Progress in Mathematics
Volume 154

Albrecht Böttcher
Yuri I. Karlovich

Carleson Curves, Muckenhoupt Weights, and Toeplitz Operators

Birkhäuser Verlag
Basel · Boston · Berlin

Authors:

Albrecht Böttcher
Fakultät für Mathematik
TU Chemnitz
D–09107 Chemnitz
Germany
e-mail: aboettch@mathematik.tu-chemnitz.de

Yuri I. Karlovich
Ukrainian Academy of Sciences
Marine Hydrophysical Institute
Hydroacoustic Department
Preobrazhenskaya Street 3
270 100 Odessa
Ukraine
e-mail: karlovic@math.ist.utl.pt

1991 Mathematics Subject Classification 47B35, 45P05

A CIP catalogue record for this book is available from the Library of Congress, Washington D.C., USA

Deutsche Bibliothek Cataloging-in-Publication Data

Böttcher, Albrecht:
Carleson curves, Muckenhaupt weights, and Toeplitz operators /
Albrecht Böttcher ; Yuri I. Karlovich. – Basel ; Boston ; Berlin :
Birkhäuser, 1997
   (Progress in mathematics ; Vol. 154)
   ISBN 3-7643-5796-7 (Basel ...)
   ISBN 0-8176-5796-7 (Boston)

© 1997 Birkhäuser Verlag, P.O. Box 133, CH-4010 Basel, Switzerland
Printed on acid-free paper produced of chlorine-free pulp. TCF ∞
Printed in Germany
ISBN 3-7643-5796-7
ISBN 0-8176-5796-7

9 8 7 6 5 4 3 2 1

Fernando Sunyer i Balaguer 1912–1967

* * *

This book has been awarded the Ferran Sunyer i Balaguer 1997 prize.

Each year, in honor of the memory of Ferran Sunyer i Balaguer, the Institut d'Estudis Catalans awards an international research prize for a mathematical monograph of expository nature. The prize-winning monographs are published in this series. Details about the prize can be found at

`http://crm.es/info/ffsb.htm`

Previous winners include

- *Alexander Lubotzky*
  Discrete Groups, Expanding
  Graphs and Invariant Measures
  (vol. 125)
- *Klaus Schmidt*
  Dynamical Systems of Algebraic
  Origin (vol. 128)

- *M. Ram Murty & V. Kumar Murty*
  Non-vanishing of *L*-functions
  and Applications (vol. 157)

Fernando Sunyer i Balaguer 1912–1967

Born in Figueras (Gerona) with an almost fully incapacitating physical disability, Fernando Sunyer i Balaguer was confined for all his life to a wheelchair he could not move himself, and was thus constantly dependent on the care of others. His father died when Don Fernando was two years old, leaving his mother, Doña Angela Balaguer, alone with the heavy burden of nursing her son. They subsequently moved in with Fernando's maternal grandmother and his cousins Maria, Angeles, and Fernando. Later, this exemplary family, which provided the environment of overflowing kindness in which our famous mathematician grew up, moved to Barcelona.

As the physician thought it advisable to keep the sickly boy away from all sorts of possible strain, such as education and teachers, Fernando was left with the option to learn either by himself or through his mother's lessons which, thanks to her love and understanding, were considered harmless to his health. Without a doubt, this education was strongly influenced by his living together with cousins who were to him much more than cousins for all his life. After a period of intense reading, arousing a first interest in astronomy and physics, his passion for mathematics emerged and dominated his further life.

In 1938, he communicated his first results to Prof. J. Hadamard of the Academy of Sciences in Paris, who published one of his papers in the Academy's "Comptes Rendus" and encouraged him to proceed in his selected course of investigation. From this moment, Fernando Sunyer i Balaguer maintained a constant interchange with the French analytical school, in particular with Mandelbrojt and his students. In the following years, his results were published regularly. The limited space here does not, unfortunately, allow for a critical analysis of his scientific achievements. In the mathematical community his work, for which he attained international recognition, is well known.

Don Fernando's physical handicap did not allow him to write down any of his papers by himself. He dictated them to his mother until her death in 1955, and when, after a period of grief and desperation, he resumed research with new vigor, his cousins took care of the writing. His working power, paired with exceptional talents, produced a number of results which were eventually recognized for their high scientific value and for which he was awarded various prizes. These honours not withstanding, it was difficult for him to reach the social and professional position corresponding to his scientific achievements. At times, his economic situation was not the most comfortable either. It wasn't until the 9th of December 1967, 18 days prior his death, that his confirmation as a scientific member was made public by the División de Ciencias, Médicas y de Naturaleza of the Council. Furthermore, he was elected only as "de entrada", in contrast to class membership.

Due to his physical constraints, the academic degrees for his official studies were granted rather belatedly. By the time he was given the Bachelor degree, he had already been honoured by several universities! In 1960 he finished his Master's degree and was awarded the doctorate after the requisite period of two years as a student. Although he had been a part-time employee of the Mathematical Seminar since 1948, he was not allowed to become a full member of the scientific staff until 1962. This despite his actually heading the department rather than just being a staff member.

His own papers regularly appeared in the journals of the Barcelona Seminar, *Collectanea Mathematica*, to which he was also an eminent reviewer and advisor. On several occasions, he was consulted by the Proceedings of the American Society of Mathematics as an advisor. He always participated in and supported guest lectures in Barcelona, many of them having been prepared or promoted by him. On the occasion of a conference in 1966, H. Mascart of Toulouse publicly pronounced his feeling of being honoured by the presence of F. Sunyer Balaguer, "the first, by far, of Spanish mathematicians".

At all times, Sunyer Balaguer felt a strong attachment to the scientific activities of his country and modestly accepted the limitations resulting from his attitude, resisting several calls from abroad, in particular from France and some institutions in the USA. In 1963 he was contracted by the US Navy, and in the following years he earned much respect for the results of his investigations. "His value to the prestige of the Spanish scientific community was outstanding and his work in mathematics of a steady excellence that makes his loss difficult to accept" (letter of condolence from T.B. Owen, Rear Admiral of the US Navy).

Twice, Sunyer Balaguer was approached by young foreign students who wanted to write their thesis under his supervision, but he had to decline because he was unable to raise the necessary scholarship money. Many times he reviewed doctoral theses for Indian universities, on one occasion as the president of a distinguished international board. The circumstances under which Sunyer attained his scientific achievements also testify to his remarkable human qualities. Indeed, his manner was friendly and his way of conversation reflected his gift for friendship as well as enjoyment of life and work which went far beyond a mere acceptance of the situation into which he had been born. His opinions were as firm as they were cautious, and at the same time he had a deep respect for the opinion and work of others. Though modest by nature, he achieved due credit for his work, but his petitions were free of any trace of exaggeration or undue self-importance. The most surprising of his qualities was, above all, his absolute lack of preoccupation with his physical condition, which can largely be ascribed to the sensible education given by his mother and can be seen as an indicator of the integration of the disabled into our society.

On December 27, 1967, still fully active, Ferran Sunyer Balaguer unexpectedly passed away. The memory of his remarkable personality is a constant source of stimulation for our own efforts.

Translated from Juan Augé: Fernando Sunyer Balaguer. *Gaceta Matematica*, 1.a Serie – Tomo XX – Nums. 3 y 4, 1968, where a complete bibliography can be found.

Contents

Preface

This book is a reasonably self-contained introduction to the spectral theory of Toeplitz operators with piecewise continuous symbols and of singular integral operators with piecewise continuous coefficients on Carleson curves with Muckenhoupt weights. For piecewise Lyapunov curves with power weights, the corresponding theory was accomplished by Gohberg and Krupnik in the seventies. Only in the eighties, after a long development and by the efforts of many mathematicians, did it become clear that the Cauchy singular integral operator S_Γ is bounded on the weighted Lebesgue space $L^p(\Gamma, w)$ $(1 < p < \infty)$ if and only if Γ is a Carleson curve and w is a Muckenhoupt weight. Extending the Gohberg-Krupnik theory to this more (and even, in a sense, "most") general setting would have been a thankless job had it turned out that by means of refined techniques the results for piecewise Lyapunov curves and power weights could essentially be carried over to Carleson curves and Muckenhoupt weights. However, in recent times it was discovered that general Carleson curves and general Muckenhoupt weights yield qualitatively new phenomena in the spectra of Toeplitz and singular integral operators. The resulting spectral theory is surprisingly rich and extremely beautiful. It is the subject of this book.

To get an idea of what is going on, let us consider the essential spectrum of the Cauchy singular integral operator S_Γ on $L^p(\Gamma, w)$ in the case where Γ is a bounded simple arc. If Γ is piecewise Lyapunov and w is a power weight, then the essential spectrum consists of two circular arcs between -1 and 1. We will show that these circular arcs metamorphose into logarithmic double spirals for more complicated curves and that in the case of general Carleson curves these double spirals may blow up to heavy sets whose boundaries are nevertheless comprised of pieces of logarithmic spirals. Proper (i.e. non-powerlike) Muckenhoupt weights may further thicken the spectrum: until some point the weights are unable to destroy the circular arcs and logarithmic spirals in the spectrum, but beyond this point some kind of interference between the curve and the weight results in a complete disappearance of spirality and the emergence of so-called leaves. In other words: when considering boundedness of S_Γ, it is only workers in the field who know of the precipice between Lyapunov curves and Carleson curves or between power weights and Muckenhoupt weights – when looking at the spectrum of S_Γ, everyone can *see* this precipice.

The problem of finding the spectrum of the Cauchy singular integral operator is, in a sense, equivalent to describing the spectrum of Toeplitz operators with piecewise continuous symbols on weighted Hardy spaces $L^p_+(\Gamma, w)$ over Jordan curves Γ. The language of Toeplitz operators is more convenient for our purposes, and therefore it is Toeplitz operators which will play the dominant part in this book. Having identified the local spectra of Toeplitz operators, we will employ local principles, an appropriate N projections theorem, and results of geometric function theory pertaining to the problem of extending Carleson curves and Muckenhoupt weights in order to construct a symbol calculus for Banach algebras of singular integral operators over composed curves.

The table of contents provides an overall view of what this book is all about. We merely want to add the following remarks.

The first three chapters are an introduction to Carleson curves and Muckenhoupt weights. Various results of these chapters are well known, but a series of concepts, methods, and results are new and are dictated by the needs of the spectral theory of Toeplitz and singular integral operators. In particular, the use of submultiplicative functions and their indices in order to characterize Muckenhoupt weights seems to be a novelty. Here, we also introduce the notions of the indicator set and of the indicator functions, which contain just the information hidden in the curve and the weight that is of relevance in the spectral theory. The spirality indices of a curve and the indices of powerlikeness of a weight are important parameters of the indicator functions.

In Chapters 4 and 5 we give a detailed proof of the theorem stating that the Cauchy singular integral operator is bounded on $L^p(\Gamma, w)$ $(1 < p < \infty)$ if and only if Γ is a Carleson curve and w is a Muckenhoupt weight.

Chapter 6 contains some background material on Toeplitz operators and exhibits two basic techniques for tackling them: localization and Wiener-Hopf factorization. Chapter 7 is the high point of the book. In this chapter we completely describe the essential spectrum and the spectrum of Toeplitz operators with piecewise continuous symbols. In a sense, Chapters 1 to 6 serve to prepare for Chapter 7, while Chapters 8 to 10 are the harvest from Chapter 7.

Harvest needs harvesting machines. The central result of Chapter 8 is an N projections theorem, whose $N = 2$ version allows us to establish a symbol calculus for algebras of singular integral operators over Jordan curves. In Chapter 9 we employ the machinery of geometric function theory in order to deal with certain problems of extending Carleson curves and Muckenhoupt weights. Thereafter, we can use the results of Chapters 7 and 8 (including the N projections theorem in its full strength) to construct a symbol calculus for singular integral operators over composed curves. Chapter 10 records some further results, which could not be incorporated into the main text for lack of space.

In the late seventies, the spectral theory of Toeplitz operators with piecewise continuous symbols was considered as round and complete. In 1990, Spitkovsky surprised the community with the spectacular discovery that in the case of Lyapunov curves with arbitrary Muckenhoupt weights the circular arcs metamorphose into horns, and again it seemed then that there remained nothing to say. We now know the spectra of Toeplitz operators with piecewise continuous symbols in the case of arbitrary Carleson curves and arbitrary Muckenhoupt weights. Is this the end of the story ? Our experience tells us that the answer to such a question must be NO. Also notice that consideration of operators with oscillating symbols or passage to higher dimensions are among the challenges of the future.

Part of the book is heavily based on results obtained only in the last three years. Thus, we are aware of the fact that several things certainly can and will be done better. We nevertheless hope that we succeeded to convey to the reader an idea of the fascinating beauty of the spectral theory of Toeplitz and singular integral operators and the mathematics behind it.

Acknowledgements. This book was written during a stay of the second author at the Technical University Chemnitz from 1993 to 1996. We are deeply indebted to the

Alfried Krupp Foundation

for supporting our joint work over these years through funds from a Förderpreis für junge Hochschullehrer. Without the support by the Krupp Foundation, this book would not exist.

We also wish to express our sincere gratitude to Sylvia Böttcher for the energy and patience she devoted to the production of the LATEX masters of this book and to Alexei Yu. Karlovich for proof-reading the entire manuscript with untiring enthusiasm and for suggesting a large number of improvements.

Chemnitz, April 1997   The authors

Chapter 1
Carleson curves

The purpose of this chapter is to acquaint the reader with some simple but basic properties of Carleson curves and to provide a sufficient supply of examples. The "oscillation" of a Carleson curve Γ at a point $t \in \Gamma$ may be measured by its Seifullayev bounds σ_t^- and σ_t^+ as well as its spirality indices δ_t^- and δ_t^+. The definition of the spirality indices requires the notion of the W transform and some facts from the theory of submultiplicative functions. In the spectral theory of Toeplitz and singular integral operators, the spirality indices will play a decisive role. We therefore compute the spirality indices for a sufficiently large class of concrete Carleson curves.

1.1 Definitions and examples

We refer to a subset Γ of the complex plane $\mathbf{C}$ as an *arc* if it is homeomorphic to a connected subset of the real line $\mathbf{R}$ which contains at least two distinct points. Equivalently, $\Gamma \subset \mathbf{C}$ is an arc if and only if Γ is homeomorphic to one of the sets $[0, 1]$, $[0, \infty)$, or $(-\infty, \infty)$. A subset Γ of $\mathbf{C}$ is referred to as a *Jordan curve* if it is homeomorphic to the complex unit circle $\mathbf{T} := \{z \in \mathbf{C} : |z| = 1\}$. Throughout what follows, we understand by a *simple curve* always an arc or a Jordan curve.

Let $\Gamma \subset \mathbf{C}$ be an arc. Then there exists a connected subset $I \subset \mathbf{R}$ containing at least two distinct points and a homeomorphism $\varphi : I \to \Gamma$. The interior of I is an interval (a, b) with $a \in \mathbf{R} \cup \{-\infty\}$ and $b \in \mathbf{R} \cup \{+\infty\}$. If

$$t_1 := \lim_{x \to a+0} \varphi(x) \quad \text{or} \quad t_2 := \lim_{x \to b-0} \varphi(x)$$

exist and are finite, they are referred to as *endpoints* of the arc Γ. Thus, an arc may have two, one, or no endpoints. An arc is said to be an *open arc* if it does not contain its endpoints.

A subset Γ of $\mathbf{C}$ is called a *composed curve* if it is connected and may be represented as the union of finitely many arcs each pair of which have at most endpoints in common.

1

Now suppose Γ is a composed curve. For a point $t \in \Gamma$ and a number $\varepsilon \in (0, \infty)$, let $\Gamma(t, \varepsilon) := \{\tau \in \Gamma : |\tau - t| < \varepsilon\}$ stand for the portion of Γ contained in the open disk of radius ε centered at t. The set $\Gamma(t, \varepsilon)$ is an at most countable union of arcs. If all these arcs are rectifiable and the sum of their lenghts is finite, we say that $\Gamma(t, \varepsilon)$ is *rectifiable*. The composed curve Γ is said to be *locally rectifiable* if $\Gamma(t, \varepsilon)$ is rectifiable for every $t \in \Gamma$ and every $\varepsilon \in (0, \infty)$. It is easily seen that the local rectifiability of Γ is equivalent to the requirement that $\Gamma \cap \{z \in \mathbf{C} : |z| < R\}$ be rectifiable for every $R > 0$ (note that the latter set is also an at most countable union of arcs).

In what follows we will only consider locally rectifiable curves. Local rectifiability is a condition which rules out many (more or less pathological) curves. For example, the arc $\Gamma = \{(1 - e^{-\theta})e^{i\theta} : \theta \geq 0\}$ is an arc homeomorphic to $[0, \infty)$, but it is not locally rectifiable. To have another example, note that the connected set

$$\Gamma = \{\sin \frac{1}{|x|} : -1 \leq x < 0\} \cup \{iy : -1 \leq y \leq 1\} \cup \{\sin \frac{1}{x} : 0 < x \leq 1\}$$

is a composed curve which, however, is not locally rectifiable. On the other hand, if $\Gamma = \{re^{ir} : r \geq 0\}$, then Γ is an arc, it is locally rectifiable, but it is not rectifiable. The arc $\Gamma = \{e^{-\theta}e^{i\theta} : \theta \geq 0\}$ is rectifiable.

A composed curve Γ is said to be *bounded* if it is a bounded subset of the plane, i.e. if there is an $R > 0$ such that $\Gamma \subset \{z \in \mathbf{C} : |z| < R\}$. Otherwise Γ is called *unbounded*. Obviously, arcs homeomorphic to $[0, 1]$ or Jordan curves are always bounded. It is also obvious that a locally rectifiable arc homeomorphic to $[0, \infty)$ or $(-\infty, \infty)$ is bounded if and only if it is rectifiable.

Let Γ be a locally rectifiable composed curve and equip Γ with Lebesgue length measure. The measure of a measurable subset $\gamma \subset \Gamma$ will be denoted by $|\gamma|$. In particular, $|\Gamma(t, \varepsilon)|$ is the sum of the lengths of the at most countably many arcs constituting $\Gamma(t, \varepsilon)$. The curve Γ is said to be a *Carleson curve* (an *Ahlfors regular curve*, a *David regular curve*, or an *Ahlfors-David curve*) if

$$C_\Gamma := \sup_{t \in \Gamma} \sup_{\varepsilon > 0} \frac{|\Gamma(t, \varepsilon)|}{\varepsilon} < \infty. \tag{1.1}$$

In other words, Γ is Carleson if and only if there is a constant C_Γ such that $|\Gamma(t, \varepsilon)| \leq C_\Gamma \varepsilon$ for all $t \in \Gamma$ and all $\varepsilon > 0$. Condition (1.1) is frequently referred to as the *Carleson condition* and the constant C_Γ defined by (1.1) is sometimes called the *Carleson constant*.

Let Γ be a composed locally rectifiable curve and let $\Gamma_1, \ldots, \Gamma_N$ be a finite number of arcs such that $\Gamma = \Gamma_1 \cup \ldots \cup \Gamma_N$. Since

$$\Gamma_j(t, \varepsilon) \subset \Gamma(t, \varepsilon) \subset \Gamma_1(t, \varepsilon) \cup \ldots \cup \Gamma_N(t, \varepsilon)$$

and thus

$$|\Gamma_j(t, \varepsilon)| \leq |\Gamma(t, \varepsilon)| \leq |\Gamma_1(t, \varepsilon)| + \ldots + |\Gamma_N(t, \varepsilon)|,$$

it follows that Γ is a Carleson curve if and only if each arc Γ_j is Carleson. Due to this observation we may henceforth focus our attention to simple curves.

When checking condition (1.1), it is often useful to take into account that *if Γ is bounded*, then (1.1) is equivalent to the condition

$$\exists \varepsilon_0 > 0 \ : \ \sup_{t \in \Gamma} \ \sup_{0 < \varepsilon < \varepsilon_0} \frac{|\Gamma(t, \varepsilon)|}{\varepsilon} < \infty. \tag{1.2}$$

Indeed, (1.2) is obviously implied by (1.1). Conversely, if $\varepsilon \geq \varepsilon_0$ then $|\Gamma(t, \varepsilon)|/\varepsilon \leq |\Gamma(t, \varepsilon)|/\varepsilon_0 \leq |\Gamma|/\varepsilon_0$, which together with (1.2) gives (1.1).

However, in the case of unbounded curves one cannot replace (1.1) by (1.2). This can be seen by constructing a "locally nice" unbounded arc Γ the lengths of whose portions Γ_n in a sequence of squares of side lengths $2n$ are comparable to n^2. For $n = 1, 2, 3, \ldots$, put

$$x_{n,k} = n^2 - n + k, \ \ y_{n,k} = x_{n,k} + 2ni \ \ (k = 0, 1, \ldots, 2n).$$

Then define $\Gamma = \cup_{n=1}^{\infty} \Gamma_n$ with

$$\Gamma_n := \bigcup_{k=1,2,\ldots,2n} [x_{n,k}, y_{n,k}] \ \cup \ \bigcup_{k=0,2,\ldots,2n-2} [x_{n,k}, x_{n,k+1}] \ \cup \ \bigcup_{k=1,3,\ldots,2n-1} [y_{n,k}, y_{n,k+1}].$$

After drawing a picture (see Fig. 1a), it is easily seen that $|\Gamma(t, \varepsilon)|/\varepsilon < 1 + \sqrt{2}$ for all $\varepsilon < 1/2$ but

$$\frac{|\Gamma(n^2 + ni, n\sqrt{2})|}{n\sqrt{2}} \geq \frac{2n + 2n \cdot 2n}{n\sqrt{2}} = \sqrt{2} + 2\sqrt{2}n \to \infty.$$

Thus (1.2) holds (with $\varepsilon_0 = 1/2$), but (1.1) is not satisfied.

Here is a second example. Let $\Gamma := [0, 2\pi] \cup \{re^{ir} : r \geq 2\pi\}$. It is readily seen that $|\Gamma(t, \varepsilon)| < 2\pi\varepsilon$ for all $t \in \Gamma$ whenever $\varepsilon < \pi$. On the other hand, if $R > 2\pi$ then

$$|\Gamma(0, R)| = 2\pi + \int_{2\pi}^{r} |d(re^{ir})| = 2\pi + \int_{2\pi}^{R} \sqrt{1 + r^2} \, dr > \int_{2\pi}^{R} r \, dr = R^2/2 - 2\pi^2,$$

whence $|\Gamma(0, R)|/R \to \infty$ as $R \to \infty$ (see Fig. 1b).

If Γ is an arc and t is an endpoint of Γ, then $|\Gamma(t, \varepsilon)| \geq \varepsilon$ whenever Γ has a point on the circle $\{z \in \mathbf{C} : |z - t| = \varepsilon\}$. This is clearly the case for all sufficiently small ε. In case t is not an endpoint of an arc Γ or Γ is a Jordan curve, we see analogously that $|\Gamma(t, \varepsilon)| \geq 2\varepsilon$ for all sufficiently small ε. It follows in particular that always $C_\Gamma \geq 2$. Further, we arrive at the conclusion that if Γ is a *bounded* curve, then Γ is a Carleson curve if and only if $\varepsilon \leq |\Gamma(t, \varepsilon)| \leq C_\Gamma \varepsilon$ for all $t \in \Gamma$ and all sufficiently small ε, and in this sense bounded Carleson curves may be characterized as bounded curves for which the measure of small portions $\Gamma(t, \varepsilon)$ is

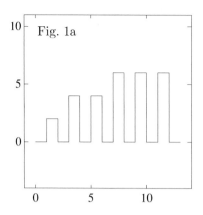

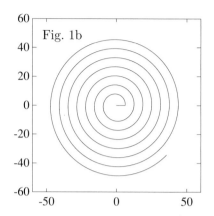

Figure 1a shows the beginning of the "comb" $\Gamma = \cup_{n=1}^{\infty}\Gamma_n$. In Figure 1b we see the segment $[0, 2\pi]$ and a piece of the Archimedian spiral $\{re^{ir} : r \geq 2\pi\}$.

comparable with the diameter of the disks $\{z \in \mathbf{C} : |z - t| < \varepsilon\}$ (uniformly with respect to $t \in \Gamma$).

Finally, we remark that a locally rectifiable composed curve Γ is Carleson if and only if

$$C'_{\Gamma} := \sup_{z\in\mathbf{C}} \sup_{\varepsilon>0} \frac{|\Gamma \cap D(z,\varepsilon)|}{\varepsilon} < \infty, \tag{1.3}$$

where $D(z, \varepsilon) := \{\zeta \in \mathbf{C} : |\zeta - z| < \varepsilon\}$, Indeed, if (1.3) holds, then (1.1) is obviously satisfied. Conversely, suppose (1.1) is valid. We claim that then (1.3) is true with $C'_{\Gamma} \leq 2C_{\Gamma}$. If $\Gamma \cap D(z, \varepsilon) = \emptyset$, then $|\Gamma \cap D(z, \varepsilon)|/\varepsilon = 0$. On the other hand, if there is a point $t \in \Gamma \cap D(z, \varepsilon)$, then $\Gamma \cap D(z, \varepsilon) \subset \Gamma \cap D(t, 2\varepsilon)$, whence $|\Gamma \cap D(z, \varepsilon)|/\varepsilon \leq C_{\Gamma}2\varepsilon$. This proves our claim.

Given real numbers $a < b$, we denote by $C[a, b]$ the continuous real-valued functions on the segment $[a, b]$, by $C^1(a, b)$ the continuously differentiable real-valued functions on the interval (a, b), and by $C^1[a, b]$ the functions in $C[a, b] \cap C^1(a, b)$ whose derivatives have finite one-sided limits at a and b.

Proposition 1.1. *If*

$$\Gamma = \{\tau \in \mathbf{C} : \tau = x + if(x), \ a \leq x \leq b\} \tag{1.4}$$

with

$$f \in C[a, b] \cap C^1(a, b) \ \text{and} \ |f'(x)| \leq M \ \text{for all} \ x \in (a, b) \tag{1.5}$$

then Γ is a Carleson curve.

Proof. If $t = x_0 + if(x_0) \in \Gamma$ and $\varepsilon > 0$, then $\Gamma(t, \varepsilon)$ is contained in the stripe of all $x + iy \in \mathbf{C}$ satisfying $\max\{a, x_0 - \varepsilon\} \leq x \leq \min\{b, x_0 + \varepsilon\}$. Consequently,

$$|\Gamma(t, \varepsilon)| = \int_{\Gamma(t,\varepsilon)} |d\tau| \leq \int_{\max\{a,x_0-\varepsilon\}}^{\min\{b,x_0+\varepsilon\}} \sqrt{1 + (f'(x))^2}\, dx$$

$$\leq \int_{x_0-\varepsilon}^{x_0+\varepsilon} \sqrt{1 + M^2}\, dx = 2\sqrt{1 + M^2}\,\varepsilon. \qquad \square$$

Example 1.2. We call a bounded arc a C^1 *arc* if, after suitable rotation, it may be parametrized as in (1.4) with $f \in C^1[a, b]$ (in which case (1.5) is automatically satisfied). A composed curve Γ is referred to as a *piecewise C^1 curve* if it may be represented as a finite union of C^1 arcs. Proposition 1.1 tells us that piecewise C^1 curves are Carleson curves. We remark that the only possible "singularities" of piecewise C^1 Jordan curves are corners and cusps. $\qquad \square$

Example 1.3. To have a less trivial example of a Carleson curve, pick $\alpha > 0$ and define Γ by (1.4) with

$$f(x) = x^\alpha \sin(1/x) \text{ for } x \in (0, 1], \ f(0) = 0.$$

Clearly, $f \in C[0, 1] \cap C^1(0, 1)$. We claim that

$$\Gamma \text{ is Carleson for } \alpha \geq 2, \tag{1.6}$$

$$\Gamma \text{ is rectifiable but not Carleson for } 1 < \alpha < 2, \tag{1.7}$$

$$\Gamma \text{ is not rectifiable for } 0 < \alpha \leq 1. \tag{1.8}$$

Since $f'(x) = \alpha x^{\alpha-1} \sin(1/x) - x^{\alpha-2} \cos(1/x)$ is bounded on $(0, 1)$ for $\alpha \geq 2$, we may deduce (1.6) from Proposition 1.1. Determine $x_n \in (1/((n+1)\pi), 1/(n\pi))$ by the equation $|\sin(1/x_n)| = 1$. Since

$$N\pi |\Gamma(0, \tfrac{1}{N\pi})| > N\pi \sum_{0 < x_n < \frac{1}{N\pi}} x_n^\alpha > N\pi \sum_{n=N}^{\infty} \left(\frac{1}{(n+1)\pi}\right)^\alpha$$

and the latter sum is ∞ for $\alpha \leq 1$ and greater than $C/N^{\alpha-1}$ with some $C > 0$ for $\alpha > 1$, we arrive at (1.8) and (1.7). $\qquad \square$

For $d > 0$, we denote by $C(0, d]$ the real-valued continuous functions on $(0, d] := \{x \in \mathbf{R} : 0 < x \leq d\}$.

Proposition 1.4. *If*

$$\Gamma = \{0\} \cup \{\tau \in \mathbf{C} : \tau = re^{i\varphi(r)},\ 0 < r \le d\} \tag{1.9}$$

with

$$\varphi \in C(0, d] \cap C^1(0, d) \quad and \quad |r\varphi'(r)| \le M \ \text{ for all } \ r \in (0, d], \tag{1.10}$$

then Γ *is a Carleson curve.*

Proof. It is clear that Γ is a bounded arc. For $t \in \Gamma$, the portion $\Gamma(t, \varepsilon)$ is a subset of the annulus between the two circles of radius $\max\{0, |t| - \varepsilon\}$ and $\min\{d, |t| + \varepsilon\}$ centered at the origin. Thus,

$$|\Gamma(t, \varepsilon)| = \int_{\Gamma(t, \varepsilon)} |d\tau| \le \int_{\max\{0, |t| - \varepsilon\}}^{\min\{d, |t| + \varepsilon\}} \sqrt{1 + r^2 \left(\varphi'(r)\right)^2}\, dr$$

$$\le \int_{|t| - \varepsilon}^{|t| + \varepsilon} \sqrt{1 + M^2}\, dr = 2\sqrt{1 + M^2}\,\varepsilon. \qquad \square$$

Let us for a moment assume that φ' extends to a function in $C(0, d]$. If $\varphi(r)$ has a finite limit as $r \to 0$, then Γ is simply a piecewise C^1 curve. Functions φ with no limit at the origin provide a lot of interesting curves. For example, if $\varphi(r)$ increases monotoneously to $+\infty$ as $r \to 0$, then $\tau = re^{i\varphi(r)}$ traces out a spiral scrolling up counter-clockwise at the origin. Clearly, the more rapidly φ increases, the less are our chances to get a Carleson curve: "analytically" this is seen from the estimate

$$\frac{|\Gamma(0, \varepsilon)|}{\varepsilon} = \frac{1}{\varepsilon} \int_0^\varepsilon \sqrt{1 + r^2 \left(\varphi'(r)\right)^2}\, dr \ge \frac{1}{\varepsilon} \int_0^\varepsilon r|\varphi'(r)|\, dr, \tag{1.11}$$

and "geometrically" this follows from the observation that $\Gamma(0, \varepsilon)$ contains the more whorls of the spiral the faster φ increases.

Example 1.5. Suppose φ has powerlike growth, i.e. $\varphi(r) = \delta/r^\alpha$ with $\delta > 0$ and $\alpha > 0$. Then $r\varphi'(r) = -\alpha\delta/r^\alpha$ and hence Proposition 1.4 is not applicable. In fact, Γ is *not* a Carleson curve: from (1.11) we infer that

$$\frac{|\Gamma(0, \varepsilon)|}{\varepsilon} \ge \frac{\alpha|\delta|}{\varepsilon} \int_0^\varepsilon r^{-\alpha}\, dr,$$

and the latter term is infinite for $\alpha \ge 1$ (implying that Γ is not even rectifiable) and equals $\alpha|\delta|/((1 - \alpha)\varepsilon^\alpha)$ for $0 < \alpha < 1$, which goes to infinity as $\varepsilon \to 0$. $\qquad \square$

Example 1.6 Let $\varphi(r) = -\delta \log r$ with some constant $\delta \in \mathbf{R}$. For $\delta = 0$, we have nothing but a line segment terminating at the origin. However, if $\delta > 0$ (resp. $\delta < 0$) then

$$\tau = re^{-i\delta \log r} = r^{1-i\delta}$$

describes a *logarithmic spiral* scrolling up counter-clockwise (resp. clockwise) at the origin as $r \to 0$. Since $r\varphi'(r) = -\delta$, we conclude from Proposition 1.4 that logarithmic spirals and connected pieces of them are Carleson arcs. □

Example 1.7. A very large class of Carleson curves emerges from (1.9) with the choice $d = 1$,

$$\varphi(r) = h\big(\log(-\log r)\big)(-\log r), \quad 0 < r < 1 \tag{1.12}$$

where

$$h \in C^1(\mathbf{R}), \quad |h(x)| \le M \text{ and } |h'(x)| \le M \text{ for all } x \in \mathbf{R}. \tag{1.13}$$

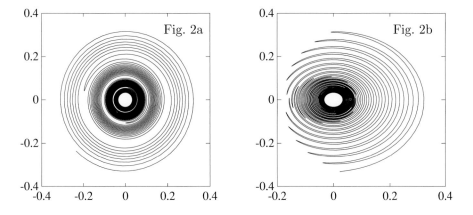

Figures 2a and 2b show pieces of two simple Carleson arcs as in Example 1.7. In order to make visible the whirl, we chose a logarithmic scale of the radius in polar coordinates.

On defining $\varphi(1) = 0$, we achieve that $\varphi \in C(0,1] \cap C^1(0,1)$. Since

$$r\varphi'(r) = -h\big(\log(-\log r)\big) - h'\big(\log(-\log r)\big), \tag{1.14}$$

condition (1.10) is satisfied if (1.13) holds. For instance, if $h(x) = \delta + \mu \sin \lambda x$, then (1.13) is obviously true. In order to demonstrate the variety of Carleson curves supplied by (1.12), (1.13), we mention the following: given any sequence $\{k_n\}_{n=1}^\infty$ of natural numbers, one can choose a function h subject to (1.13) such that the moduli of the differences of successive local extremal values of $\varphi(r)$ constitute just the sequence $\{2\pi k_n\}_{n=1}^\infty$ as r decreases from 1 to 0. In other words, $\tau = re^{i\varphi(r)}$ approaches the origin making first k_1 whorls counter-clockwise, then k_2 whorls clockwise, then k_3 whorls counter-clockwise etc. □

1.2   Growth of the argument

Let Γ be an arbitrary simple Carleson curve and fix $t \in \Gamma$. We then have

$$\tau - t = |\tau - t| e^{i \arg(\tau - t)} \quad \text{for} \quad \tau \in \Gamma \setminus \{t\}, \tag{1.15}$$

and the argument $\arg(\tau - t)$ may be chosen to be a continuous function of τ in $\Gamma \setminus \{t\}$. We fix any continuous branch of $\arg(\tau - t)$. For an arc $\gamma \subset \Gamma \setminus \{t\}$ with the endpoints t_1 and t_2, we denote by

$$\big\{ \arg(\tau - t) \big\}_{\tau \in \gamma} := \arg(t_2 - t) - \arg(t_1 - t)$$

the increment of $\arg(\tau - t)$ as τ moves along γ from t_1 to t_2.

By virtue of the Carleson condition, one expects that $\Delta := \{\arg(\tau - t)\}_{\tau \in \gamma}$ cannot be very large unless $|\tau - t|$ decays a sufficient amount along γ. Put $|t_1 - t| =: R$, $|t_2 - t| =: r$, and let $R \geq r$. For the sake of simplicity, assume γ is entirely contained in the annulus $\{z \in \mathbf{C} : r \leq |z - t| \leq R\}$. Then clearly $|\gamma| \geq |\Delta| r$, while (1.1) implies that $|\gamma| \leq |\Gamma(t, R)| \leq C_\Gamma R$. Thus, $|\Delta| \leq C_\Gamma R / r$, which is a precise estimate confirming our expectation. The following theorem provides such an estimate for the general case.

Theorem 1.8. *Let Γ be a simple Carleson curve and $t \in \Gamma$. Let $\gamma \subset \Gamma \setminus \{t\}$ be any arc whose endpoints lie on the concentric circles $\{z \in \mathbf{C} : |z - t| = R\}$ and $\{z \in \mathbf{C} : |z - t| = r\}$ with $R \geq r > 0$. Then*

$$\big| \big\{ \arg(\tau - t) \big\}_{\tau \in \gamma} \big| < 2\pi C_\Gamma R / r, \tag{1.16}$$

where C_Γ is the Carleson constant (1.1).

Proof. Abbreviate $\{\arg(\tau - t)\}_{\tau \in \gamma}$ to $\Delta(\gamma)$ and put $K(y) := \{z \in \mathbf{C} : |z - t| = y\}$. The set $\Gamma \cap K(y)$ consists of at most countable many singletons $p_i(y)$ $(i \in I)$ and at most countable many arcs $q_l(y)$ $(l \in L)$. Denote by $q'_l(y)$ and $q''_l(y)$ the endpoints of the arc $q_l(y)$ and define

$$M(y) := \bigcup_{i \in I} \{p_i(y)\} \cup \bigcup_{l \in L} \{q'_l(y), q''_l(y)\}.$$

The set $\gamma \setminus (M(r) \cup M(R))$ is the union of at most countable many open arcs $\gamma_j \subset \gamma$ $(j \in J)$. Let t'_j and t''_j denote the endpoints of γ_j. We divide the collection of the arcs γ_j into three pairwise disjoint classes:

$$\begin{aligned}
\mathcal{N}_1 &:= \big\{ \gamma_j : \{t'_j, t''_j\} \subset K(r) \text{ and } \gamma_j \subset \Gamma(t, r) \big\}, \\
\mathcal{N}_2 &:= \big\{ \gamma_j : \{t'_j, t''_j\} \subset K(r) \cup K(R) \text{ and } \gamma_j \subset \Gamma(t, 2R) \setminus \Gamma(t, r) \big\}, \\
\mathcal{N}_3 &:= \big\{ \gamma_j : \{t'_j, t''_j\} \subset K(R) \text{ and } \gamma_j \not\subset \Gamma(t, 2R) \big\}.
\end{aligned}$$

Let first $\gamma_j \in \mathcal{N}_1$. We denote by $\tilde{\gamma}_j \subset K(r)$ the arc whose endpoints are t'_j and t''_j and which is uniquely determined by the requirement that t belongs to the unbounded component of $\mathbf{C} \setminus (\gamma_j \cup \tilde{\gamma}_j)$. Clearly, $0 < |\Delta(\gamma_j)| < 2\pi$. If $0 < |\Delta(\gamma_j)| < \pi$, then

$$|\gamma_j| \geq |t''_j - t'_j| > (2/\pi)|\tilde{\gamma}_j| = (2r/\pi)|\Delta(\gamma_j)|,$$

while if $\pi \leq |\Delta(\gamma_j)| < 2\pi$, we have

$$|\gamma_j| \geq 2r = (r/\pi)2\pi > (r/\pi)|\Delta(\gamma_j)|.$$

Thus, for every $\gamma_j \in \mathcal{N}_1$ the inequality

$$|\gamma_j| > (r/\pi)|\Delta(\gamma_j)| \tag{1.17}$$

holds.

If $\gamma_j \in \mathcal{N}_2$ then $\gamma_j \subset \mathbf{C} \setminus \Gamma(t, r)$, whence

$$|\gamma_j| \geq r|\Delta(\gamma_j)| > (r/\pi)|\Delta(\gamma_j)|. \tag{1.18}$$

If $\gamma_j \in \mathcal{N}_3$, then the endpoints of γ_j lie on $K(R)$ and γ_j has a point outside $K(2R)$. This implies that $|\gamma_j \cap \Gamma(t, 2R)| \geq 2R$. On the other hand, because again $0 < |\Delta(\gamma_j)| < 2\pi$, we obtain

$$|\gamma_j \cap \Gamma(t, 2R)| \geq 2R \geq 2r = (r/\pi)2\pi > (r/\pi)|\Delta(\gamma_j)|. \tag{1.19}$$

We now put the things together. Since $\gamma_j \subset \Gamma(t, 2R)$ for $\gamma_j \in \mathcal{N}_1 \cup \mathcal{N}_2$, we have

$$|\Gamma(t, 2R)| \geq \sum_{\gamma_j \in \mathcal{N}_1} |\gamma_j| + \sum_{\gamma_j \in \mathcal{N}_2} |\gamma_j| + \sum_{\gamma_j \in \mathcal{N}_3} |\gamma_j \cap \Gamma(t, 2R)|,$$

and taking into account (1.17) to (1.19) we so get

$$|\Gamma(t, 2R)| > (r/\pi) \sum_{\gamma_j \in \mathcal{N}_1 \cup \mathcal{N}_2 \cup \mathcal{N}_3} |\Delta(\gamma_j)| \geq (r/\pi)|\Delta(\gamma)|. \tag{1.20}$$

From (1.1) we know that $|\Gamma(t, 2R)| \leq C_\Gamma 2R$, and thus (1.20) gives

$$|\Delta(\gamma)| < (\pi/r)C_\Gamma 2R = 2\pi C_\Gamma R/r. \qquad \square$$

The following result essentially sharpens the previous theorem in the case where R/r is large.

Theorem 1.9. *Under the hypotheses of Theorem 1.8 we have*

$$\left|\left\{\arg(\tau - t)\right\}_{\tau \in \gamma}\right| < 2\pi e C_\Gamma \left(\log \frac{R}{r} + 1\right) \tag{1.21}$$

Proof. Denote the endpoints of γ by τ_1 and τ_2. Suppose $|\tau_1 - t| = R$ and $|\tau_2 - t| = r$, and give γ the orientation from τ_1 to τ_2. There is a unique integer $n \geq 0$ such that

$$R/e^{n+1} < r \leq R/e^n. \tag{1.22}$$

Let $t_0 = \tau_1$, and for $k \in \{1, 2, \ldots, n\}$, let t_k be the last point on the (oriented) arc γ which lies on the circle $\{z \in \mathbf{C} : |z - t| = R/e^k\}$. We have $\gamma = \gamma_1 \cup \ldots \cup \gamma_{n+1}$ where

$$\gamma_k = \gamma(t_{k-1}, t_k) \quad (1 \leq k \leq n), \quad \gamma_{n+1} = \gamma(t_n, \tau_2)$$

and $\gamma(\alpha, \beta)$ stands for the piece of γ between α, β. Denote by Δ and Δ_k ($1 \leq k \leq n + 1$) the increment of $\arg(\tau - t)$ along γ and γ_k, respectively. Clearly, $|\Delta| \leq |\Delta_1| + |\Delta_2| + \ldots + |\Delta_{n+1}|$.

From Theorem 1.8 we infer that

$$|\Delta_k| < 2\pi C_\Gamma \frac{|t_{k-1} - t|}{|t_k - t|} = 2\pi C_\Gamma \frac{R/e^{k-1}}{R/e^k} = 2\pi e C_\Gamma$$

for $1 \leq k \leq n$ and that

$$|\Delta_{n+1}| < 2\pi C_\Gamma \frac{|t_n - t|}{|\tau_2 - t|} = 2\pi C_\Gamma \frac{R/e^n}{r} < 2\pi e C_\Gamma,$$

the last inequality resulting from (1.22). Consequently,

$$|\Delta| \leq \sum_{k=1}^{n+1} |\Delta_k| < 2\pi e C_\Gamma (n + 1),$$

and since $n \leq \log(R/r)$ by virtue of (1.22), we arrive at the desired inequality (1.21). $\qquad\square$

The next theorem is a straightforward consequence of Theorem 1.9 and was first proved in the late seventies by Seifullayev with the help of other (and less elementary) methods.

Theorem 1.10 (Seifullayev). *If Γ is a simple Carleson curve and $t \in \Gamma$, then*

$$\arg(\tau - t) = O(-\log|\tau - t|) \quad \text{as } \tau \to t. \tag{1.23}$$

Proof. Fix any point $\tau_0 \in \Gamma \setminus \{t\}$. From Theorem 1.9 we obtain that

$$|\arg(\tau - t) - \arg(\tau_0 - t)| < 2\pi e C_\Gamma \left(\log|\tau_0 - t| - \log|\tau - t| + 1\right)$$

and thus $|\arg(\tau - t)| \leq M(-\log|\tau - t|)$ with some constant $M < \infty$ as $\tau \to t$. $\qquad\square$

Notice that immediate application of Theorem 1.8 gives only the estimate $\arg(\tau - t) = O(1/|\tau - t|)$ as $\tau \to t$.

At the whirl point $t = 0$ of the logarithmic spiral of Example 1.6 we have $\arg \tau = -\delta \log |\tau|$, and the Carleson curves considered in Example 1.7 deliver plenty of arguments oscillating between $-\delta_1 \log |\tau|$ and $-\delta_2 \log |\tau|$ as $\tau \to 0$. This shows that (1.23) cannot be improved.

1.3   Seifullayev bounds

By virtue of Theorem 1.10, we may with each point t of a simple Carleson curve associate the two numbers

$$\sigma_t^- := \liminf_{\tau \to t} \frac{\arg(\tau - t)}{-\log |\tau - t|}, \quad \sigma_t^+ := \limsup_{\tau \to t} \frac{\arg(\tau - t)}{-\log |\tau - t|}. \tag{1.24}$$

Clearly, $-\infty < \sigma_t^- \leq \sigma_t^+ < +\infty$. We call σ_t^- and σ_t^+ the *lower* and *upper Seifullayev bounds* of Γ at t. In a sense, the Seifullayev bounds measure the spirality of the curve at the given point.

A point $t \in \Gamma$ is said to be *nonhelical* if $\arg(\tau - t) = O(1)$ as $\tau \to t$; otherwise it is called *helical*. For instance, all points of the curves of Examples 1.2 and 1.3 are nonhelical. If $t \in \Gamma$ is nonhelical then $\sigma_t^- = \sigma_t^+ = 0$. We refer to helical points t at which $\arg(\tau - t) = o(-\log |\tau - t|)$ as *hidden whirl points*. For example, if Γ is given by (1.9), (1.12) with $h(x) = (1/\sqrt{x^2 + 1}) \sin x$, then the origin is a hidden whirl point of Γ. When plotting a Carleson curve, hidden whirl points are difficult to detect – they look like nonhelical points. Obviously, $\sigma_t^- = \sigma_t^+ = 0$ if and only if t is nonhelical or a hidden whirl point.

A logarithmic spiral scrolling up at the origin is given by $\tau = re^{-i\delta \log r} e^{i\beta}$ with $0 \leq \beta < 2\pi$. In this case $\arg \tau = -\delta \log |\tau| + \beta$ and thus $\sigma_0^- = \sigma_0^+ = -\delta$. Now consider two logarithmic spirals:

$$\tau = re^{-i\delta_1 \log r} e^{i\beta_1} \quad \text{and} \quad \tau = re^{-i\delta_2 \log r} e^{i\beta_2}. \tag{1.25}$$

If $\delta_1 \neq \delta_2$, then every interval $(0, r_0)$ contains infinitely many r for which

$$-\delta_1 \log r + \beta_1 \in -\delta_2 \log r + \beta_2 + 2\pi \mathbf{Z},$$

therefore the two spirals (1.25) intersect at infinitely many points, and thus, their union is not a simple (nor even a composed) curve. However, we are given a simple Carleson curve in case $\delta_1 = \delta_2 = \delta$ and $\beta_1 \neq \beta_2$. We then have $\arg \tau = -\delta \log |\tau| + \beta(\tau)$ where $\beta(\tau)$ is constant on the two connected components of $\Gamma \backslash \{0\}$, whence $\sigma_0^- = \sigma_0^+ = -\delta$.

We remark that for a general simple Carleson curve the Seifullayev bounds coincide, $\sigma_0^- = \sigma_0^+ = -\delta$, if and only if

$$\arg(\tau - t) = -\delta \log |\tau - t| + o(\log |\tau - t|) \quad \text{as } \tau \to t,$$

that is, if and only if t is nonhelical, a hidden whirl point, or Γ is a "hidden perturbation" of one or two logarithmic spirals in a neighborhood of t.

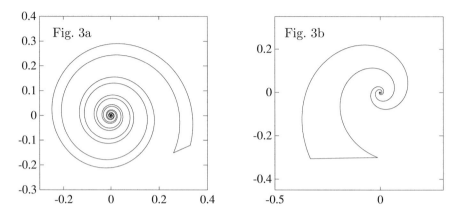

Figures 3a and 3b show Carleson Jordan curves comprised of a line segment and two logarithmic spirals (with $\delta = 10$ in Figure 3a and $\delta = -3$ in Figure 3b).

For further reference, we still single out two more general cases.

Example 1.11. Define arcs Γ_1 and Γ_2 by

$$\Gamma_1 = \{0\} \cup \{\tau \in \mathbf{C} : \tau = re^{i\varphi(r)},\ 0 < r \leq 1\},$$
$$\Gamma_2 = \{0\} \cup \{\tau \in \mathbf{C} : \tau = re^{i(\varphi(r)+b(r))},\ 0 < r \leq 1\},$$

where φ and b are functions in $C(0,1] \cap C^1(0,1)$ satisfying

$$|r\varphi'(r)| \leq M \ \text{ and } \ |r(\varphi'(r) + b'(r))| \leq M \ \text{ for all } \ r \in (0,1),$$
$$b(1) = 0 \ \text{ and } \ 0 < b(r) < 2\pi \ \text{ for all } \ r \in (0,1).$$

From Proposition 1.4 we see that Γ_1 and Γ_2 and thus also $\Gamma_1 \cup \Gamma_2$ are Carleson curves. Notice that $\Gamma_1 \cup \Gamma_2$ is a Jordan curve. Since $b(r)/(-\log r) \to 0$ as $r \to 0$, it follows that $\Gamma_1, \Gamma_2, \Gamma_1 \cup \Gamma_2$ have the same Seifullayev bounds σ_0^- and σ_0^+ at the origin:

$$\sigma_0^- = \liminf_{r \to 0} \big(\varphi(r)/(-\log r)\big), \ \ \sigma_0^+ = \limsup_{r \to 0} \big(\varphi(r)/(-\log r)\big). \qquad (1.26)$$

$\square$

Example 1.12. Let $\Gamma_1, \Gamma_2, \Gamma_1 \cup \Gamma_2$ be as in previous example. In addition, suppose

$$\varphi(r) = h\big(\log(-\log r)\big)(-\log r)$$

where $h \in C^1(\mathbf{R})$ and h as well as h' are bounded on $\mathbf{R}$. Taking into account the special form of φ, we obtain from (1.26) that

$$\sigma_0^- = \liminf_{x \to +\infty} h(x), \quad \sigma_0^+ = \limsup_{x \to +\infty} h(x). \tag{1.27}$$

$\square$

It turns out that in the spectral theory of Toeplitz and singular integral operators on Carleson curves it is not the Seifullayev bounds but rather two other numbers, the spirality indices, which play the decisive role. These indices will be defined in Section 1.6, where we will also discuss their connection with the Seifullayev bounds.

1.4 Submultiplicative functions

In what follows we will repeatedly make use of some simple and well known properties of submultiplicative functions. These properties are assembled in this section.

We call a function $\varrho : (0, \infty) \to (0, \infty]$ *regular* if it is bounded from above in some open neighborhood of the point 1. A function $\varrho : (0, \infty) \to (0, \infty]$ is said to be *submultiplicative* if

$$\varrho(x_1 x_2) \leq \varrho(x_1)\varrho(x_2) \quad \text{for all} \quad x_1, x_2 \in (0, \infty). \tag{1.28}$$

If $\varrho : (0, \infty) \to (0, \infty]$ is both regular and submultiplicative, then $\varrho(x)$ is bounded away from zero in some open neighborhood of 1, because $\varrho(1) \leq \varrho(x)\varrho(x^{-1})$ for all $x > 1$. Obviously, a submultiplicative function ϱ is regular if and only if $\log \varrho$ is bounded in some open neighborhood of 1. Moreover, if $\varrho : (0, \infty) \to (0, \infty]$ is regular and submultiplicative, then $\varrho(x)$ is finite for all $x \in (0, \infty)$ and $\log \varrho$ is bounded on every bounded segment $[a, b] \subset (0, \infty)$. Indeed, if $|\log \varrho(x)| \leq M$ for all $x \in [x_0^{-1}, x_0]$ with some $x_0 > 1$, then, by the submultiplicativity, $|\log \varrho(x)| \leq nM$ for all $x \in [x_0^{-n}, x_0^n]$.

Given a regular submultiplicative function $\varrho : (0, \infty) \to (0, \infty)$, one defines

$$\alpha(\varrho) := \sup_{x \in (0,1)} \frac{\log \varrho(x)}{\log x}, \quad \beta(\varrho) := \inf_{x \in (1,\infty)} \frac{\log \varrho(x)}{\log x}.$$

Clearly, $-\infty < \alpha(\varrho)$ and $\beta(\varrho) < +\infty$. We call $\alpha(\varrho)$ and $\beta(\varrho)$ the *lower* and *upper* *indices* of ϱ, respectively.

Theorem 1.13. *If ϱ is regular and submultiplicative, then*

$$\alpha(\varrho) = \lim_{x \to 0} \frac{\log \varrho(x)}{\log x}, \quad \beta(\varrho) = \lim_{x \to \infty} \frac{\log \varrho(x)}{\log x}, \tag{1.29}$$

and $-\infty < \alpha(\varrho) \leq \beta(\varrho) < +\infty$.

Proof. Assume first that $\beta := \beta(\varrho) > -\infty$. Given any $\varepsilon > 0$, pick any $x_0 > 1$ so that $\log \varrho(x_0)/\log x_0 < \beta + \varepsilon$. For $x > x_0$, choose a natural number n so that $x_0^n \le x < x_0^{n+1}$. Then, by (1.28),

$$\log \varrho(x) = \log \varrho(x x_0^{-n} x_0^n) \le \log \varrho(x x_0^{-n}) + n \log \varrho(x_0)$$

and hence

$$\beta \le \frac{\log \varrho(x)}{\log x} \le \frac{\log \varrho(x x_0^{-n})}{\log x} + \frac{n \log x_0}{\log x} \frac{\log \varrho(x_0)}{\log x_0}. \qquad (1.30)$$

Since $x x_0^{-n} \in [1, x_0]$, it follows that $\log \varrho(x x_0^{-n})$ remains bounded and that therefore the first term on the right of (1.30) goes to zero as $x \to \infty$. Because $n/(n+1) < n \log x_0/\log x \le 1$, the second term on the right of (1.30) tends to $\log \varrho(x_0)/\log x_0 < \beta + \varepsilon$ as $x \to \infty$. Consequently, $\beta \le \log \varrho(x)/\log x < \beta + \varepsilon$ for all sufficiently large x, which proves the second equality of (1.29).

If $\beta = -\infty$, we may for every $M > 0$ find an $x_0 > 1$ such that $\log \varrho(x_0)/\log x_0 < -M$. Repeating the argument of the previous paragraph we get the inequality $\log \varrho(x)/\log x < -M$ for all sufficiently large x, which proves the second equality of (1.29) in case $\beta = -\infty$.

If ϱ is regular and submultiplicative, then so also is the function $\tilde{\varrho}$ given by $\tilde{\varrho}(x) = \varrho(x^{-1})$. We have $\alpha(\varrho) = -\beta(\tilde{\varrho})$ and thus, by what has already been proved,

$$\alpha(\varrho) = -\beta(\tilde{\varrho}) = - \lim_{x \to \infty} \frac{\log \varrho(x^{-1})}{\log x} = \lim_{x \to 0} \frac{\log \varrho(x)}{\log x}.$$

This completes the proof of (1.29).

Finally, since $\varrho(1) \le \varrho(x)\varrho(x^{-1})$, we get

$$\frac{\log \varrho(x)}{\log x} \ge \frac{\log \varrho(1)}{\log x} + \frac{\log \varrho(x^{-1})}{\log x^{-1}} \quad \text{for all } x > 1.$$

Passing to the limit $x \to \infty$ and taking into account (1.29), we arrive at the inequality $\beta(\varrho) \ge \alpha(\varrho)$. $\qquad \square$

Corollary 1.14. *Let ϱ be regular and submultiplicative. Then $\varrho(x) \ge x^{\alpha(\varrho)}$ for all $x \in (0, 1)$ and $\varrho(x) \ge x^{\beta(\varrho)}$ for all $x \in (1, \infty)$. Furthermore, given any $\varepsilon > 0$, there is an $x_0 > 1$ such that $\varrho(x) \le x^{\alpha(\varrho)-\varepsilon}$ for all $x \in (0, x_0^{-1})$ and $\varrho(x) \le x^{\beta(\varrho)+\varepsilon}$ for all $x \in (x_0, \infty)$.*

Proof. By the definition of $\beta(\varrho)$, we have $\log \varrho(x)/\log x \ge \beta(\varrho)$ and thus $\varrho(x) \ge x^{\beta(\varrho)}$ for all $x > 1$. Theorem 1.13 implies that $\log \varrho(x)/\log x \le \beta(\varrho) + \varepsilon$ and therefore $\varrho(x) \le x^{\beta(\varrho)+\varepsilon}$ whenever $x > 1$ is sufficiently large. The assertions with $\alpha(\varrho)$ can be shown in the same way. $\qquad \square$

1.5 The W transform

Let Γ be a bounded simple Carleson curve and fix $t \in \Gamma$. We are now going to associate two functions $W_t\psi, W_t^0\psi : (0, \infty) \to (0, \infty]$ with every continuous function $\psi : \Gamma \setminus \{t\} \to (0, \infty)$, which will be called the W *transforms* of ψ.

Given a continuous function $\psi : \Gamma \setminus \{t\} \to (0, \infty)$, we define

$$(W_t\psi)(x) := \begin{cases} \sup_{R>0} \left(\max_{|\tau-t|=xR} \psi(\tau) / \min_{|\tau-t|=R} \psi(\tau) \right) & \text{for } x \in (0, 1] \\ \sup_{R>0} \left(\max_{|\tau-t|=R} \psi(\tau) / \min_{|\tau-t|=x^{-1}R} \psi(\tau) \right) & \text{for } x \in [1, \infty); \end{cases} \tag{1.31}$$

here and in the following we use the abbreviations

$$\sup_{R>0} := \sup_{0<R\leq d_t} \quad \text{where } d_t := \max_{\tau\in\Gamma} |\tau - t|, \tag{1.32}$$

$$\max_{|\tau-t|=y} := \max_{\{\tau\in\Gamma:|\tau-t|=y\}}, \quad \min_{|\tau-t|=y} := \min_{\{\tau\in\Gamma:|\tau-t|=y\}}. \tag{1.33}$$

The function $W_t^0\psi$ is defined by (1.31) with $\sup_{R>0}$ replaced by $\limsup_{R\to 0}$. A little thought shows that for all $x \in (0, \infty)$ we may write

$$(W_t^0\psi)(x) = \limsup_{R\to 0} \left(\max_{|\tau-t|=xR} \psi(\tau) / \min_{|\tau-t|=R} \psi(\tau) \right) \tag{1.34}$$

$$= \limsup_{R\to 0} \left(\max_{|\tau-t|=R} \psi(\tau) / \min_{|\tau-t|=x^{-1}R} \psi(\tau) \right). \tag{1.35}$$

To avoid huge expressions, we also define

$$M(\psi, R_1, R_2) := \max_{|\tau-t|=R_1} \psi(\tau) / \min_{|\tau-t|=R_2} \psi(\tau) \tag{1.36}$$

for $R_1, R_2 \in (0, d_t]$. With this notation, (1.31) reads

$$(W_t\psi)(x) = \begin{cases} \sup_{R>0} M(\psi, xR, R) & \text{for } x \in (0, 1] \\ \sup_{R>0} M(\psi, R, x^{-1}R) & \text{for } x \in [1, \infty). \end{cases}$$

Lemma 1.15. *The function $W_t\psi$ is submultiplicative for every continuous function $\psi : \Gamma \setminus \{t\} \to (0, \infty)$.*

Proof. Clearly, $(W_t\psi)(x) > 0$ for all $x \in (0, \infty)$. If $x = x_1x_2$ with x_1 and x_2 in $(0, 1]$, we have

$$(W_t\psi)(x) \leq \sup_{0<R\leq d_t} M(\psi, xR, x_2R,) \sup_{0<R\leq d_t} M(\psi, x_2R, R)$$

$$= \sup_{0<R'\leq x_2d_t} M(\psi, x_1R', R') \sup_{0<R\leq d_t} M(\psi, x_2R, R)$$

$$\leq \sup_{0<R\leq d_t} M(\psi, x_1R, R) \sup_{0<R\leq d_t} M(\psi, x_2R, R) = (W_t\psi)(x_1)(W_t\psi)(x_2).$$

If $x = x_1 x_2$ with $x_1 \in (0,1]$, $x_2 \in (1, \infty)$, $x \in (0,1]$ then

$$
\begin{aligned}
(W_t \psi)(x) &\leq \sup_{0 < R \leq d_t} M(\psi, xR, x_1 R) \sup_{0 < R \leq d_t} M(\psi, x_1 R, R) \\
&= \sup_{0 < R' \leq xd_t} M(\psi, R', x_2^{-1} R') \sup_{0 < R \leq d_t} M(\psi, x_1 R, R) \\
&\leq \sup_{0 < R \leq d_t} M(\psi, R, x_2^{-1} R) \sup_{0 < R \leq d_t} M(\psi, x_1 R, R) = (W_t \psi)(x_2)(W_t \psi)(x_1).
\end{aligned}
$$

The remaining cases can be treated similarly. This proves the submultiplicativity of $W_t \psi$. $\qquad \square$

The problems of deciding whether $W_t^0 \psi$ is submultiplicative and whether $W_t \psi$ and $W_t^0 \psi$ are regular are in general difficult, but we have at least the following result.

Lemma 1.16. *Let* $\psi : \Gamma \setminus \{t\} \to (0, \infty)$ *be a continuous function. If* $W_t \psi$ *is regular, then* $W_t^0 \psi$ *is regular and submultiplicative and both functions have the same lower and upper indices:*

$$
\alpha(W_t \psi) = \alpha(W_t^0 \psi), \quad \beta(W_t \psi) = \beta(W_t^0 \psi). \tag{1.37}
$$

Proof. Suppose $W_t \psi$ is regular. Since $(W_t^0 \psi)(x) \leq (W_t \psi)(x)$ for all $x \in (0, \infty)$, the function $W_t^0 \psi$ is bounded from above in a neighborhood of the point 1 if $W_t \psi$ is bounded from above there.

Abbreviate $\alpha(W_t \psi) =: \alpha$ and $\beta(W_t \psi) =: \beta$. By Corollary 1.14, there is a function $\varepsilon : (0,1) \to [0, \infty)$ such that $\varepsilon(x) \to 0$ as $x \to 0$ and

$$
x^\alpha \leq (W_t \psi)(x) \leq x^{\alpha - \varepsilon(x)} \quad \text{for} \quad x \in (0,1). \tag{1.38}
$$

Fix any $x \in (0,1)$. Then for every integer $m \geq 1$ there exists a number $R_m \in (0, d_t]$ such that $(W_t \psi)(x^m) \leq M(\psi, x^m R_m, R_m) x^{-1}$. Hence, by (1.38),

$$
M(\psi, x^m R_m, R_m) \geq x^{m\alpha + 1}. \tag{1.39}
$$

On the other hand, we infer from (1.38) that

$$
M(\psi, x^m R_{2m}, R_{2m}) \leq (W_t \psi)(x^m) \leq x^{m\alpha - m\varepsilon(x^m)}. \tag{1.40}
$$

We have

$$
\begin{aligned}
\prod_{n=m}^{2m-1} M(\psi, x^{n+1} R_{2m}, x^n R_{2m}) &\geq M(\psi, x^{2m} R_{2m}, x^m R_{2m}) \\
&\geq M(\psi, x^{2m} R_{2m}, R_{2m}) / M(\psi, x^m R_{2m}, R_{2m})
\end{aligned}
$$

and hence by (1.39) and (1.40),

$$\prod_{n=m}^{2m-1} M(\psi, x^{n+1} R_{2m}, x^n R_{2m}) \geq \frac{x^{2m\alpha+1}}{x^{m\alpha-m\varepsilon(x^m)}} = x^{m\alpha+1+m\varepsilon(x^m)}. \tag{1.41}$$

Choose $n(m)$ so that

$$M(\psi, x^{n(m)+1} R_{2m}, x^{n(m)} R_{2m})$$
$$= \max \{ M(\psi, x^{n+1} R_{2m}, x^n R_{2m}) : m \leq n \leq 2m-1 \}.$$

Then

$$\left[M(\psi, x^{n(m)+1} R_{2m}, x^{n(m)} R_{2m}) \right]^m \geq \prod_{n=m}^{2m-1} M(\psi, x^{n+1} R_{2m}, x^n R_{2m})$$

and (1.41) implies that $M(\psi, x^{n(m)+1} R_{2m}, x^{n(m)} R_{2m}) \geq x^{\alpha+1/m+\varepsilon(x^m)}$. Because $x^{n(m)} R_{2m} \leq x^m d_t = o(1)$ and $\varepsilon(x^m) = o(1)$ as $m \to \infty$, it follows that

$$(W_t^0 \psi)(x) = \limsup_{R \to 0} M(\psi, xR, R)$$
$$\geq \limsup_{m \to \infty} M(\psi, x^{n(m)+1} R_{2m}, x^{n(m)} R_{2m}) \geq x^\alpha. \tag{1.42}$$

Since (1.42) holds for every $x \in (0,1)$, we see that $(W_t^0 \psi)(x) > 0$ for all $x \in (0,1)$. It can be shown analogously that $(W_t^0 \psi)(x) \geq x^\beta$ for $x \in (1,\infty)$, where $\beta := \beta(W_t \psi)$, which implies that $(W_t^0 \psi)(x) > 0$ for all $x \in (1,\infty)$. As obviously $(W_t^0 \psi)(1) \geq 1$, we finally obtain that $W_t^0 \psi$ maps $(0,\infty)$ to $(0,\infty)$. Thus, since $W_t^0 \psi$ is bounded from above, it is regular. The submultiplicativity of $W_t^0 \psi$ can now be verified as in the proof of Lemma 1.15.

From (1.42) we deduce that $x^\alpha \leq (W_t^0 \psi)(x) \leq (W_t \psi)(x)$ for $x \in (0,1)$. Thus

$$\alpha \geq \frac{\log(W_t^0 \psi)(x)}{\log x} \geq \frac{\log(W_t \psi)(x)}{\log x},$$

whence

$$\alpha \geq \alpha(W_t^0 \psi) := \sup_{x \in (0,1)} \frac{\log(W_t^0 \psi)(x)}{\log x} \geq \sup_{x \in (0,1)} \frac{\log(W_t \psi)(x)}{\log x} = \alpha,$$

which proves that $\alpha = \alpha(W_t^0 \psi)$. In a similar way we obtain the equality $\beta = \beta(W_t^0 \psi)$. $\qquad \square$

We now specify ψ to the function $\eta_t : \Gamma \setminus \{t\} \to (0,\infty)$ defined by

$$\eta_t(\tau) := e^{-\arg(\tau-t)}, \tag{1.43}$$

where $\arg(\tau - t)$ denotes any continuous branch of the argument on $\Gamma \setminus \{t\}$.

Lemma 1.17. *If η_t is given by (1.43) then*

$$e^{-4\pi C_\Gamma} \leq (W_t \eta_t)(x) \leq e^{4\pi C_\Gamma} \quad \text{for } x \in [1/2, 2]$$

and hence, $W_t \eta_t$ is regular.

Proof. Let $x \in [1/2, 1]$ and $R \in (0, d_t]$. Since η_t cannot be constant on arcs contained in circles centered at t, there are uniquely determined points $\tau', \tau'' \in \Gamma$ such that

$$|\tau' - t| = xR, \quad \eta_t(\tau') = \max_{|\tau - t| = xR} \eta_t(\tau), \quad |\tau'' - t| = R, \quad \eta_t(\tau'') = \min_{|\tau - t| = R} \eta_t(\tau).$$

We have

$$\left| \log M(\eta_t, xR, R) \right| = \left| \log \left(\eta_t(\tau') / \eta_t(\tau'') \right) \right| = \left| -\log \eta_t(\tau'') + \log \eta_t(\tau') \right|$$
$$= \left| \arg(\tau'' - t) - \arg(\tau' - t) \right| = \left| \{ \arg(\tau - t) \}_{t \in \gamma} \right|$$

where $\gamma \subset \Gamma$ is the arc with the endpoints τ' and τ''. Thus, by Theorem 1.8,

$$\left| \log M(\eta_t, xR, R) \right| \leq 2\pi C_\Gamma R / (xR) = 2\pi C_\Gamma x^{-1} \leq 4\pi C_\Gamma,$$

whence $e^{-4\pi C_\Gamma} \leq M(\eta_t, xR, R) \leq e^{4\pi C_\Gamma}$ for all $x \in [1/2, 1]$ and all $R \in (0, d_t]$. The same estimate for $M(\eta_t, R, x^{-1}R)$ and $x \in [1, 2]$ can be shown analogously. $\qquad \square$

Theorem 1.18. *Let Γ be a bounded simple Carleson curve, fix $t \in \Gamma$, and define η_t by (1.43). Then $W_t \eta_t$ and $W_t^0 \eta_t$ are regular submultiplicative functions with the same lower and upper indices.*

Proof. Combine the preceding three lemmas. $\qquad \square$

1.6 Spirality indices

For a bounded simple Carleson curve Γ, we define the *lower* and *upper spirality indices* δ_t^- and δ_t^+ at a point $t \in \Gamma$ by

$$\delta_t^- := \alpha(W_t \eta_t) = \alpha(W_t^0 \eta_t), \quad \delta_t^+ := \beta(W_t \eta_t) = \beta(W_t^0 \eta_t). \tag{1.44}$$

Theorem 1.18 tells us that these indices are well-defined and that

$$-\infty < \delta_t^- \leq \delta_t^+ < +\infty.$$

The curve Γ is said to be *spiralic at the point* t if $\delta_t^- = \delta_t^+$ and is called a *spiralic curve* if $\delta_t^- = \delta_t^+$ for all $t \in \Gamma$.

To begin with an easy example, suppose Γ is a logarithmic spiral with the whirl point t. Then $\arg(\tau - t) = -\delta \log |\tau - t| + \beta$, hence

$$\eta_t(\tau) = e^{\delta \log |\tau - t| - \beta} = e^{-\beta} |\tau - t|^\delta,$$

and thus

$$(W_t^0 \eta_t)(x) = \limsup_{R \to 0} \left(e^{-\beta} (xR)^\delta / (e^{-\beta} R^\delta) \right) = x^\delta,$$

which implies that $\delta_t^- = \delta_t^+ = \delta$. If Γ consists of two logarithmic spirals Γ_1 and Γ_2 scrolling up at the same point t, then

$$\arg(\tau - t) = \begin{cases} -\delta \log |\tau - t| + \beta_1 & \text{for } \tau \in \Gamma_1, \\ -\delta \log |\tau - t| + \beta_2 & \text{for } \tau \in \Gamma_2 \end{cases}$$

(recall the discussion after (1.25)). Without loss of generality assume $\beta_1 < \beta_2$. We have

$$\max_{|\tau - t| = xR} e^{-\arg(\tau - t)} = \max \left\{ e^{-\beta_1} (xR)^\delta, e^{-\beta_2} (xR)^\delta \right\} = e^{-\beta_1} (xR)^\delta,$$

$$\min_{|\tau - t| = R} e^{-\arg(\tau - t)} = \min \left\{ e^{-\beta_1} R^\delta, e^{-\beta_2} R^\delta \right\} = e^{-\beta_2} R^\delta,$$

and therefore $(W_t^0 \eta_t)(x) = e^{\beta_2 - \beta_1} x^\delta$. Again it results that $\delta_t^- = \delta_t^+ = \delta$.

In case t is a nonhelical point, which means that $\arg(\tau - t) = O(1)$, we obtain that $\eta_t(\tau) = e^{O(1)}$ and consequently, $C_1 \leq (W_t^0 \eta_t)(x) \leq C_2$ with certain positive constants C_1 and C_2. Thus, $\delta_t^- = \delta_t^+ = 0$.

The following proposition provides us with a variety of less trivial examples.

Proposition 1.19. *Let $\Gamma_1, \Gamma_2, \Gamma = \Gamma_1 \cup \Gamma_2$ be as in Example 1.12. In addition, suppose $h \in C^2(\mathbf{R})$ and $|h''(x)| \leq M$ for $x \in \mathbf{R}$. Then $\Gamma_1, \Gamma_2, \Gamma$ have the same spirality indices at the origin, which are given at $t = 0$ by*

$$\delta_t^- = \liminf_{r \to 0} \left(- r\varphi'(r) \right) = \liminf_{x \to +\infty} \left(h(x) + h'(x) \right), \tag{1.45}$$

$$\delta_t^+ = \limsup_{r \to 0} \left(- r\varphi'(r) \right) = \limsup_{x \to +\infty} \left(h(x) + h'(x) \right). \tag{1.46}$$

Proof. We only consider $\Gamma = \Gamma_1 \cup \Gamma_2$, in which case

$$\arg(\tau - t) = \begin{cases} \varphi(|\tau - t|) & \text{for } \tau \in \Gamma_1 \setminus \{0\}, \\ \varphi(|\tau - t|) + b(|\tau - t|) & \text{for } \tau \in \Gamma_2 \setminus \{0\}. \end{cases}$$

Hence, if xR and R are in $(0, 1]$,

$$\max_{|\tau - t| = xR} e^{-\arg(\tau - t)} = \max \left\{ e^{-\varphi(xR)}, e^{-\varphi(xR) - b(R)} \right\} = e^{-\varphi(xR)},$$

$$\min_{|\tau - t| = R} e^{-\arg(\tau - t)} = \min \left\{ e^{-\varphi(R)}, e^{-\varphi(R) - b(R)} \right\} = e^{-\varphi(R) - b(R)},$$

and it follows that

$$(W_t^0 \eta_t)(x) = \limsup_{R \to 0} e^{\varphi(R) - \varphi(xR) + b(R)} = a(x) e^{\limsup_{R \to 0}(\varphi(R) - \varphi(xR))} \tag{1.47}$$

where, because of the inequalities $0 \le b(R) \le 2\pi$,

$$1 \le a(x) := (W_t^0 \eta_t)(x) e^{-\limsup_{R \to 0}(\varphi(R) - \varphi(xR))} \le e^{2\pi}. \tag{1.48}$$

We are so left with computing $\limsup_{R \to 0}(\varphi(R) - \varphi(xR))$. Put $y = -\log R$ and $\varepsilon = -\log x$. Then

$$\varphi(R) - \varphi(xR)$$
$$= h\big(\log(-\log R)\big)(-\log R) - h\big(\log(-\log(xR))\big)\big(-\log(xR)\big)$$
$$= h(\log y)y - h\big(\log(y + \varepsilon)\big)(y + \varepsilon)$$
$$= h(\log y)y - h\big(\log y + \log(1 + \varepsilon/y)\big)(y + \varepsilon)$$
$$= \big(h(\log y) - h(\log y + \log(1 + \varepsilon/y))\big)(y + \varepsilon) - \varepsilon h(\log y)$$
$$= h'\big(\xi(y)\big)\big(-\log(1 + \varepsilon/y)\big)(y + \varepsilon) - \varepsilon h(\log y)$$

with some $\xi(y) \in (\log y, \log y + \log(1 + \varepsilon/y))$. Since $\log(1 + \varepsilon/y) = \varepsilon/y + O(1/y^2)$, we get

$$\varphi(R) - \varphi(xR) = -\varepsilon h'\big(\xi(y)\big) - \varepsilon h(\log y) + O(1/y)$$
$$= -\varepsilon h'\big(\log y + O(1/y)\big) - \varepsilon h(\log y) + O(1/y)$$
$$= -\varepsilon h'(\log y) - \varepsilon h''\big(\eta(y)\big)O(1/y) - \varepsilon h(\log y) + O(1/y)$$

with $\eta(y) \in (\log y, \log y + O(1/y))$. Taking into account that h'' is bounded, we so obtain

$$\varphi(R) - \varphi(xR) = -\varepsilon\big(h(\log y) + h'(\log y)\big) + O(1/y)$$

and thus,

$$\limsup_{R \to 0}\big(\varphi(R) - \varphi(xR)\big) = \limsup_{y \to +\infty}\Big(\log x\big(h(\log y) + h'(\log y)\big)\Big).$$
$$= \begin{cases} \log x \liminf_{z \to +\infty}\big(h(z) + h'(z)\big) & \text{for } x \in (0, 1] \\ \log x \limsup_{z \to +\infty}\big(h(z) + h'(z)\big) & \text{for } x \in [1, \infty). \end{cases}$$

From (1.47) we now see that

$$\frac{\log(W_t^0 \eta_t)(x)}{\log x} = \begin{cases} \frac{a(x)}{\log x} + \liminf_{z \to +\infty}\big(h(z) + h'(z)\big) & \text{for } x \in (0, 1] \\ \frac{a(x)}{\log x} + \limsup_{z \to +\infty}\big(h(z) + h'(z)\big) & \text{for } x \in [1, \infty). \end{cases}$$

which in view of (1.48) gives the most right expressions for δ_t^- and δ_t^+ in (1.45) and (1.46). Because

$$-r\varphi'(r) = -r\Big(h\big(\log(-\log r)\big)(-\log r)\Big)' = h\big(\log(-\log r)\big) + h'\big(\log(-\log r)\big),$$

we finally arrive at the equalities

$$\liminf_{r\to 0} \left(-r\varphi'(r) \right) = \liminf_{z\to+\infty} \left(h(z) + h'(z) \right),$$

$$\limsup_{r\to 0} \left(-r\varphi'(r) \right) = \limsup_{z\to+\infty} \left(h(z) + h'(z) \right). \qquad \square$$

Comparing (1.26), (1.27) and (1.45), (1.46) we arrive at the observation that the Seifullayev bounds need not coincide with the spirality indices. For example, if $h(x) = \delta + \mu \sin \lambda x$, then

$$\sigma_t^- = \delta - |\mu|, \quad \sigma_t^+ = \delta + |\mu|,$$

while $h(x) + h'(x) = \delta + \mu(\lambda \cos \lambda x + \sin \lambda x)$ and thus

$$\delta_t^- = \delta - |\mu| \max_{x\in\mathbf{R}}(\lambda \cos \lambda x + \sin \lambda x) = \delta - |\mu|\sqrt{\lambda^2 + 1},$$

$$\delta_t^+ = \delta + |\mu| \max_{x\in\mathbf{R}}(\lambda \cos \lambda x + \sin \lambda x) = \delta + |\mu|\sqrt{\lambda^2 + 1}.$$

Moreover, this example provides us with Carleson curves which have arbitrarily prescribed spirality indices $\delta_t^- \le \delta_t^+$ at the point t.

Proposition 1.20. *If Γ is a bounded simple Carleson curve and $t \in \Gamma$, then*

$$-\infty < \delta_t^- \le \sigma_t^- \le \sigma_t^+ \le \delta_t^+ < +\infty. \qquad (1.49)$$

Proof. Put $\mu(y) := \max_{|\tau - t| = y} \eta_t(\tau)$ and $\nu(y) := \min_{|\tau - t| = y} \eta_t(\tau)$. Then

$$\delta_t^+ = \lim_{x\to\infty} \frac{\log(W_t \eta_t)(x)}{\log x} = \lim_{x\to\infty} \sup_{R>0} \frac{\log \mu(R) - \log \nu(x^{-1} R)}{\log x} \qquad (1.50)$$

and hence, given any $\varepsilon > 0$, there is an $x_0 > 1$ such that

$$\delta_t^+ + \varepsilon > \frac{\log \mu(R) - \log \nu(x^{-1} R)}{\log x}$$

$$= \frac{\log \mu(R)}{\log x} + \frac{\log \nu(x^{-1} R)}{\log(x^{-1} R)} \cdot \frac{-\log(x^{-1} R)}{\log x} \qquad (1.51)$$

for all $R \in (0, d_t]$ and all $x \in (x_0, \infty)$. Fix $R \in (0, d_t]$. Since $\log \mu(R)/\log x \to 0$ and $(-\log(x^{-1} R))/\log x \to 1$ as $x \to \infty$, we see from (1.51) that

$$\limsup_{x\to\infty} \frac{\log \nu(x^{-1} R)}{\log(x^{-1} R)} \le \delta_t^+ + \varepsilon.$$

Therefore

$$\limsup_{r\to 0} \frac{\log \nu(r)}{\log r} \le \delta_t^+ + \varepsilon$$

or equivalently,

$$\limsup_{r\to 0} \left(\log \min_{|\tau - t| = r} e^{-\arg(\tau - t)} / \log r \right) \le \delta_t^+ + \varepsilon. \qquad (1.52)$$

Because

$$\log \min_{|\tau - t| = r} e^{-\arg(\tau - t)} = \min_{|\tau - t| = r} \big(-\arg(\tau - t) \big) = - \max_{|\tau - t| = r} \arg(\tau - t),$$

it follows from (1.52) that

$$\sigma_t^+ := \limsup_{\tau \to t} \frac{\arg(\tau - t)}{-\log|\tau - t|} \le \delta_t^+ + \varepsilon.$$

As $\varepsilon > 0$ was arbitrary, we arrive at the inequality $\sigma_t^+ \le \delta_t^+$. The inequality $\sigma_t^- \ge \delta_t^-$ can be shown analogously. The proof of the second part of the proposition is left to the reader. $\square$

The previous proposition implies in particular that if $\delta_t^- = \delta_t^+ =: \delta$, then $\sigma_t^- = \sigma_t^+ = \delta$. Notice that for curves as in Proposition 1.19 this is also immediate from L' Hospital's rule and formulas (1.26), (1.45), (1.46):

$$\lim_{r \to 0} \frac{\varphi(r)}{-\log r} = \lim_{r \to 0} \frac{\varphi'(r)}{-1/r} = \lim_{r \to 0} \big(-r\varphi'(r) \big)$$

whenever $-r\varphi'(r)$ has a limit as $r \to 0$. The first half of the following proposition is a curious converse of the observation just made.

Proposition 1.21. *Let $\Gamma = \Gamma_1$ be as in Proposition 1.19. If $\sigma_0^- = \sigma_0^+ =: \sigma_0$, then $\delta_0^- = \delta_0^+ = \sigma_0$. On the other hand, given any numbers $\delta_t^\pm, \sigma_t^\pm$ such that*

$$-\infty < \delta_t^- \le \sigma_t^- < \sigma_t^+ \le \delta_t^+ < +\infty, \tag{1.53}$$

there exists a Carleson curve as in Proposition 1.19 whose spirality indices $\delta_0^\pm$ and Seifullayev bounds $\sigma_0^\pm$ are the numbers $\delta_t^\pm$ and $\sigma_t^\pm$, respectively.

Proof. Suppose $\sigma_0^- = \sigma_0^+ = \sigma_0$. Then $h(\log(-\log r)) = \sigma_0 + o(1)$ as $r \to 0$. Put $f(x) := h(x) - \sigma_0$. Since $f(x) \to 0$ as $x \to +\infty$ and f'' is bounded, we deduce from the inequality

$$|f'(x)|^2 \le 4 \sup_{y \ge x} |f(y)| \sup_{y \ge x} |f''(y)|$$

(see [179, Exercise 5.14]) that $f'(x) \to 0$ as $x \to +\infty$. Hence $h(x) \to \sigma_0$ and $h'(x) \to 0$ as $x \to +\infty$, and Proposition 1.19 therefore implies that $\delta_0^- = \delta_0^+ = \sigma_0$.

To find a curve whose Seifullayev and spirality indices satisfy (1.53), let first a, b, c, d be any real numbers such that $c > a > b > d$. We construct real numbers $x_1 < 0 < x_2$ and a real-valued function $f \in C^2[x_1, x_2]$ such that

$$c = \max_{x \in [x_1, x_2]} \big(f(x) + f'(x) \big), \quad a = \max_{x \in [x_1, x_2]} f(x),$$

$$b = \min_{x \in [x_1, x_2]} f(x), \quad d = \min_{x \in [x_1, x_2]} \big(f(x) + f'(x) \big),$$

and, in addition,

$$f(x_1) = f(x_2) = b, \quad f'(x_1) = f'(x_2) = f''(x_1) = f''(x_2) = 0.$$

To do this, consider the function $\varphi(\lambda) := \lambda - \arctan \lambda$. The function φ increases strictly monotoneously from 0 to ∞ as λ moves from 0 to ∞. Hence, there are uniquely determined numbers $\lambda_1 > 0, \lambda_2 > 0$ such that

$$\lambda_1 - \arctan \lambda_1 = \pi(c-a)/\omega, \quad \lambda_2 - \arctan \lambda_2 = \pi(b-d)/\omega, \tag{1.54}$$

where $\omega := a - b$. Put $x_1 := -2\pi/\lambda_1, \ x_2 := 2\pi/\lambda_2$ and define

$$f(x) := \begin{cases} \frac{\omega\lambda_1}{2\pi}\left(\lambda_1 x - \sin(\lambda_1 x)\right) + a & \text{for } x \in [x_1, 0] \\ -\frac{\omega\lambda_2}{2\pi}\left(\lambda_2 x - \sin(\lambda_2 x)\right) + a & \text{for } x \in [0, x_2]. \end{cases}$$

We claim that f is the function we are looking for. It is easily seen that f is in $C^2[x_1, x_2]$,

$$f(0) = a, \quad f(x_1) = f(x_2) = b,$$

and

$$f'(x_1) = f'(0) = f'(x_2) = f''(x_1) = f''(0) = f''(x_2) = 0.$$

Since f increases on $[x_1, 0]$ and decreases on $[0, x_2]$, it is clear the max $f = a$ and min $f = b$. Further,

$$f'(x) + f''(x) = \begin{cases} \frac{\omega\lambda_1}{\pi} \sin \frac{\lambda_1 x}{2}\left(\sin \frac{\lambda_1 x}{2} + \lambda_1 \cos \frac{\lambda_1 x}{2}\right) & \text{for } x \in [x_1, 0] \\ -\frac{\omega\lambda_2}{\pi} \sin \frac{\lambda_2 x}{2}\left(\sin \frac{\lambda_2 x}{2} + \lambda_2 \cos \frac{\lambda_2 x}{2}\right) & \text{for } x \in [0, x_2]. \end{cases}$$

Thus, $f' + f''$ assumes its maximum at some point $y_1 \in [x_1, 0]$ and its minimum at some point $y_2 \in [0, x_2]$, and we have $\tan(\lambda_k y_k/2) = -\lambda_k$ for $k = 1, 2$. It follows that

$$\lambda_1 y_1/2 = -\arctan \lambda_1, \quad \lambda_2 y_2/2 = \pi - \arctan \lambda_2,$$

which together with (1.54) gives

$$\begin{aligned}
f(y_1) + f'(y_1) &= \frac{\omega}{2\pi}\left(\lambda_1 y_1 - \sin(\lambda_1 y_1) + \lambda_1\left(1 - \cos(\lambda_1 y_1)\right)\right) + a \\
&= \frac{\omega}{\pi}\left(\frac{\lambda_1 y_1}{2} + \cos^2 \frac{\lambda_1 y_1}{2}\left(-\tan \frac{\lambda_1 y_1}{2} + \lambda_1 \tan^2 \frac{\lambda_1 y_1}{2}\right)\right) + a \\
&= \frac{\omega}{\pi}\left(-\arctan \lambda_1 + \frac{1}{1 + \lambda^2}(\lambda_1 + \lambda_1^3)\right) + a = \frac{\omega}{\pi}\pi\frac{c-a}{\omega} + a = a.
\end{aligned}$$

Analogously one can show that $f(y_2) + f'(y_2) = d$.

Now let $\sigma_t^{\pm}, \delta_t^{\pm}$ be numbers subject to (1.53). Looking for a Carleson curve as in Proposition 1.19 whose spirality indices and Seifullayev bounds at the origin are just these numbers amounts to finding a function $h \in C^2(\mathbf{R})$ for which h, h', h'' are bounded on $\mathbf{R}$ and for which

$$\liminf_{x \to +\infty} h(x) = \sigma_t^-, \qquad \limsup_{x \to +\infty} h(x) = \sigma_t^+,$$
$$\liminf_{x \to +\infty} \big(h(x) + h'(x)\big) = \delta_t^-, \quad \limsup_{x \to +\infty} \big(h(x) + h'(x)\big) = \delta_t^+.$$

If $\delta_t^- < \sigma_t^- < \sigma_t^+ < \delta_t^+$, we put $a = \sigma_t^+, b = \sigma_t^-, c = \delta_t^+, d = \delta_t^-$ and extend the corresponding function f constructed above periodically to a function h on all of $\mathbf{R}$. Let us assume that $\delta_t^- < \sigma_t^-$ but $\delta_t^+ = \sigma_t^+$. We then put $a = \sigma_t^+, b = \sigma_t^-, d = \delta_t^-$ and for $n \in \{1, 2, \ldots\}$, we let $c_n = a + 1/n$. For each quadruple (a, b, c_n, d) we construct a function $f_n \in C^2[x_{1,n}, x_{2,n}]$ as above and then we define h by

$$h(x) := \begin{cases} f_n(x + x_{1,n} - z_{1,n}) & \text{for } x \in I_n \ (n = 1, 2, \ldots) \\ b & \text{for } x \in \mathbf{R} \setminus \bigcup_{n=1}^{\infty} I_n \end{cases}$$

where $z_{1,n} := x_{1,1} + \sum_{k=1}^{n-1} \ell_k$, $I_n := [z_{1,n}, z_{1,n} + \ell_n]$, $\ell_n := x_{2,n} - x_{1,n}$. It is easily seen that h is the desired function. In the remaining cases one can proceed similarly. $\qquad\square$

Remark 1.22. In the first part of Proposition 1.21, the restriction to curves as in Proposition 1.19 is essential. To see this, we will give an example of a Carleson arc for which $\delta_0^- = -1$, $\sigma_0^- = \sigma_0^+ = 0$, and $\delta_0^+ = 1$.

To construct the desired arc, we first notice that Proposition 1.19 remains true with the requirement that h'' be bounded replaced by the weaker assumption that

$$h''(x) = o(e^x) \text{ as } x \to +\infty. \tag{1.55}$$

Indeed, in the proof of Proposition 1.19 we only have to replace the conclusion

$$h''(x) = O(1) \implies h''(\eta(y))O(1/y) = O(1/y)$$

by the conclusion that (1.55) implies

$$h''(\eta(y))O(1/y) = o\Big(e^{\log y + O(1/y)}\Big)O(1/y) = o(1) \text{ as } y \to +\infty.$$

Now define $\Gamma = \Gamma_1$ as in Example 1.12 with a function $h \in C^2(\mathbf{R})$ such that h, h', h'' are bounded on $(-\infty, 1]$ and

$$h(x) = xe^{-x} \sin \int_1^x \frac{e^u du}{u} \quad \text{for } x > 1.$$

For $x > 1$ we then have

$$h'(x) = (1-x)e^{-x}\sin\int_1^x \frac{e^u\,du}{u} + \cos\int_1^x \frac{e^u\,du}{u},$$

$$h''(x) = \left((x-2)e^{-x} - \frac{e^x}{x}\right)\sin\int_1^x \frac{e^u\,du}{u} + \frac{1-x}{x}\cos\int_1^x \frac{e^u\,du}{u}.$$

Thus, h and h' are bounded on $\mathbf{R}$, while h'' satisfies satisfies (1.55). We therefore can apply the modified form of Proposition 1.19 to obtain

$$\sigma_0^{\pm} = \lim_{x\to+\infty} h(x) = 0,$$

$$\delta_0^- = \liminf_{x\to+\infty}\left(h(x)+h'(x)\right) = \liminf_{x\to+\infty}\cos\int_1^x \frac{e^u\,du}{u} = -1,$$

$$\delta_0^+ = \limsup_{x\to+\infty}\left(h(x)+h'(x)\right) = \limsup_{x\to+\infty}\cos\int_1^x \frac{e^u\,du}{u} = +1. \qquad \square$$

We conclude this section with a final remark about the relation between the spirality indices and the Seifullayev bounds. Suppose Γ is as in Proposition 1.4. Then Theorem 1.9 implies that $(\varphi(x^{-1}R)-\varphi(R))/\log x$ remains bounded as (R,x) ranges over $(0, d_t] \times (1, \infty)$. Consider the two iterated upper limits

$$\limsup_{x\to\infty}\limsup_{R\to0} \frac{\varphi(x^{-1}R) - \varphi(R)}{\log x} \tag{1.56}$$

and

$$\limsup_{R\to0}\limsup_{x\to\infty} \frac{\varphi(x^{-1}R) - \varphi(R)}{\log x}. \tag{1.57}$$

The first of these iterated upper limits equals

$$\limsup_{x\to\infty}\left(\frac{1}{\log x}\log\limsup_{R\to0}\left(e^{-\varphi(R)}/e^{-\varphi(x^{-1}R)}\right)\right)$$

$$= \limsup_{x\to\infty}\frac{\log(W_0^0\eta_0)(x)}{\log x} = \lim_{x\to\infty}\frac{\log(W_0^0\eta_0)(x)}{\log x} = \beta(W_0^0\eta_0) = \delta_0^+,$$

while the second iterated upper limit is equal to

$$\limsup_{R\to0}\limsup_{x\to\infty}\frac{\varphi(x^{-1}R)}{\log x} = \limsup_{R\to0}\limsup_{x\to\infty}\frac{\varphi(x^{-1}R)}{-\log(x^{-1}R)}$$

$$= \limsup_{R\to0}\limsup_{r\to0}\frac{\varphi(r)}{-\log r} = \limsup_{R\to0}\sigma_0^+ = \sigma_0^+.$$

The finiteness of the two numbers (1.56) and (1.57) is one mystery of Carleson curves, the fact that in the spectral theory of Toeplitz and singular integral operators the number in (1.56) is the deciding quantity is another one.

1.7   Notes and comments

1.1. Our terminology follows Dynkin and Osilenker [62], [60], [61] who call a locally rectifiable composed curve Γ a *Carleson curve* if it satisfies (1.3), which is equivalent to (1.1). Carleson curves first appeared in the paper [1] by Ahlfors and have gained permanently increasing attention in mathematical analysis since David's papers [46], [47]. A non-negative Borel measure μ in the plane is referred to as a *Carleson measure* if $\sup_D \mu(D)/\mathrm{diam}\,(D) < \infty$, the supremum over all open disks, and it is clear that the measure μ given by $\mu(D) := |D \cap \Gamma|$ is a Carleson measure if and only if (1.3) holds, i.e. if and only if Γ is a Carleson curve.

We here restrict ourselves to connected curves. The major part of the results of this book is also true for curves with a finite number of connected components. Notice that if Γ has infinitely many connected components, then condition (1.3) is still necessary for the Cauchy singular integral S_Γ to be bounded on $L^p(\Gamma)$ but no longer sufficient; see, for example, the book by David and Semmes [50, Chapter 1.1] and the paper by Mattila, Melnikov, and Verdera [143].

1.2–1.3. Theorem 1.8 was established in our paper [16] and Theorem 1.9 seems to appear here for the first time, but it is very likely that these theorems are known to specialists. A result very close to Theorem 1.10 was obtained by Seifullayev [186] with the help of other methods. He also introduced the numbers $\sigma_t^{\pm}$ we call the Seifullayev bounds.

1.4. The theory of submultiplicative functions and their indices is well presented in the books by Bennett and Sharpley [8], S.G. Krein, Petunin, and Semenov [131], and Hille and Phillips [105]. The results and proofs of Section 1.4 are taken from these books.

1.5–1.6. The main results of these sections are from our paper [16]. There we also introduced the notion of the W-transform and the spirality indices. After the appearance of our papers [16], [18] we learned that functions of the form $h(\log(-\log x))\log x$ and especially the function $\sin(\log(-\log x))\log x$ are good friends of workers in Orlicz spaces. In particular, such functions have already been employed by Lindberg [135] and Maligranda [141] for other purposes. Lemma 1.16 is an analogue of Lemma 2(a) of Boyd [28] (also see Theorem 8.18 of [8]). The $c_n = a + 1/n$ trick in the proof of Proposition 1.21 is due to Alexei Yu. Karlovich.

Chapter 2
Muckenhoupt weights

This chapter aims at developing a certain feeling for Muckenhoupt weights. Power weights are a simple class of Muckenhoupt weights, but we have a long way to go in finding a sufficient supply of nontrivial Muckenhoupt weights. We have in particular to prove the so-called reverse Hölder inequality, which implies that if w is a weight in $A_p(\Gamma)$ then $w^{1+\varepsilon}$ belongs to $A_p(\Gamma)$ for all sufficiently small $|\varepsilon|$ ("stability" of Muckenhoupt weights). For weights with only one singularity, we establish a criterion for their membership in $A_p(\Gamma)$ in terms of the W transform. It is this criterion which will enable us to identify plenty of oscillating weights as Muckenhoupt weights.

2.1 Definitions

Let Γ be a composed locally rectifiable curve. A measurable function $w : \Gamma \to [0, \infty]$ is called a *weight* if the preimage $w^{-1}(\{0, \infty\})$ has measure zero. Thus, a weight is finite and nonzero almost everywhere. In the following chapters we will be concerned with the weighted Lebesgue spaces $L^p(\Gamma, w)$ $(1 < p < \infty)$ with the norm

$$\|f\|_{p,w} := \left(\int_\Gamma |f(\tau)|^p w(\tau)^p |d\tau| \right)^{1/p}.$$

In case $w(\tau) = 1$ for all $\tau \in \Gamma$, we abbreviate $L^p(\Gamma, w)$ to $L^p(\Gamma)$ and $\| \cdot \|_{p,w}$ to $\| \cdot \|_p$. Let $L^p_{\mathrm{loc}}(\Gamma)$ stand for the measurable functions on Γ belonging to $L^p(\gamma)$ for every bounded measurable subset $\gamma \subset \Gamma$. Equivalently, w is in $L^p_{\mathrm{loc}}(\Gamma)$ if and only if $w \in L^p(\Gamma(t, \varepsilon))$ for every $t \in \Gamma$ and every $\varepsilon \in (0, \infty)$.

Fix $p \in (1, \infty)$. Throughout what follows, we define $q \in (1, \infty)$ by $1/p + 1/q = 1$. We denote by $A_p(\Gamma)$ the set of all weights $w : \Gamma \to [0, \infty]$ such that

$w \in L^p_{\mathrm{loc}}(\Gamma)$, $w^{-1} \in L^q_{\mathrm{loc}}(\Gamma)$, and

$$C_w := \sup_{t \in \Gamma} \sup_{\varepsilon > 0} \left(\frac{1}{\varepsilon} \int\limits_{\Gamma(t,\varepsilon)} w(\tau)^p |d\tau| \right)^{1/p} \left(\frac{1}{\varepsilon} \int\limits_{\Gamma(t,\varepsilon)} w(\tau)^{-q} |d\tau| \right)^{1/q} < \infty, \qquad (2.1)$$

where, as in Chapter 1, $\Gamma(t,\varepsilon) := \{\tau \in \Gamma : |\tau - t| < \varepsilon\}$. Condition (2.1) is referred to as the *Muckenhoupt condition* and weights in $\bigcup_{1 < p < \infty} A_p(\Gamma)$ are called *Muckenhoupt weights* on Γ.

Let us begin with a few simple remarks about the Muckenhoupt condition. If $w \in A_p(\Gamma)$, then, by Hölder's inequality,

$$\frac{|\Gamma(t,\varepsilon)|}{\varepsilon} = \frac{1}{\varepsilon} \int\limits_{\Gamma(t,\varepsilon)} |d\tau| \leq \left(\frac{1}{\varepsilon} \int\limits_{\Gamma(t,\varepsilon)} w(\tau)^p |d\tau| \right)^{1/p} \left(\frac{1}{\varepsilon} \int\limits_{\Gamma(t,\varepsilon)} w(\tau)^{-q} |d\tau| \right)^{1/q} \leq C_w$$

and hence, the Muckenhoupt condition (2.1) implies the Carleson condition (1.1) with $C_\Gamma \leq C_w$. In other words, if the class $A_p(\Gamma)$ is non-empty, then Γ must necessarily be a Carleson curve. Conversely, if Γ is Carleson then constant weights belong to $A_p(\Gamma)$. Therefore, when dealing with Muckenhoupt weights we will henceforth always assume that the underlying curve is a Carleson curve.

In connection with what was said in the preceding paragraph the following remark seems to be in order. The Carleson condition is a condition for solely the curve Γ. However, the Muckenhoupt condition (2.1) is a condition for the triple (Γ, p, w), and it is only by psychological reasons that (2.1) is viewed as a condition for the weight w.

If Γ is bounded, then condition (2.1) is equivalent to the condition

$$\exists\, \varepsilon_0 > 0 : \sup_{t \in \Gamma} \sup_{\varepsilon < \varepsilon_0} \left(\frac{1}{\varepsilon} \int\limits_{\Gamma(t,\varepsilon)} w(\tau)^p |d\tau| \right)^{1/p} \left(\frac{1}{\varepsilon} \int\limits_{\Gamma(t,\varepsilon)} w(\tau)^{-q} |d\tau| \right)^{1/q} < \infty. \qquad (2.2)$$

Indeed, it is clear that (2.2) follows from (2.1). To get the reverse implication, suppose (2.2) holds. Then if $\varepsilon \geq \varepsilon_0$,

$$\frac{1}{\varepsilon} \int\limits_{\Gamma(t,\varepsilon)} w(\tau)^p |d\tau| \leq \frac{1}{\varepsilon_0} \int\limits_{\Gamma(t,\varepsilon)} w(\tau)^p |d\tau| \leq \frac{1}{\varepsilon_0} \int\limits_{\Gamma} w(\tau)^p |d\tau| = \frac{1}{\varepsilon_0} \|w\|_p^p,$$

and an analogous estimate holds with w^p replaced by w^{-q}. Consequently,

$$\sup_{t \in \Gamma} \sup_{\varepsilon \geq \varepsilon_0} \left(\frac{1}{\varepsilon} \int\limits_{\Gamma(t,\varepsilon)} w(\tau)^p |d\tau| \right)^{1/p} \left(\frac{1}{\varepsilon} \int\limits_{\Gamma(t,\varepsilon)} w(\tau)^{-q} |d\tau| \right)^{1/q} \leq \frac{1}{\varepsilon_0} \|w\|_p \|w^{-1}\|_q,$$

which proves the assertion. Since constant weights belong to $A_p(\Gamma)$ if and only if Γ is Carleson, we deduce from Section 1.1 that (2.1) and (2.2) are not equivalent in case Γ is unbounded.

If Γ is a bounded Carleson curve, then $\varepsilon \leq |\Gamma(t,\varepsilon)| \leq C_\Gamma \varepsilon$ for all sufficiently small $\varepsilon > 0$ by virtue of the Carleson condition, hence condition (2.2) is equivalent to the condition

$$\exists \varepsilon_0 > 0 : \sup_{t \in \Gamma} \sup_{\varepsilon < \varepsilon_0} \left(\frac{1}{|\Gamma(t,\varepsilon)|} \int_{\Gamma(t,\varepsilon)} w(\tau)^p |d\tau| \right)^{1/p} \left(\frac{1}{|\Gamma(t,\varepsilon)|} \int_{\Gamma(t,\varepsilon)} w(\tau)^{-q} |d\tau| \right)^{1/q} < \infty,$$

(2.3)

and since $\Gamma \supset \Gamma(t,\varepsilon) \supset \Gamma(t,\varepsilon_0)$ for $\varepsilon \geq \varepsilon_0$, we obtain as above that (2.3) is in turn equivalent to

$$c_w := \sup_{t \in \Gamma} \sup_{\varepsilon > 0} \left(\frac{1}{|\Gamma(t,\varepsilon)|} \int_{\Gamma(t,\varepsilon)} w(\tau)^p |d\tau| \right)^{1/p} \left(\frac{1}{|\Gamma(t,\varepsilon)|} \int_{\Gamma(t,\varepsilon)} w(\tau)^{-q} |d\tau| \right)^{1/q} < \infty.$$

(2.4)

If Γ is an unbounded Carleson curve, then $\varepsilon \leq |\Gamma(t,\varepsilon)| \leq C_\Gamma \varepsilon$ for all $\varepsilon > 0$, which implies that (2.1) is the same as (2.4). Thus, independently of whether Γ is bounded or not, we arrive at the conclusion that a weight w belongs to $A_p(\Gamma)$ if and only if Γ is a Carleson curve, $w \in L^p_{\mathrm{loc}}(\Gamma)$, $w^{-1} \in L^q_{\mathrm{loc}}(\Gamma)$, and

$$c_w = \sup_Q \left(\frac{1}{|Q|} \int_Q w(\tau)^p |d\tau| \right)^{1/p} \left(\frac{1}{|Q|} \int_Q w(\tau)^{-q} |d\tau| \right)^{1/q} < \infty, \qquad (2.5)$$

where Q ranges over all portions $\Gamma(t,\varepsilon)$ $(t \in \Gamma, \varepsilon > 0)$.

To simplify notation, we will in the following frequently employ the abbreviation

$$\Delta_t(f,\varepsilon) := \frac{1}{|\Gamma(t,\varepsilon)|} \int_{\Gamma(t,\varepsilon)} f(\tau) |d\tau|. \qquad (2.6)$$

With this notation, (2.4) reads

$$\sup_{t \in \Gamma} \sup_{\varepsilon > 0} \left(\Delta_t(w^p, \varepsilon) \right)^{1/p} \left(\Delta_t(w^{-q}, \varepsilon) \right)^{1/q} < \infty. \qquad (2.7)$$

For a measurable subset γ of Γ, we denote by $L^\infty(\gamma)$ the Banach algebra of all essentially bounded functions on γ and by $GL^\infty(\gamma)$ the group of invertible elements of $L^\infty(\gamma)$, that is, the functions $f \in L^\infty(\gamma)$ for which $1/f$ also belongs to $L^\infty(\gamma)$.

Proposition 2.1. *Let Γ be a bounded composed Carleson curve and $1 < p < \infty$.*

(a) *If $w_1, w_2 \in A_p(\Gamma)$ and there exists an $\varepsilon_0 > 0$ such that for each $t \in \Gamma$ at least one of the functions w_1, w_2 belongs to $GL^\infty(\Gamma(t,\varepsilon_0))$, then $w_1 w_2 \in A_p(\Gamma)$.*

(b) *If $w_1, w_2 \in A_p(\Gamma)$ then $w_1^\theta w_2^{1-\theta} \in A_p(\Gamma)$ for every $\theta \in [0,1]$.*

Proof. (a) Choose points $t_i \in \Gamma$ $(i = 1, \ldots, n)$ so that $\Gamma \subset \bigcup_{i=1}^{n} \Gamma(t_i, \varepsilon_0/2)$. Clearly, each portion $\Gamma(t, \varepsilon)$ with $\varepsilon < \varepsilon_0/2$ is entirely contained in some portion $\Gamma(t_i, \varepsilon_0)$. Denote by $w_j \in \{w_1, w_2\}$ the function in $GL^{\infty}(\Gamma(t_i, \varepsilon_0))$ and by $w_k \in \{w_1, w_2\}$ the other function. Then there is an $M_i \in (0, \infty)$ such that $|w_j(\tau)| \leq M_i$ and $|w_j^{-1}(\tau)| \leq M_i$ for almost all $\tau \in \Gamma(t_i, \varepsilon_0)$. We have

$$\sup_{t \in \Gamma} \sup_{\varepsilon < \varepsilon_0/2} \left(\Delta_t(w_1^p w_2^p, \varepsilon)\right)^{1/p} \left(\Delta_t(w_1^{-q} w_2^{-q}, \varepsilon)\right)^{1/q}$$

$$\leq \max_{1 \leq i \leq n} \sup_{\varepsilon < \varepsilon_0} \left(\Delta_{t_i}(w_j^p w_k^p, \varepsilon)\right)^{1/p} \left(\Delta_{t_i}(w_j^{-q} w_k^{-q}, \varepsilon)\right)^{1/q}$$

$$\leq \max_{1 \leq i \leq n} \sup_{\varepsilon < \varepsilon_0} M_i \left(\Delta_{t_i}(w_k^p, \varepsilon)\right)^{1/p} \left(\Delta_{t_i}(w_k^{-q}, \varepsilon)\right)^{1/q},$$

and the latter quantity is finite since $w_k \in \{w_1, w_2\} \subset A_p(\Gamma)$.

(b) The assertion is trivial for $\theta = 0$ or $\theta = 1$. If $0 < \theta < 1$, we may apply Hölder's inequality to get

$$\left(\Delta_t(w_1^{\theta p} w_2^{(1-\theta)p}, \varepsilon)\right)^{1/p} \leq \left(\Delta_t(w_1^p, \varepsilon)\right)^{\theta/p} \left(\Delta_t(w_2^p, \varepsilon)\right)^{(1-\theta)/p},$$

$$\left(\Delta_t(w_1^{-\theta q} w_2^{-(1-\theta)q}, \varepsilon)\right)^{1/q} \leq \left(\Delta_t(w_1^{-q}, \varepsilon)\right)^{\theta/q} \left(\Delta_t(w_2^{-q}, \varepsilon)\right)^{(1-\theta)/q},$$

which implies that $w := w_1^{\theta} w_2^{1-\theta}$ satisfies (2.7). $\square$

Throughout what follows in this chapter, we restrict ourselves to simple Carleson curves. We will return to composed Carleson curves in Chapter 9.

2.2   Power weights

We are now in a position to provide a first class of nonconstant Muckenhoupt weights.

Theorem 2.2. *Let Γ be a bounded simple Carleson curve, $1 < p < \infty$, and $1/p + 1/q = 1$. Let further $t_1, \ldots, t_n$ be distinct points on Γ and let $\lambda_1, \ldots, \lambda_n$ be real numbers. Then the weight*

$$w(\tau) := \prod_{j=1}^{n} |\tau - t_j|^{\lambda_j} \quad (\tau \in \Gamma) \tag{2.8}$$

belongs to $A_p(\Gamma)$ if and only if

$$-1/p < \lambda_j < 1/q \text{ for all } j. \tag{2.9}$$

Proof. If $w \in A_p(\Gamma)$, then necessarily $w \in L^p(\Gamma)$ and $w^{-1} \in L^q(\Gamma)$, which implies (2.9). To prove the reverse implication, we consider the special case where $w_0(\tau) := |\tau - t|^\lambda$ with $-1/p < \lambda < 1/q$ and show that $w_0 \in A_p(\Gamma)$. By Proposition 2.1(a), this implies that the weight (2.8) belongs to $A_p(\Gamma)$ whenever (2.9) is valid.

Assume first that $0 < \lambda < 1/q$. Let $0 < \varepsilon \le d_t$. Since $|\tau - t|^{\lambda p}$ increases together with $|\tau - t|$, we have

$$\left(\frac{1}{\varepsilon} \int_{\Gamma(t,\varepsilon)} |\tau - t|^{\lambda p} |d\tau| \right)^{1/p} \le \left(\frac{\varepsilon^{\lambda p}}{\varepsilon} \int_{\Gamma(t,\varepsilon)} |d\tau| \right)^{1/p}$$

$$= \left(\frac{\varepsilon^{\lambda p}}{\varepsilon} |\Gamma(t,\varepsilon)| \right)^{1/p} \le C_\Gamma^{1/p} \varepsilon^\lambda. \tag{2.10}$$

Further, we may write

$$\frac{1}{\varepsilon} \int_{\Gamma(t,\varepsilon)} |\tau - t|^{-\lambda q} |d\tau| = \frac{1}{\varepsilon} \sum_{n=1}^\infty \int_{\Gamma(t,\varepsilon/2^{n-1}) \setminus \Gamma(t,\varepsilon/2^n)} |\tau - t|^{-\lambda q} |d\tau|,$$

and because $|\tau - t|^{-\lambda q}$ decreases as $|\tau - t|$ increases, we have

$$\int_{\Gamma(t,\varepsilon/2^{n-1}) \setminus \Gamma(t,\varepsilon/2^n)} |\tau - t|^{-\lambda q} |d\tau| \le \left(\frac{\varepsilon}{2^n} \right)^{-\lambda q} \left| \Gamma\left(t, \frac{\varepsilon}{2^{n-1}} \right) \setminus \Gamma\left(t, \frac{\varepsilon}{2^n} \right) \right|.$$

The Carleson condition implies that $|\Gamma(t, \varepsilon/2^{n-1})| \le C_\Gamma \varepsilon/2^{n-1}$, and it is clear that $|\Gamma(t, \varepsilon/2^n)| \ge \varepsilon/2^n$ for $0 < \varepsilon \le d_t$. Hence

$$\left| \Gamma\left(t, \frac{\varepsilon}{2^{n-1}} \right) \setminus \Gamma\left(t, \frac{\varepsilon}{2^n} \right) \right| \le (2C_\Gamma - 1) \frac{\varepsilon}{2^n},$$

and since $1 - \lambda q > 0$, it follows that

$$\left(\frac{1}{\varepsilon} \int_{\Gamma(t,\varepsilon)} |\tau - t|^{-\lambda q} |d\tau| \right)^{1/q} \le \left(\frac{1}{\varepsilon} (2C_\Gamma - 1) \sum_{n=1}^\infty \left(\frac{\varepsilon}{2^n} \right)^{1-\lambda q} \right)^{1/q}$$

$$= (2C_\Gamma - 1)^{1/q} (2^{1-\lambda q} - 1)^{-1/q} \varepsilon^{-\lambda}. \tag{2.11}$$

Combining (2.10) and (2.11) we arrive at the estimate

$$\sup_{\varepsilon > 0} \frac{1}{\varepsilon} \left(\int_{\Gamma(t,\varepsilon)} |\tau - t|^{\lambda p} |d\tau| \right)^{1/p} \left(\int_{\Gamma(t,\varepsilon)} |\tau - t|^{-\lambda q} |d\tau| \right)^{1/q} < M$$

with some constant $M < \infty$. This estimate can also be proved at points $t_0 \neq t$. We omit the proof here, because the precise argument will be given later in a more general situation (see the proofs of (2.67) and (2.70)). In summary, $w_0 \in A_p(\Gamma)$.

The case where $-1/p < \lambda < 0$ may be disposed of analogously, and the case $\lambda = 0$ is trivial. $\qquad \square$

Weights of the form (2.8) are usually referred to as *power weights*.

We remark that if Γ is unbounded, we may also proceed as in the above proof to see that $|\tau - t|^\lambda$ is a weight in $A_p(\Gamma)$ if and only if $-1/p < \lambda < 1/q$.

2.3 The logarithm of a Muckenhoupt weight

The logarithm of the power weight (2.8) is

$$\log w(\tau) = \sum_{j=1}^{n} \lambda_j \log |\tau - t_j| \qquad (2.12)$$

and thus the simplest nontrivial example of a function in $BMO(\Gamma)$. The purpose of this section is to prove Theorem 2.5, which shows that the logarithm of every Muckenhoupt weight belongs to $BMO(\Gamma)$. This provides a nice necessary condition for membership in the Muckenhoupt class, which, however, is far from being a sufficient condition: for example, the function (2.12) is in $BMO(\Gamma)$ for every choice of the real numbers λ_j, while it is the logarithm of a Muckenhoupt weight in $A_p(\Gamma)$ if and only if the numbers λ_j are subject to the restrictions (2.9).

Lemma 2.3. *Let Γ be a simple locally rectifiable curve, $1 < p < \infty$, and suppose $w : \Gamma \to [0, \infty]$ is a weight. If $w \in L^p_{\mathrm{loc}}(\Gamma)$ and $w^{-1} \in L^q_{\mathrm{loc}}(\Gamma)$, then $\log w \in L^1_{\mathrm{loc}}(\Gamma)$.*

Proof. Let $\gamma \subset \Gamma$ be a bounded measurable set. Put $\gamma^+ := \{\tau \in \gamma : 1 \le w(\tau) < \infty\}$, $\gamma^- := \{\tau \in \gamma : 0 < w(\tau) < 1\}$. Then

$$|\log w(\tau)| < w(\tau) \text{ for } \tau \in \gamma^+, \ |\log w(\tau)| < 1/w(\tau) \text{ for } \tau \in \gamma^-$$

and hence,

$$\int_\gamma |\log w(\tau)|\,|d\tau| = \int_{\gamma^+} |\log w(\tau)|\,|d\tau| + \int_{\gamma^-} |\log w(\tau)|\,|d\tau|$$

$$< \int_{\gamma^+} w(\tau)\,|d\tau| + \int_{\gamma^-} w(\tau)^{-1}|d\tau| \le \int_\gamma w(\tau)\,|d\tau| + \int_\gamma w(\tau)^{-1}|d\tau|$$

$$\le |\gamma|^{1/q}\left(\int_\gamma w(\tau)^p|d\tau|\right)^{1/p} + |\gamma|^{1/p}\left(\int_\gamma w(\tau)^{-q}|d\tau|\right)^{1/q} < \infty. \qquad \square$$

A function $f : \Gamma \to [-\infty, \infty]$ is said to have *bounded mean oscillation at a point* $t \in \Gamma$, $f \in BMO(\Gamma, t)$, if $f \in L^1_{\mathrm{loc}}(\Gamma)$ and

$$\|f\|_{*,t} := \sup_{\varepsilon > 0} \frac{1}{|\Gamma(t,\varepsilon)|} \int_{\Gamma(t,\varepsilon)} |f(\tau) - \Delta_t(f,\varepsilon)|\,|d\tau| < \infty \qquad (2.13)$$

where $\Delta_t(f, \varepsilon)$ is defined by (2.6). Of course, we may write (2.13) also in the form

$$\|f\|_{*,t} := \sup_{\varepsilon>0} \Delta_t\left(\left|f - \Delta_t(f, \varepsilon)\right|, \varepsilon\right) < \infty. \tag{2.14}$$

One says that a function $f : \Gamma \to [-\infty, \infty]$ is of *bounded mean oscillation on* Γ, $f \in BMO(\Gamma)$, if $f \in BMO(\Gamma, t)$ for all $t \in \Gamma$ and

$$\|f\|_* := \sup_{t\in\Gamma} \|f\|_{*,t} < \infty. \tag{2.15}$$

A beautiful presentation of BMO theory (at least in the case of "nice" curves Γ) is in Garnett's book [74] and therefore we may be concise here. First of all we remark that $\|\cdot\|_*$ is a semi-norm on $BMO(\Gamma)$ with the property that $\|f\|_* = 0$ if and only if f is constant (a.e.).

Let us recall *Jensen's inequality*: if Γ is a simple rectifiable curve, f is a real-valued function in $L^1(\Gamma)$, and φ is a convex function on $\mathbf{R}$, then

$$\varphi\left(\frac{1}{|\gamma|} \int_\gamma f(\tau)\,|d\tau|\right) \leq \frac{1}{|\gamma|} \int_\gamma \varphi(f(\tau))\,|d\tau| \tag{2.16}$$

for every measurable subset γ of Γ. In particular, taking $\varphi(x) = e^{sx}$ we obtain that

$$\exp\left(\frac{s}{|\gamma|} \int_\gamma f(\tau)\,|d\tau|\right) \leq \frac{1}{|\gamma|} \int_\gamma \exp\left(sf(\tau)\right)\,|d\tau|$$

or, with the notation (2.6),

$$\exp\left(s\Delta_t(f, \varepsilon)\right) \leq \Delta_t\left(\exp(sf), \varepsilon\right).$$

Putting $f = \log w$, the latter inequality yields

$$\exp\left(s\Delta_t(\log w, \varepsilon)\right) \leq \Delta_t(w^s, \varepsilon) \quad \text{for all } s \in \mathbf{R}. \tag{2.17}$$

If φ is a concave function on $\mathbf{R}$, then (2.16) holds with "$\leq$" replaced by "$\geq$".

Given a simple Carleson curve Γ, $1 < p < \infty$, and a point $t \in \Gamma$, we denote by $A_p(\Gamma, t)$ the collection of all weights $w : \Gamma \to [0, \infty]$ such that $w \in L^p_{loc}(\Gamma)$, $w^{-1} \in L^q_{loc}(\Gamma)$, and

$$C_{w,t} := \sup_{\varepsilon>0} \left(\frac{1}{\varepsilon} \int_{\Gamma(t,\varepsilon)} w(\tau)^p|d\tau|\right)^{1/p} \left(\frac{1}{\varepsilon} \int_{\Gamma(t,\varepsilon)} w(\tau)^{-q}|d\tau|\right)^{1/q} < \infty. \tag{2.18}$$

Clearly, $w \in A_p(\Gamma)$ if and only if $w \in A_p(\Gamma, t)$ for all $t \in \Gamma$ and $\sup_{t\in\Gamma} C_{w,t} < \infty$.

Proposition 2.4. *If Γ is a simple Carleson curve, $t \in \Gamma$, and w is in $A_p(\Gamma, t)$ $(1 < p < \infty)$, then $\log w \in BMO(\Gamma, t)$.*

Proof. Lemma 2.3 implies that $\log w \in L^1_{\text{loc}}(\Gamma)$ and hence we may consider $\Delta_t(\varepsilon) := \Delta_t(\log w, \varepsilon)$. In the same way we showed the equivalence of (2.1) and (2.5), we see that (2.18) is equivalent to the requirement

$$c_{w,t} := \sup_{\varepsilon > 0} \left(\Delta_t(w^p, \varepsilon) \right)^{1/p} \left(\Delta_t(w^{-q}, \varepsilon) \right)^{1/q} < \infty.$$

Thus,

$$c_{w,t} = \sup_{\varepsilon > 0} \left(e^{-p\Delta_t(\varepsilon)} \Delta_t(w^p, \varepsilon) \right)^{1/p} \left(e^{q\Delta_t(\varepsilon)} \Delta_t(w^{-q}, \varepsilon) \right)^{1/q}. \tag{2.19}$$

By Jensen's inequality (2.17),

$$e^{s\Delta_t(\varepsilon)} = e^{s\Delta_t(\log w, \varepsilon)} \le \Delta_t(w^s, \varepsilon)$$

whenever $s \in \mathbf{R}$ and $w^s \in L^1_{\text{loc}}(\Gamma)$. Letting $s = p$ and $s = -q$, we therefore get

$$e^{-p\Delta_t(\varepsilon)} \Delta_t(w^{-p}, \varepsilon) \ge 1, \quad e^{q\Delta_t(\varepsilon)} \Delta_t(w^{-q}, \varepsilon) \ge 1.$$

These inequalities together with (2.19) imply that

$$\sup_{\varepsilon > 0} e^{-\Delta_t(\varepsilon)} \left(\Delta_t(w^p, \varepsilon) \right)^{1/p} \le c_{w,t} \tag{2.20}$$

and

$$\sup_{\varepsilon > 0} e^{\Delta_t(\varepsilon)} \left(\Delta_t(w^{-q}, \varepsilon) \right)^{1/q} \le c_{w,t}. \tag{2.21}$$

Put $\Gamma^+(t, \varepsilon) := \{ \tau \in \Gamma(t, \varepsilon) : \log w(\tau) \ge \Delta_t(\varepsilon) \}$ and $\Gamma^-(t, \varepsilon) := \{ \tau \in \Gamma(t, \varepsilon) : \log w(\tau) < \Delta_t(\varepsilon) \}$. Using the Jensen and Hölder inequalities, we deduce from (2.20) and (2.21) that

$$\exp \left(\Delta_t(|\log w - \Delta_t(\varepsilon)|, \varepsilon) \right) \le \Delta_t \left(\exp(|\log w - \Delta_t(\varepsilon)|), \varepsilon \right)$$

$$= \frac{1}{|\Gamma(t, \varepsilon)|} \int_{\Gamma^+(t,\varepsilon)} e^{\log w(\tau) - \Delta_t(\varepsilon)} |d\tau| + \frac{1}{|\Gamma(t, \varepsilon)|} \int_{\Gamma^-(t,\varepsilon)} e^{-(\log w(\tau) - \Delta_t(\varepsilon))} |d\tau|$$

$$\le \left(\Delta_t(e^{p(\log w - \Delta_t(\varepsilon))}, \varepsilon) \right)^{1/p} + \left(\Delta_t(e^{-q(\log w - \Delta_t(\varepsilon))}, \varepsilon) \right)^{1/q}$$

$$= e^{-\Delta_t(\varepsilon)} \left(\Delta_t(w^p, \varepsilon) \right)^{1/p} + e^{\Delta_t(\varepsilon)} \left(\Delta_t(w^{-q}, \varepsilon) \right)^{1/q} \le 2 c_{w,t}.$$

Thus

$$\| \log w \|_{*,t} \le 2 c_{w,t}, \tag{2.22}$$

whence $\log w \in BMO(\Gamma, t)$. $\square$

Theorem 2.5. *If Γ is a simple Carleson curve and $w \in A_p(\Gamma)$ $(1 < p < \infty)$ then $\log w \in BMO(\Gamma)$.*

Proof. Immediate from (2.22) and (2.5). $\square$

Example 2.6. As a first application of Theorem 2.5, we show that Muckenhoupt weights cannot simply be "sticked together" to produce a new Muckenhoupt weight. In order to avoid unnecessary technical complications, assume Γ is the segment $[-1, 1]$ and $w(x) = |x|^\lambda$ for $x < 0$, $w(x) = x^\mu$ for $x > 0$. We claim that $w \in A_p(\Gamma)$ if and only if $\lambda = \mu \in (-1/p, 1/q)$.

The "if" portion follows from Theorem 2.2. To prove the "only if" part, we invoke Theorem 2.5 and verify that if $\lambda \neq \mu$, then $\log w \notin BMO(\Gamma)$. By symmetry, it suffices to consider the case where $\mu > 0$ and $\lambda \in [-\mu, \mu)$. We have

$$\frac{1}{2\varepsilon} \int_{-\varepsilon}^{\varepsilon} \log w(x)\, dx = m(\log \varepsilon - 1)$$

with $m := (\mu + \lambda)/2 \geq 0$. Further, as $\mu > 0$,

$$\frac{1}{2\varepsilon} \int_{-\varepsilon}^{\varepsilon} \big| \log w(x) - m(\log \varepsilon - 1) \big|\, dx \geq \frac{1}{2\varepsilon} \int_{0}^{\varepsilon} \big| \mu \log x - m(\log \varepsilon - 1) \big|\, dx$$

$$= (1/2)|\mu \log \xi - m \log \varepsilon + m| \quad \text{with some } \xi \in (0, \varepsilon)$$

$$\geq (1/2)\big(\mu |\log \xi| - m|\log \varepsilon| \big) - (1/2)m. \tag{2.23}$$

Because $|\log \xi| \geq |\log \varepsilon|$ and $\mu > m$, it follows that (2.23) goes to infinity as $\varepsilon \to 0$. Thus, $\|\log w\|_{*,0} = \infty$ and hence $\log w \notin BMO(\Gamma)$. $\qquad\square$

2.4 Symmetric and periodic reproduction

In Example 2.6, the weight patched up was not a Muckenhoupt weight because something was wrong at the seam. However, given a weight $w_0 \in A_p([-1, 0])$, one can easily verify that the weight w defined on $[-1, 1]$ by

$$w(x) := \begin{cases} w_0(x) & \text{for } x \in [-1, 0) \\ w_0(-x) & \text{for } x \in (0, 1] \end{cases}$$

really belongs to $A_p([-1, 1])$. This simple observation is the foundation of the following construction.

Let $\Delta_0, \Delta_1, \Delta_2, \ldots$ be a finite or countable number of consecutive segments of the real line of positive lengths $l_0, l_1, l_2, \ldots$. Without loss of generality assume

$$\Delta_0 = [0, l_0], \quad \Delta_1 = [l_0, l_0 + l_1], \quad \Delta_2 = [l_0 + l_1, l_0 + l_1 + l_2], \quad \ldots.$$

Let further w_0 be a given weight on Δ_0. We call w_0 the *mother weight* and extend it to the union $\Delta_0 \cup \Delta_1 \cup \Delta_2 \cup \ldots$ as follows:

$$w_1(l_0 + x l_1) := w_0(l_0 - x l_0) \quad (x \in (0, 1]),$$

$$w_2(l_0 + l_1 + x l_2) := w_1(l_0 + l_1 - x l_1) \quad (x \in (0, 1]),$$

$$\ldots.$$

The weight w defined on $\Delta := \Delta_0 \cup \Delta_1 \cup \Delta_2 \cup \dots$ by $w|\Delta_j := w_j$ is referred to as the weight resulting by *symmetric reproduction* from the mother weight w_0.

Theorem 2.7. *If the mother weight w_0 belongs to $A_p(\Delta_0)$ and*

$$\frac{1}{K} \le \frac{l_{k+1}}{l_k} \le K \quad for \ k = 0, 1, 2, \dots \tag{2.24}$$

with some constant $K \in (0, \infty)$ independent of k, then the weight w resulting from w_0 by symmetric reproduction is a weight in $A_p(\Delta)$.

Proof. By the construction of w, there are functions $\varphi_k : \Delta_k \to \Delta_0$ such that $w(x) = w_0(\varphi_k(x))$ for $x \in \Delta_k$. It is easily seen that φ_k is of the form

$$\varphi_k(x) = c_k + (-1)^k \frac{l_0}{l_k} x \quad (x \in \Delta_k)$$

with some constant c_k. Thus, if I is a subset of Δ_k, then

$$\int_I w(x)^p dx = \int_I w_0\big(\varphi_k(x)\big)^p dx = \frac{l_k}{l_0} \int_{\varphi_k(I)} w_0(y)^p dy, \tag{2.25}$$

$$\int_I w(x)^{-q} dx = \int_I w_0\big(\varphi_k(x)\big)^{-q} dx = \frac{l_k}{l_0} \int_{\varphi_k(I)} w_0(y)^{-q} dy, \tag{2.26}$$

$$|I| = (l_k/l_0)|\varphi_k(I)|. \tag{2.27}$$

Take any interval $I \subset \Delta$. Suppose first that $I \subset \Delta_k$ for some k. Then, by (2.25) to (2.27),

$$\frac{1}{|I|} \left(\int_I w(x)^p dx \right)^{1/p} \left(\int_I w(x)^{-q} dx \right)^{1/q}$$

$$= \frac{l_0}{l_k} \frac{1}{|\varphi_k(I)|} \left(\frac{l_k}{l_0} \int_{\varphi_k(I)} w_0(y)^p dy \right)^{1/p} \left(\frac{l_k}{l_0} \int_{\varphi_k(I)} w_0(y)^{-q} dy \right)^{1/q}$$

$$= \frac{1}{|\varphi_k(I)|} \left(\int_{\varphi_k(I)} w_0(y)^p dy \right)^{1/p} \left(\int_{\varphi_k(I)} w_0(y)^{-q} dy \right)^{1/q} \le c_{w_0}$$

(recall (2.4)). Now suppose $I \subset \Delta_k \cup \Delta_{k+1}$ and

$$|I \cap \Delta_k| = \varepsilon l_k > 0, \ |I \cap \Delta_{k+1}| = \delta l_{k+1} > 0.$$

Put $E_k := \varphi_k(I \cap \Delta_k)$ and $E_{k+1} := \varphi_{k+1}(I \cap \Delta_{k+1})$. By (2.27),

$$|E_k| = \frac{l_0}{l_k}|I \cap \Delta_k| = \varepsilon l_0, \ |E_{k+1}| = \frac{l_0}{l_{k+1}}|I \cap \Delta_{k+1}| = \delta l_0.$$

Assume $\varepsilon \geq \delta$. Then $E_k \supset E_{k+1}$ and from (2.25) and (2.27) we infer that

$$\int_I w(x)^p dx = \int_{I \cap \Delta_k} w(x)^p dx + \int_{I \cap \Delta_{k+1}} w(x)^p dx$$

$$= \frac{l_k}{l_0} \int_{E_k} w_0(y)^p dy + \frac{l_{k+1}}{l_0} \int_{E_{k+1}} w_0(y)^p dy$$

$$\leq \frac{l_k}{l_0} \int_{E_k} w_0(y)^p dy + \frac{l_{k+1}}{l_0} \int_{E_k} w_0(y)^p dy = (l_k + l_{k+1})\varepsilon \frac{1}{|E_k|} \int_{E_k} w_0(y)^p dy.$$

Analogously, $\quad \int_I w(x)^{-q} dx \leq (l_k + l_{k+1})\varepsilon \frac{1}{|E_k|} \int_{E_k} w_0(y)^{-q} dy.$

Thus,

$$\frac{1}{|I|} \left(\int_I w(x)^p dx \right)^{1/p} \left(\int_I w(x)^{-q} dx \right)^{1/q}$$

$$\leq \frac{(l_k + l_{k+1})\varepsilon}{|I|} \frac{1}{|E_k|} \left(\int_{E_k} w_0(y)^p dy \right)^{1/p} \left(\int_{E_k} w_0(y)^{-q} dy \right)^{1/q},$$

and this is uniformly bounded because $w_0 \in A_p(\Delta_0)$ and

$$\frac{(l_k + l_{k+1})\varepsilon}{|I|} = \frac{(l_k + l_{k+1})\varepsilon}{\varepsilon l_k + \delta l_{k+1}} < \frac{l_k + l_{k+1}}{l_k} \leq 1 + K$$

by virtue of (2.24). In the same way one can dispose of the case where $\delta \geq \varepsilon$.

Finally, suppose I contains $n \geq 1$ of the segments $\Delta_0, \Delta_1, \ldots$ and is contained in $n + 2$ of them:

$$\Delta_{k+1} \cup \ldots \cup \Delta_{k+n} \subset I \subset \Delta_k \cup \Delta_{k+1} \cup \ldots \cup \Delta_{k+n} \cup \Delta_{k+n+1}. \tag{2.28}$$

Then, again by (2.25) to (2.27),

$$\frac{1}{|I|} \left(\int_I w(x)^p dx \right)^{1/p} \left(\int_I w(x)^{-q} dx \right)^{1/q}$$

$$\leq \frac{1}{|I|} \left(\sum_{j=k}^{k+n+1} \int_{\Delta_j} w(x)^p dx \right)^{1/p} \left(\sum_{j=k}^{k+n+1} \int_{\Delta_j} w(x)^{-q} dx \right)^{1/q}$$

$$= \frac{1}{|I|} \left(\sum_{j=k}^{k+n+1} \frac{l_j}{l_0} \int_{\Delta_0} w_0(y)^p dy \right)^{1/p} \left(\sum_{j=k}^{k+n+1} \frac{l_j}{l_0} \int_{\Delta_0} w_0(y)^{-q} dy \right)^{1/q}$$

$$= \frac{1}{|I|} \left(\sum_{j=k}^{k+n+1} l_j \right) \frac{1}{l_0} \left(\int_{\Delta_0} w_0(y)^p dy \right)^{1/p} \left(\int_{\Delta_0} w_0(y)^{-q} dy \right)^{1/q}. \tag{2.29}$$

Since $|I| \geq l_{k+1} + \ldots + l_{k+n}$, we have

$$\frac{1}{|I|}\left(\sum_{j=k}^{k+n+1} l_j\right) \leq \frac{l_k + l_{k+1} + \ldots + l_{k+n} + l_{k+n+1}}{l_{k+1} + \ldots + l_{k+n}} < \frac{l_k}{l_{k+1}} + 1 + \frac{l_{k+n+1}}{l_{k+n}} \leq K + 1 + K$$

due to (2.24). This shows that (2.29) is uniformly bounded. $\qquad\square$

Example 2.8. Let $\Delta_0 = [0, 2]$ and suppose the mother weight is a power weight, say $w_0(x) = |x - 1|^\lambda$ with $-1/p < \lambda < 1/q$. From Theorems 2.2 and 2.7 we learn that the weight given by

$$w(x) := |x - 2k - 1|^\lambda \quad \text{for } x \in \Delta_k := [2k, 2k + 2]$$

belongs to $A_p([0, \infty))$ $(k \in \mathbf{Z}_+)$ and $A_p(\mathbf{R})$ $(k \in \mathbf{Z})$, respectively. $\qquad\square$

Example 2.9. Let $\Delta_0 = [0, 1]$ and let $w_0 \in A_p(\Delta_0)$ be a power weight. Extend w_0 by symmetric reproduction with $l_k = 1/2^k$ to a weight v_0 on the segment

$$\left[0, 1 + \sum_{k=1}^\infty (1/2^k)\right] = [0, 2].$$

Thus, if w_0 has both a pole and a zero, then v_0 is a weight in $A_p([0, 2])$ such that

$$\limsup_{x \to 2} v_0(x) = +\infty, \quad \liminf_{x \to 2} v_0(x) = 0.$$

Now take $v_0 \in A_p([0, 2])$ as the mother weight and extend it by symmetric reproduction with $l_k = 1/2^k$ to a Muckenhoupt weight u_0 on the segment

$$\left[0, 2 + \sum_{k=1}^\infty (1/2^k)\right] = [0, 3].$$

The strange singularity of v_0 at the point $x = 2$ is then repeated a countable number of times by u_0. $\qquad\square$

The following result provides another way of constructing new Muckenhoupt weights from given Muckenhoupt weights.

Theorem 2.10. *Let I be a closed and connected subset of $\mathbf{R}$ containing at least two points. If $w_0 \in A_p(\mathbf{T})$, then the weight w given on I by $w(x) := w_0(e^{ix})$ belongs to $A_p(I)$.*

Proof. A simple computation gives the assertion for the case where $|I| \leq 2\pi$ (notice that w has no "critical seams" since $w_0 \in A_p(\mathbf{T})$). If $|I| > 2\pi$, one may proceed as in the cases $I \subset \Delta_k \cup \Delta_{k+1}$ and (2.28) of the proof of Theorem 2.7. $\qquad\square$

Theorem 2.10 says in particular that after transplanting a weight in $A_p(\mathbf{T})$ via the map $[0, 2\pi] \to \mathbf{T}$, $x \mapsto e^{ix}$ to a weight on $[0, 2\pi]$ and subsequently extending the resulting weight periodically, we get a weight in $A_p(\mathbf{R})$.

The function $x \mapsto e^{ix}$ is a simple example of an inner function on $\mathbf{R}$. Let $H^\infty(\mathbf{C}_+)$ denote the Hardy space of all analytic and bounded function in the upper half-plane $\mathbf{C}_+ := \{z \in \mathbf{C} : \operatorname{Im} z > 0\}$. Fatou's theorem states that functions in $H^\infty(\mathbf{C}_+)$ have nontangential limits a.e. on $\mathbf{R} = \partial \mathbf{C}_+$. A function $u : \mathbf{R} \to \mathbf{T}$ is said to be an *inner function* if there exists a function $v \in H^\infty(\mathbf{C}_+)$ such that u coincides with the nontangential limits of v a.e. on $\mathbf{R}$. Since $z \mapsto e^{iz} = e^{i(x+iy)} = e^{ix}e^{-y}$ is analytic and bounded in $\mathbf{C}_+$, it follows that $x \mapsto e^{ix}$ is inner.

Let now $u : \mathbf{R} \to \mathbf{T}$ be any inner function. Given a weight $w_0 \in A_p(\mathbf{T})$, define a weight w on $\mathbf{R}$ by $w(x) = w_0(u(x))$. Question: does w belong to $A_p(\mathbf{R})$?

Theorem 2.11. *If $u : \mathbf{R} \to \mathbf{T}$ is inner and $w_0 \in A_2(\mathbf{R})$ then $w_0 \circ u \in A_2(\mathbf{T})$. If $p \neq 2$, then there exist inner functions $u : \mathbf{R} \to \mathbf{T}$ and weights $w_0 \in A_p(\mathbf{T})$ such that $w_0 \circ u \notin A_p(\mathbf{R})$.*

As we will not make use of this theorem, we renounce to give a proof.

In summary, Theorems 2.7 and 2.10 allow us to construct plenty of nontrivial Muckenhoupt weights provided we have a mother weight in $A_p([0, 1])$ or $A_p(\mathbf{T})$. However, all these weights are, in sense, repetitions of the mother weight. And the only mother weights we know until the present moment are (repetitions of) power weights ...

2.5 Portions versus arcs

Let Γ be a simple rectifiable curve. For τ and t on Γ, let $\ell(\tau, t)$ stand for the length of the shortest subarc of Γ with the endpoints τ and t. Given $t \in \Gamma$ and $\varepsilon > 0$, we put $\Gamma[t, \varepsilon] := \{\tau \in \Gamma : \ell(\tau, t) < \varepsilon\}$. Thus, if t is not an endpoint of Γ, then at least for small $\varepsilon > 0$ the set $\Gamma[t, \varepsilon]$ is the arc of length 2ε centered at t. It is clear that the arc $\Gamma[t, \varepsilon]$ is contained in the portion $\Gamma(t, \varepsilon) := \{\tau \in \Gamma : |\tau - t| < \varepsilon\}$.

For $p \in (1, \infty)$, we define $W_p(\Gamma)$ as the set of all weights $w : \Gamma \to [0, \infty]$ for which $w \in L^p(\Gamma)$, $w^{-1} \in L^q(\Gamma)$, and

$$\sup_{t \in \Gamma} \sup_{\varepsilon > 0} \left(\frac{1}{\varepsilon} \int_{\Gamma[t,\varepsilon]} w(\tau)^p |d\tau| \right)^{1/p} \left(\frac{1}{\varepsilon} \int_{\Gamma[t,\varepsilon]} w(\tau)^{-q} |d\tau| \right)^{1/q} < \infty. \tag{2.30}$$

In (2.30) we may again the $\sup_{\varepsilon > 0}$ replace by $\sup_{\varepsilon < \varepsilon_0}$ where $\varepsilon_0 > 0$ is any fixed number. We also remark that (2.30) is equivalent to (2.5) with Q ranging over all subarcs of Γ or over all subarcs of Γ whose length is at most ε_0.

Since $\Gamma[t, \varepsilon] \subset \Gamma(t, \varepsilon)$, we see that always $A_p(\Gamma) \subset W_p(\Gamma)$. If Γ is not Carleson, then $A_p(\Gamma) = \emptyset$ (recall Section 2.1), while $W_p(\Gamma)$ contains at least the constants. Thus, in general $W_p(\Gamma)$ may be properly larger than $A_p(\Gamma)$. The question whether $A_p(\Gamma)$ and $W_p(\Gamma)$ coincide for a given Carleson curve is very delicate.

Theorem 2.12 (Simonenko). *If Γ is a simple piecewise Lyapunov curve having a cusp, then $A_p(\Gamma) \neq W_p(\Gamma)$.*

A full *proof* of this theorem is in [197]. As we will not make use of this result in the following, we confine ourselves to a few remarks.

We call an arc is a *Lyapunov arc* if, after appropriate rotation, it may be given by (1.4) with a function f whose derivative satisfies a Hölder condition. A simple curve Γ which may be represented as a finite union of Lyapunov arcs is called a *piecewise Lyapunov curve*. Clearly, piecewise Lyapunov curves are piecewise C^1 curves in the sense of Example 1.2.

The proof of Simonenko's theorem may be reduced to the case where

$$\Gamma = [0,1] \cup \{\tau \in \mathbf{C} : \tau = x + if(x), \ 0 \le x \le 1\}$$

with some function $f \in C^1[0,1]$ satisfying the following conditions: $f(0) = 0$, $f(x) > 0$ for $x \in (0,1]$, $f'(0) = 0$, and f' satisfies a Hölder condition on $[0,1]$.

Given such a curve Γ, Simonenko constructed a weight as follows. For a segment $l = [a,b]$ on the real line and an integer $n \ge 1$, put

$$\psi_{l,n}(x) := \prod_{k=1}^{n} \frac{|x - x_{2k-1}|^2}{|x - x_{2k-2}||x - x_{2k}|}$$

where $x_m := a + (b-a)m/(2n)$ $(m = 0, 1, \ldots, 2n)$. Then, for $j \in \{1, 2, \ldots\}$, let $l_j := [1/2^j, 1/2^{j-1}]$ and define

$$\psi(x) := \prod_{j=1}^{\infty} \psi_{l_j, j^2}(x).$$

One can show that this infinite product converges for all $x \in \mathbf{R}$ at which its factors are finite and nonzero. Finally, define a weight $\varrho : \Gamma \to [0,\infty]$ by $\varrho(x) := \psi(x)$ and $\varrho(x + if(x)) := 1$. Simonenko [197] showed that $\varrho^\mu \notin A_p(\Gamma)$ for all $\mu > 0$ and all $1 < p < \infty$ and that for every $p \in (1,\infty)$ there exists a $\mu > 0$ such that $\varrho^\mu \in W_p(\Gamma)$. We remark that Simonenko did not prove that $\varrho^\mu \notin A_p(\Gamma)$ by showing that the Muckenhoupt condition (2.1) is violated for $w := \varrho^\mu$; he rather proved that the Cauchy singular integral operator S_Γ is unbounded on the space $L^p(\Gamma, \varrho^\mu)$. □

Working with weights in $W_p(\Gamma)$ is, in a sense, more convenient than with weights in $A_p(\Gamma)$. For example, almost all the proofs given in [74] or [127] for the cases where $\Gamma = \mathbf{R}$ and $\Gamma = \mathbf{T}$ may be easily extended to curves Γ for which $W_p(\Gamma) = A_p(\Gamma)$. Since every weight in $A_p(\Gamma)$ belongs to $W_p(\Gamma)$, the reader might ask where the problem is. The problem consists, for instance, in the following. One of the deepest results in the theory of Muckenhoupt weights says that if $w \in A_p(\Gamma)$, then $w^{1+\varepsilon} \in A_p(\Gamma)$ for all sufficiently small $\varepsilon > 0$. The proofs of this fact given in

[74] and [127] (and other books) for $\Gamma = \mathbf{R}$ and $\Gamma = \mathbf{T}$ may without difficulty be adapted to show that

$$w \in W_p(\Gamma) \Longrightarrow w^{1+\varepsilon} \in W_p(\Gamma) \quad \forall \varepsilon \in [0, \varepsilon_0)$$

As $A_p(\Gamma) \subset W_p(\Gamma)$, this gives the implication

$$w \in A_p(\Gamma) \Longrightarrow w^{1+\varepsilon} \in W_p(\Gamma) \quad \forall \varepsilon \in [0, \varepsilon_0)$$

which, however, does not tell us anything concerning the implication

$$w \in A_p(\Gamma) \Longrightarrow w^{1+\varepsilon} \in A_p(\Gamma) \quad \forall \varepsilon \in [0, \varepsilon_0).$$

Much of what follows in this and the next three sections serves the proof of the latter implication.

We call a simple bounded Carleson curve Γ *arclike* if there exists a $K > 0$ such that every portion $\Gamma(t, \varepsilon)$ is contained in some subarc of Γ whose length is at most $K\varepsilon$. Equivalently, Γ is arclike if and only if there are $\varepsilon_0 > 0$ and $K > 0$ such that for every $\varepsilon < \varepsilon_0$ the portion $\Gamma(t, \varepsilon)$ is a subset of some subarc of length $K\varepsilon$ (notice that for $\varepsilon \geq \varepsilon_0$ we may take $K = |\Gamma|/\varepsilon_0$).

If Γ is an arclike Carleson curve then $A_p(\Gamma) = W_p(\Gamma)$. Indeed, since $\Gamma(t, \varepsilon)$ is always contained in $\Gamma[s, K\varepsilon/2]$ for some $s \in \Gamma$, it follows that

$$\sup_{t \in \Gamma} \sup_{\varepsilon > 0} \left(\frac{1}{\varepsilon} \int_{\Gamma(t,\varepsilon)} w(\tau)^p |d\tau| \right)^{1/p} \left(\frac{1}{\varepsilon} \int_{\Gamma(t,\varepsilon)} w(\tau)^{-q} |d\tau| \right)^{1/q}$$

$$\leq \sup_{s \in \Gamma} \sup_{\varepsilon > 0} \left(\frac{1}{\varepsilon} \int_{\Gamma[s,K\varepsilon/2]} w(\tau)^p |d\tau| \right)^{1/p} \left(\frac{1}{\varepsilon} \int_{\Gamma[s,K\varepsilon/2]} w(\tau)^{-q} |d\tau| \right)^{1/q}$$

$$= \frac{K}{2} \sup_{s \in \Gamma} \sup_{\delta > 0} \left(\frac{1}{\delta} \int_{\Gamma[s,\delta]} w(\tau)^p |d\tau| \right)^{1/p} \left(\frac{1}{\delta} \int_{\Gamma[s,\delta]} w(\tau)^{-q} |d\tau| \right)^{1/q},$$

which proves the desired inclusion $W_p(\Gamma) \subset A_p(\Gamma)$.

Example 2.13. If Γ is as in Proposition 1.1 or 1.4, then Γ is arclike and hence $W_p(\Gamma) = A_p(\Gamma)$. To see this, we note that in the case at hand the portion $\Gamma(t, \varepsilon)$ is contained in an arc of length at most $2\varepsilon\sqrt{1 + M^2}$. $\qquad\square$

Example 2.14. Let $\Gamma = \Gamma_1 \cup \Gamma_2$ be the Carleson Jordan curve of Example 1.11. If $b(r) = o(1)$ as $r \to 0$, then Γ is not arclike. Indeed, the distance between $re^{i\varphi(r)}$ and $re^{i(\varphi(r)+b(r))}$ is $2r \sin(b(r)/2)$, which is certainly less than $r/2$, the half distance between $re^{i\varphi(r)}$ and the origin, if only r is sufficiently small. Hence, if r is small enough, then the portion $\Gamma(re^{i\varphi(r)}, 4r \sin(b(r)/2))$ is disconnected and the minimal arc containing this portion passes through the origin. The length of this arc is at least r, and so the assumption that Γ be arclike implies that

$$r \leq 4Kr\sin\left(b(r)/2\right) = 4Kro(1) \text{ as } r \to 0,$$

which is a contradiction. In particular, if $\varphi(r) = 0$ and $b(r) = r^\alpha$ ($\alpha > 0$), then Γ has a cusp at the origin and is not arclike.

On the other hand, if $\varphi(r) = -\delta \log r$ and $b(r)$ is a nonzero constant for sufficiently small r, then Γ is arclike. For $\delta = 0$ this is obvious, for $\delta \neq 0$ this requires a little computation. $\square$

Although the curves considered in Example 2.14 need not be arclike, we have the following result.

Proposition 2.15. *Let $\Gamma = \Gamma_1 \cup \Gamma_2$ be as in Example 1.11, let v be a weight on $I = [0,1]$, and define a weight w on Γ by $w(\tau) = v(|\tau|)$. Then the following are equivalent:*

(i) *$w \in A_p(\Gamma)$;*
(ii) *$w \in W_p(\Gamma)$*
(iii) *$v \in A_p(I)\left(= W_p(I)\right)$.*

Proof. (i) $\Longrightarrow$ (ii). This is clear since always $A_p(\Gamma) \subset W_p(\Gamma)$.

(ii) $\Longrightarrow$ (iii). Pick a segment $\sigma = [R - \varepsilon, R + \varepsilon] \subset I$ and denote by $\gamma \subset \Gamma$ the arc $\{re^{i\varphi(r)} : r \in \sigma\}$. We have

$$\frac{1}{|\gamma|}\int_\gamma w(\tau)^p|d\tau| = \int_{R-\varepsilon}^{R+\varepsilon} v(r)^p\sqrt{1+r^2(\varphi'(r))^2}\,dr \Big/ \int_{R-\varepsilon}^{R-\varepsilon}\sqrt{1+r^2(\varphi'(r))^2}\,dr$$

$$\geq \int_{R-\varepsilon}^{R+\varepsilon} v(r)^p\,dr \Big/ \int_{R-\varepsilon}^{R+\varepsilon}\sqrt{1+M^2}\,dr$$

$$= \int_\sigma v(r)^p\,dr \Big/ \left(2\varepsilon\sqrt{1+M^2}\right) = \frac{1}{|\sigma|}\int_\sigma v(r)^p\,dr \Big/ \sqrt{1+M^2}$$

and analogously,

$$\frac{1}{|\gamma|}\int_\gamma w(\tau)^{-q}|d\tau| \geq \frac{1}{|\sigma|}\int_\sigma v(r)^{-q}\,dr \Big/ \sqrt{1+M^2}.$$

This proves that $v \in A_p(I)$ whenever $w \in W_p(\Gamma)$.

(iii) $\implies$ (i). Let $t \in \Gamma$, put $R = |t|$, and suppose $0 \leq R - \varepsilon < R + \varepsilon \leq 1$. Then

$$\Gamma(t,\varepsilon) \subset \{\tau \in \Gamma : R - \varepsilon \leq |\tau| \leq R + \varepsilon\} =: \gamma_1 \cup \gamma_2$$

where γ_1 and γ_2 are arcs. Hence,

$$\frac{1}{\varepsilon} \int\limits_{\Gamma(t,\varepsilon)} w(\tau)^p |d\tau|$$

$$\leq \frac{|\gamma_1|}{\varepsilon} \frac{1}{|\gamma_1|} \int\limits_{\gamma_1} w(\tau)^p |d\tau| + \frac{|\gamma_2|}{\varepsilon} \frac{1}{|\gamma_2|} \int\limits_{\gamma_2} w(\tau)^p |d\tau| \qquad (2.31)$$

$$\leq \frac{\sqrt{1+M^2}\, 2\varepsilon}{\varepsilon} \left(\frac{1}{|\gamma_1|} \int\limits_{\gamma_1} w(\tau)^p |d\tau| + \frac{1}{|\gamma_2|} \int\limits_{\gamma_2} w(\tau)^p |d\tau| \right)$$

$$\leq 2\sqrt{1+M^2} \left(\left(\frac{1}{|\gamma_1|} \int\limits_{\gamma_1} w(\tau)^p |d\tau| \right)^{1/p} + \left(\frac{1}{|\gamma_2|} \int\limits_{\gamma_2} w(\tau)^p |d\tau| \right)^{1/p} \right)^p . \qquad (2.32)$$

In the same way we get

$$\frac{1}{\varepsilon} \int\limits_{\Gamma(t,\varepsilon)} w(\tau)^{-q} |d\tau|$$

$$\leq 2\sqrt{1+M^2} \left(\left(\frac{1}{|\gamma_1|} \int\limits_{\gamma_1} w(\tau)^{-q} |d\tau| \right)^{1/q} + \left(\frac{1}{|\gamma_2|} \int\limits_{\gamma_2} w(\tau)^{-q} |d\tau| \right)^{1/q} \right)^q .$$

Consequently,

$$\left(\frac{1}{\varepsilon} \int\limits_{\Gamma(t,\varepsilon)} w(\tau)^p |d\tau| \right)^{1/p} \left(\frac{1}{\varepsilon} \int\limits_{\Gamma(t,\varepsilon)} w(\tau)^{-q} |d\tau| \right)^{1/q}$$

$$\leq 2\sqrt{1+M^2} \sum_{j,k=1}^{2} \left(\frac{1}{|\gamma_j|} \int\limits_{\gamma_j} w(\tau)^p |d\tau| \right)^{1/p} \left(\frac{1}{|\gamma_k|} \int\limits_{\gamma_k} w(\tau)^{-q} |d\tau| \right)^{1/q} . \qquad (2.33)$$

For $\gamma \in \{\gamma_1, \gamma_2\}$, we have

$$\frac{1}{|\gamma|} \int\limits_{\gamma} w(\tau)^p |d\tau| = \int\limits_{R-\varepsilon}^{R+\varepsilon} v(r)^p \sqrt{1 + r^2 (\psi'(r))^2}\, dr \Big/ \int\limits_{R-\varepsilon}^{R+\varepsilon} \sqrt{1 + r^2 (\psi'(r))^2}\, dr$$

with $\psi \in \{\varphi, \varphi + b\}$ and thus,

$$\frac{1}{|\gamma|} \int\limits_{\gamma} w(\tau)^p |d\tau| \leq \sqrt{1+M^2} \int\limits_{R-\varepsilon}^{R+\varepsilon} v(r)^p dr \Big/ \int\limits_{R-\varepsilon}^{R+\varepsilon} dr = \frac{\sqrt{1+M^2}}{2\varepsilon} \int\limits_{R-\varepsilon}^{R+\varepsilon} v(r)^p dr .$$

An analogous estimate holds for w^{-q} and therefore

$$
\left(\frac{1}{|\gamma_j|}\int_{\gamma_j} w(\tau)^p |d\tau|\right)^{1/p}\left(\frac{1}{|\gamma_k|}\int_{\gamma_k} w(\tau)^{-q}|d\tau|\right)^{1/q}
$$

$$
\leq \frac{\sqrt{1+M^2}}{2\varepsilon}\left(\int_{R-\varepsilon}^{R+\varepsilon} v(r)^p dr\right)^{1/p}\left(\int_{R-\varepsilon}^{R+\varepsilon} v(r)^{-q}dr\right)^{1/q}
$$

$$
\leq \sqrt{1+M^2}\sup_{\sigma\subset[0,1]}\left(\frac{1}{|\sigma|}\int_\sigma v(r)^p\right)^{1/p}\left(\frac{1}{|\sigma|}\int_\sigma v(r)^{-q}dr\right)^{1/q}=:B<\infty.
$$

We so have proved that the left-hand side of (2.33) is at most $8B\sqrt{1+M^2}$ if $0\leq |t|-\varepsilon<|t|+\varepsilon\leq 1$.

If $R-\varepsilon<0<R+\varepsilon\leq 1$, then $\Gamma(t,\varepsilon)\subset\Gamma(0,R+\varepsilon)$ and the set $\{\tau\in\Gamma:|\tau|\leq R+\varepsilon\}$ is the union of two arcs γ_1 and γ_2 meeting at the origin. Again (2.31) is true, and since now $R<\varepsilon$, we have

$$
\max\{|\gamma_1|,|\gamma_2|\}\leq\sqrt{1+M^2}\,(R+\varepsilon)<\sqrt{1+M^2}\,2\varepsilon,
$$

which gives (2.32) and then (2.33). Replacing in the rest of the above reasoning $R-\varepsilon$ by 0, we see that the left-hand side of (2.33) does not exceed $8B\sqrt{1+M^2}$ for $|t|-\varepsilon<0<|t|+\varepsilon\leq 1$, too. The cases $0\leq R-\varepsilon<1<R+\varepsilon$ and $R-\varepsilon<0<1<R+\varepsilon$ can be settled analogously.   □

2.6   The maximal operator

Our next objective is to prove Theorem 2.31. For this purpose, we need some auxiliary results.

Let Γ be a simple Carleson curve. For a function $f\in L^1_{loc}(\Gamma)$, we define the so-called *maximal function* Mf of f on Γ by

$$
(Mf)(t):=\sup_{\varepsilon>0}\frac{1}{\varepsilon}\int_{\Gamma(t,\varepsilon)}|f(\tau)||d\tau|,\ \ t\in\Gamma.
$$

The map $M:f\mapsto Mf$ is referred to as the *maximal operator*.

The length measure on Γ makes Γ to a metric space: the distance between $\tau,t\in\Gamma$ is $\ell(\tau,t)$. A function $g:\Gamma\to[0,+\infty]$ is said to be *lower semi-continuous at a point* $t\in\Gamma$ if for each $\varepsilon>0$ there exists a $\delta>0$ such that $g(\tau)>g(t)-\varepsilon$ for all τ in the open arc $\Gamma[t,\delta]$. The function g is said to be *lower semi-continuous on the curve* Γ if it is lower semi-continuous at each point $t\in\Gamma$. Equivalently, g is lower semi-continuous on Γ if and only if the set $\{t\in\Gamma:g(t)>\lambda\}$ is an open subset of Γ for each $\lambda>0$. It follows in particular that lower semi-continuous functions on Γ are always measurable.

Lemma 2.16. *If $f \in L^1_{loc}(\Gamma)$, then $Mf : \Gamma \rightarrow [0, +\infty]$ is lower semi-continuous and thus measurable on Γ.*

Proof. Fix $t_0 \in \Gamma$. If $(Mf)(t_0) = 0$, then Mf is obviously lower semi-continuous at t_0. So let $(Mf)(t_0) > 0$ and pick $\lambda \in (0, (Mf)(t_0))$. For every $\lambda_0 \in (\lambda, (Mf)(t_0))$ there exists an $\varepsilon_0 > 0$ such that

$$\frac{1}{\varepsilon_0} \int_{\Gamma(t_0, \varepsilon_0)} |f(\tau)| \, |d\tau| > \lambda_0.$$

Put $\varkappa := \lambda_0/\lambda - 1$ and choose $\delta \in (0, \varkappa\varepsilon_0)$. Since $\Gamma(t_0, \varepsilon_0) \subset \Gamma(t, \varepsilon_0 + |t - t_0|)$, we obtain that for $t \in \Gamma(t_0, \delta)$,

$$\frac{1}{\varepsilon_0 + |t - t_0|} \int_{\Gamma(t, \varepsilon_0 + |t - t_0|)} |f(\tau)| \, |d\tau|$$

$$\geq \frac{\varepsilon_0}{\varepsilon_0 + |t - t_0|} \frac{1}{\varepsilon_0} \int_{\Gamma(t_0, \varepsilon_0)} |f(\tau)| \, |d\tau| > \frac{\lambda_0}{1 + \varkappa} = \lambda.$$

Consequently, $(Mf)(t) > \lambda$ for all $t \in \Gamma(t_0, \delta)$ and thus, all the more, for all $t \in \Gamma[t_0, \delta]$. $\qquad\square$

The *weak L^1 space* over Γ is defined as the collection of all measurable functions $g : \Gamma \rightarrow \mathbf{C} \cup \{\infty\}$ for which

$$\|g\|_\sim := \sup_{\lambda > 0} \lambda \big| \{t \in \Gamma : |g(t)| > \lambda\} \big| < \infty.$$

We denote this space by $L^\sim(\Gamma)$. Clearly, $L^\sim(\Gamma)$ is a linear space. If $g \in L^1(\Gamma)$, then

$$\lambda \big| \{t \in \Gamma : |g(t)| > \lambda\} \big| \leq \int_{|g(t)| > \lambda} |g(t)| \, |dt| \leq \|g\|_1, \qquad (2.34)$$

which implies that $L^1(\Gamma) \subset L^\sim(\Gamma)$. The function $g(t) := 1/|t - t_0|$ $(t_0 \in \Gamma)$ is easily seen to belong to $L^\sim(\Gamma) \setminus L^1(\Gamma)$ whenever Γ is bounded.

A map $T : L^1(\Gamma) \rightarrow L^\sim(\Gamma)$ is said to be a *sublinear operator* if

$$|T(\alpha f + \beta g)| \leq |\alpha| \, |Tf| + |\beta| \, |Tg| \quad \text{a.e. on } \Gamma$$

for all $\alpha, \beta \in \mathbf{C}$ and all $f, g \in L^1(\Gamma)$. A sublinear operator $T : L^1(\Gamma) \rightarrow L^\sim(\Gamma)$ is called an *operator of weak type $(1,1)$* if there exists a constant $C < \infty$ such that $\|Tf\|_\sim \leq C\|f\|$ for all $f \in L^1(\Gamma)$, or equivalently, such that

$$\big| \{t \in \Gamma : |(Tf)(t)| > \lambda\} \big| \leq \frac{C}{\lambda} \int_\Gamma |f(\tau)| \, |d\tau|$$

for all $\lambda > 0$ and all $f \in L^1(\Gamma)$.

Theorem 2.17 (Besicovitch). *For each integer $n \geq 1$, there exists a number θ_n with the following property. If A is a bounded subset of $\mathbf{R}^n$ and for each $x \in A$ we are given an open ball $B(x, r(x)) \subset \mathbf{R}^n$ of radius $r(x) > 0$ centered at x, then the set $\{B(x, r(x))\}_{x \in A}$ contains an at most countable subset $\{B_k\} := \{B(x_k, r(x_k))\}$ such that*

(i) $A \subset \bigcup_k B_k$,

(ii) *every point $x \in \mathbf{R}^n$ belongs to at most θ_n balls of $\{B_k\}$.*

A *proof* of this well known result may be found in [95, Chapter 1].   □

Theorem 2.18. *Let Γ be a simple Carleson curve. Then the maximal operator M is of weak type $(1, 1)$, i.e.*

$$\left| \{t \in \Gamma : (Mf)(t) > \lambda\} \right| \leq \frac{C}{\lambda} \int_\Gamma |f(\tau)| \, |d\tau| \tag{2.35}$$

for all $\lambda > 0$ and $f \in L^1(\Gamma)$ with $C := C_\Gamma \theta_2$, which is a constant depending only on Γ.

Proof. Let $f \in L^1(\Gamma)$ and $\lambda > 0$. By Lemma 2.16, the set $A_\lambda := \{t \in \Gamma : (Mf)(t) > \lambda\}$ is an open subset of Γ. Hence, $A_{\lambda,m} := A_\lambda \cap \{z \in \mathbf{C} : |z| < m\}$ is an open subset of Γ and a bounded subset of $\mathbf{C}$. For each $t \in A_\lambda$, pick an open disk $D(t, \varepsilon(t))$ centered at t such that

$$\frac{1}{\varepsilon(t)} \int_{\Gamma(t, \varepsilon(t))} |f(\tau)| \, |d\tau| > \lambda. \tag{2.36}$$

Theorem 2.17 implies that the set $\{D(t, \varepsilon(t))\}_{t \in A_{\lambda,m}}$ contains an at most countable subset $\{D(t_k, \varepsilon_k)\}$ such that $A_{\lambda,m} \subset \bigcup_k \Gamma(t_k, \varepsilon_k)$ and $\sum_k \chi_k \leq \theta_2$, where χ_k denotes the characteristic function of the portion $\Gamma(t_k, \varepsilon_k)$. Taking into account the Carleson condition (1.1) and the inequality (2.36), we get

$$|A_{\lambda,m}| \leq \sum_k |\Gamma(t_k, \varepsilon_k)| \leq C_\Gamma \sum_k \varepsilon_k < C_\Gamma \sum_k \frac{1}{\lambda} \int_{\Gamma(t_k, \varepsilon_k)} |f(\tau)| \, |d\tau|$$

$$= C_\Gamma \frac{1}{\lambda} \int_\Gamma \sum_k \chi_k(\tau) |f(\tau)| \, |d\tau| \leq C_\Gamma \theta_2 \frac{1}{\lambda} \int_\Gamma |f(\tau)| \, |d\tau|.$$

Thus,

$$|A_\lambda| = \lim_{m \to \infty} |A_{\lambda,m}| \leq C_\Gamma \theta_2 \frac{1}{\lambda} \int_\Gamma |f(\tau)| \, |d\tau|,$$

which is (2.35) with $C := C_\Gamma \theta_2$. Finally, it is clear that M is sublinear.   □

The following theorem extends Lebesgue's differentiability theorem to the case of Carleson curves.

Theorem 2.19. *Let Γ be a simple Carleson curve and $f \in L^1_{\mathrm{loc}}(\Gamma)$. Then*

$$\lim_{\varepsilon \to 0} \frac{1}{|\Gamma(t,\varepsilon)|} \int_{\Gamma(t,\varepsilon)} f(\tau)\,|d\tau| = f(t)$$

for almost all $t \in \Gamma$.

Proof. We may clearly assume that Γ is bounded. A moment's thought reveals that it suffices to show that for every $\lambda > 0$ the set

$$P_\lambda := \left\{ t \in \Gamma : \limsup_{\varepsilon \to 0} \left| \Delta_t(f,\varepsilon) - f(t) \right| > \lambda \right\}$$

has measure zero (recall the notation (2.6)).

Fix $\varepsilon > 0$ and choose a function $g \in C(\Gamma)$ such that $\|f - g\|_1 < \varepsilon$. Given a point $t \in \Gamma$ and a proper open arc $\gamma \subset \Gamma$ containing t, let $r := (1/2)\min_{\tau \in \Gamma \setminus \gamma} |\tau - t|$. Clearly, $r > 0$ and $\Gamma(t,\varepsilon) \subset \gamma$ for all $\varepsilon < r$. Since

$$\left| \int_{\Gamma(t,\varepsilon)} g(\tau)\,|d\tau| - g(t)|\Gamma(t,\varepsilon)| \right| \leq \int_{\Gamma(t,\varepsilon)} |g(\tau) - g(t)|\,|d\tau| \leq \|g - g(t)\|_{L^\infty(\gamma)} |\Gamma(t,\varepsilon)|,$$

it follows that $\lim_{\varepsilon \to 0} \Delta_t(g,\varepsilon) = g(t)$ for all $t \in \Gamma$.

Consequently, if we let $h := f - g$, then

$$P_\lambda = \left\{ t \in \Gamma : \limsup_{\varepsilon \to 0} \left| \Delta_t(h,\varepsilon) - h(t) \right| > \lambda \right\}.$$

The latter set is contained in $P^1_\lambda \cup P^2_\lambda$ where

$$P^1_\lambda := \left\{ t \in \Gamma : \limsup_{\varepsilon \to 0} \left| \Delta_t(h,\varepsilon) \right| > \lambda/2 \right\}, \quad P^2_\lambda := \left\{ t \in \Gamma : h(t) > \lambda/2 \right\}.$$

As $|\Gamma(t,\varepsilon)| \geq \varepsilon$ for small $\varepsilon > 0$, we have

$$\limsup_{\varepsilon \to 0} \left| \Delta_t(h,\varepsilon) \right| \leq \limsup_{\varepsilon \to 0} \frac{1}{\varepsilon} \int_{\Gamma(t,\varepsilon)} |h(\tau)|\,|d\tau| \leq (Mh)(t).$$

Thus, letting $|P^1_\lambda|_e$ denoting the exterior Lebesgue measure of P^1_λ, we infer from (2.35) that

$$|P^1_\lambda|_e \leq \left| \{ t \in \Gamma : (Mh)(t) > \lambda/2 \} \right| \leq (2/\lambda)C\|h\|_1 < 2C\varepsilon/\lambda.$$

Furthermore, by (2.34),

$$|P^2_\lambda| = \left| \{ t \in \Gamma : |h(t)| > \lambda/2 \} \right| \leq (2/\lambda)\|h\|_1 < 2\varepsilon/\lambda.$$

In summary,

$$|P_\lambda| \leq |P^1_\lambda|_e + |P^2_\lambda| \leq 2(C+1)\varepsilon/\lambda,$$

and as $\varepsilon > 0$ can be chosen as small as desired, we get $|P_\lambda| = 0$. $\qquad \square$

2.7   The reverse Hölder inequality

This section is devoted to the proof of Theorem 2.29. Our reasoning essentially
follows Calderón's paper [31].

Let Γ be a simple Carleson curve, $1 < p < \infty$, and $w \in A_p(\Gamma)$. Then w^p is in
$L^1_{\mathrm{loc}}(\Gamma)$ and we may consider the measure $d\nu(\tau) := w(\tau)^p|d\tau|$ on Γ. For a subset
E of Γ measurable with respect to Lebesgue length measure, we put

$$|E| := \int_E |d\tau|, \; |E|_\nu := \int_E w(\tau)^p|d\tau|.$$

Lemma 2.20. *If E is a measurable subset of the portion $Q := \Gamma(t, \varepsilon)$ ($t \in \Gamma, \varepsilon > 0$),*
then

$$\frac{|E|_\nu}{|Q|_\nu} \geq c_w^{-p}\left(\frac{|E|}{|Q|}\right)^p \tag{2.37}$$

where c_w is given by (2.5).

Proof. We have

$$|E| = \int_E w(\tau)w(\tau)^{-1}|d\tau| \leq \left(\int_E w(\tau)^p|d\tau|\right)^{1/p}\left(\int_E w(\tau)^{-q}|d\tau|\right)^{1/q}$$

$$\leq |E|_\nu^{1/p}\left(\int_Q w(\tau)^{-q}|d\tau|\right)^{1/q} \leq |E|_\nu^{1/p}c_w|Q| \, |Q|_\nu^{-1/p},$$

which yields (2.37).   □

Lemma 2.21. *If E is a measurable subset of the portion $Q := \Gamma(t, \varepsilon)$ ($t \in \Gamma, \varepsilon > 0$)*
and $|E| \leq \alpha|Q|$ with $\alpha \in (0, 1)$, then $|E|_\nu \leq \beta|Q|_\nu$ where $\beta \in (0, 1)$ is given by
$\beta = 1 - c_w^{-p}(1 - \alpha)^p$.

Proof. Applying Lemma 2.20 to the set $E' = Q \setminus E$ we obtain

$$\frac{|E|_\nu}{|Q|_\nu} = 1 - \frac{|E'|_\nu}{|Q|_\nu} \leq 1 - c_w^{-p}\left(\frac{|E'|}{|Q|}\right)^p$$

$$= 1 - c_w^{-p}\left(1 - \frac{|E|}{|Q|}\right)^p \leq 1 - c_w^{-p}(1 - \alpha)^p,$$

and as $c_w \in [1, \infty)$ and $\alpha \in (0, 1)$, it is clear that $\beta := 1 - c_w^{-p}(1 - \alpha)^p \in (0, 1)$.
                                                                                □

Lemma 2.22. *For each $t \in \Gamma$, the function $\Phi_t : (0, \infty) \to (0, \infty)$ given by*

$$\Phi_t(\varepsilon) := \frac{1}{|\Gamma(t, \varepsilon)|} \int_{\Gamma(t, \varepsilon)} w(\tau)^p |d\tau|$$

is continuous from the left.

Proof. We may write $\Gamma(t, \varepsilon) = \Gamma(t, \varepsilon/2) \cup \bigcup_{n=1}^{\infty} R_n$ where

$$R_n := \{\tau \in \Gamma : \varepsilon(1 - 1/2^n) \leq |\tau - t| < \varepsilon(1 - 1/2^{n+1})\}.$$

Since

$$|\Gamma(t, \varepsilon)| = |\Gamma(t, \varepsilon/2)| + \sum_{n=1}^{\infty} |R_n|, \quad |\Gamma(t, \varepsilon)|_\nu = |\Gamma(t, \varepsilon/2)|_\nu + \sum_{n=1}^{\infty} |R_n|_\nu,$$

it follows that

$$\sum_{n=N}^{\infty} |R_n| \to 0 \text{ and } \sum_{n=N}^{\infty} |R_n|_\nu \to 0 \text{ as } N \to \infty.$$

If $\delta > \varepsilon(1 - 1/2^N)$ then

$$|\Gamma(t, \delta)| \geq |\Gamma(t, \varepsilon)| - \sum_{n=N}^{\infty} |R_n|, \quad |\Gamma(t, \delta)|_\nu \geq |\Gamma(t, \varepsilon)|_\nu - \sum_{n=N}^{\infty} |R_n|_\nu,$$

and hence $|\Gamma(t, \delta)| \to |\Gamma(t, \varepsilon)|$, $|\Gamma(t, \delta)|_\nu \to |\Gamma(t, \varepsilon)|_\nu$ as $\delta \to \varepsilon - 0$. Consequently,

$$\Phi_t(\delta) = |\Gamma(t, \delta)|_\nu / |\Gamma(t, \delta)| \to |\Gamma(t, \varepsilon)|_\nu / |\Gamma(t, \varepsilon)| = \Phi_t(\varepsilon)$$

as $\delta \to \varepsilon - 0$. $\square$

Lemma 2.23. *Let $\mathcal{F}$ be a set $\{\Gamma(t, \varepsilon(t))\}$ of portions whose centers lie on Γ and whose radii $\varepsilon(t)$ are uniformly bounded, $\varepsilon(t) \leq d < \infty$ for all t. Then there exists an at most countable subset $\mathcal{F}' = \{\Gamma(t_j, \varepsilon(t_j))\}$ of $\mathcal{F}$ such that*

(i) $\Gamma(t_i, \varepsilon(t_i)) \cap \Gamma(t_j, \varepsilon(t_j)) = \emptyset$ *for $i \neq j$,*

(ii) *each portion $\Gamma(t, \varepsilon(t)) \in \mathcal{F}$ is contained in some portion $\Gamma(t_i, 5\varepsilon(t_i))$ with $\Gamma(t_i, \varepsilon(t_i)) \in \mathcal{F}'$.*

Proof. Let $\mathcal{F}_1 := \{\Gamma(t_{j,1}, \varepsilon(t_{j,1}))\} \subset \mathcal{F}$ be any set of pairwise disjoint portions such that $d/2 < \varepsilon(t_{j,1}) \leq d$ and $\Gamma(t, \varepsilon(t)) \cap \Gamma(t_{j,1}, \varepsilon(t_{j,1})) \neq \emptyset$ for some j whenever $\Gamma(t, \varepsilon(t)) \in \mathcal{F}$ and $d/2 < \varepsilon(t) \leq d$. Given $\mathcal{F}_1, \ldots, \mathcal{F}_{k-1}$, denote by $\mathcal{F}_k := \{\Gamma(t_{j,k}, \varepsilon(t_{j,k}))\} \subset \mathcal{F}$ any set of pairwise disjoint portions such that $d/2^k < \varepsilon(t_{j,k}) \leq d/2^{k-1}$,

$$\Gamma(t_{j,k}, \varepsilon(t_{j,k})) \cap Q = \emptyset \text{ for all } Q \in \mathcal{F}_1 \cup \ldots \mathcal{F}_{k-1},$$

and $\Gamma(t, \varepsilon(t)) \cap \Gamma(t_{j,l}, \varepsilon(t_{j,l})) \neq \emptyset$ for some j and $l \in \{1, \ldots, k\}$ if $\Gamma(t, \varepsilon(t)) \in \mathcal{F}$ and $d/2^k < \varepsilon(t) \leq d/2^{k-1}$. We claim that $\mathcal{F}' := \bigcup_{k=1}^{\infty} \mathcal{F}_k$ has the desired properties.

Since each $\mathcal{F}_k$ is at most countable, the set $\mathcal{F}'$ is also at most countable. It is also clear that (i) is satisfied. To show (ii), pick any $\Gamma(t, \varepsilon(t)) \in \mathcal{F}$ and suppose $d/2^k < \varepsilon(t) \leq d/2^{k-1}$. Then $\Gamma(t, \varepsilon(t))$ must hit a portion $\Gamma(t_{j,l}, \varepsilon(t_{j,l})) \in \mathcal{F}_l$ with $l \in \{1, \ldots, k\}$. Thus, if $z \in \Gamma(t, \varepsilon(t))$ and $y \in \Gamma(t_{j,l}, \varepsilon(t_{j,l})) \cap \Gamma(t, \varepsilon(t))$, then

$$|z - t_{j,l}| \leq |z - t| + |t - y| + |y - t_{j,l}| < \varepsilon(t) + \varepsilon(t) + \varepsilon(t_{j,l})$$
$$\leq 2\,d/2^{k-1} + \varepsilon(t_{j,l}) \leq 4\,d/2^l + \varepsilon(t_{j,l}) < 5\,\varepsilon(t_{j,l}),$$

implying that $z \in \Gamma(t_{j,l}, 5\,\varepsilon(t_{j,l}))$. $\square$

Now fix $t_0 \in \Gamma$, $\varepsilon_0 > 0$, and put

$$Q := \Gamma(t_0, \varepsilon_0), \quad \lambda := \left(\frac{1}{|Q|} \int_Q w(\tau)^p |d\tau| \right)^{1/p}. \tag{2.38}$$

Let Φ_t be as in Lemma 2.22. Suppose further we are given a real number $a > 1$. For every natural number $m \geq 1$, define

$$A_m := \Big\{ t \in Q : \Phi_t(\varepsilon) \geq \lambda^p a^{mp} \text{ for some } \varepsilon \in (0, \varepsilon_0] \Big\}.$$

Since $a > 1$, it is clear that $A_1 \supset A_2 \supset \ldots \supset A_m \supset \ldots$. If $A_m \neq \emptyset$ and $t \in A_m$, let

$$\varepsilon_m(t) := \max \Big\{ \varepsilon \in (0, \varepsilon_0] : \Phi_t(\varepsilon) \geq \lambda^p a^{mp} \Big\}; \tag{2.39}$$

the existence of $\varepsilon_m(t)$ is guaranteed by Lemma 2.22. Applying Lemma 2.23 to the set $\mathcal{F} := \{ \Gamma(t, \varepsilon_m(t)) \}_{t \in A_m}$, we see that there is an at most countable subset $\mathcal{F}' := \{ \Gamma(t_i, \varepsilon_m(t_i)) \} \subset \mathcal{F}$ of pairwise disjoint portions such that every portion $\Gamma(t, \varepsilon_m(t))$ is contained in $\Gamma(t_i, 5\,\varepsilon_m(t_i))$ with some $\Gamma(t_i, \varepsilon_m(t_i)) \in \mathcal{F}'$. Finally, put

$$\Gamma_{m,i} := \Gamma(t_i, \varepsilon_m(t_i)), \quad \tilde{\Gamma}_{m,i} := \Gamma(t_i, 5\,\varepsilon_m(t_i)), \quad E_m := \bigcup_i \tilde{\Gamma}_{m,i}$$

in case $A_m \neq \emptyset$, and let $E_m = \emptyset$ in case $A_m = \emptyset$.

Lemma 2.24. *Let $C_\Gamma (\geq 1)$ be the Carleson constant of Γ and let $c_w (\geq 1)$ be the constant given by (2.5). If*

$$a > 22\,C_\Gamma c_w, \tag{2.40}$$

then $\Phi_t(\varepsilon) < \lambda^p a^p$ for all $t \in Q$ and all $\varepsilon \in [\varepsilon_0, 11\,\varepsilon_0]$.

Proof. First notice that if $t_1, t_2 \in \Gamma$ and $0 < \varepsilon_1 \leq \varepsilon_2$, then

$$\frac{|\Gamma(t_2, \varepsilon_2)|}{|\Gamma(t_1, \varepsilon_1)|} \leq C_\Gamma \frac{\varepsilon_2}{\varepsilon_1}. \tag{2.41}$$

Indeed, if Γ is bounded, then

$$\frac{|\Gamma(t_2, \varepsilon_2)|}{|\Gamma(t_1, \varepsilon_1)|} \leq \begin{cases} C_\Gamma \varepsilon_2 / \varepsilon_1 & \text{if } \varepsilon_1 \leq d_{t_1}, \\ |\Gamma|/|\Gamma| \leq C_\Gamma \varepsilon_2 / \varepsilon_1 & \text{if } \varepsilon_1 > d_{t_1}, \end{cases}$$

where $d_t := \max_{\tau \in \Gamma} |\tau - t|$, and it is clear that (2.41) always holds in case Γ is unbounded. If $t \in Q$ and $\varepsilon \geq \varepsilon_0$, then Q is a subset of the portion $\Gamma(t, 2\varepsilon)$. Hence, (2.41) and Lemma 2.20 imply that

$$\begin{aligned}
\Phi_t(\varepsilon) &= \frac{|\Gamma(t, \varepsilon)|_\nu}{|\Gamma(t, \varepsilon)|} \leq \frac{|\Gamma(t, 2\varepsilon)|}{|\Gamma(t, \varepsilon)|} \cdot \frac{|\Gamma(t, 2\varepsilon)|_\nu}{|\Gamma(t, 2\varepsilon)|} \leq 2\, C_\Gamma \frac{|\Gamma(t, 2\varepsilon)|_\nu}{|\Gamma(t, 2\varepsilon)|} \\
&= 2\, C_\Gamma \frac{|\Gamma(t, 2\varepsilon)|_\nu}{|\Gamma(t_0, \varepsilon_0)|_\nu} \cdot \frac{|\Gamma(t_0, \varepsilon_0)|}{|\Gamma(t, 2\varepsilon)|} \cdot \frac{|\Gamma(t_0, \varepsilon_0)|_\nu}{|\Gamma(t_0, \varepsilon_0)|} \\
&\leq 2\, C_\Gamma c_w^p \left(\frac{|\Gamma(t, 2\varepsilon)|}{|\Gamma(t_0, \varepsilon_0)|} \right)^p \frac{|\Gamma(t_0, \varepsilon_0)|}{|\Gamma(t, 2\varepsilon)|} \cdot \frac{|\Gamma(t_0, \varepsilon_0)|_\nu}{|\Gamma(t_0, \varepsilon_0)|} \\
&= 2\, C_\Gamma c_w^p \left(\frac{|\Gamma(t, 2\varepsilon)|}{|\Gamma(t_0, \varepsilon_0)|} \right)^{p-1} \lambda^p \leq 2\, C_\Gamma c_w^p \left(\frac{2\, C_\Gamma \varepsilon}{\varepsilon_0} \right)^{p-1} \lambda^p,
\end{aligned}$$

and this is less than $\lambda^p a^p$ if $\varepsilon \leq 11\,\varepsilon_0$ and (2.40) holds. $\qquad\square$

Lemma 2.25. *We have $w(t) \leq \lambda a^m$ for almost all $t \in Q \setminus E_m$.*

Proof. If $A_m = \emptyset$ then $\Phi_t(\varepsilon) < \lambda^p a^{mp}$ for all $t \in Q$ and all $\varepsilon \in (0, \varepsilon_0]$. If $A_m \neq \emptyset$ then $E_m \supset \bigcup_{t \in A_m} \Gamma(t, \varepsilon_m(t)) \supset A_m$ by construction and hence $\Phi_t(\varepsilon) < \lambda^p a^{mp}$ for all $t \in Q \setminus E_m \subset Q \setminus A_m$ and all $\varepsilon \in (0, \varepsilon_0]$. Thus, in either case we deduce from Theorem 2.19 that

$$w(t)^p = \lim_{\varepsilon \to 0} \Phi_t(\varepsilon) \leq \lambda^p a^{mp}$$

for almost all $t \in Q \setminus E_m$. $\qquad\square$

Lemma 2.26. *If a satisfies (2.40) and*

$$a > (10\, C_\Gamma{}^2)^{1/p} \tag{2.42}$$

then $\Gamma(t_0, 6\,\varepsilon_0) \supset E_1 \supset E_2 \supset \ldots \supset E_m \supset \ldots$.

Proof. If $A_{m+1} = \emptyset$ then $E_{m+1} = \emptyset \subset E_m$. So suppose $A_{m+1} \neq \emptyset$ and $\tau \in A_{m+1}$. Since $\Phi_\tau(\varepsilon_0) < \lambda^p a^p \leq \lambda^p a^{mp}$ due to Lemma 2.24, we have $\varepsilon_m(t) < \varepsilon_0$. If $\varepsilon_m(t) < \varepsilon \leq \varepsilon_0$, then, by (2.39), $\Phi_\tau(\varepsilon) < \lambda^p a^{mp} < \lambda^p a^{(m+1)p}$, whence, again by (2.39), $\varepsilon_{m+1}(\tau) \leq \varepsilon_m(\tau)$. Pick a portion $\Gamma_{m+1,j}$ and let $\Gamma_{m+1,j} =: \Gamma(t, r)$. Then $r = \varepsilon_{m+1}(t)$, and since $\varepsilon_{m+1}(t) \leq \varepsilon_m(t)$, it follows that $\Gamma_{m+1,j}$ is contained in $\Gamma(t, \varepsilon_m(t)) =: \Gamma(t, s)$.

We have $|\Gamma(t, 2s)|_\nu / |\Gamma(t, 2s)| = \Phi_t(2s) = \Phi_t(2\varepsilon_m(t))$. If $2\varepsilon_m(t) \leq \varepsilon_0$ then $\Phi_t(2\varepsilon_m(t)) < \lambda^p a^{mp}$ due to (2.39), while if $\varepsilon_0 \leq 2\varepsilon_m(t) < 2\varepsilon_0 < 11\varepsilon_0$, we see that $\Phi_t\left(2\,\varepsilon_m(t)\right) < \lambda^p a^p \le \lambda^p a^{mp}$ from Lemma 2.24. Thus, in either case we obtain

$$\frac{|\Gamma(t,2\,s)|_\nu}{|\Gamma(t,2\,s)|} < \lambda^p a^{mp} \le \Phi_t\left(\varepsilon_m(t)\right) = \Phi_t\left(s\right) = \frac{|\Gamma(t,s)|_\nu}{|\Gamma(t,s)|}.$$

Therefore, by (2.41),

$$\lambda^p a^{mp} \le \frac{|\Gamma(t,s)|_\nu}{|\Gamma(t,s)|} \le \frac{|\Gamma(t,2\,s)|}{|\Gamma(t,s)|} \cdot \frac{|\Gamma(t,2\,s)|_\nu}{|\Gamma(t,2\,s)|} < 2\,C_\Gamma \lambda^p a^{mp}. \qquad (2.43)$$

Since $r = \varepsilon_{m+1}(t)$, it follows analogously that

$$\lambda^p a^{(m+1)p} \le \frac{|\Gamma(t,r)|_\nu}{|\Gamma(t,r)|} < 2\,C_\Gamma \lambda^p a^{(m+1)p}. \qquad (2.44)$$

Combining (2.43) and (2.44) and taking into account (2.41) we get

$$1 \le \frac{|\Gamma(t,s)|_\nu}{|\Gamma(t,r)|_\nu} < \frac{2\,C_\Gamma \lambda^p a^{mp}|\Gamma(t,s)|}{\lambda^p a^{(m+1)p}|\Gamma(t,r)|} \le \frac{2\,C_\Gamma^{\,2}}{a^p}\frac{s}{r},$$

that is, $s/r > a^p/(2\,C_\Gamma^{\,2})$, whence, by (2.42), $s > 5\,r$. Thus,

$$\tilde{\Gamma}_{m+1,j} := \Gamma(t,5\,r) \subset \Gamma(t,s) = \Gamma(t,\varepsilon_m(t)) \subset E_m.$$

Since $\Gamma_{m+1,j}$ was arbitrary, we arrive at the conclusion that $E_{m+1} = \bigcup_j \tilde{\Gamma}_{m+1,j}$ is a subset of E_m.

Finally, if $t \in A_1 \subset \Gamma(t_0,\varepsilon_0)$ then $\varepsilon_1(t) \le \varepsilon_0$ and $|t - t_0| < \varepsilon_0$, which implies that

$$E_1 = \bigcup_j \tilde{\Gamma}_{1,j} \subset \bigcup_{t\in A_1} \Gamma(t,5\,\varepsilon_1(t)) \subset \Gamma(t_0,6\,\varepsilon_0). \qquad \square$$

Lemma 2.27. *If*

$$a > \max\left\{22\,C_\Gamma c_w,\ (60\,C_\Gamma)^{1/p},\ (40\,C_\Gamma^{\,3})^{1/p}\right\}, \qquad (2.45)$$

then there is a constant $\alpha \in (0,1)$ depending only on the numbers C_Γ, c_w, p such that $|E_{m+1} \cap \Gamma_{m,i}| \le \alpha|\Gamma_{m,i}|$ for all m and i.

Proof. Let $\Gamma_{m,i} =: \Gamma(t,\varepsilon_m(t))$ and pick any portion $\tilde{\Gamma}_{m+1,j} =: \Gamma(\tau,r)$. Suppose first that $\tilde{\Gamma}_{m+1,j} \cap \Gamma_{m,i} \ne \emptyset$ but $\tilde{\Gamma}_{m+1,j}$ is not contained in $\Gamma_{m,i}$. Then there is a $y \in \tilde{\Gamma}_{m+1,j} \cap \Gamma_{m,i}$, and for every $z \in \tilde{\Gamma}_{m+1,j}$ we have $|y - z| < 2\,r$ and thus

$$|t - z| \le |t - y| + |y - z| < \varepsilon_m(t) + 2\,r =: s,$$

whence $\tilde{\Gamma}_{m+1,j} = \Gamma(\tau,r) \subset \Gamma(t,s)$. Since

$$|t - \tau| \le |t - y| + |y - \tau| < \varepsilon_m(t) + r < s,$$

it follows that $\Gamma(t,s) \subset \Gamma(\tau, s+s) = \Gamma(\tau, 2\,s)$. We have $r = 5\,\varepsilon_{m+1}(\tau) \le 5\,\varepsilon_0$ and thus, $s \le \varepsilon_0 + 10\,\varepsilon_0 = 11\,\varepsilon_0$. Consequently, $\Phi_t(s) < \lambda^p a^p \le \lambda^p a^{mp}$ by virtue of (2.39) and Lemma 2.24. Therefore,

$$\lambda^p a^{mp} > \Phi_t(s) = \frac{|\Gamma(t,s)|_\nu}{|\Gamma(t,s)|} \ge \frac{|\tilde\Gamma_{m+1,j}|_\nu}{|\Gamma(t,s)|} \ge \frac{|\Gamma_{m+1,j}|_\nu}{|\Gamma(t,s)|}$$

$$= \frac{|\Gamma_{m+1,j}|_\nu}{|\Gamma_{m+1,j}|} \cdot \frac{|\Gamma_{m+1,j}|}{|\Gamma(t,s)|} = \Phi_t\big(\varepsilon_{m+1}(\tau)\big) \frac{|\Gamma_{m+1,j}|}{|\Gamma(t,s)|}$$

$$\ge \lambda^p a^{(m+1)p} \frac{|\Gamma_{m+1,j}|}{|\Gamma(t,s)|} \ge \lambda^p a^{(m+1)p} \frac{|\Gamma_{m+1,j}|}{|\Gamma(\tau,2\,s)|},$$

and hence, again by (2.41),

$$\big|\Gamma(\tau,2\,s)\big| \ge a^p |\Gamma_{m+1,j}| = a^p \big|\Gamma(\tau,r/5)\big| \ge \frac{a^p}{10\,C_\Gamma} \frac{r}{s} \big|\Gamma(\tau,2\,s)\big|.$$

It follows that

$$r \le 10\,C_\Gamma a^{-p} s = 10\,C_\Gamma a^{-p}\big(\varepsilon_m(t) + 2\,r\big),$$

and taking into account that $a^p > 60\,C_\Gamma$ by (2.45), we so get

$$r < 10\,C_\Gamma (60\,C_\Gamma)^{-1}\big(\varepsilon_m(t) + 2\,r\big) = \varepsilon_m(t)/6 + 2\,r/6,$$

whence $r < \varepsilon_m(t)/4$. Since $\tilde\Gamma_{m+1,j} \setminus \Gamma_{m,i} \ne \emptyset$, the latter inequality implies that for every $z \in \tilde\Gamma_{m+1,j}$,

$$|t - z| \ge \varepsilon_m(t) - 2\,r > \varepsilon_m(t)/2.$$

Thus, $\tilde\Gamma_{m+1,j} \cap \Gamma\big(t, \varepsilon_m(t)/2\big) = \emptyset$ for all $j \in J_1$ where

$$J_1 := \{j : \tilde\Gamma_{m+1,j} \cap \Gamma_{m,i} \ne \emptyset,\ \tilde\Gamma_{m+1,j} \not\subset \Gamma_{m,i}\}.$$

Consequently, if $j \in J_1$ then

$$\Big|\bigcup_{j \in J_1} \tilde\Gamma_{m+1,j} \cap \Gamma_{m,i}\Big| \le \big|\Gamma_{m,i} \setminus \Gamma(t, \varepsilon_m(t)/2)\big|. \tag{2.46}$$

We always have

$$\frac{|\Gamma_{m+1,j}|_\nu}{|\Gamma_{m+1,j}|} = \Phi_t\big(\varepsilon_{m+1}(\tau)\big) \ge \lambda^p a^{(m+1)p},$$

$$\frac{|\Gamma_{m,i}|_\nu}{|\Gamma_{m,i}|} = \frac{|\Gamma(t,\varepsilon_m(t))|_\nu}{|\Gamma(t,\varepsilon_m(t))|} \le \frac{|\Gamma(t,2\,\varepsilon_m(t))|_\nu}{|\Gamma(t,2\,\varepsilon_m(t))|} \cdot \frac{|\Gamma(t,2\,\varepsilon_m(t))|}{|\Gamma(t,\varepsilon_m(t))|}$$

$$= \Phi_t\big(2\,\varepsilon_m(t)\big) \frac{|\Gamma(t,2\,\varepsilon_m(t))|}{|\Gamma(t,\varepsilon_m(t))|} < \lambda^p a^{mp} 2\,C_\Gamma$$

(recall (2.41), (2.39) and Lemma 2.24 for the last inequality). Hence, letting j range over the set $J_2 := \{j : \tilde{\Gamma}_{m+1,j} \subset \Gamma_{m,i}\}$, we obtain that

$$\sum_j |\tilde{\Gamma}_{m+1,j}| \le 5\, C_\Gamma \sum_j |\Gamma_{m+1,j}| \le 5\, C_\Gamma \lambda^{-p} a^{-(m+1)p} \sum_j |\Gamma_{m+1,j}|_\nu$$
$$\le 5\, C_\Gamma \lambda^{-p} a^{-(m+1)p} |\Gamma_{m,i}|_\nu < 10\, C_\Gamma{}^2 a^{-p} |\Gamma_{m,i}|$$
$$= 10\, C_\Gamma{}^2 a^{-p} |\Gamma(t, \varepsilon_m(t))| \le 20\, C_\Gamma{}^3 a^{-p} |\Gamma(t, \varepsilon_m(t)/2)|.$$

Since $a > (40\, C_\Gamma{}^3)^{1/p}$ due to (2.45), we get

$$\sum_{j \in J_2} |\tilde{\Gamma}_{m+1,j}| < (1/2) |\Gamma(t, \varepsilon_m(t)/2)|. \tag{2.47}$$

Combining (2.46) and (2.47) we finally arrive at the conclusion that

$$|E_{m+1} \cap \Gamma_{m,i}| = \left| \bigcup_j \tilde{\Gamma}_{m+1,j} \cap \Gamma_{m,i} \right|$$
$$\le \left| \bigcup_{j \in J_1} \tilde{\Gamma}_{m+1,j} \cap \Gamma_{m,i} \right| + \left| \bigcup_{j \in J_2} \tilde{\Gamma}_{m+1,j} \right|$$
$$\le \left| \Gamma_{m,i} \setminus \Gamma(t, \varepsilon_m(t)/2) \right| + (1/2) |\Gamma(t, \varepsilon_m(t)/2)|$$
$$= |\Gamma_{m,i}| - (1/2) |\Gamma(t, \varepsilon_m(t)/2)| \le |\Gamma_{m,i}| - (1/(4\, C_\Gamma)) |\Gamma_{m,i}|,$$

which gives the assertion with $\alpha = 1 - 1/(4\, C_\Gamma)$. $\square$

Lemma 2.28. *If a satisfies (2.45), then $|E_{m+1}|_\nu \le \beta |E_m|_\nu$ for all $m = 1, 2, \ldots$ with some constant $\beta \in (0,1)$ depending only on C_Γ, c_w, p.*

Proof. Lemmas 2.27 and 2.21 show that

$$|E_{m+1} \cap \Gamma_{m,i}|_\nu \le \left(1 - c_w^{-p}(1 - \alpha)^p\right) |\Gamma_{m,i}|_\nu$$

with some $\alpha \in (0,1)$ depending only on C_Γ, c_w, p, and since the portions $\Gamma_{m,i}$ are pairwise disjoint, we see that

$$\left| E_{m+1} \cap \left(\bigcup_i \Gamma_{m,i} \right) \right|_\nu \le \left(1 - c_w^{-p}(1 - \alpha)^p\right) \left| \bigcup_i \Gamma_{m,i} \right|_\nu. \tag{2.48}$$

On the other hand, from Lemma 2.20 and (2.41) we deduce that

$$\frac{|\tilde{\Gamma}_{m,i}|_\nu}{|\Gamma_{m,i}|_\nu} \le c_w^p \left(\frac{|\tilde{\Gamma}_{m,i}|}{|\Gamma_{m,i}|} \right)^p \le c_w^p (5\, C_\Gamma)^p,$$

whence

$$|E_m|_\nu = \left| \bigcup_i \tilde{\Gamma}_{m,i} \right|_\nu \le \sum_i |\tilde{\Gamma}_{m,i}|_\nu$$
$$\le c_w^p (5\, C_\Gamma)^p \sum_i |\Gamma_{m,i}|_\nu = c_w^p (5\, C_\Gamma)^p \left| \sum_i \Gamma_{m,i} \right|_\nu. \tag{2.49}$$

Since $E_{m+1} \subset E_m$ by Lemma 2.26, the estimates (2.48) and (2.49) yield that

$$
|E_{m+1}|_\nu \leq \left| E_{m+1} \cap \left(\bigcup_i \Gamma_{m,i} \right) \right|_\nu + \left| E_m \setminus \bigcup_i \Gamma_{m,i} \right|_\nu
$$

$$
\leq \left| \bigcup_i \Gamma_{m,i} \right|_\nu - c_w^{-p}(1-\alpha)^p \left| \bigcup_i \Gamma_{m,i} \right|_\nu + \left| E_m \setminus \bigcup_i \Gamma_{m,i} \right|_\nu
$$

$$
= |E_m|_\nu - c_w^{-p}(1-\alpha)^p \left| \bigcup_i \Gamma_{m,i} \right|_\nu
$$

$$
\leq |E_m|_\nu - c_w^{-2p}(1-\alpha)^p(5\,C_\Gamma)^{-p}|E_m|_\nu = \beta|E_m|_\nu
$$

with $\beta := 1 - c_w^{-2p}(1-\alpha)^p(5\,C_\Gamma)^{-p} \in (0,1)$. $\qquad\square$

Now we are in a position to establish the main result of this section, the so-called *reverse Hölder inequality*.

Theorem 2.29. *Let Γ be a simple Carleson curve, $1 < p < \infty$, and $w \in A_p(\Gamma)$. Then there exist constants $\delta \in (0,\infty)$ and $C \in (0,\infty)$ depending only on C_Γ, c_w, p such that*

$$
\left(\frac{1}{|\Gamma(t,\varepsilon)|} \int_{\Gamma(t,\varepsilon)} w(\tau)^{p(1+\delta)}|d\tau| \right)^{1/(1+\delta)} \leq C \left(\frac{1}{|\Gamma(t,\varepsilon)|} \int_{\Gamma(t,\varepsilon)} w(\tau)^p|d\tau| \right)
$$

for all $t \in \Gamma$ and all $\varepsilon > 0$.

Proof. Fix $t_0 \in \Gamma$ and $\varepsilon_0 > 0$, define Q and λ by (2.38), choose any a satisfying (2.45), and construct $E_1 \supset E_2 \supset \ldots \supset E_m \supset \ldots$ as above. From Lemma 2.28 we infer that $|\bigcap_m E_m|_\nu = 0$, and since $w^{-1}(0)$ has measure zero, this implies that $|\bigcap_m E_m| = 0$. Let β be the constant delivered by Lemma 2.28 and let $\delta > 0$ be any number such that $0 < a^{p\delta}\beta < 1$. Taking into account Lemmas 2.25 and 2.28 we get

$$
\int_Q w(\tau)^{p(1+\delta)}|d\tau| = \int_{Q\setminus E_1} w(\tau)^{p(1+\delta)}|d\tau| + \sum_{m=1}^{\infty} \int_{Q\cap(E_m\setminus E_{m+1})} w(\tau)^{p(1+\delta)}|d\tau|
$$

$$
\leq (\lambda a)^{p\delta} \int_{Q\setminus E_1} w(\tau)^p|d\tau| + \sum_{m=1}^{\infty} (\lambda a^{m+1})^{p\delta} \int_{Q\cap(E_m\setminus E_{m+1})} w(\tau)^p|d\tau|
$$

$$
\leq (\lambda a)^{p\delta} \int_Q w(\tau)^p|d\tau| + \sum_{m=1}^{\infty} (\lambda a^{m+1})^{p\delta} \int_{E_m} w(\tau)^p|d\tau|
$$

$$
= (\lambda a)^{p\delta}|Q|_\nu + (\lambda a)^{p\delta} \sum_{m=1}^{\infty} a^{mp\delta}|E_m|_\nu
$$

$$
\leq (\lambda a)^{p\delta} \left(|Q|_\nu + |E_1|_\nu \sum_{m=1}^{\infty} (a^{p\delta}\beta)^m \beta^{-1} \right),
$$

and $\sum_{m=1}^{\infty}(a^{p\delta}\beta)^m < \infty$ since $0 < a^{p\delta}\beta < 1$. Because (2.45) implies (2.42) and because $E_1 \subset \Gamma(t_0, 6\,\varepsilon_0)$ by Lemma 2.26, we infer from Lemma 2.20 that

$$\frac{|E_1|_\nu}{|Q|_\nu} \leq \frac{|\Gamma(t_0, 6\,\varepsilon_0)|_\nu}{|\Gamma(t_0, \varepsilon_0)|_\nu} \leq c_w^p \left(\frac{|\Gamma(t_0, 6\,\varepsilon_0)|}{|\Gamma(t_0, \varepsilon_0)|} \right)^p \leq c_w^p (6\,C_\Gamma)^p.$$

Thus,

$$\int_Q w(\tau)^{p(1+\delta)} |d\tau| \leq K \lambda^{p\delta} |Q|_\nu \quad \text{with} \quad K := a^{p\delta} \left(1 + c_w^p (6C_\Gamma)^p \beta^{-1} \sum_{m=1}^{\infty}(a^{p\delta}\beta)^m \right).$$

Since $\lambda^p = |Q|_\nu / |Q|$, this implies that

$$\int_Q w(\tau)^{p(1+\delta)} |d\tau| \leq K \left(\frac{|Q|_\nu}{|Q|} \right)^\delta |Q|_\nu = K|Q| \left(\frac{1}{|Q|} \int_Q w(\tau)^p |d\tau| \right)^{1+\delta},$$

which gives the assertion with $C = K^{1/(1+\delta)}$. $\square$

2.8   Stability of Muckenhoupt weights

Let Γ be a simple Carleson curve and let $w : \Gamma \to [0, \infty]$ be a weight. This section is concerned with the two sets

$$G := G(w, \Gamma) := \{ (p, \varrho) \in (1, \infty) \times \mathbf{R} : w^\varrho \in A_p(\Gamma) \},$$
$$\tilde{G} := \tilde{G}(w, \Gamma) := \{ (p, \varrho) \in (1, \infty) \times \mathbf{R} : w^{\varrho/p} \in A_p(\Gamma) \}.$$

Since Γ is supposed to be a Carleson curve, both sets contain the half-line $(1, \infty) \times \{0\}$. If w is a weight for which $\log w$ does not belong to $BMO(\Gamma)$ (recall Example 2.6), then, by Theorem 2.5, there are no $p \in (1, \infty)$ and no $\varrho \neq 0$ such that $w^\varrho \in A_p(\Gamma)$. Hence, in this case both G and $\tilde{G}$ coincide with $(1, \infty) \times \{0\}$. From Theorem 2.2 we deduce that if w is the power weight $w(\tau) = |\tau - t|^\lambda$, then

$$G = \begin{cases} \{(p, \varrho) \in (1, \infty) \times \mathbf{R} : -1/(\lambda p) < \varrho < 1/\lambda - 1/(\lambda p)\} & \text{if } \lambda > 0, \\ (1, \infty) \times \mathbf{R} \text{ if } \lambda = 0, \\ \{(p, \varrho) \in (1, \infty) \times \mathbf{R} : 1/\lambda - 1/(\lambda p) < \varrho < -1/(\lambda p)\} & \text{if } \lambda < 0, \end{cases}$$

$$\tilde{G} = \begin{cases} \{(p, \varrho) \in (1, \infty) \times \mathbf{R} : -1/\lambda < \varrho < -1/\lambda + p/\lambda\} & \text{if } \lambda > 0, \\ (1, \infty) \times \mathbf{R} \text{ if } \lambda = 0, \\ \{(p, \varrho) \in (1, \infty) \times \mathbf{R} : -1/\lambda + p/\lambda < \varrho < -1/\lambda\} & \text{if } \lambda < 0, \end{cases}$$

Figures 4a and 4b show G and $\tilde{G}$ for $\lambda = 1$.

Proposition 2.30. *Let Γ be a simple Carleson curve and let $w : \Gamma \to [0, \infty]$ be a weight. Then $\tilde{G}$ is convex and, moreover, if $(p_0, \varrho_0) \in \tilde{G}$ then $(p, \varrho_0) \in \tilde{G}$ for all $p \geq p_0$.*

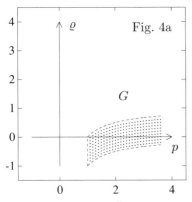
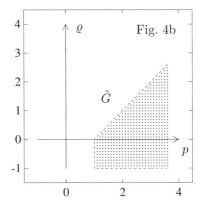

Proof. Let $(p_1, \varrho_1) \in \tilde{G}$, $(p_2, \varrho_2) \in \tilde{G}$ and put $p(\theta) := (1 - \theta)p_1 + \theta p_2$, $\varrho(\theta) :=$ $(1 - \theta)\varrho_1 + \theta\varrho_2$. We know that $w^{\varrho_1/p_1} \in A_{p_1}(\Gamma)$, $w^{\varrho_2/p_2} \in A_{p_2}(\Gamma)$, and we have to show that $w^{\varrho(\theta)/p(\theta)} \in A_{p(\theta)}(\Gamma)$ for all $\theta \in (0, 1)$. Let $\gamma \subset \Gamma$ be any portion. Hölder's inequality with $\alpha = 1/(1 - \theta)$ and $\beta = 1/\theta$ gives

$$\left(\frac{1}{|\gamma|} \int_\gamma (w^{\varrho(\theta)/p(\theta)})^{p(\theta)} |d\tau| \right)^{1/p(\theta)} = \left(\frac{1}{|\gamma|} \int_\gamma w^{\varrho_1(1-\theta)+\varrho_2\theta} |d\tau| \right)^{1/p(\theta)}$$

$$\leq \left(\frac{1}{|\gamma|} \int_\gamma w^{\varrho_1} |d\tau| \right)^{(1-\theta)/p(\theta)} \left(\frac{1}{|\gamma|} \int_\gamma w^{\varrho_2} |d\tau| \right)^{\theta/p(\theta)}$$

$$= \left(\frac{1}{|\gamma|} \int_\gamma (w^{\varrho_1/p_1})^{p_1} |d\tau| \right)^{\frac{1}{p_1} \frac{(1-\theta)p_1}{p(\theta)}} \left(\frac{1}{|\gamma|} \int_\gamma (w^{\varrho_2/p_2})^{p_2} |d\tau| \right)^{\frac{1}{p_2} \frac{\theta p_2}{p(\theta)}}. \quad (2.50)$$

Define q_j and $q(\theta)$ by $1/p_j + 1/q_j = 1$ and $1/p(\theta) + 1/q(\theta) = 1$. Applying Hölder's inequality with

$$\alpha = \frac{p(\theta)}{q(\theta)} \frac{q_1}{p_1} \frac{1}{1 - \theta}, \quad \beta = \frac{p(\theta)}{q(\theta)} \frac{q_2}{p_2} \frac{1}{\theta}$$

(note that $1/\alpha + 1/\beta = 1$), we obtain

$$\left(\frac{1}{|\gamma|} \int_\gamma (w^{-\varrho(\theta)/p(\theta)})^{q(\theta)} |d\tau| \right)^{1/q(\theta)}$$

$$= \left(\frac{1}{|\gamma|} \int_\gamma w^{-(q(\theta)/p(\theta))(\varrho_1(1-\theta)+\varrho_2\theta)} |d\tau| \right)^{1/q(\theta)}$$

$$\leq \left(\frac{1}{|\gamma|} \int_\gamma w^{-\varrho_1 q_1/p_1} |d\tau| \right)^{1/(\alpha q(\theta))} \left(\frac{1}{|\gamma|} \int_\gamma w^{-\varrho_2 q_2/p_2} |d\tau| \right)^{1/(\beta q(\theta))}$$

$$= \left(\frac{1}{|\gamma|} \int_{\gamma} (w^{-\varrho_1/p_1})^{q_1} |d\tau| \right)^{\frac{1}{q_1} \frac{(1-\theta)p_1}{p(\theta)}} \left(\frac{1}{|\gamma|} \int_{\gamma} (w^{-\varrho_2/p_2})^{q_2} |d\tau| \right)^{\frac{1}{q_2} \frac{\theta p_2}{p(\theta)}}. \quad (2.51)$$

Multiplication of (2.50) and (2.51) shows that $w^{\varrho(\theta)/p(\theta)} \in A_{p(\theta)}(\Gamma)$.

Now let $(p_0, \varrho_0) \in \tilde{G}$, i.e. suppose $w^{\varrho_0/p_0} \in A_{p_0}(\Gamma)$. For $v := w^{\varrho_0/p}$ and every portion $\gamma \subset \Gamma$ we then have

$$\left(\frac{1}{|\gamma|} \int_{\gamma} v^p |d\tau| \right)^{1/p} \left(\frac{1}{|\gamma|} \int_{\gamma} v^{-q} |d\tau| \right)^{1/q}$$

$$= \left(\frac{1}{|\gamma|} \int_{\gamma} w^{\varrho_0} |d\tau| \right)^{1/p} \left(\frac{1}{|\gamma|} \int_{\gamma} w^{-\varrho_0/(p-1)} |d\tau| \right)^{1/q}. \quad (2.52)$$

If $p \geq p_0$, we may invoke Jensen's inequality (2.16) with $\varphi(x) = x^{(p-1)/(p_0-1)}$ to see that (2.52) is at most

$$\left(\frac{1}{|\gamma|} \int_{\gamma} w^{\varrho_0} |d\tau| \right)^{\frac{1}{p}} \left(\frac{1}{|\gamma|} \int_{\gamma} w^{-\varrho_0/(p_0-1)} |d\tau| \right)^{\frac{p_0-1}{p-1} \frac{1}{q}}$$

$$= \left(\left(\frac{1}{|\gamma|} \int_{\gamma} (w^{\varrho_0/p_0})^{p_0} |d\tau| \right)^{1/p_0} \left(\frac{1}{|\gamma|} \int_{\gamma} (w^{-\varrho_0/p_0})^{q_0} |d\tau| \right)^{1/q_0} \right)^{p_0/p}.$$

This shows that $v \in A_p(\Gamma)$ and thus that $(p, \varrho) \in \tilde{G}$. $\qquad \square$

The following theorem is the main result of this section. It makes precise what is meant by stability of Muckenhoupt weights.

Theorem 2.31. *Let Γ be a simple Carleson curve, $1 < p_0 < \infty$, and $w \in A_{p_0}(\Gamma)$. Then there exists an $\varepsilon_0 > 0$ such that $w^{1+\varepsilon} \in A_p(\Gamma)$ for all $\varepsilon \in (-\varepsilon_0, \varepsilon_0)$ and all $p \in (p_0 - \varepsilon_0, p_0 + \varepsilon_0)$.*

Proof. By Theorem 2.29, there exist constants $\delta \in (0, \infty)$ and $C \in (0, \infty)$ such that

$$\left(\frac{1}{|\gamma|} \int_{\gamma} w(\tau)^{p_0(1+\delta)} |d\tau| \right)^{1/(1+\delta)} \leq \frac{C}{|\gamma|} \int_{\gamma} w(\tau)^{p_0} |d\tau|$$

for every portion $\gamma \subset \Gamma$. If $w \in A_{p_0}(\Gamma)$, then $w^{-1} \in A_{q_0}(\Gamma)$ where $1/p_0 + 1/q_0 = 1$. Hence, again by Theorem 2.29, we may find constants $\eta \in (0, \infty)$ and $D \in (0, \infty)$ such that

$$\left(\frac{1}{|\gamma|} \int_{\gamma} w(\tau)^{-q_0(1+\eta)} |d\tau| \right)^{1/(1+\eta)} \leq \frac{D}{|\gamma|} \int_{\gamma} w(\tau)^{-q_0} |d\tau|$$

for all portions $\gamma \subset \Gamma$. If $-1 < \lambda_1 \leq \lambda_2$, then $\varphi(x) = x^{(1+\lambda_2)/(1+\lambda_1)}$ is convex and we may apply Jensen's inequality (2.16) to deduce that

$$\left(\frac{1}{|\gamma|} \int\limits_\gamma |f(\tau)|^{1+\lambda_1} |d\tau| \right)^{1/(1+\lambda_1)} \leq \left(\frac{1}{|\gamma|} \int\limits_\gamma |f(\tau)|^{1+\lambda_2} |d\tau| \right)^{1/(1+\lambda_2)}$$

for every $f \in L^1_{\mathrm{loc}}(\Gamma)$. Thus, letting $\lambda := \min\{\delta, \eta\}$, we obtain that $w^{1+\lambda} \in A_{p_0}(\Gamma)$. Consequently, $(p_0, (1+\lambda)p_0) \in \tilde{G}$. Because, by Proposition 2.30, the set $\tilde{G}$ is convex and contains the half-line $(1, \infty) \times \{0\}$, it follows that (p_0, p_0) is an inner point of $\tilde{G}$, which is equivalent to the assertion of the theorem. □

Corollary 2.32. *Let Γ be a simple Carleson curve and let w be a weight on Γ.*

(a) *The set G is either the half-line $(1, \infty) \times \{0\}$ or an open and connected set with the property*

$$(p_0, \varrho_0) \in G \Longrightarrow (p_0, \varrho) \in G \text{ for } 0 \leq \varrho/\varrho_0 \leq 1 \text{ and } (p, \varrho_0/p) \in G \text{ for } p \geq p_0.$$

(b) *The set $\tilde{G}$ is either the half-line $(1, \infty) \times \{0\}$ or an open and convex set enjoying the property*

$$(p_0, \varrho_0) \in \tilde{G} \Longrightarrow (p_0, p_0\varrho) \in \tilde{G} \text{ for } 0 \leq \varrho/\varrho_0 \leq 1 \text{ and } (p, \varrho_0) \in \tilde{G} \text{ for } p \geq p_0.$$

Proof. Suppose $\tilde{G}$ and G are not $(1, \infty) \times \{0\}$. Theorem 2.31 implies that G and $\tilde{G}$ are open, Proposition 2.30 says that $\tilde{G}$ is convex, and since $\psi : G \to \tilde{G}$, $(p, \varrho) \mapsto (p, p\varrho)$ is a homeomorphism, it follows that G is connected. If $(p_0, \varrho_0) \in G$, then $w^{\varrho_0} \in A_{p_0}(\Gamma)$ and thus, by Proposition 2.1(b) with $w_1 = w^{\varrho_0}$ and $w_2 = 1$, we have $w^\varrho \in A_{p_0}(\Gamma)$ whenever $0 \leq \varrho/\varrho_0 \leq 1$. Proposition 2.30 tells us that $(p, \varrho_0) \in \tilde{G}$ for every $p \geq p_0$ if only $(p_0, \varrho_0) \in \tilde{G}$. Finally, because $(p, \varrho) \in G$ if and only if $(p, p\varrho) \in \tilde{G}$, we obtain the remaining two assertions. □

2.9 Muckenhoupt condition and W transform

We now consider weights w on Γ which are continuous and nonzero on Γ minus a point t. For such weights we may consider the W transforms $W_t w$ and $W^0_t w$ introduced in Section 1.5. We there showed that if $W_t w$ is regular, then so also is $W^0_t w$ and both functions have the same lower and upper indices:

$$\alpha(W_t w) = \alpha(W^0_t w) \leq \beta(W^0_t w) = \alpha(W_t w). \tag{2.53}$$

The following theorem provides us with a very useful tool for checking the Muckenhoupt condition once these indices are available.

Theorem 2.33. *Let Γ be a bounded simple Carleson curve and $t \in \Gamma$. Suppose $w : \Gamma \setminus \{t\} \to (0, \infty)$ is a continuous function and $W_t w$ is regular. Then w is in $A_p(\Gamma)$ $(1 < p < \infty)$ if and only if*

$$-1/p < \alpha(W_t^0 w) \le \beta(W_t^0 w) < 1/q. \tag{2.54}$$

Proof. Put $\alpha_t := \alpha(W_t w)$ and $\beta_t := \beta(W_t w)$. First suppose that (2.54) holds. Then, by (2.53), $-1/p < \alpha_t \le \beta_t < 1/q$, and we may find an $\varepsilon > 0$ such that

$$-1/p < \alpha_t - \varepsilon < \beta_t + \varepsilon < 1/q. \tag{2.55}$$

Since $W_t w$ is regular, we infer from Corollary 1.14 that there are $x_0 \in (0, 1)$ and $C_t \in (0, \infty)$ such that

$$(W_t w)(x) \le x^{\alpha_t - \varepsilon} \text{ if } x \in (0, x_0), \ (W_t w)(x) \le x^{\beta_t + \varepsilon} \text{ if } x \in (x_0^{-1}, \infty), \tag{2.56}$$
$$(W_t w)(x) \le C_t \text{ if } x \in [x_0, x_0^{-1}]. \tag{2.57}$$

From the definitions (1.31) and (1.36) we see that

$$\sup_{R>0} M(w^{-1}, xR, R) = \sup_{R>0} M(w, R, xR) = (W_t w)(x^{-1})$$

for all $x \in (0, 1)$. The latter equality and (1.31) show that if $x \in (0, 1)$, $\tau_0 \in \Gamma$, $|\tau_0 - t| = xR$, then

$$w(\tau_0) \le \max_{|\tau - t| = xR} w(\tau) \le (W_t w)(x) \min_{|\tau - t| = R} w(\tau), \tag{2.58}$$

$$w(\tau_0)^{-1} \le \max_{|\tau - t| = xR} w(\tau)^{-1} \le (W_t w)(x^{-1}) \min_{|\tau - t| = R} w(\tau)^{-1}. \tag{2.59}$$

Using (2.56) we also obtain for $n > 1$ that

$$\sup_{x \in [x_0^n, x_0^{n-1})} (W_t w)(x) \le \sup_{x \in [x_0^n, x_0^{n-1})} x^{\alpha_t - \varepsilon}$$
$$\le \max \left\{ x_0^{n(\alpha_t - \varepsilon)}, x_0^{(n-1)(\alpha_t - \varepsilon)} \right\} = c_1 x_0^{(n-1)(\alpha_t - \varepsilon)}, \tag{2.60}$$

$$\sup_{x \in [x_0^n, x_0^{n-1})} (W_t w)(x^{-1}) \le \sup_{x \in [x_0^n, x_0^{n-1})} x^{-(\beta_t + \varepsilon)}$$
$$\le \max \left\{ x_0^{-n(\beta_t + \varepsilon)}, x_0^{-(n-1)(\beta_t + \varepsilon)} \right\} = c_2 x_0^{-(n-1)(\beta_t + \varepsilon)}, \tag{2.61}$$

where $c_1 := \max\{1, x_0^{\alpha_t - \varepsilon}\}$ and $c_2 := \max\{1, x_0^{-(\beta_t + \varepsilon)}\}$.

By virtue of (2.55), $1 + p(\alpha_t - \varepsilon) > 0$ and $1 - q(\beta_t + \varepsilon) > 0$. Consequently,

$$0 < x_0^{1 + p(\alpha_t - \varepsilon)} < 1, \ 0 < x_0^{1 - q(\beta_t + \varepsilon)} < 1. \tag{2.62}$$

Further, by the Carleson condition (1.1),

$$\left| \Gamma(t, x_0^{n-1} R) \right| - \left| \Gamma(t, x_0^n R) \right| \le C_\Gamma x_0^{n-1} R - x_0^n R = c_0 x_0^{n-1} R \tag{2.63}$$

with $c_0 := C_\Gamma - x_0 > 0$. Taking into account (2.58), (2.63), (2.57), (2.60), (2.62) we get

$$
\int_{\Gamma(t,R)} w(\tau)^p |d\tau| = \sum_{n=1}^{\infty} \int_{\Gamma(t,x_0^{n-1}R)\setminus\Gamma(t,x_0^n R)} w(\tau)^p |d\tau|
$$

$$
\leq \sum_{n=1}^{\infty} \sup_{x\in[x_0^n,x_0^{n-1})} (W_t w)(x)^p \left(\min_{|\tau-t|=R} w(\tau)^p \right) \left(\left|\Gamma(t,x_0^{n-1}R)\right| - \left|\Gamma(t,x_0^n R)\right| \right)
$$

$$
\leq \left(C_t^p c_0 R + \sum_{n=2}^{\infty} c_1^p x_0^{p(n-1)(\alpha_t - \varepsilon)} c_0 x_0^{n-1} R \right) \min_{|\tau-t|=R} w(\tau)^p
$$

$$
= c_0 \left(C_t^p + c_1^p \frac{x_0^{1+p(\alpha_t - \varepsilon)}}{1 - x_0^{1+p(\alpha_t - \varepsilon)}} \right) R \min_{|\tau-t|=R} w(\tau)^p, \tag{2.64}
$$

while (2.59), (2.63), (2.57), (2.59), (2.62) yield

$$
\int_{\Gamma(t,R)} w(\tau)^{-q} |d\tau| = \sum_{n=1}^{\infty} \int_{\Gamma(t,x_0^{n-1}R)\setminus\Gamma(t,x_0^n R)} w(\tau)^{-q} |d\tau|
$$

$$
\leq \sum_{n=1}^{\infty} \sup_{x\in[x_0^n,x_0^{n-1})} (W_t w)(x^{-1})^q
$$

$$
\times \left(\min_{|\tau-t|=R} w(\tau)^{-q} \right) \left(\left|\Gamma(t,x_0^{n-1}R)\right| - \left|\Gamma(t,x_0^n R)\right| \right)
$$

$$
\leq \left(C_t^q c_0 R + \sum_{n=2}^{\infty} c_2^q x_0^{-q(n-1)(\beta_t + \varepsilon)} c_0 x_0^{n-1} R \right) \min_{|\tau-t|=R} w(\tau)^{-q}
$$

$$
= c_0 \left(C_t^q + c_2^q \frac{x_0^{1-q(\beta_t + \varepsilon)}}{1 - x_0^{1-q(\beta_t + \varepsilon)}} \right) R \min_{|\tau-t|=R} w(\tau)^{-q}. \tag{2.65}
$$

Put

$$
c_3 := c_0 \left(C_t^p + c_1^p \frac{x_0^{1+p(\alpha_t - \varepsilon)}}{1 - x_0^{1+p(\alpha_t - \varepsilon)}} \right), \quad c_4 := c_0 \left(C_t^q + c_2^q \frac{x_0^{1-q(\beta_t + \varepsilon)}}{1 - x_0^{1-q(\beta_t + \varepsilon)}} \right).
$$

Then (2.64) and (2.65) read

$$
\left(\frac{1}{R} \int_{\Gamma(t,R)} w(\tau)^p |d\tau| \right)^{1/p} \leq c_3^{1/p} \min_{|\tau-t|=R} w(\tau),
$$

$$
\left(\frac{1}{R} \int_{\Gamma(t,R)} w(\tau)^{-q} |d\tau| \right)^{1/q} \leq c_4^{1/q} \min_{|\tau-t|=R} w(\tau)^{-1} \leq c_4^{1/q} \left(\min_{|\tau-t|=R} w(\tau) \right)^{-1},
$$

and multiplication of these two inequalities shows that $C_{w,t}$ equals

$$\sup_{R>0}\left(\frac{1}{R}\int_{\Gamma(t,R)}w(\tau)^p|d\tau|\right)^{1/p}\left(\frac{1}{R}\int_{\Gamma(t,R)}w(\tau)^{-q}|d\tau|\right)^{1/q}\le c_3^{1/p}c_4^{1/q}<\infty.\qquad(2.66)$$

Let now $t_0\in\Gamma\setminus\{t\}$. Suppose first that $R\ge|t-t_0|/2$. Then $(R+|t-t_0|)/R$ is at most 3, and since always $\Gamma(t_0,R)\subset\Gamma(t,R+|t-t_0|)$, we obtain

$$\frac{1}{R}\int_{\Gamma(t_0,R)}w(\tau)^p|d\tau|\le\frac{R+|t-t_0|}{R}\frac{1}{R+|t-t_0|}\int_{\Gamma(t,R+|t-t_0|)}w(\tau)^p|d\tau|$$

$$\le\frac{3}{R+|t-t_0|}\int_{\Gamma(t,R+|t-t_0|)}w(\tau)^p|d\tau|$$

and analogously,

$$\frac{1}{R}\int_{\Gamma(t_0,R)}w(\tau)^{-q}|d\tau|\le\frac{3}{R+|t-t_0|}\int_{\Gamma(t,R+|t-t_0|)}w(\tau)^{-q}|d\tau|.$$

Consequently, by (2.66),

$$\sup_{R\ge|t-t_0|/2}\left(\frac{1}{R}\int_{\Gamma(t_0,R)}w(\tau)^p|d\tau|\right)^{1/p}\left(\frac{1}{R}\int_{\Gamma(t_0,R)}w(\tau)^{-q}|d\tau|\right)^{1/q}$$

$$\le 3\sup_{r>0}\left(\frac{1}{r}\int_{\Gamma(t,r)}w(\tau)^p|d\tau|\right)^{1/p}\left(\frac{1}{r}\int_{\Gamma(t,r)}w(\tau)^{-q}|d\tau|\right)^{1/q}=3\,C_{w,t}<\infty.\qquad(2.67)$$

Now suppose $0<R<|t-t_0|/2$. Put

$$R_0:=|t-t_0|-R,\quad R_1:=\min\big\{|t-t_0|+R,d_t\big\},$$

where $d_t:=\max\{|\tau-t|:\tau\in\Gamma\}$. The estimates (2.58) and (2.59) show that

$$\left(\frac{1}{R}\int_{\Gamma(t_0,R)}w(\tau)^p|d\tau|\right)^{1/p}\le C_\Gamma^{1/p}\sup_{\tau\in\Gamma(t_0,R)}w(\tau)$$

$$\le C_\Gamma^{1/p}\sup_{x\in[R_0,R_1]}\max_{|\tau-t|=x}w(\tau)$$

$$\le C_\Gamma^{1/p}\sup_{x\in[R_0,R_1]}(W_tw)\left(\frac{x}{R_1}\right)\min_{|\tau-t|=R_1}w(\tau),$$

$$\left(\frac{1}{R}\int_{\Gamma(t_0,R)}w(\tau)^{-q}|d\tau|\right)^{1/q}\le C_\Gamma^{1/q}\sup_{\tau\in\Gamma(t_0,R)}w(\tau)^{-1}$$

$$\le C_\Gamma^{1/q} \sup_{x\in[R_0,R_1]} \max_{|\tau-t|=x} w(\tau)^{-1}$$

$$\le C_\Gamma^{1/q} \sup_{x\in[R_0,R_1]} (W_t w)\left(\frac{R_1}{x}\right) \min_{|\tau-t|=R_1} w(\tau)^{-1}.$$

Multiplying the latter two inequalities we arrive at the estimate

$$\sup_{0<R<|t-t_0|/2} \left(\frac{1}{R}\int_{\Gamma(t_0,R)} w(\tau)^p|d\tau|\right)^{1/p} \left(\frac{1}{R}\int_{\Gamma(t_0,R)} w(\tau)^{-q}|d\tau|\right)^{1/q}$$

$$\le C_\Gamma \sup_{0<R<|t-t_0|/2} \sup_{x\in[R_0,R_1]} (W_t w)\left(\frac{x}{R_1}\right)(W_t w)\left(\frac{R_1}{x}\right). \tag{2.68}$$

We now consider

$$\frac{x}{R_1} = \frac{x}{\min\{|t-t_0|+R, d_t\}}$$

for $x \in [R_0, R_1] = [|t-t_0| - R, \min\{|t-t_0|+R, d_t\}]$ and $0 < R < |t-t_0|/2$. If $R \le d_t - |t-t_0|$, then

$$1 \ge \frac{x}{\min\{|t-t_0|+R, d_t\}} \ge \frac{|t-t_0|-R}{|t-t_0|+R} \ge \frac{|t-t_0|-|t-t_0|/2}{|t-t_0|+|t-t_0|/2} = \frac{1}{3},$$

while if $R > d_t - |t-t_0|$, we have $d_t - |t-t_0| < R < |t-t_0|/2$, whence $d_t < (3/2)|t-t_0|$ and thus,

$$1 \ge \frac{x}{\min\{|t-t_0|+R, d_t\}} \ge \frac{|t-t_0|-R}{d_t} \ge \frac{|t-t_0|/2}{(3/2)|t-t_0|} = \frac{1}{3}. \tag{2.69}$$

Consequently, in either case $1/3 \le x/R_1 \le 1$. This estimate and (2.68) together with (2.56), (2.57) give

$$\sup_{0<R<|t-t_0|/2} \left(\frac{1}{R}\int_{\Gamma(t_0,R)} w(\tau)^p|d\tau|\right)^{1/p} \left(\frac{1}{R}\int_{\Gamma(t_0,R)} w(\tau)^{-q}|d\tau|\right)^{1/q}$$

$$\le C_\Gamma \sup_{y\in[1/3,3]} (W_t w)(y)^2 < \infty. \tag{2.70}$$

From (2.66), (2.67), (2.70) we finally infer that $w \in A_p(\Gamma)$, which proves the "if" portion of the theorem.

To show the "only if" part, assume that $\alpha_t < -1/p$. Since $W_t w$ is regular, Corollary 1.14 yields that

$$(W_t w)(x) \ge x^{\alpha_t} \quad \text{for } x \in (0,1). \tag{2.71}$$

The definitions (1.31) and (1.36) imply that if $c \in (0,1]$ and $R \in (0, d_t]$, then

$$\min_{|\tau-t|=cR} w(\tau) = \big(M(w,R,cR)\big)^{-1} \max_{|\tau-t|=R} w(\tau)$$

$$\geq \frac{1}{(W_t w)(c^{-1})} \max_{|\tau-t|=R} w(\tau) \tag{2.72}$$

$$\min_{|\tau-t|=cR} w(\tau)^{-1} = \big(M(w^{-1}, R, cR)\big)^{-1} \max_{|\tau-t|=R} w(\tau)^{-1}$$

$$= \big(M(w, cR, R)\big)^{-1} \max_{|\tau-t|=R} w(\tau)^{-1} \geq \frac{1}{(W_t w)(c)} \max_{|\tau-t|=R} w(\tau)^{-1}. \tag{2.73}$$

Fix $\varepsilon \in (0,1)$ and $\kappa \in (0,1)$. Since $W_t w$ is positive on $(0,\infty)$, we see from definition (1.31) that for each integer $n \geq 1$ there is an $R_n \in (0, d_t]$ such that

$$\max_{|\tau-t|=\kappa^n R_n} w(\tau) \geq (1-\varepsilon)(W_t w)(\kappa^n) \min_{|\tau-t|=R_n} w(\tau). \tag{2.74}$$

If $c \in [\kappa, 1)$, we obtain from (2.72), (2.74), (2.71) that

$$\min_{|\tau-t|=c\kappa^n R_n} w(\tau) \geq \frac{1}{(W_t w)(c^{-1})} \max_{|\tau-t|=\kappa^n R_n} w(\tau)$$

$$\geq \frac{1-\varepsilon}{(W_t w)(c^{-1})} (W_t w)(\kappa^n) \min_{|\tau-t|=R_n} w(\tau) \geq \frac{1-\varepsilon}{(W_t w)(c^{-1})} \kappa^{n\alpha_t} \min_{|\tau-t|=R_n} w(\tau). \tag{2.75}$$

By (2.73), we have for $c \in [\kappa, 1)$

$$\min_{|\tau-t|=cR_n} w(\tau)^{-1} \geq \frac{1}{(W_t w)(c)} \max_{|\tau-t|=R_n} w(\tau)^{-1}. \tag{2.76}$$

For $m \geq 0$ and $n \geq 1$, put $\Gamma_{m,n} := \Gamma(t, \kappa^m R_n) \setminus \Gamma(t, \kappa^{m+1} R_n)$. Clearly, $|\Gamma_{m,n}| \geq (1-\kappa)\kappa^m R_n$. This together with (2.75) and (2.76) implies that

$$\left(\frac{1}{R_n} \int_{\Gamma(t,R_n)} w(\tau)^p |d\tau|\right)^{1/p} \geq \left(\frac{|\Gamma_{n,n}|}{R_n} \frac{1}{|\Gamma_{n,n}|} \int_{\Gamma_{n,n}} w(\tau)^p |d\tau|\right)^{1/p}$$

$$\geq \left(\frac{(1-\kappa)\kappa^n R_n}{R_n}\right)^{1/p} \inf_{c \in [\kappa,1)} \min_{|\tau-t|=c\kappa^n R_n} w(\tau)$$

$$\geq (1-\kappa)^{1/p} \kappa^{n/p} \frac{1-\varepsilon}{\sup_{c\in[\kappa,1)} (W_t w)(c^{-1})} \kappa^{n\alpha_t} \min_{|\tau-t|=R_n} w(\tau), \tag{2.77}$$

$$\left(\frac{1}{R_n} \int_{\Gamma(t,R_n)} w(\tau)^{-q} |d\tau|\right)^{1/q} \geq \left(\frac{|\Gamma_{0,n}|}{R_n} \frac{1}{|\Gamma_{0,n}|} \int_{\Gamma_{0,n}} w(\tau)^{-q} |d\tau|\right)^{1/q}$$

$$\geq \left(\frac{(1-\kappa)R_n}{R_n}\right)^{1/q} \inf_{c \in [\kappa,1)} \min_{|\tau-t|=cR_n} w(\tau)^{-1}$$

$$\geq (1-\kappa)^{1/q} \frac{1}{\sup_{c\in[\kappa,1)} (W_t w)(c)} \max_{|\tau-t|=R_n} w(\tau)^{-1}. \tag{2.78}$$

Taking into account that

$$D_t := \sup_{c \in [\kappa, 1]} (W_t w)(c) \sup_{c \in [\kappa, 1]} (W_t w)(c^{-1}) < \infty$$

due to the regularity of $W_t w$, we conclude from (2.77) and (2.78) that

$$
\begin{aligned}
C_{w,t} &:= \sup_{R>0} \left(\frac{1}{R} \int_{\Gamma(t,R)} w(\tau)^p |d\tau| \right)^{1/p} \left(\frac{1}{R} \int_{\Gamma(t,R)} w(\tau)^{-q} |d\tau| \right)^{1/q} \\
&\geq \left(\frac{1}{R_n} \int_{\Gamma(t,R_n)} w(\tau)^p |d\tau| \right)^{1/p} \left(\frac{1}{R_n} \int_{\Gamma(t,R_n)} w(\tau)^{-q} |d\tau| \right)^{1/q} \\
&\geq (1-\kappa)(1-\varepsilon) D_t^{-1} \kappa^{n(1/p+\alpha_t)}.
\end{aligned}
$$

Since $\kappa \in (0,1)$ and $1/p + \alpha_t < 0$, we see that $\kappa^{n(1/p+\alpha_t)} \to \infty$ as $n \to \infty$. Hence $C_{w,t} = \infty$ and therefore w cannot belong to $A_p(\Gamma)$.

It can be shown analogously that $C_{w,t} = \infty$ and thus $w \notin A_p(\Gamma)$ in case $\beta_t > 1/q$.

Suppose finally that $\alpha_t = -1/p$ or $\beta_t = 1/q$ and assume $w \in A_p(\Gamma)$. Then, by Theorem 2.31, there exists an $\varepsilon > 0$ such that $\tilde{w} := w^{1+\varepsilon} \in A_p(\Gamma)$. We have $W_t \tilde{w} = (W_t w)^{1+\varepsilon}$, so $W_t \tilde{w}$ is regular and

$$\alpha(W_t \tilde{w}) = (1+\varepsilon)\alpha(W_t w) = (1+\varepsilon)\alpha_t, \ \beta(W_t \tilde{w}) = (1+\varepsilon)\beta(W_t w) = (1+\varepsilon)\beta_t.$$

Consequently,

$$\alpha(W_t \tilde{w}) = -(1+\varepsilon)(1/p) < -1/p \ \text{ or } \ \beta(W_t \tilde{w}) = (1+\varepsilon)(1/q) > 1/q.$$

By what was already shown, this is impossible if $\tilde{w} \in A_p(\Gamma)$. This contradiction completes the proof. $\qquad\square$

Example 2.34. Let Γ be a simple Carleson curve, fix $t \in \Gamma$, and define $\eta_t(\tau) := e^{-\arg(\tau-t)}$ as in Section 1.2. Then η_t is a weight on Γ which is continuous and nonvanishing on $\Gamma \setminus \{t\}$. From Lemma 1.17 we know that $W_t \eta_t$ is regular and hence Theorem 2.33 is applicable. Since $\alpha(W_t^0 \eta_t)$ and $\beta(W_t^0 \eta_t)$ are nothing but the spirality indices δ_t^- and δ_t^+ (recall (1.44)), we arrive at the conclusion that $\eta_t \in A_p(\Gamma)$ if and only if $-1/p < \delta_t^- \leq \delta_t^+ < 1/q$.

We say that two weights w_1 and w_2 are *equivalent* if $w_1 = w_2 c$ with some function $c \in GL^\infty(\Gamma)$, i.e. some function c which is bounded and bounded away from zero. Since $L^p(\Gamma, w_1) = L^p(\Gamma, w_2)$ whenever w_1 is equivalent to w_2, we will not regard equivalent weights as different weights.

If t is a nonhelical point of Γ (recall Section 1.3), then $\arg(\tau - t)$ is bounded for $\tau \in \Gamma \setminus \{t\}$ and hence η_t is equivalent to the weight which is identically 1. In case

$\arg(\tau - t) = -\delta \log |\tau - t| + O(1)$, we get $\eta_t(\tau) = |\tau - t|^\delta e^{O(1)}$ and consequently, η_t is equivalent to a power weight. Thus, in these cases the consideration of η_t does not provide Muckenhoupt weights beyond power weights.

However, for curves as in Proposition 1.19 with $\delta_t^- \neq \delta_t^+$ the above construction gives us Muckenhoupt weights which are not equivalent to power weights. If, for example, $h(x) = \varepsilon \sin x$, then

$$\eta_0(\tau) = e^{\varepsilon \sin(\log(-\log |\tau|)) \log |\tau|},$$

and since $\delta_0^- = -|\varepsilon|\sqrt{2}$, $\delta_0^+ = |\varepsilon|\sqrt{2}$, it follows that $\eta_0 \in A_p(\Gamma)$ if and only if $|\varepsilon|\sqrt{2} < \min\{1/p, 1/q\}$. Because

$$\liminf_{\tau \to 0} \eta_0(\tau) = 0 \quad \text{and} \quad \limsup_{\tau \to 0} \eta_0(\tau) = \infty,$$

it is clear that η_0 is not equivalent to a power weight. $\square$

2.10   Oscillating weights

Using Theorem 2.33 and modifying the construction of Example 2.34 we can find plenty of Muckenhoupt weights on arbitrary simple bounded Carleson curves.

Lemma 2.35. *Let Γ be a simple bounded Carleson curve, $t \in \Gamma$, and without loss of generality suppose $d_t := \max\{|\tau - t| : \tau \in \Gamma\} = 1$. Let*

$$w(\tau) = e^{F(|\tau - t|)} \quad \text{for } \tau \in \Gamma \setminus \{t\}$$

where $F : (0, 1] \to \mathbf{R}$ is a function in $C(0, 1] \cap C^1(0, 1)$. If $rF'(r)$ is bounded for $r \in (0, 1)$, then both $W_t w$ and $W_t^0 w$ are regular submultiplicative functions.

Proof. Suppose $|rF'(r)| \leq C < \infty$ for $r \in (0, 1)$. If $x \in (0, 1)$, then

$$M(w, xR, R) := \frac{\max\limits_{|\tau - t| = xR} w(\tau)}{\min\limits_{|\tau - t| = R} w(\tau)} = e^{F(xR) - F(R)} = e^{F'(\xi R)(x - 1)R} \tag{2.79}$$

with some $\xi \in (x, 1)$. Clearly

$$\left| F'(\xi R)(x - 1)R \right| = \left| (\xi R) F'(\xi R)(x - 1)/\xi \right| \leq C x^{-1}$$

and thus

$$e^{-Cx^{-1}} \leq M(w, xR, R) \leq e^{Cx^{-1}} \tag{2.80}$$

whenever $x \in (0, 1)$ and $R > 0$. Since $M(w, R, R) = 1$, the estimate also holds for $x = 1$. Consequently, by the definition of $W_t w$ and $W_t^0 w$, both functions are positive on $(0, 1]$ and bounded from above on $[1/2, 1]$, for example. Analogously one can show that they are positive on $[1, \infty)$ and bounded from above on $(1, 2]$. Hence, $W_t w$ and $W_t^0 w$ are regular. Their submultiplicativity follows from Lemmas 1.15 and 1.16. $\square$

Theorem 2.36. *Let* Γ *and* w *be as in Lemma* 2.35. *In addition, suppose*

$$F(r) = f\big(\log(-\log r)\big) \log r \quad for \ r \in (0,1) \tag{2.81}$$

where

$$f \in C^2(\mathbf{R}) \quad and \ f, f', f'' \ are \ bounded \ on \ \mathbf{R}. \tag{2.82}$$

Then

$$\alpha(W_t^0 w) = \liminf_{r \to 0} \big(r F'(r)\big) = \liminf_{x \to +\infty} \big(f(x) + f'(x)\big), \tag{2.83}$$

$$\beta(W_t^0 w) = \limsup_{r \to 0} \big(r F'(r)\big) = \limsup_{x \to +\infty} \big(f(x) + f'(x)\big). \tag{2.84}$$

Furthermore, $w \in A_p(\Gamma)$ $(1 < p < \infty)$ *if and only if* (2.54) *holds.*

Comment. Note that if F is of the form (2.81), (2.82), then the requirement $F(1) = 0$ makes F continuous on $(0, 1]$, while the identity

$$rF'(r) = f\big(\log(-\log r)\big) + f'\big(\log(-\log r)\big)$$

implies that $rF'(r)$ is automatically bounded and also gives the second equalities in (2.83) and (2.84).

Proof. Lemma 2.35 tells us that $W_t w$ and $W_t^0 w$ are regular. The last assertion of the theorem is therefore immediate from Theorem 2.33. As in the proof of Proposition 1.19 we get

$$\limsup_{R \to 0} \big(F(xR) - F(R)\big) = (\log x) \liminf_{z \to +\infty} \big(f(z) + f'(z)\big) \ .$$

Hence, by (2.79), for $x \in (0, 1)$ we have

$$\log(W_t^0 w)(x)/\log x = \liminf_{z \to +\infty} \big(f(z) + f'(z)\big).$$

This proves (2.83). In the same way one can prove (2.84). $\qquad\square$

Example 2.37. Let Γ be any simple Carleson curve, $t \in \Gamma$, and suppose $d_t = 1$. Put $w(\tau) = e^{F(|\tau - t|)}$ with F given by (2.81) and (2.82).

If $f(x) = \lambda$, then $F(r) = \lambda \log r$ and hence, $w(\tau) = |\tau - t|^\lambda$ is a power weight. From (2.83) and (2.84) we deduce that $\alpha(W_t^0 w) = \beta(W_t^0 w) = \lambda$. Thus, Theorem 2.36 tells us that $w \in A_p(\Gamma)$ if and only if $-1/p < \lambda < 1/q$, which is nothing but Theorem 2.2.

Now choose $f(x) = \lambda + \varepsilon \sin(\eta x)$. In this case Theorem 2.36 along with the computation after the proof of Proposition 1.19 shows that $w \in A_p(\Gamma)$ if and only if

$$-1/p < \lambda - |\varepsilon|\sqrt{\eta^2 + 1} \leq \lambda + |\varepsilon|\sqrt{\eta^2 + 1} < 1/q. \tag{2.85}$$

In particular, if $f(x) = \varepsilon \sin x$ then

$$\liminf_{\tau \to t} w(\tau) = 0, \quad \limsup_{\tau \to t} w(\tau) = \infty,$$

if $f(x) = |\varepsilon|(1 - \sin x)$ then

$$\liminf_{\tau \to t} w(\tau) = 0, \quad \limsup_{\tau \to t} w(\tau) = 1 < \infty,$$

and if $f(x) = |\varepsilon|(-1 + \sin x)$ then

$$\liminf_{\tau \to t} w(\tau) = 1 > 0, \quad \limsup_{\tau \to t} w(\tau) = \infty.$$

In either case we infer from (2.85) that the "oscillating weight" w belongs to $A_p(\Gamma)$ for all sufficiently small ε.   □

2.11   Notes and comments

2.1. The Muckenhoupt condition (2.1) with $\Gamma = \mathbf{R}$ appeared in Muckenhoupt's paper [148] for the first time, where it was shown to be equivalent to the boundedness on $L^p(\mathbf{R})$ $(1 < p < \infty)$ of the Hardy-Littlewood maximal operator $\mathcal{M}$, given by

$$(\mathcal{M}f)(t) := \sup_{t \in I} \frac{1}{|I|} \int_I |f(t)|\,|d\tau|, \quad t \in \mathbf{R},$$

the supremum over all intervals $I \subset \mathbf{R}$ containing t. Clearly, in this case the portions $\Gamma(t, \varepsilon)$ in (2.1) are nothing but the intervals $(t - \varepsilon, t + \varepsilon) \subset \mathbf{R}$. A little bit later, Hunt, Muckenhoupt, and Wheeden [107] proved that just this condition is also necessary and sufficient for the Cauchy singular integral S_Γ to be bounded on $L^p(\Gamma, w)$ $(1 < p < \infty)$ in case $\Gamma = \mathbf{R}$ or $\Gamma = \mathbf{T}$.

In 1983, Simonenko [197] as well as Dynkin and Osilenker [62] (note that the Russian original of [62] and [197] appeared at the same time) observed that the results of the preceding paragraph are no longer true for more general curves Γ if the Muckenhoupt condition is written in the form (2.30), i.e. with arcs in place of portions. After David's preprint [46], Dynkin and Osilenker [62] (also see [60], [61]) understood that (2.1) is the right generalization of Muckenhoupt's original condition.

Standard texts on Muckenhoupt weights include the surveys by Dynkin, Osilenker [62], Dynkin [60], [61], and the monographs by Garcia-Cuerva, Rubio de Francia [73], Garnett [74], Journé [109], and Koosis [127].

2.2. Theorem 2.2 is well known in case Γ is a piecewise Lyapunov curve (see, e.g., [89]). For general bounded simple Carleson curves it was established by Danilov [44, Theorem 2.6]. The proof given in the text is Danilov's.

2.3. These results and their proofs are well known: see Hunt, Muckenhoupt, Wheeden [107], Garnett [74], Dynkin and Osilenker [62], or Dynkin [60] for the case of nice curves.

2.4. Symmetric and periodic reproduction are certainly constructions which are well known to specialists, but we have not found a reference in the literature.

Theorem 2.10 was obtained by Grudsky and Khevelev [94] using different arguments. The assertion concerning $p = 2$ in Theorem 2.11 is a simple consequence of the Helson-Szegö theorem (Theorem 4.18). In 1992, I. Spitkovsky also observed that Theorem 4.19 implies that $A_p(\mathbf{T}) \cap A_q(\mathbf{T})$ $(1/p + 1/q = 1)$ is preserved under composition by inner functions $u : \mathbf{T} \to \mathbf{T}$ (private communication). The $p \neq 2$ part of Theorem 2.11 was established by Grudsky and one of the authors in [14].

2.5. The nontrivial piece of this section, Theorem 2.12, is Simonenko's [197]. See also the notes and comments to Section 2.1.

2.6. The maximal function

$$(Mf)(t) := \sup_{\varepsilon>0} \frac{1}{\varepsilon} \int\limits_{\Gamma(t,\varepsilon)} |f(\tau)| \, |d\tau|, \quad t \in \Gamma,$$

is the analogue of the "centered" Hardy-Littlewood maximal function

$$(Mf)(t) := \sup_{\varepsilon>0} \frac{1}{2\varepsilon} \int\limits_{t-\varepsilon}^{t+\varepsilon} |f(\tau)| \, d\tau, \quad t \in \mathbf{R},$$

for simple Carleson curves. Several properties of the maximal function and the corresponding maximal operator can be found in the books by Garcia-Cuerva, Rubio de Francia [73], Garnett [74], Guzman [95], Stein [202], Strömberg, Torchinsky [203] and the surveys by Dynkin, Osilenker [62], [60], [61]. A proof of Besicovitch's covering theorem, (Theorem 2.17) is in [9] and Chapter 1 of [95]. Our proofs of Theorems 2.18 and 2.19 follow [95, Chapter 2].

2.7. The reverse Hölder inequality saying that there exist positive constants C and δ such

$$\left(\frac{1}{|Q|} \int\limits_Q w(\tau)^{1+\delta} |d\tau| \right)^{1/(1+\delta)} \leq C \left(\frac{1}{|Q|} \int\limits_Q w(\tau) |d\tau| \right)$$

for all cubes (or balls) of the $\mathbf{R}^n$ was introduced by R.R. Coifman and C. Fefferman [36]. They proved it with the help of the Calderón-Zygmund decomposition for the $\mathbf{R}^n$. In Garnett's book [74, Chapter 6], the reverse Hölder inequality for the real line $\mathbf{R}$ is proved using an inequality by Gehring [76]. These proofs do not work for general so-called spaces of homogeneous type. For the theory of these spaces see Calderón [31], Coifman, Weiss [39], [40], Folland, Stein [69], and Strömberg, Torchinsky [203], for example.

A.P. Calderón [31] gave a completely different proof of the reverse Hölder inequality, which was applicable to general spaces of homogeneous type. Loosely speaking, he replaced the Calderón-Zygmund decomposition by a sequence of such decompositions corresponding to "telescopically" embedded coverings. A similar strategy is also pursued by Strömberg and Torchinsky [203]. Our exposition in Section 2.7 basically follows Calderón [31]. It should be noted that, in contrast to the situation encountered in [31] and [203], in our case the function Φ_t of Lemma 2.22 need not be continuous. However, it is continuous from the left, which in conjunction with some modifications of Calderón's arguments saves the proof.

2.8. For nice curves, $\Gamma = \mathbf{R}$ or $\Gamma = \mathbf{T}$, Theorem 2.31 is already in the papers by Muckenhoupt [148] and Hunt, Muckenhoupt, Wheeden [107]. The $\mathbf{R}^n$ analogue of Theorem 2.31 was first proved by Coifman and Fefferman [36], who also realized the relevance of the reverse Hölder inequality in this connection. A result very close to Theorem 2.31 was established by Calderón [31] and Strömberg, Torchinsky [203] (notice that the hypothesis (v) of [31] and the continuity hypothesis of [203, Chapter 1, Theorems 17 and 18] are not satisfied for every Carleson curve). As far as we know, Simonenko [198] was the first to prove Theorem 2.31 in full generality. When dealing with the stability of Muckenhoupt weights, Simonenko did not yet know of the results of David [46], [47]. He denoted by $\mathcal{R}$ the class of all curves Γ for which the Cauchy singular integral S_Γ is bounded on $L^2(\Gamma)$ (and thus also on $L^p(\Gamma)$) and showed Theorem 2.31 with the hypothesis that Γ be a Carleson curve replaced by the hypothesis that Γ be in $\mathcal{R}$. We also remark that Simonenko did not use a reverse Hölder inequality. The idea of his proof is to rewrite the condition $S_\Gamma \in \mathcal{B}(L^p(\Gamma, w))$ in terms of the invertibility of a certain operator A (this can be managed by having recourse to Theorem 6.32) and subsequently to make use of a result by Shneiberg [190] which says that if the operator A is invertible on L^{p_0}, then it is invertible on L^p for all p close to p_0.

2.9–2.10. Theorem 2.33, which enables us to decide in terms of the indices of the W-transform whether a continuous function $\varphi : \Gamma \setminus \{t\} \to (0, \infty)$ belongs to $A_p(\Gamma)$, was established in our paper [16]. We remark that this theorem can also be derived from Theorem 3.13, which provides us with such a criterion for arbitrary weights φ in terms of the indices of the U-transform: one has only to set $w = 1$ in the latter theorem.

The examples given in Section 2.10 nicely illustrate the usefulness of results like Theorem 2.33. It may be that such examples are well known to specialists, but we have not seen them in the literature.

Chapter 3
Interaction between curve and weight

Chapter 1 dealt with curves only, and in Chapter 2 we focussed our attention on weights, although weights cannot exist without a curve they live on. We now turn to some problems arising when considering a curve and a weight on this curve as a whole.

The spirality indices δ_t^- and δ_t^+ of a simple Carleson curve Γ at a point $t \in \Gamma$ are intrinsic characteristics of the curve. In this chapter we introduce two numbers μ_t and ν_t which measure the powerlikeness of a weight $w \in A_p(\Gamma)$ at $t \in \Gamma$. Of course, since w lives on Γ, the numbers μ_t and ν_t are in fact characteristics of the pair (Γ, w).

Sole knowledge of the numbers $\delta_t^-, \delta_t^+, \mu_t, \nu_t$, though important and useful in many instances, does not suffice for our further purposes. We have to measure more carefully. It turns out that the appropriate "measuring instrument" is the weight $|(\tau - t)^\gamma|$ with a complex number γ and that the information we are interested in is completely contained in the so-called indicator set $N_t(\Gamma, p, w)$, which is defined as the set of all $\gamma \in \mathbf{C}$ for which $|(\tau - t)^\gamma| w(\tau)$ is a weight in $A_p(\Gamma)$. If the curve Γ or the weight w are sufficiently nice at t, then the indicator set may be described in terms of the numbers $\delta_t^-, \delta_t^+, \mu_t, \nu_t$. In general, however, the interaction between Γ and w, which, in a sense, may be understood as some kind of interference between the oscillation of the curve and the oscillation of the weight, leads to indicator sets of more intricate shapes. We will give estimates of the indicator sets via the numbers $\delta_t^-, \delta_t^+, \mu_t, \nu_t$ and will present final results concerning the shape of the indicator set.

3.1 Moduli of complex powers

Let Γ be a bounded simple Carleson curve and $t \in \Gamma$. For a complex number γ, we define the weight $\varphi_{t,\gamma} : \Gamma \to [0, \infty]$ by

$$\varphi_{t,\gamma}(\tau) := |(\tau - t)^\gamma| = |\tau - t|^{\operatorname{Re}\gamma} e^{-\operatorname{Im}\gamma \arg(\tau - t)} \quad \text{for } \tau \in \Gamma \setminus \{t\} \qquad (3.1)$$

where $\arg(\tau - t)$ is any continuous argument on $\Gamma \setminus \{t\}$. Given $p \in (1, \infty)$ and a weight $w \in A_p(\Gamma)$, we define the *indicator set* of Γ, p, w at t by

$$N_t := N_t(\Gamma, p, w) := \{\gamma \in \mathbf{C} : \varphi_{t,\gamma} w \in A_p(\Gamma)\}. \tag{3.2}$$

Since $w \in A_p(\Gamma)$, the indicator set always contains the origin. In this section we identify the indicator set in the case where w is constant.

Clearly, $\varphi_{t,\gamma}$ is equivalent (in the sense of Example 2.34) to a power weight if and only if γ is real or $\arg(\tau - t) = -\delta \log|\tau - t| + O(1)$ with some $\delta \in \mathbf{R}$. In the general case, $\varphi_{t,\gamma}$ is the product of a power weight and (the power of) a weight as in Example 2.34 : $\varphi_{t,\gamma} = \varphi_{t,\mathrm{Re}\,\gamma} \eta_t^{\mathrm{Im}\,\gamma}$. From Theorem 2.36 with $f(x) = \mathrm{Re}\,\gamma$ we infer that

$$\alpha(W_t^0 \varphi_{t,\mathrm{Re}\,\gamma}) = \beta(W_t^0 \varphi_{t,\mathrm{Re}\,\gamma}) = \mathrm{Re}\,\gamma. \tag{3.3}$$

From Example 2.34 we know that

$$\alpha(W_t^0 \eta_t) = \delta_t^-, \quad \beta(W_t^0 \eta_t) = \delta_t^+ \tag{3.4}$$

are the spirality indices of Γ at t. The following proposition gives us $\alpha(W_t^0 \varphi_{t,\gamma})$ and $\beta(W_t^0 \varphi_{t,\gamma})$ as a mixture of the numbers (3.3) and (3.4).

Proposition 3.1. *Let Γ be a bounded simple Carleson curve and $t \in \Gamma$. Then the functions $W_t \varphi_{t,\gamma}$ and $W_t^0 \varphi_{t,\gamma}$ are regular for every $\gamma \in \mathbf{C}$, we have*

$$\alpha(W_t \varphi_{t,\gamma}) = \alpha(W_t^0 \varphi_{t,\gamma}) = \mathrm{Re}\,\gamma + \min\{\delta_t^- \mathrm{Im}\,\gamma, \delta_t^+ \mathrm{Im}\,\gamma\}, \tag{3.5}$$
$$\beta(W_t \varphi_{t,\gamma}) = \beta(W_t^0 \varphi_{t,\gamma}) = \mathrm{Re}\,\gamma + \max\{\delta_t^- \mathrm{Im}\,\gamma, \delta_t^+ \mathrm{Im}\,\gamma\}, \tag{3.6}$$

and $\varphi_{t,\gamma} \in A_p(\Gamma)$ if and only if $-1/p < \alpha(W_t \varphi_{t,\gamma}) \le \beta(W_t \varphi_{t,\gamma}) < 1/q$.

Proof. Obviously,

$$\varphi_{t,\gamma}(\tau) = |\tau - t|^{\mathrm{Re}\,\gamma}(\eta_t(\tau)^{\mathrm{sign}\,\mathrm{Im}\,\gamma})^{|\mathrm{Im}\,\gamma|}. \tag{3.7}$$

If $x \in (0, 1]$, then

$$(W_t \eta_t)(x^{-1}) = \sup_{R>0} M(\eta_t, R, xR) = \sup_{R>0} M(\eta_t^{-1}, xR, R),$$

while if $x \in [1, \infty)$, we have

$$(W_t \eta_t)(x^{-1}) = \sup_{R>0} M(\eta_t, x^{-1}R, R) = \sup_{R>0} M(\eta_t^{-1}, R, x^{-1}R).$$

Thus, by (3.7) and (1.31),

$$(W_t \varphi_{t,\gamma})(x) = x^{\mathrm{Re}\,\gamma}\left((W_t \eta_t)(x^{\mathrm{sign}\,\mathrm{Im}\,\gamma})\right)^{|\mathrm{Im}\,\gamma|} \quad \text{for all} \quad x \in (0, \infty). \tag{3.8}$$

We know from Theorem 1.18 that $W_t\eta_t$ and $W_t^0\eta_t$ are regular. Hence, by (3.8), so also are $W_t\varphi_{t,\gamma}$ and $W_t^0\varphi_{t,\gamma}$. Lemma 1.16 implies that $W_t\varphi_{t,\gamma}$ and $W_t^0\varphi_{t,\gamma}$ have the same lower and upper indices. From (3.8) we get

$$
\alpha(W_t w_t \varphi_{t,\gamma}) = \lim_{x \to 0} \left((\log W_t\varphi_{t,\gamma})(x)/\log x \right)
$$

$$
= \begin{cases}
\operatorname{Re}\gamma + |\operatorname{Im}\gamma| \lim_{x \to 0} \left((\log W_t\eta_t)(x)/\log x \right) & \text{for } \operatorname{Im}\gamma \geq 0 \\
\operatorname{Re}\gamma + |\operatorname{Im}\gamma| \lim_{x \to \infty} \left((\log W_t\eta_t)(x)/(-\log x) \right) & \text{for } \operatorname{Im}\gamma < 0
\end{cases}
$$

$$
= \begin{cases}
\operatorname{Re}\gamma + \alpha(W_t\eta_t)\operatorname{Im}\gamma & \text{for } \operatorname{Im}\gamma \geq 0 \\
\operatorname{Re}\gamma + \beta(W_t\eta_t)\operatorname{Im}\gamma & \text{for } \operatorname{Im}\gamma < 0,
\end{cases}
$$

which in conjunction with (3.4) and the inequality $\delta_t^- \leq \delta_t^+$ gives (3.5). The proof of (3.6) is analogous. The last assertion of the proposition follows from Theorem 2.33. $\qquad\square$

Proposition 3.1 implies that if $w(\tau) = 1$ for all $\tau \in \Gamma$, then the indicator set N_t is the set of all $\gamma \in \mathbf{C}$ for which

$$
-\frac{1}{p} < \operatorname{Re}\gamma + \min\left\{\delta_t^- \operatorname{Im}\gamma, \delta_t^+ \operatorname{Im}\gamma\right\} \leq \operatorname{Re}\gamma + \max\left\{\delta_t^- \operatorname{Im}\gamma, \delta_t^+ \operatorname{Im}\gamma\right\} < \frac{1}{q}.
$$

Letting $\gamma = x + iy$, we so may draw a picture of N_t in the x, y plane. However, in what follows it will be more convenient to write $\gamma = y + ix$ and then to represent N_t in the x, y plane. Thus, we have

$$
N_t = \left\{ y + ix \in \mathbf{C} : -\frac{1}{p} < y + \min\left\{\delta_t^- x, \delta_t^+ x\right\} \leq y + \max\left\{\delta_t^- x, \delta_t^+ x\right\} < \frac{1}{q} \right\}. \quad (3.9)
$$

If $\delta_t^- = \delta_t^+ = \delta_t$, then (3.9) is the open stripe in the x, y plane between the straight lines of slope $-\delta_t$ passing through the points $(0, -1/p)$ and $(0, 1/q)$. If $\delta_t^- < \delta_t^+$, then (3.9) is the open parallelogram in the x, y plane two vertices of which are $(0, -1/p)$ and $(0, 1/q)$ and the sides of which have the slopes $-\delta_t^-$ and $-\delta_t^+$.

3.2 U and V transforms

In the previous section we identified the indicator set $N_t(\Gamma, p, w)$ for constant weights w. Employing the W transform one can also tackle general weights w which are continuous and nonzero on $\Gamma \setminus \{t\}$. However, for arbitrary weights w the W transform is not the appropriate tool. We therefore introduce two other transforms, the U and V transforms.

Let Γ be a bounded simple Carleson curve and $1 < p < \infty$. Fix $t \in \Gamma$. For $0 \leq R_1 < R_2 \leq d_t := \max_{\tau \in \Gamma} |\tau - t|$, we define

$$
\Gamma(t, R_1, R_2) := \left\{ \tau \in \Gamma : R_1 \leq |\tau - t| < R_2 \right\}.
$$

Thus, $\Gamma(t, 0, R)$ is nothing else than the portion $\Gamma(t, R)$ of Γ in the disk $\{z \in \mathbf{C} : |z - t| < R\}$, and if $R_1 > 0$ then $\Gamma(t, R_1, R_2)$ is the portion of Γ in the annulus $\{z \in \mathbf{C} : R_1 \leq |z - t| < R_2\}$. For a function f which is integrable on $\Gamma(t, R_1, R_2)$ we put

$$\Delta_t(f, R_1, R_2) := \frac{1}{|\Gamma(t, R_1, R_2)|} \int_{\Gamma(t, R_1, R_2)} f(\tau) \, |d\tau|.$$

Now let $\psi : \Gamma \to [0, \infty]$ be a weight. In the subsequent applications, ψ is usually the weight $\varphi_{t,\gamma} w$ where $\varphi_{t,\gamma}$ is given by (3.1) and w is a fixed weight in $A_p(\Gamma)$. In dependence on some additional properties of ψ, we define four functions

$$U_t \psi, U_t^0 \psi, V_t \psi, V_t^0 \psi : (0, \infty) \to [0, \infty].$$

Suppose first that $\psi \in L_{\mathrm{loc}}^p(\Gamma \setminus \{t\})$ and $\psi^{-1} \in L_{\mathrm{loc}}^q(\Gamma \setminus \{t\})$, which means that $\psi \in L^p(\Gamma \setminus \Gamma(t, R))$ and $\psi^{-1} \in L^q(\Gamma \setminus \Gamma(t, R))$ for all $R > 0$. Fix a number $\kappa \in (0, 1)$. For example, in all what follows we may suppose that $\kappa := 1/2$; notice, however, that the formulas don't become shorter when writing $1/2$ in place of κ. We define $(U_t \psi)(x)$ by

$$\begin{cases} \sup_{R>0} \left(\Delta_t(\psi^p, \kappa x R, x R) \right)^{1/p} \left(\Delta_t(\psi^{-q}, \kappa R, R) \right)^{1/q} & \text{for } x \in (0, 1] \\ \sup_{R>0} \left(\Delta_t(\psi^p, \kappa R, R) \right)^{1/p} \left(\Delta_t(\psi^{-q}, \kappa x^{-1} R, x^{-1} R) \right)^{1/q} & \text{for } x \in [1, \infty) \end{cases} \tag{3.10}$$

where, as usual, $1/p + 1/q = 1$ and where, according to (1.32), $\sup_{R>0}$ means $\sup_{0<R\leq d_t}$. The dependence of $U_t \psi$ on κ will be suppressed. We define $U_t^0 \psi$ by (3.10) with $\sup_{R>0}$ replaced by $\limsup_{R\to 0}$. We remark that $U_t^0 \psi$ can alternatively also be given as follows (recall (1.34) and (1.35)):

$$\begin{aligned} (U_t^0 \psi)(x) &= \limsup_{R \to 0} \left(\Delta_t(\psi^p, \kappa x R, x R) \right)^{1/p} \left(\Delta_t(\psi^{-q}, \kappa R, R) \right)^{1/q} \tag{3.11} \\ &= \limsup_{R \to 0} \left(\Delta_t(\psi^p, \kappa R, R) \right)^{1/p} \left(\Delta_t(\psi^{-q}, \kappa x^{-1} R, x^{-1} R) \right)^{1/q} \end{aligned}$$

If $\log \psi \in L^1(\Gamma)$, we put

$$(V_t \psi)(x) := \begin{cases} \sup_{R>0} \dfrac{\exp(\Delta_t(\log \psi, 0, xR))}{\exp(\Delta_t(\log \psi, 0, R))} & \text{for } x \in (0, 1] \\ \sup_{R>0} \dfrac{\exp(\Delta_t(\log \psi, 0, R))}{\exp(\Delta_t(\log \psi, 0, x^{-1} R))} & \text{for } x \in [1, \infty), \end{cases} \tag{3.12}$$

and we denote by $V_t^0 \psi$ the function given by (3.12) with $\limsup_{R\to 0}$ in place of $\sup_{R>0}$. Notice that in (3.12) only means over portions in disks occur. Thus, we could also have recourse to the slightly shorter notation (2.6), but for the sake of uniformity, we now write $\Delta_t(f, 0, R)$ for $\Delta_t(f, R)$.

The following lemma provides uniform lower and upper estimates for $U_t\psi$, $U_t^0\psi$, $V_t\psi$, $V_t^0\psi$ in a neighborhood of the point 1. We remark that, obviously,

$$(U_t^0\psi)(x) \le (U_t\psi)(x), \quad (V_t^0\psi)(x) \le (V_t\psi)(x) \quad \text{for all } x \in (0, \infty). \tag{3.13}$$

Also recall the definitions of $BMO(\Gamma, t)$ $(\subset L^1(\Gamma))$ and $A_p(\Gamma, t)$ $(\subset L^p(\Gamma) \cap \{w : w^{-1} \in L^q(\Gamma)\})$ from Section 2.3.

Lemma 3.2. *Let Γ be a bounded simple Carleson curve, let $\psi : \Gamma \to [0, \infty]$ be a weight, let $t \in \Gamma$, and let $x_0 \in (1, \infty)$ be an arbitrary point.*

(a) *If $\log \psi \in BMO(\Gamma, t)$ then*

$$0 < 1/C_t(x_0) \le \inf_{x \in [x_0^{-1}, x_0]} (V_t^0\psi)(x) \le \sup_{x \in [x_0^{-1}, x_0]} (V_t\psi)(x) \le C_t(x_0) < \infty$$

where $C_t(x_0) := \exp(C_\Gamma x_0 \| \log \psi \|_{*, t})$ and C_Γ is the Carleson constant of Γ.

(b) *If $\psi \in A_p(\Gamma, t)$ then*

$$0 < C_t^{(1)}(x_0) \le \inf_{x \in [x_0^{-1}, x_0]} (U_t^0\psi)(x) \le \sup_{x \in [x_0^{-1}, x_0]} (U_t\psi)(x) \le C_t^{(2)}(x_0) < \infty$$

where

$$C_t^{(1)}(x_0) := \frac{1}{C_t(x_0)} \exp\left(-\frac{2C_\Gamma}{1 - \kappa} \| \log \psi \|_{*, t}\right), \quad C_t^{(2)} = \frac{C_\Gamma C_{\psi, t}^2}{1 - \kappa} C_t(x_0),$$

and $C_{\psi, t}$ is given by (2.18) with ψ in place of w.

Proof. For brevity, put $\Delta_t(\varepsilon) := \Delta_t(\log \psi, 0, \varepsilon)$.

(a) By the Carleson condition (1.1), we have for $x \in [x_0^{-1}, 1]$ the estimate

$$|\Delta_t(xR) - \Delta_t(R)| = \frac{1}{|\Gamma(t, xR)|} \left| \int\limits_{\Gamma(t, xR)} (\log \psi(\tau) - \Delta_t(R)) |d\tau| \right|$$

$$\le \frac{|\Gamma(t, R)|}{|\Gamma(t, xR)|} \frac{1}{|\Gamma(t, R)|} \int\limits_{\Gamma(t, R)} |\log \psi(\tau) - \Delta_t(R)| |d\tau|$$

$$\le (C_\Gamma / x) \| \log \psi \|_{*, t} \le C_\Gamma x_0 \| \log \psi \|_{*, t} \tag{3.14}$$

and thus,

$$(V_t^0\psi)(x) = \limsup_{R \to 0} \exp\left(\Delta_t(xR) - \Delta_t(R)\right)$$

$$\ge \exp\left(-C_\Gamma x_0 \| \log \psi \|_{*, t}\right) =: 1/C_t(x_0).$$

Analogously, if $x \in [1, x_0]$ then

$$\left| \Delta_t(R) - \Delta_t(x^{-1}R) \right| \leq C_\Gamma x \| \log \psi \|_{*,t} \leq C_\Gamma x_0 \| \log \psi \|_{*,t}, \tag{3.15}$$

whence

$$\begin{aligned}
(V_t^0 \psi)(x) &= \limsup_{R \to 0} \exp \left(\Delta_t(R) - \Delta_t(x^{-1}R) \right) \\
&\geq \exp \left(- C_\Gamma x_0 \| \log \psi \|_{*,t} \right) = 1/C_t(x_0).
\end{aligned}$$

From (3.14) and (3.15) we also see that

$$(V_t \psi)(x) \leq \exp \left(C_\Gamma x_0 \| \log \psi \|_{*,t} \right) = C_t(x_0)$$

for all $x \in [x_0^{-1}, x_0]$.

(b) Since $\psi \in L^p(\Gamma)$ and $\psi^{-1} \in L^q(\Gamma)$, we deduce from Lemma 2.3 that $\log \psi$ is in $L^1(\Gamma)$. Hence, $\Delta_t(R)$ is well-defined for $\psi \in A_p(\Gamma, t)$. We have

$$\begin{aligned}
& \left| \frac{1}{\Gamma(t, \kappa R, R)} \int_{\Gamma(t, \kappa R, R)} \log \psi(\tau) |d\tau| - \Delta_t(R) \right| \\
&= \left| \frac{1}{|\Gamma(t, \kappa R, R)|} \int_{\Gamma(t, \kappa R, R)} \left(\log \psi(\tau) - \Delta_t(R) \right) |d\tau| \right| \\
&\leq \frac{|\Gamma(t, R)|}{|\Gamma(t, \kappa R, R)|} \frac{1}{|\Gamma(t, R)|} \int_{\Gamma(t, R)} \left| \log \psi(\tau) - \Delta_t(R) \right| |d\tau|,
\end{aligned}$$

and since $\log \psi \in BMO(\Gamma, t)$ by Proposition 2.4, $|\Gamma(t, R)| \leq C_\Gamma R$ by the Carleson condition, and $|\Gamma(t, \kappa R, R)| \geq (1 - \kappa) R$, we get

$$\left| \Delta_t(\log \psi, \kappa R, R) - \Delta_t(R) \right| \leq \frac{C_\Gamma}{1 - \kappa} \| \log \psi \|_{*,t}. \tag{3.16}$$

Jensen's inequality (2.17) so shows that

$$\begin{aligned}
e^{\Delta_t(R)} &= \exp \left(\Delta_t(\log \psi, \kappa R, R) \right) \exp \left(- \Delta_t(\log \psi, \kappa R, R) + \Delta_t(R) \right) \\
&\leq \exp \left(\Delta_t(\log \psi, \kappa R, R) \right) \exp \left| \Delta_t(\log \psi, \kappa R, R) - \Delta_t(R) \right| \\
&\leq \left(\Delta_t(\psi^p, \kappa R, R) \right)^{1/p} \exp \left(\frac{C_\Gamma}{1 - \kappa} \| \log \psi \|_{*,t} \right). \tag{3.17}
\end{aligned}$$

On the other hand, since $c_{\psi,t} \leq C_{\psi,t}$, we infer from (2.20) that

$$\begin{aligned}
\left(\Delta_t(\psi^p, \kappa R, R) \right)^{1/p} &\leq \left(\frac{|\Gamma(t, R)|}{|\Gamma(t, \kappa R, R)|} \right)^{1/p} \left(\Delta_t(\psi^p, R) \right)^{1/p} \\
&\leq \left(\frac{C_\Gamma}{1 - \kappa} \right)^{1/p} C_{\psi,t} e^{\Delta_t(R)}. \tag{3.18}
\end{aligned}$$

Combining (3.17) and (3.18) we obtain

$$D_1 e^{\Delta_t(R)} \leq \left(\Delta_t(\psi^p, \kappa R, R)\right)^{1/p} \leq D_2 e^{\Delta_t(R)} \qquad (3.19)$$

with

$$D_1 := \exp\left(-\frac{C_\Gamma}{1-\kappa}\|\log\psi\|_{*,t}\right), \quad D_2 := \left(\frac{C_\Gamma}{1-\kappa}\right)^{1/p} C_{\psi,t}.$$

Analogously one can show that

$$D_1 e^{-\Delta_t(R)} \leq \left(\Delta_t(\psi^{-q}, \kappa R, R)\right)^{1/q} \leq D_3 e^{\Delta_t(R)} \qquad (3.20)$$

with $D_3 := (C_\Gamma/(1-\kappa))^{1/q} C_{\psi,t}$. Consequently, if $x \in [x_0^{-1}, 1]$ then

$$
\begin{aligned}
(U_t^0 \psi)(x) &= \limsup_{R \to 0} \left(\Delta_t(\psi^p, \kappa x R, xR)\right)^{1/p} \left(\Delta_t(\psi^{-q}, \kappa R, R)\right)^{1/q} \\
&\geq \limsup_{R \to 0} D_1^2 e^{\Delta_t(xR) - \Delta_t(R)},
\end{aligned}
$$

which, by (3.14), gives

$$(U_t^0 \psi)(x) \geq D_1^2 e^{-C_\Gamma x_0 \|\log\psi\|_{*,t}} =: C_t^{(1)}(x_0). \qquad (3.21)$$

Using (3.15) one analogously gets (3.21) for $x \in [1, x_0]$. Finally, in the same way one can show that

$$(U_t \psi)(x) \leq D_2 D_3 e^{C_\Gamma x_0 \|\log\psi\|_{*,t}} = C_t^{(2)}(x_0) \text{ for } x \in [x_0^{-1}, x_0]. \qquad \square$$

Theorem 3.3. *Let Γ be a bounded simple Carleson curve, let $\psi : \Gamma \to [0, \infty]$ be a weight, and let $t \in \Gamma$.*
(a) *If $\log \psi \in BMO(\Gamma, t)$ then $V_t \psi$ and $V_t^0 \psi$ are regular functions mapping $(0, \infty)$ to $(0, \infty)$.*
(b) *If $\psi \in A_p(\Gamma, t)$ $(1 < p < \infty)$ then $U_t \psi, U_t^0 \psi, V_t \psi$, and $V_t^0 \psi$ are regular functions mapping $(0, \infty)$ to $(0, \infty)$.*
(c) *If one of the functions $U_t \psi, U_t^0 \psi, V_t \psi, V_t^0 \psi$ is regular, then it is submultiplicative.*

Proof. The regularity and finiteness of the functions under the conditions of (a) and (b) follow from Lemma 3.2. We now show that $U_t \psi$ is submultiplicative. Suppose $x_1, x_2 \in (0, \infty)$ and $x_1 x_2 \in (0, 1]$. We then have

$$
\begin{aligned}
(U_t \psi)(x_1 x_2) &= \sup_{R>0} \left(\Delta_t(\psi^p, \kappa x_1 x_2 R, x_1 x_2 R)\right)^{1/p} \left(\Delta_t(\psi^{-q}, \kappa R, R)\right)^{1/q} \\
&\leq \sup_{R>0} \left(\left(\Delta_t(\psi^p, \kappa x_1 x_2 R, x_1 x_2 R)\right)^{1/p} \left(\Delta_t(\psi^{-q}, \kappa x_2 R, x_2 R)\right)^{1/q}\right) \\
&\quad \times \sup_{R>0} \left(\left(\Delta_t(\psi^p, \kappa x_2 R, x_2 R)\right)^{1/p} \left(\Delta_t(\psi^{-q}, \kappa R, R)\right)^{1/q}\right) \quad (3.22)
\end{aligned}
$$

If $x_1, x_2 \in (0, 1]$, then the first factor on the right of (3.22) equals

$$
\sup_{0<R\leq x_2 d_t} \left(\left(\Delta_t(\psi^p, \kappa x_1 R, x_1 R) \right)^{1/p} \left(\Delta_t(\psi^{-q}, \kappa R, R) \right)^{1/q} \right)
$$

$$
\leq \sup_{0<R\leq d_t} \left(\left(\Delta_t(\psi^p, \kappa x_1 R, x_1 R) \right)^{1/p} \left(\Delta_t(\psi^{-q}, \kappa R, R) \right)^{1/q} \right) = (U_t\psi)(x_1),
$$

and hence, the right-hand side of (3.22) is not greater than $(U_t\psi)(x_1)(U_t\psi)(x_2)$. Suppose $x_1 \in (1, \infty)$ and $x_2 \in (0, 1]$. Then the first factor on the right of (3.22) is

$$
\sup_{0<R\leq x_1 x_2 d_t} \left(\left(\Delta_t(\psi^p, \kappa R, R) \right)^{1/p} \left(\Delta_t(\psi^{-q}, \kappa x_1^{-1} R, x_1^{-1} R) \right)^{1/q} \right)
$$

$$
\leq \sup_{0<R\leq d_t} \left(\left(\Delta_t(\psi^p, \kappa R, R) \right)^{1/p} \left(\Delta_t(\psi^{-q}, \kappa x_1^{-1} R, x_1^{-1} R) \right)^{1/q} \right) = (U_t\psi)(x_1),
$$

i.e. again $(U_t\psi)(x_1 x_2) \leq (U_t\psi)(x_1)(U_t\psi)(x_2)$. This settles the case $x_1 x_2 \in (0, 1]$. For $x_1 x_2 \in (1, \infty)$ the proof is analogous. Thus, $U_t\psi$ is submultiplicative.

The submultiplicativity of $U_t^0\psi, V_t\psi$, and $V_t^0\psi$ can be verified similarly. $\square$

Given a weight $\psi : \Gamma \to [0, \infty]$ such that $\psi \in L^p(\Gamma)$ and $\psi^{-1} \in L^q(\Gamma)$, we may define two functions $U\psi, U^0\psi : (0, \infty) \to [0, \infty]$ by

$$
(U\psi)(x) := \sup_{t\in\Gamma}(U_t\psi)(x), \quad (U^0\psi)(x) := \sup_{t\in\Gamma}(U_t^0\psi)(x).
$$

If ψ is a weight for which $\log\psi \in L^1(\Gamma)$, we denote by $V\psi, V^0\psi : (0, \infty) \to [0, \infty]$ the functions given by

$$
(V\psi)(x) := \sup_{t\in\Gamma}(V_t\psi)(x), \quad (V^0\psi)(x) := \sup_{t\in\Gamma}(V_t^0\psi)(x).
$$

Theorem 3.4. *Let Γ be a bounded simple Carleson curve and let $\psi : \Gamma \to [0, \infty]$ be a weight.*
  (a) *If $\log\psi \in BMO(\Gamma)$ then $V_t\psi, V_t^0\psi$ $(t \in \Gamma)$ and $V\psi, V^0\psi$ are regular submultiplicative functions mapping $(0, \infty)$ to $(0, \infty)$.*
  (b) *If $\psi \in A_p(\Gamma)$ $(1 < p < \infty)$ then $U_t\psi, U_t^0\psi, V_t\psi, V_t^0\psi$ $(t \in \Gamma)$ as well as $U\psi, U^0\psi, V\psi, V^0\psi$ are regular submultiplicative functions mapping $(0, \infty)$ to $(0, \infty)$.*

Proof. (a) Since $\|f\|_* = \sup_{t\in\Gamma} \|f\|_{*,t}$ for $f \in BMO(\Gamma)$, the assertion follows from Lemma 3.2(a) and Theorem 3.3(a),(c).

(b) If $\psi \in A_p(\Gamma)$, then $C_\psi := \sup_{t\in\Gamma} C_{\psi,t}$ where C_ψ and $C_{\psi,t}$ are defined in accordance with (2.1) and (2.18). This together with Theorem 2.5, part (a) of the present theorem, Lemma 3.2(b), and Theorem 3.3(b),(c) gives the assertion. $\square$

3.3 Muckenhoupt condition and U transform

Under the hypotheses of Theorems 3.3 and 3.4 we may define the lower and upper indices of the corresponding regular submultiplicative functions by the formulas (1.29):

$$\alpha(U_t\psi) := \lim_{x\to 0} \frac{\log(U_t\psi)(x)}{\log x}, \quad \beta(U_t\psi) := \lim_{x\to\infty} \frac{\log(U_t\psi)(x)}{\log x}, \quad \dots$$

Lemma 3.5. *Let Γ be a bounded simple Carleson curve, let $\psi : \Gamma \to [0,\infty]$ be a weight, and let $t \in \Gamma$.*

(a) *If $\log\psi \in L^1(\Gamma)$ and the function $V_t\psi$ is regular, then $V_t\psi$ and $V_t^0\psi$ are regular submultiplicative functions and*

$$\alpha(V_t\psi) = \alpha(V_t^0\psi) \le \beta(V_t^0\psi) = \beta(V_t\psi). \tag{3.23}$$

(b) *If $\psi \in L^p_{\mathrm{loc}}(\Gamma \setminus \{t\})$ and $\psi^{-1} \in L^q_{\mathrm{loc}}(\Gamma \setminus \{t\})$ and if $U_t\psi$ is regular, then $U_t\psi$ and $U_t^0\psi$ are regular submultiplicative functions and*

$$\alpha(U_t\psi) = \alpha(U_t^0\psi) \le \beta(U_t^0\psi) = \beta(U_t\psi). \tag{3.24}$$

Proof. This can almost literally be shown by the argument of the proof of Lemma 1.16. One has only to replace W by V or U and $M(\psi, R_1, R_2)$ by

$$\exp\left(\Delta_t(\log\psi, 0, R_1)\right) / \exp\left(\Delta_t(\log\psi, 0, R_2)\right)$$

and

$$\left(\Delta_t(\psi^p, \kappa R_1, R_1)\right)^{1/p} \left(\Delta_t(\psi^{-q}, \kappa R_2, R_2)\right)^{1/q},$$

respectively. $\qquad\square$

Lemma 3.6. *Let Γ be a bounded simple Carleson curve, let $\psi : \Gamma \to [0,\infty]$ be a weight, and let $t \in \Gamma$.*

(a) *If $\psi \in L^p_{\mathrm{loc}}(\Gamma \setminus \{t\}), \psi^{-1} \in L^q_{\mathrm{loc}}(\Gamma \setminus \{t\})$, $\log\psi \in BMO(\Gamma, t)$, and if the function $U_t\psi$ is regular, then*

$$\alpha(U_t\psi) \le \alpha(V_t\psi) \le \beta(V_t\psi) \le \beta(U_t\psi). \tag{3.25}$$

(b) *If $\psi \in A_p(\Gamma, t)$ then*

$$\alpha(U_t\psi) = \alpha(V_t\psi) \le \beta(V_t\psi) = \beta(U_t\psi). \tag{3.26}$$

(c) *If $\psi \in A_p(\Gamma)$ then*

$$\alpha(U\psi) = \alpha(V\psi) \le \beta(V\psi) = \beta(U\psi). \tag{3.27}$$

Remark. After Lemma 3.15 we will show that in part (a) the additional requirement that $\log\psi$ be in $BMO(\Gamma, t)$ can actually be removed.

Proof. (a) Since $U_t \psi$ is submultiplicative by virtue of Theorem 3.3(c) and since $V_t \psi$ is regular and submultiplicative due to Theorem 3.3(a),(c), each of the four indices in (3.25) is well-defined by (1.29).

Let $x \in (0, 1]$. By Jensen's inequality (2.17) and by (3.16), we obtain

$$
\begin{aligned}
&\left(\Delta_t(\psi^p, \kappa x R, x R)\right)^{1/p} \left(\Delta_t(\psi^{-q}, \kappa R, R)\right)^{1/q} \\
&\geq \exp\left(\Delta_t(\log \psi, \kappa x R, x R)\right) \exp\left(\Delta_t(\log \psi^{-1}, \kappa R, R)\right) \\
&= \exp\left(\Delta_t(\log \psi, 0, x R) - \Delta_t(\log \psi, 0, R)\right) \\
&\quad \times \exp\left(\Delta_t(\log \psi, \kappa x R, x R) - \Delta_t(\log \psi, 0, x R)\right) \\
&\quad \times \exp\left(-\Delta_t(\log \psi, \kappa R, R) + \Delta_t(\log \psi, 0, R)\right) \\
&\geq \exp\left(\Delta_t(\log \psi, 0, x R) - \Delta_t(\log \psi, 0, R)\right) \exp\left(-\frac{2\,C_\Gamma}{1 - \kappa} \|\log \psi\|_{*,t}\right),
\end{aligned}
$$

whence

$$
(U_t \psi)(x) \geq C(V_t \psi)(x) \ \text{ for } \ x \in (0, 1] \tag{3.28}
$$

with $C := \exp(-2\,C_\Gamma \|\log \psi\|_{*,t}/(1 - \kappa))$. It can be shown analogously that (3.28) also holds for $x \in (1, \infty)$. Thus,

$$
\frac{\log(U_t \psi)(x)}{\log x} \leq \frac{\log C}{\log x} + \frac{\log(V_t \psi)(x)}{\log x} \ \text{ for } \ x \in (0, 1)
$$

and

$$
\frac{\log(U_t \psi)(x)}{\log x} \geq \frac{\log C}{\log x} + \frac{\log(V_t \psi)(x)}{\log x} \ \text{ for } \ x \in (1, \infty).
$$

Consequently, by (1.29), $\alpha(U_t \psi) \leq \alpha(V_t \psi)$ and $\beta(U_t \psi) \geq \beta(V_t \psi)$. This proves (3.25).

(b) The four indices are well-defined by virtue of Theorem 3.3(b),(c). We have

$$
\alpha(V_t \psi) = \lim_{x \to 0} \frac{1}{\log x} \sup_{R > 0} \left(\Delta_t(x R) - \Delta_t(R)\right)
$$

where $\Delta_t(R) := \Delta_t(\log \psi, 0, R)$, and

$$
\alpha(U_t \psi) = \lim_{x \to 0} \frac{\sup\limits_{R > 0} \left(\frac{1}{p} \log \Delta_t(\psi^p, \kappa x R, x R) + \frac{1}{q} \log \Delta_t(\psi^{-q}, \kappa R, R)\right)}{\log x}.
$$

The equality $\alpha(V_t \psi) = \alpha(U_t \psi)$ is therefore immediate from (3.19) and (3.20). In the same way one gets the equality $\beta(V_t \psi) = \beta(U_t \psi)$.

(c) The proof is the one of part (b) with $\sup_{t \in \Gamma} \sup_{R > 0}$ in place of $\sup_{R > 0}$. $\square$

In particular, the preceding two lemmas in conjunction with Theorem 3.4(b) imply that if $\psi \in A_p(\Gamma)$, then

$$\alpha(U_t\psi) = \alpha(U_t^0\psi) = \alpha(V_t\psi) = \alpha(V_t^0\psi), \tag{3.29}$$
$$\alpha(U\psi) = \alpha(U^0\psi) = \alpha(V\psi) = \alpha(V^0\psi), \tag{3.30}$$

and the same equalities hold with α replaced by β. The relation between the indices of $U_t\psi$ and $U\psi$ as well as between the indices of $V_t\psi$ and $V\psi$ is more complicated. Since for every $t \in \Gamma$,

$$\alpha(U\psi) = \lim_{x\to 0} \frac{\log(U\psi)(x)}{\log x} \leq \lim_{x\to 0} \frac{\log(U_t\psi)(x)}{\log x} = \alpha(U_t\psi),$$
$$\beta(U\psi) = \lim_{x\to \infty} \frac{\log(U\psi)(x)}{\log x} \geq \lim_{x\to \infty} \frac{\log(U_t\psi)(x)}{\log x} = \beta(U_t\psi),$$

we have at least the estimates

$$\alpha(U\psi) \leq \inf_{t\in\Gamma} \alpha(U_t\psi) \leq \sup_{t\in\Gamma} \beta(U_t\psi) \leq \beta(U\psi), \tag{3.31}$$
$$\alpha(V\psi) \leq \inf_{t\in\Gamma} \alpha(V_t\psi) \leq \sup_{t\in\Gamma} \beta(V_t\psi) \leq \beta(V\psi), \tag{3.32}$$

for $\psi \in A_p(\Gamma)$.

The following theorem provides a sufficient condition for the membership of a weight in $A_p(\Gamma, t)$. It will be used to prove Theorem 3.13. The main result of this section is Theorem 3.8.

Theorem 3.7. *Let Γ be a bounded simple Carleson curve, let $t \in \Gamma$, $p \in (1, \infty)$, and let $\psi : \Gamma \to [0, \infty]$ be a weight. Suppose that $\psi \in L_{loc}^p(\Gamma \setminus \{t\})$, $\psi^{-1} \in L_{loc}^q(\Gamma \setminus \{t\})$, and the function $U_t\psi$ is regular. If*

$$-1/p < \alpha(U_t\psi) \leq \beta(U_t\psi) < 1/q, \tag{3.33}$$

then $\psi \in A_p(\Gamma, t)$. Conversely, if $\psi \in A_p(\Gamma, t)$, then the function $U_t\psi$ is regular and

$$-1/p \leq \alpha(U_t\psi) \leq \beta(U_t\psi) \leq 1/q, \tag{3.34}$$

Proof. Put $\alpha := \alpha(U_t\psi)$ and $\beta := \beta(U_t\psi)$. By (3.33), there is an $\varepsilon > 0$ such that $-1/p < \alpha - \varepsilon < \beta + \varepsilon < 1/q$ and hence

$$1 + p(\alpha - \varepsilon) > 0, \ 1 - q(\beta + \varepsilon) > 0. \tag{3.35}$$

Recall that $U_t\psi$ involves the parameter $\kappa \in (0, 1)$. From Corollary 1.14 we deduce that there is an $n_0 \geq 0$ such that

$$(U_t\psi)(\kappa^n) \leq \kappa^{n(\alpha-\varepsilon)} \text{ for } n \geq n_0, \ (U_t\psi)(\kappa^n) \leq \kappa^{n(\beta+\varepsilon)} \text{ for } n \leq -n_0. \tag{3.36}$$

The regularity and submultiplicativity of $U_t\psi$ imply that

$$(U_t\psi)(\kappa^n) \le M := \max_{-n_0 < n < n_0} (U_t\psi)(\kappa^n) < \infty \quad \text{for} \quad -n_0 < n < n_0. \qquad (3.37)$$

Taking into account the Carleson condition (1.1), we get

$$\begin{aligned}
\left|\Gamma(t, \kappa^{n+1}R, \kappa^n R)\right| &= \left|\Gamma(t, \kappa^n R)\right| - \left|\Gamma(t, \kappa^{n+1}R)\right| \\
&\le C_\Gamma \kappa^n R - \kappa^{n+1}R = c_0 \kappa^n R
\end{aligned} \qquad (3.38)$$

with $c_0 := C_\Gamma - \kappa > 0$. From definition (3.10) we obtain

$$\begin{aligned}
&\Delta_t(\psi^p, \kappa^{n+1}R, \kappa^n R) \\
&\qquad \le \left((U_t\psi)(\kappa^n)\right)^p \left(\Delta_t(\psi^{-q}, \kappa R, R)\right)^{-p/q} \quad \text{for} \quad n \ge 0, \qquad (3.39) \\
&\Delta_t(\psi^{-q}, \kappa^{n+1}R, \kappa^n R) \\
&\qquad \le \left((U_t\psi)(\kappa^{-n})\right)^q \left(\Delta_t(\psi^p, \kappa R, R)\right)^{-q/p} \quad \text{for} \quad n \ge 0. \qquad (3.40)
\end{aligned}$$

Since $0 < \kappa^{1+p(\alpha-\varepsilon)} < 1$ and $0 < \kappa^{1-q(\beta+\varepsilon)} < 1$ by (3.35), we see from (3.36) to (3.40) that

$$\begin{aligned}
\int_{\Gamma(t,R)} \psi(\tau)^p |d\tau| &= \sum_{n=0}^{\infty} \left|\Gamma(t, \kappa^{n+1}R, \kappa^n R)\right| \Delta_t(\psi^p, \kappa^{n+1}R, \kappa^n R) \\
&\le \sum_{n=0}^{\infty} c_0 \kappa^n R \left((U_t\psi)(\kappa^n)\right)^p \left(\Delta_t(\psi^{-q}, \kappa R, R)\right)^{-p/q} \\
&\le c_0 R \left(\sum_{n=n_0}^{\infty} \kappa^{n(1+p(\alpha-\varepsilon))} + \sum_{n=0}^{n_0-1} \kappa^n M^p\right) \left(\Delta_t(\psi^{-q}, \kappa R, R)\right)^{-p/q} \\
&=: R c_1 \left(\Delta_t(\psi^{-q}, \kappa R, R)\right)^{-p/q} \quad \text{with} \quad c_1 < \infty
\end{aligned} \qquad (3.41)$$

and analogously,

$$\int_{\Gamma(t,R)} \psi(\tau)^{-q} |d\tau| \le R c_2 \left(\Delta_t(\psi^p, \kappa R, R)\right)^{-q/p} \qquad (3.42)$$

with

$$c_2 := c_0 \left(\sum_{n=n_0}^{\infty} \kappa^{n(1-q(\beta+\varepsilon))} + \sum_{n=0}^{n_0-1} \kappa^n M^q\right) < \infty.$$

Multiplying (3.41) and (3.42) and applying Hölder's inequality, we get

$$\begin{aligned}
&\left(\frac{1}{R} \int_{\Gamma(t,R)} \psi(\tau)^p |d\tau|\right)^{1/p} \left(\frac{1}{R} \int_{\Gamma(t,R)} \psi(\tau)^{-q} |d\tau|\right)^{1/q} \\
&\qquad \le c_1^{1/p} c_2^{1/q} \left(\Delta_t(\psi^{-q}, \kappa R, R)\right)^{-1/q} \left(\Delta_t(\psi^p, \kappa R, R)\right)^{-1/p} \le c_1^{1/p} c_2^{1/q}. \qquad (3.43)
\end{aligned}$$

This proves that $\psi \in A_p(\Gamma, t)$ (including that $\psi \in L^p(\Gamma)$ and $\psi^{-1} \in L^q(\Gamma)$).

Now suppose that $\psi \in A_p(\Gamma, t)$. Then, by Theorem 3.3(b), the function $U_t\psi$ is regular. Since

$$\Delta_t(\psi^p, \kappa x R, x R) = \frac{1}{|\Gamma(t, \kappa x R, x R)|} \int\limits_{\Gamma(t,\kappa x R,x R)} \psi(\tau)^p |d\tau|$$

$$\leq \frac{1}{(1-\kappa)xR} \int\limits_{\Gamma(t,R)} \psi(\tau)^p |d\tau| \quad \text{for } x \in (0,1]$$

and

$$\Delta_t(\psi^{-q}, \kappa x^{-1} R, x^{-1} R) = \frac{1}{|\Gamma(t, \kappa x^{-1} R, x^{-1} R)|} \int\limits_{\Gamma(t,\kappa x^{-1}R,x^{-1}R)} \psi(\tau)^{-q} |d\tau|$$

$$\leq \frac{1}{(1-\kappa)x^{-1}R} \int\limits_{\Gamma(t,R)} \psi(\tau)^{-q} |d\tau| \quad \text{for } x \in [1,\infty),$$

we obtain from (3.10) and (2.18) the estimates

$$(U_t\psi)(x) \leq \frac{x^{-1/p}}{1-\kappa} \sup_{\varepsilon>0} \left(\frac{1}{\varepsilon} \int\limits_{\Gamma(t,\varepsilon)} \psi(\tau)^p |d\tau|\right)^{1/p} \left(\frac{1}{\varepsilon} \int\limits_{\Gamma(t,\varepsilon)} \psi(\tau)^{-q} |d\tau|\right)^{1/q}$$

$$= \frac{x^{-1/p}}{1-\kappa} C_{\psi,t} \quad \text{for } x \in (0,1], \tag{3.44}$$

$$(U_t\psi)(x) \leq \frac{x^{1/q}}{1-\kappa} C_{\psi,t} \quad \text{for } x \in [1,\infty). \tag{3.45}$$

These two estimates and (1.29) give (3.34). $\qquad\square$

Here now is an analogue of Theorem 2.33.

Theorem 3.8. *Let Γ be a bounded simple Carleson curve and let $\psi : \Gamma \to [0,\infty]$ be a weight such that $\psi \in L^p(\Gamma)$ and $\psi^{-1} \in L^q(\Gamma)$. Then $\psi \in A_p(\Gamma)$ $(1 < p < \infty)$ if and only if*
  (i) *the function $U\psi$ is regular,*
  (ii) $-1/p < \alpha(U\psi) \leq \beta(U\psi) < 1/q$.

Proof. Suppose (i) and (ii) hold. Repeating the proof of Theorem 3.7 with $U_t\psi$ replaced by $U\psi$, we arrive at the estimate (3.43) with certain constants c_1, c_2 independent of t and R, which shows that $\psi \in A_p(\Gamma)$.

Conversely, suppose $\psi \in A_p(\Gamma)$. The condition (i) follows from Theorem 3.4(b). Put $\alpha := \alpha(U\psi)$ and $\beta := \beta(U\psi)$. By virtue of (3.44) and (3.45),

$$(U\psi)(x) \leq \frac{C_\psi}{1-\kappa} \begin{cases} x^{-1/p} & \text{for } x \in (0,1] \\ x^{1/q} & \text{for } x \in [1,\infty), \end{cases}$$

whence $-1/p \leq \alpha \leq \beta \leq 1/q$.

Finally, assume $\psi \in A_p(\Gamma)$ but $\alpha = -1/p$ or $\beta = 1/q$. By Theorem 2.31, there exists and $\varepsilon > 0$ such that $\psi^{1+\varepsilon} \in A_p(\Gamma)$. From Lemma 3.6(c) we deduce that

$$\alpha(U\psi^{1+\varepsilon}) = \alpha(V\psi^{1+\varepsilon}) = (1+\varepsilon)\alpha(V\psi) = (1+\varepsilon)\alpha(U\psi) = (1+\varepsilon)\alpha \qquad (3.46)$$

and, analogously, $\beta(U\psi^{1+\varepsilon}) = (1+\varepsilon)\beta$. Since $(1+\varepsilon)\alpha < -1/p$ or $(1+\varepsilon)\beta > 1/q$, we obtain from what was already proved that $\psi^{1+\varepsilon} \notin A_p(\Gamma)$. This contradiction completes the proof. □

Here are a few remarks about the previous theorems.

First, in contrast to Theorem 3.8, we only have the following partial converse of Theorem 3.7: if $\psi \in A_p(\Gamma, t)$ then $U_t\psi$ is regular and $-1/p \leq \alpha(U_t\psi) \leq \beta(U_t\psi) \leq 1/q$. Since we do not know a "local version" of Theorem 2.31, i.e. a result saying that $\psi^{1+\varepsilon} \in A_p(\Gamma, t)$ for all sufficiently small $\varepsilon > 0$ whenever $\psi \in A_p(\Gamma, t)$, we have not been able to establish a full converse of Theorem 3.7.

Secondly, we have not been able to prove a version of Theorem 3.8 in which U is replaced by V. On the other hand, we cannot do (3.46) without an intermediate passage to $V\psi$. Thus, at the present moment, we could dispense with Theorem 2.31 and with the V transform if we were satisfied by the partial converses of Theorems 2.33 and 3.8 which say that $-1/p \leq \alpha \leq \beta \leq 1/q$ whenever the weight is in $A_p(\Gamma)$. Until now, we used Theorem 2.31 and the V transform solely to show that actually $-1/p < \alpha \leq \beta < 1/q$.

3.4   Indicator set and U transform

Let Γ be a bounded simple Carleson curve, fix a point $t \in \Gamma$ and a weight $w \in A_p(\Gamma)$ $(1 < p < \infty)$. The purpose of this section is to describe the indicator set $N_t := N_t(\Gamma, p, w)$ given by (3.2) in terms of the U transform.

Of course, we would like to make use of Theorem 3.8. This confronts us with two questions: first, for which γ is $U(\varphi_{t,\gamma}w)$ regular and secondly, if this is the case, what are $\alpha(U(\varphi_{t,\gamma}w))$ and $\beta(U(\varphi_{t,\gamma}w))$? All we know so far with respect to the first question is that $U(\varphi_{t,\gamma}w)$ is regular for $\gamma = 0$ (Theorem 3.4(b)).

In what follows we use the notations $U_t\varphi w := U_t(\varphi w)$, $U\varphi w := U(\varphi w), \ldots$

Lemma 3.9. Let $\varphi : \Gamma \to [0, \infty]$ be a weight which is continuous and nonvanishing on $\Gamma \setminus \{t\}$ and let $w : \Gamma \to [0, \infty]$ be a weight such that $w \in L^p_{\mathrm{loc}}(\Gamma \setminus \{t\})$ and $w^{-1} \in L^q_{\mathrm{loc}}(\Gamma \setminus \{t\})$. If $W_t\varphi$ and U_tw are regular, then $U_t\varphi w$ is a regular submultiplicative function and

$$\alpha(U_tw) + \alpha(W_t\varphi) \leq \alpha(U_t\varphi w) \leq \min\left\{\alpha(U_tw) + \beta(W_t\varphi), \beta(U_tw) + \alpha(W_t\varphi)\right\},$$

$$\max\left\{\alpha(U_tw) + \beta(W_t\varphi), \beta(U_tw) + \alpha(W_t\varphi)\right\} \leq \beta(U_t\varphi w) \leq \beta(U_tw) + \beta(W_t\varphi).$$

Proof. We first show that

$$(U_t \varphi w)(x) \leq D_t (W_t \varphi)(x)(U_t w)(x) \quad \text{for all } x \in (0, \infty) \tag{3.47}$$

where

$$D_t := \left(\sup_{c \in [\kappa, 1)} (W_t \varphi)(c) \right) \left(\sup_{c \in [\kappa, 1)} (W_t \varphi)(c^{-1}) \right). \tag{3.48}$$

Note that $D_t \in (0, \infty)$ due to the regularity of $W_t \varphi$.

For $x \in (0, 1]$,

$$(W_t \varphi^{-1})(x) = \sup_{R > 0} M(\varphi^{-1}, xR, R) = \sup_{R > 0} M(\varphi, R, xR) = (W_t \varphi)(x^{-1})$$

and if $x \in [1, \infty)$, then

$$(W_t \varphi^{-1})(x) = \sup_{R > 0} M(\varphi^{-1}, R, x^{-1}R) = \sup_{R > 0} M(\varphi, x^{-1}R, R) = (W_t \varphi)(x^{-1}).$$

Thus, for every $R \in (0, d_t]$ and every $c \in (0, 1]$ we have

$$\max_{|\tau - t| = cR} \varphi(\tau) \leq (W_t \varphi)(c) \min_{|\tau - t| = R} \varphi(\tau), \tag{3.49}$$

$$\max_{|\tau - t| = cR} \varphi(\tau)^{-1} \leq (W_t \varphi)(c^{-1}) \min_{|\tau - t| = R} \varphi(\tau)^{-1}. \tag{3.50}$$

Using the latter two inequalities and Lemma 1.15 we obtain for every $x \in (0, 1]$,

$$\left(\Delta_t(\varphi^p w^p, \kappa x R, xR) \right)^{1/p} \leq \left(\sup_{c \in [\kappa, 1)} \max_{|\tau - t| = cxR} \varphi(\tau) \right) \left(\Delta_t(w^p, \kappa x R, xR) \right)^{1/p}$$

$$\leq \left(\sup_{c \in [\kappa, 1)} (W_t \varphi)(cx) \right) \left(\min_{|\tau - t| = R} \varphi(\tau) \right) \left(\Delta_t(w^p, \kappa x R, xR) \right)^{1/p}$$

$$\leq \left(\sup_{c \in [\kappa, 1)} (W_t \varphi)(c) \right) (W_t \varphi)(x) \left(\min_{|\tau - t| = R} \varphi(\tau) \right) \left(\Delta_t(w^p, \kappa x R, xR) \right)^{1/p},$$

$$\left(\Delta_t(\varphi^{-q} w^{-q}, \kappa R, R) \right)^{1/q} \leq \left(\sup_{c \in [\kappa, 1)} \max_{|\tau - t| = cR} \varphi(\tau)^{-1} \right) \left(\Delta_t(w^{-q}, \kappa R, R) \right)^{1/q}$$

$$\leq \left(\sup_{c \in [\kappa, 1)} (W_t \varphi)(c^{-1}) \right) \left(\min_{|\tau - t| = R} \varphi(\tau)^{-1} \right) \left(\Delta_t(w^{-q}, \kappa R, R) \right)^{1/q}.$$

Multiplication of these inequalities gives

$$\left(\Delta_t(\varphi^p w^p, \kappa x R, xR) \right)^{1/p} \left(\Delta_t(\varphi^{-q} w^{-q}, \kappa R, R) \right)^{1/q}$$

$$\leq \left(\sup_{c \in [\kappa, 1)} (W_t \varphi)(c) \right) \left(\sup_{c \in [\kappa, 1)} (W_t \varphi)(c^{-1}) \right) (W_t \varphi)(x)$$

$$\times \left(\Delta_t(w^p, \kappa x R, xR) \right)^{1/p} \left(\Delta_t(w^{-q}, \kappa R, R) \right)^{1/q},$$

and taking the supremum over $R \in (0, d_t]$ we arrive at the estimate (3.47) for $x \in (0, 1]$. Analogously one can prove (3.47) for $x \in [1, \infty)$. Since $W_t\varphi$ and $U_t w$ are regular, we infer from (3.47) that $U_t\varphi w$ is bounded from above in some open neighborhood of the point 1.

To estimate $U_t\varphi w$ from below, let again $R \in (0, d_t]$ and $c \in (0, 1]$. As above, we get

$$\min_{|\tau-t|=cR} \varphi(\tau) \geq \left((W_t\varphi)(c^{-1})\right)^{-1} \max_{|\tau-t|=R} \varphi(\tau), \tag{3.51}$$

$$\min_{|\tau-t|=cR} \varphi(\tau)^{-1} \geq \left((W_t\varphi)(c)\right)^{-1} \max_{|\tau-t|=R} \varphi(\tau)^{-1}. \tag{3.52}$$

Hence, taking into account Lemma 1.15, we see that for every $x \in (0, 1]$,

$$\left(\Delta_t(\varphi^p w^p, \kappa x R, x R)\right)^{1/p} \geq \left(\inf_{c\in[\kappa,1)} \min_{|\tau-t|=cxR} \varphi(\tau)\right)\left(\Delta_t(w^p, \kappa x R, x R)\right)^{1/p}$$

$$\geq \left(\inf_{c\in[\kappa,1)} \left((W_t\varphi)(c^{-1}x^{-1})\right)^{-1}\right)\left(\max_{|\tau-t|=R}\varphi(\tau)\right)\left(\Delta_t(w^p, \kappa x R, x R)\right)^{1/p}$$

$$= \left(\sup_{c\in[\kappa,1)} (W_t\varphi)(c^{-1}x^{-1})\right)^{-1}\left(\max_{|\tau-t|=R}\varphi(\tau)\right)\left(\Delta_t(w^p, \kappa x R, x R)\right)^{1/p}$$

$$\geq \left(\sup_{c\in[\kappa,1)} (W_t\varphi)(c^{-1})\right)^{-1}\left((W_t\varphi)(x^{-1})\right)^{-1}\left(\max_{|\tau-t|=R}\varphi(\tau)\right)$$

$$\times \left(\Delta_t(w^p, \kappa x R, x R)\right)^{1/p},$$

$$\left(\Delta_t(\varphi^{-q} w^{-q}, \kappa R, R)\right)^{1/q} \geq \left(\inf_{c\in[\kappa,1)} \min_{|\tau-t|=cR} \varphi(\tau)^{-1}\right)\left(\Delta_t(w^{-q}, \kappa R, R)\right)^{1/q}$$

$$\geq \left(\inf_{c\in[\kappa,1)} \left((W_t\varphi)(c)\right)^{-1}\right)\left(\max_{|\tau-t|=R}\varphi(\tau)^{-1}\right)\left(\Delta_t(w^{-q}, \kappa R, R)\right)^{1/q}$$

$$= \left(\sup_{c\in[\kappa,1)} (W_t\varphi)(c)\right)^{-1}\left(\max_{|\tau-t|=R}\varphi(\tau)^{-1}\right)\left(\Delta_t(w^{-q}, \kappa R, R)\right)^{1/q}.$$

Multiplying the latter two inequalities and taking the supremum over $R \in (0, d_t]$ shows that $(U_t\varphi w)(x)$ is at least

$$\left(\sup_{c\in[\kappa,1)} (W_t\varphi)(c^{-1})\right)^{-1}\left(\sup_{c\in[\kappa,1)} (W_t\varphi)(c)\right)^{-1}(U_t w)(x)\Big/(W_t\varphi)(x^{-1})$$

for all $x \in (0, 1]$. Similarly one can prove this estimate for $x \in [1, \infty)$. Thus,

$$(U_t\varphi w)(x) \geq D_t^{-1}(U_t w)(x)/(W_t\varphi)(x^{-1}) \tag{3.53}$$

for all $x \in (0, \infty)$ where D_t is given by (3.48). Because the functions $W_t\varphi$ and $U_t w$ are regular and submultiplicative (Lemma 1.15 and Theorem 3.3(c)), they map

$(0, \infty)$ to $(0, \infty)$. Consequently, by (3.53), the function $U_t \varphi w$ is positive on $(0, \infty)$. This completes the proof of the regularity of $U_t \varphi w$. From Theorem 3.3(c) we infer that $U_t \varphi w$ is submultiplicative.

Our next goal is to prove another estimate of $U_t \varphi w$ from below, namely, the estimate

$$(U_t \varphi w)(x) \geq D_t^{-1}(W_t \varphi)(x)/(U_t w)(x^{-1}) \tag{3.54}$$

for all $x \in (0, \infty)$. Using Hölder's inequality, we get

$$(U_t w)(x^{-1}) \geq \left(\Delta_t(w^p, \kappa R, R)\right)^{1/p} \left(\Delta_t(w^{-q}, \kappa x R, x R)\right)^{1/q}$$

$$\geq \left(\Delta_t(w^{-q}, \kappa R, R)\right)^{-1/q} \left(\Delta_t(w^p, \kappa x R, x R)\right)^{-1/p} \tag{3.55}$$

for $x \in (0, 1]$ and

$$(U_t w)(x^{-1}) \geq \left(\Delta_t(w^p, \kappa x^{-1} R, x^{-1} R)\right)^{1/p} \left(\Delta_t(w^{-q}, \kappa R, R)\right)^{1/q}$$

$$\geq \left(\Delta_t(w^{-q}, \kappa x^{-1} R, x^{-1} R)\right)^{-1/q} \left(\Delta_t(w^p, \kappa R, R)\right)^{-1/p} \tag{3.56}$$

for $x \in [1, \infty)$. Let now $x \in (0, 1]$. From (3.51) and (3.52) we obtain

$$\left(\Delta_t(\varphi^p w^p, \kappa x R, x R)\right)^{1/p} \geq \left(\inf_{c \in [\kappa, 1)} \min_{|\tau - t| = c x R} \varphi(\tau)\right)\left(\Delta_t(w^p, \kappa x R, x R)\right)^{1/p}$$

$$\geq \left(\inf_{c \in [\kappa, 1)} \left((W_t \varphi)(c^{-1})\right)^{-1}\right)\left(\max_{|\tau - t| = x R} \varphi(\tau)\right)\left(\Delta_t(w^p, \kappa x R, x R)\right)^{1/p}$$

$$= \left(\sup_{c \in [\kappa, 1)} (W_t \varphi)(c^{-1})\right)^{-1} \left(\max_{|\tau - t| = x R} \varphi(\tau)\right)\left(\Delta_t(w^p, \kappa x R, x R)\right)^{1/p},$$

$$\left(\Delta_t(\varphi^{-q} w^{-q}, \kappa R, R)\right)^{1/q} \geq \left(\inf_{c \in [\kappa, 1)} \min_{|\tau - t| = c R} \varphi(\tau)^{-1}\right)\left(\Delta_t(w^{-q}, \kappa R, R)\right)^{1/q}$$

$$\geq \left(\inf_{c \in [\kappa, 1)} \left((W_t \varphi)(c)\right)^{-1}\right)\left(\max_{|\tau - t| = R} \varphi(\tau)^{-1}\right)\left(\Delta_t(w^{-q}, \kappa R, R)\right)^{1/q}$$

$$= \left(\sup_{c \in [\kappa, 1)} (W_t \varphi)(c)\right)^{-1} \left(\min_{|\tau - t| = R} \varphi(\tau)\right)^{-1}\left(\Delta_t(w^{-q}, \kappa R, R)\right)^{1/q}.$$

Multiplying these two inequalities and recalling (1.31) and (3.55) we get

$$\left(\Delta_t(\varphi^p w^p, \kappa x R, x R)\right)^{1/p}\left(\Delta_t(\varphi^{-q} w^{-q}, \kappa R, R)\right)^{1/q}$$

$$\geq D_t^{-1}\left(\max_{|\tau - t| = x R} \varphi(\tau) \Big/ \min_{|\tau - t| = R} \varphi(\tau)\right) \times \left(\Delta_t(w^p, \kappa x R, x R)\right)^{1/p}\left(\Delta_t(w^{-q}, \kappa R, R)\right)^{1/q}$$

$$\geq D_t^{-1}\left(\max_{|\tau - t| = x R} \varphi(\tau) \Big/ \min_{|\tau - t| = R} \varphi(\tau)\right)\Big/(U_t w)(x^{-1}).$$

Taking the supremum over $R \in (0, d_t]$ we see that (3.54) is true for all $x \in (0, 1]$. Analogously and with the help of (3.56) one can show that (3.54) holds for all $x \in [1, \infty)$.

We now proceed to proving the inequalities asserted by the lemma. If x is in $(0, 1)$ then $\log x < 0$ and hence, by (3.47),

$$\frac{\log(U_t \varphi w)(x)}{\log x} \geq \frac{\log D_t}{\log x} + \frac{\log(W_t \varphi)(x)}{\log x} + \frac{\log(U_t w)(x)}{\log x}. \tag{3.57}$$

Passage to the limit $x \to 0$ in (3.57) gives $\alpha(U_t w) + \alpha(W_t \varphi) \leq \alpha(U_t \varphi w)$, which is the first of the asserted inequalities.

Similarly, for $x \in (0, 1)$, the inequalities (3.53) and (3.54) imply

$$\frac{\log(U_t \varphi w)(x)}{\log x} \leq -\frac{\log D_t}{\log x} + \frac{\log(U_t w)(x)}{\log x} + \frac{\log(W_t \varphi)(x^{-1})}{\log x^{-1}},$$

$$\frac{\log(U_t \varphi w)(x)}{\log x} \leq -\frac{\log D_t}{\log x} + \frac{\log(U_t w)(x^{-1})}{\log x^{-1}} + \frac{\log(W_t \varphi)(x)}{\log x}.$$

Passing to the limit $x \to 0$, we arrive at the inequality

$$\alpha(U_t \varphi w) \leq \min\left\{\alpha(U_t w) + \beta(W_t \varphi), \beta(U_t w) + \alpha(W_t \varphi)\right\},$$

which is the second of the asserted inequalities. The remaining inequalities can be verified analogously. $\qquad\square$

Corollary 3.10. *If the hypotheses of Lemma 3.9 are satisfied and $\alpha(W_t \varphi) = \beta(W_t \varphi)$ or $\alpha(U_t w) = \beta(U_t w)$, then*

$$\alpha(U_t \varphi w) = \alpha(W_t \varphi) + \alpha(U_t w), \quad \beta(U_t \varphi w) = \beta(W_t \varphi) + \beta(U_t w). \tag{3.58}$$

Proof. Immediate from Lemma 3.9. $\qquad\square$

Corollary 3.11. *Let $\varphi : \Gamma \to [0, \infty]$ be a weight which is continuous and nonvanishing on $\Gamma \setminus \{t\}$. If $W_t \varphi$ is regular, then $U_t \varphi$ is also regular and*

$$\alpha(W_t \varphi) = \alpha(U_t \varphi), \quad \beta(W_t \varphi) = \beta(U_t \varphi). \tag{3.59}$$

Proof. Since $U_t 1 = 1$, the function $U_t 1$ is regular and $\alpha(U_t 1) = \beta(U_t 1) = 0$. The assertion is so a straightforward consequence of Corollary 3.10. $\qquad\square$

We now come back to the case where $\varphi = \varphi_{t,\gamma}$ is given by (3.1). Recall that $\eta_t := e^{-\arg(\tau - t)}$.

Corollary 3.12. *Let* Γ *be a bounded simple Carleson curve. Then the function* $U_t\varphi_{t,\gamma}w$ *is regular for every* $\gamma \in \mathbf{C}$ *and every* $w \in A_p(\Gamma)$, *and we have*

$$\alpha(U_t\varphi_{t,\gamma}w) = \operatorname{Re}\gamma + \alpha(U_t\eta_t^{\operatorname{Im}\gamma}w), \tag{3.60}$$

$$\beta(U_t\varphi_{t,\gamma}w) = \operatorname{Re}\gamma + \beta(U_t\eta_t^{\operatorname{Im}\gamma}w). \tag{3.61}$$

Proof. Proposition 3.1 tells us that $W_t\varphi_{t,\gamma}$ is regular and from Theorem 3.4(b) we know that U_tw is regular. We therefore deduce from Lemma 3.9 that $U_t\varphi_{t,\gamma}w$ is also regular. Since $\varphi_{t,\gamma} = \varphi_{t,\operatorname{Re}\gamma}\eta_t^{\operatorname{Im}\gamma}$ and, by (3.5) and (3.6),

$$\alpha(W_t\varphi_{t,\operatorname{Re}\gamma}) = \beta(W_t\varphi_{t,\operatorname{Re}\gamma}) = \operatorname{Re}\gamma, \tag{3.62}$$

we obtain (3.60) and (3.61) from Corollary 3.10. $\square$

The following result is something between Theorems 2.33 and 3.8.

Theorem 3.13. *Let* Γ *be a bounded simple Carleson curve and* $w \in A_p(\Gamma)$ $(1 < p < \infty)$. *Let* $t \in \Gamma$ *and suppose* $\varphi : \Gamma \to [0, \infty]$ *is a weight which is continuous and nonvanishing on* $\Gamma \setminus \{t\}$. *Also suppose that* $W_t\varphi$ *is regular. Then* $\varphi w \in A_p(\Gamma)$ *if and only if*

$$-1/p < \alpha(U_t\varphi w) \le \beta(U_t\varphi w) < 1/q. \tag{3.63}$$

Proof. The function $U_t\varphi w$ is regular and submultiplicative due to Theorem 3.4(b) and Lemma 3.9. If (3.63) holds, then Theorem 3.7 implies that $C_{\varphi w,t} < \infty$ (recall (2.18)). Let $t_0 \in \Gamma \setminus \{t\}$. From (2.67) we see that

$$\sup_{R \ge |t-t_0|/2} \left(\frac{1}{R} \int_{\Gamma(t_0,R)} \varphi^p w^p |d\tau| \right)^{1/p} \left(\frac{1}{R} \int_{\Gamma(t_0,R)} \varphi^{-q} w^{-q} |d\tau| \right)^{1/q} \le 3\, C_{\varphi w,t}. \tag{3.64}$$

So let $0 < R < |t-t_0|/2$. Put $R_0 := |t - t_0| - R$ and $R_1 := \min\{|t - t_0| + R, d_t\}$. From (3.49) and (3.50) we obtain

$$\left(\frac{1}{R} \int_{\Gamma(t_0,R)} \varphi^p w^p |d\tau| \right)^{1/p} \le \left(\sup_{x \in [R_0,R_1]} \max_{|\tau-t|=x} \varphi(\tau) \right) \left(\frac{1}{R} \int_{\Gamma(t_0,R)} w(\tau)^p |d\tau| \right)^{1/p}$$

$$\le \left(\sup_{x \in [R_0,R_1]} (W_t\varphi)\left(\frac{x}{R_1}\right) \right) \left(\min_{|\tau-t|=R_1} \varphi(\tau) \right) \left(\frac{1}{R} \int_{\Gamma(t_0,R)} w(\tau)^p |d\tau| \right)^{1/p},$$

$$\left(\frac{1}{R} \int_{\Gamma(t_0,R)} \varphi^{-q} w^{-q} |d\tau| \right)^{1/q} \le \left(\sup_{x \in [R_0,R_1]} \max_{|\tau-t|=x} \varphi(\tau)^{-1} \right) \left(\frac{1}{R} \int_{\Gamma(t_0,R)} w(\tau)^{-q} |d\tau| \right)^{1/q}$$

$$\le \left(\sup_{x \in [R_0,R_1]} (W_t\varphi)\left(\frac{R_1}{x}\right) \right) \left(\min_{|\tau-t|=R_1} \varphi(\tau)^{-1} \right) \left(\frac{1}{R} \int_{\Gamma(t_0,R)} w(\tau)^{-q} |d\tau| \right)^{1/q},$$

whence

$$\sup_{0<R<|t-t_0|/2} \left(\frac{1}{R} \int_{\Gamma(t_0,R)} \varphi^p w^p |d\tau| \right)^{1/p} \left(\frac{1}{R} \int_{\Gamma(t_0,R)} \varphi^{-q} w^{-q} |d\tau| \right)^{1/q}$$

$$\leq \sup_{0<R<|t-t_0|/2} \left(\left(\sup_{x \in [R_0,R_1]} (W_t\varphi) \left(\frac{x}{R_1} \right) \right) \left(\sup_{x \in [R_0,R_1]} (W_t\varphi) \left(\frac{R_1}{x} \right) \right) \right) C_{w,t_0}. \quad (3.65)$$

From (2.26) we know that $1/3 \leq R_0/R_1 \leq x/R_1 \leq 1$. Thus, combining (3.64) and (3.65) we arrive at the estimate

$$C_{\varphi w, t_0} \leq \max \left\{ 3\, C_{\varphi w, t},\ \sup_{c \in [1/3,1]} (W_t\varphi)(c) \sup_{c \in [1,3]} (W_t\varphi)(c) C_{w,t_0} \right\}.$$

Since $W_t\varphi$ is regular and $w \in A_p(\Gamma)$, it follows that $\sup_{t_0 \in \Gamma \setminus \{t\}} C_{\varphi w, t_0} < \infty$. This proves that $\varphi w \in A_p(\Gamma)$.

Conversely, if $\varphi w \in A_p(\Gamma)$, then $U_t \varphi w$ and $U\varphi w$ are regular submultiplicative functions due to Theorem 3.4(b), and we have $-1/p < \alpha(U\varphi w) \leq \beta(U\varphi w) < 1/q$ by virtue of Theorem 3.8. By (3.31), $\alpha(U\varphi w) \leq \alpha(U_t\varphi w) \leq \beta(U_t\varphi w) \leq \beta(U\varphi w)$ for all $t \in \Gamma$. This gives (3.63). $\qquad \square$

We remark that Theorem 3.13 with $w = 1$ and Corollary 3.11 imply Theorem 2.33.

Corollary 3.14. *Let Γ be a bounded simple Carleson curve, $1 < p < \infty$, $w \in A_p(\Gamma)$, and $t \in \Gamma$. Then the indicator set $N_t(\Gamma, p, w) =: N_t$ equals*

$$N_t = \left\{ \gamma \in \mathbf{C} : -\frac{1}{p} < \operatorname{Re}\gamma + \alpha(U_t \eta_t^{\operatorname{Im}\gamma} w) \leq \operatorname{Re}\gamma + \beta(U_t \eta_t^{\operatorname{Im}\gamma} w) < \frac{1}{q} \right\}. \quad (3.66)$$

Proof. By definition, $N_t := \{\gamma \in \mathbf{C} : \varphi_{t,\gamma} w \in A_p(\Gamma)\}$. The assertion is so a consequence of Theorem 3.13 with $\varphi := \varphi_{t,\gamma}$ and of Corollary 3.12. $\qquad \square$

3.5   Indicator functions

Throughout the rest of this chapter we will assume that Γ is a bounded simple Carleson curve.

The separation of $\alpha(U_t\varphi_{t,\gamma} w)$ and $\beta(U_t\varphi_{t,\gamma} w)$ into a function depending only on the real part of γ and a function of solely the imaginary part of γ (Corollary 3.12) motivates defining

$$\alpha_t^*(x) := \alpha(U_t \eta_t^x w) \quad \text{and} \quad \beta_t^*(x) := \beta(U_t \eta_t^x w) \quad \text{for } x \in \mathbf{R} \quad (3.67)$$

and rewriting (3.66) in the form

$$N_t = \left\{ y + ix \in \mathbf{C} : -\frac{1}{p} - \alpha_t^*(x) < y < \frac{1}{q} - \beta_t^*(x) \right\}. \quad (3.68)$$

We call α_t^* and β_t^* the *outer indicator functions of the triple* (Γ, p, w) *at the point* t. Clearly, in view of (3.68) we know everything about the indicator set once the outer indicator functions are available.

Unfortunately, the U transform is very unwieldy and therefore the computation of $\alpha_t^*(x)$ and $\beta_t^*(x)$ via (3.67) is difficult in concrete cases. It turns out that things are easier for the V_t^0 transform. Hence, we define

$$\alpha_t(x) := \alpha(V_t^0 \eta_t^x w) \quad \text{and} \quad \beta_t(x) := \beta(V_t^0 \eta_t^x w) \quad \text{for} \quad x \in \mathbf{R} \qquad (3.69)$$

and call α_t and β_t simply the *indicator functions of the triple* (Γ, p, w) *at the point* t. Our first concern is to show that α_t and β_t are well-defined.

Lemma 3.15. *Let* $\varphi : \Gamma \to [0, \infty]$ *be a weight which is continuous and nonvanishing on* $\Gamma \setminus \{t\}$ *and let* $w \in A_p(\Gamma)$. *Suppose* $W_t \varphi$ *is regular. Then* $\log \varphi \in BMO(\Gamma)$, *the functions* $U_t \varphi w$ *and* $V_t^0 \varphi w$ *are regular, and*

$$\alpha(U_t \varphi w) \le \alpha(V_t^0 \varphi w) \le \beta(V_t^0 \varphi w) \le \beta(U_t \varphi w). \qquad (3.70)$$

Proof. The regularity of $U_t \varphi w$ follows from Lemma 3.9. From Lemma 1.16 we see that $W_t^0 \varphi$ is also regular. Since, obviously, $W_t^0(\varphi^\varepsilon) = (W_t^0 \varphi)^\varepsilon$ for every $\varepsilon > 0$, we conclude that $W_t^0(\varphi^\varepsilon)$ is regular and that $\alpha(W_t^0(\varphi^\varepsilon)) = \varepsilon \alpha(W_t^0 \varphi)$, $\beta(W_t^0(\varphi^\varepsilon)) = \varepsilon \beta(W_t^0 \varphi)$ for all $\varepsilon > 0$. If $\varepsilon > 0$ is small enough, then

$$-1/p < \varepsilon \alpha(W_t^0 \varphi) \le \varepsilon \beta(W_t^0 \varphi) < 1/q$$

and hence $\varphi^\varepsilon \in A_p(\Gamma)$ due to Theorem 2.33. Now Theorem 2.5 implies that $\log \varphi = (1/\varepsilon) \log \varphi^\varepsilon$ belongs to $BMO(\Gamma)$. The same theorem also shows that $\log w \in BMO(\Gamma)$. Thus, $\log(\varphi w) \in BMO(\Gamma)$ and consequently, $V_t \varphi w$ and $V_t^0 \varphi w$ are regular by virtue of Theorem 3.4(a). The inequalities (3.70) now follow from Lemmas 3.6(a) and 3.5(a) with $\psi := \varphi w$. $\qquad \square$

Lemma 3.15 indicates that in Lemma 3.6(a) the assumption that $\log \psi$ be in $BMO(\Gamma, t)$ is redundant. This is indeed the case. To see this, let $\psi : \Gamma \to [0, \infty]$ be a weight such that $\psi \in L_{\mathrm{loc}}^p(\Gamma \setminus \{t\})$ and $\psi^{-1} \in L_{\mathrm{loc}}^q(\Gamma \setminus \{t\})$. We claim that $\log \psi \in BMO(\Gamma, t)$ if $U_t \psi$ is regular.

Let $\varepsilon \in (0, 1)$. By Hölder's inequality,

$$\begin{aligned}
\left(\Delta_t(\psi^{\varepsilon p}, \kappa x R, x R)\right)^{1/p} &\le \left(\Delta_t(\psi^p, \kappa x R, x R)\right)^{\varepsilon/p} \left(\Delta_t(1, \kappa x R, x R)\right)^{(1-\varepsilon)/p} \\
&= \left(\Delta_t(\psi^p, \kappa x R, x R)\right)^{\varepsilon/p}, \\
\left(\Delta_t(\psi^{-\varepsilon q}, \kappa R, R)\right)^{1/q} &\le \left(\Delta_t(\psi^{-q}, \kappa R, R)\right)^{\varepsilon/q} \left(\Delta_t(1, \kappa R, R)\right)^{(1-\varepsilon)/q} \\
&= \left(\Delta_t(\psi^{-q}, \kappa R, R)\right)^{\varepsilon/q},
\end{aligned}$$

which implies that $(U_t(\psi^\varepsilon))(x) \le ((U_t\psi)(x))^\varepsilon$ for all $x \in (0,1]$. The same estimate for $x \in [1,\infty)$ can be shown analogously. Thus, if $U_t\psi$ is regular, then so is $U_t(\psi^\varepsilon)$, and we have

$$\varepsilon\alpha(U_t\psi) \le \alpha(U_t(\psi^\varepsilon)) \le \beta(U_t(\psi^\varepsilon)) \le \varepsilon\beta(U_t\psi).$$

Hence, if $\varepsilon \in (0,1)$ is small enough, then

$$-1/p < \alpha(U_t(\psi^\varepsilon)) \le \beta(U_t(\psi^\varepsilon)) < 1/q.$$

Taking into account that $\psi^\varepsilon \in L^p_{\mathrm{loc}}(\Gamma \setminus \{t\})$ and $\psi^{-\varepsilon} \in L^q_{\mathrm{loc}}(\Gamma \setminus \{t\})$, we deduce from Theorem 3.7 that $\psi^\varepsilon \in A_p(\Gamma, t)$. Now Proposition 2.4 shows that $\log\psi = (1/\varepsilon)\log\psi^\varepsilon$ belongs to $BMO(\Gamma, t)$.

Lemma 3.16. *Let the weight φ be continuous and nonvanishing on $\Gamma \setminus \{t\}$ and suppose $W_t\varphi$ is regular. Then the functions $W_t^0\varphi$, $U_t\varphi$ and $V_t^0\varphi$ are regular and*

$$\alpha(W_t^0\varphi) = \alpha(W_t\varphi) = \alpha(U_t\varphi) = \alpha(V_t^0\varphi),$$
$$\beta(W_t^0\varphi) = \beta(W_t\varphi) = \beta(U_t\varphi) = \beta(V_t^0\varphi).$$

Proof. Corollary 3.11 implies that $U_t\varphi$ is regular and that $\alpha(W_t\varphi) = \alpha(U_t\varphi)$, $\beta(U_t\varphi) = \beta(W_t\varphi)$. From Lemma 3.15 with $w = 1$ we deduce that $V_t^0\varphi$ is regular and that $\alpha(U_t\varphi) \le \alpha(V_t^0\varphi)$, $\beta(V_t^0\varphi) \le \beta(U_t\varphi)$. Thus,

$$\alpha(W_t\varphi) = \alpha(U_t\varphi) \le \alpha(V_t^0\varphi), \quad \beta(V_t^0\varphi) \le \beta(U_t\varphi) = \beta(W_t\varphi). \qquad (3.71)$$

In the proof of Lemma 3.15 we saw that $W_t(\varphi^\varepsilon)$ is regular, that $\varphi^\varepsilon \in A_p(\Gamma)$ for all sufficiently small $\varepsilon > 0$ and that

$$\varepsilon\alpha(W_t\varphi) = \alpha(W_t(\varphi^\varepsilon)), \quad \varepsilon\beta(W_t\varphi) = \beta(W_t(\varphi^\varepsilon)). \qquad (3.72)$$

Again by Corollary 3.11,

$$\alpha(W_t(\varphi^\varepsilon)) = \alpha(U_t(\varphi^\varepsilon)), \quad \beta(W_t(\varphi^\varepsilon)) = \beta(U_t(\varphi^\varepsilon)). \qquad (3.73)$$

Since $V_t^0(\varphi^\varepsilon) = (V_t^0\varphi)^\varepsilon$ and thus $\alpha(V_t^0\varphi^\varepsilon) = \varepsilon\alpha(V_t^0\varphi)$ whenever $V_t^0\varphi$ and $V_t^0(\varphi^\varepsilon)$ are regular, we get from Lemmas 3.6(b) and 3.5(a) that

$$\varepsilon\alpha(V_t^0\varphi) = \alpha(V_t^0(\varphi^\varepsilon)) = \alpha(U_t(\varphi^\varepsilon)), \quad \varepsilon\beta(V_t^0\varphi) = \beta(V_t^0(\varphi^\varepsilon)) = \beta(U_t(\varphi^\varepsilon)). \quad (3.74)$$

The equalities (3.72) to (3.74) imply that $\alpha(W_t\varphi) = \alpha(V_t^0\varphi)$ and $\beta(W_t\varphi) = \beta(V_t^0\varphi)$. Hence, by (3.71),

$$\alpha(W_t\varphi) = \alpha(U_t\varphi) = \alpha(V_t^0\varphi), \quad \beta(V_t^0\varphi) = \beta(U_t\varphi) = \beta(W_t\varphi).$$

Lemma 1.16 completes the proof. $\qquad\qquad\qquad\qquad\qquad\qquad\qquad\qquad\qquad\qquad\square$

The next lemma is a certain counterpart to Lemma 3.9.

Lemma 3.17. *Let $\varphi : \Gamma \to [0, \infty]$ be a weight which is continuous and nonvanishing on $\Gamma \setminus \{t\}$ and let $\log w \in BMO(\Gamma, t)$. If $W_t\varphi$ is regular then $V_t^0\varphi w$ is regular and*

$$\alpha(V_t^0 w) + \alpha(W_t\varphi) \leq \alpha(V_t^0\varphi w) \leq \min\Big\{\alpha(V_t^0 w) + \beta(W_t\varphi), \beta(V_t^0 w) + \alpha(W_t\varphi)\Big\},$$

$$\max\Big\{\alpha(V_t^0 w) + \beta(W_t\varphi), \beta(V_t^0 w) + \alpha(W_t\varphi)\Big\} \leq \beta(V_t^0\varphi w) \leq \beta(V_t^0 w) + \beta(W_t\varphi).$$

Proof. Lemma 3.15 says that $\log\varphi \in BMO(\Gamma)$, and since $\log w \in BMO(\Gamma, t)$ by assumption, we obtain that $\log(\varphi w) \in BMO(\Gamma, t)$, which, by Theorem 3.3(a), implies that $V_t^0\varphi w$ is regular. Fix $x \in (0, 1)$. Then $\log(V_t^0\varphi w)(x)$ equals

$$\limsup_{R \to 0} \Big(\Delta_t(\log\varphi, 0, xR) - \Delta_t(\log\varphi, 0, R) + \Delta_t(\log w, 0, xR) - \Delta_t(\log w, 0, R)\Big). \tag{3.75}$$

From (3.75) it is clear that

$$\log(V_t^0\varphi w) \leq \log(V_t^0\varphi)(x) + \log(V_t^0 w)(x). \tag{3.76}$$

On the other hand, (3.75) implies that

$$\begin{aligned}
\log(V_t^0\varphi w)(x) \geq\ & \limsup_{R \to 0}\Big(\Delta_t(\log\varphi, 0, xR) - \Delta_t(\log\varphi, 0, R)\Big) \\
& - \limsup_{R \to 0}\Big(\Delta_t(\log w, 0, R) - \Delta_t(\log w, 0, xR)\Big) \\
=\ & \log(V_t^0\varphi)(x) - \log(V_t^0 w)(x^{-1}).
\end{aligned} \tag{3.77}$$

Analogously, by symmetry,

$$\log(V_t^0\varphi w)(x) \geq \log(V_t^0 w)(x) - \log(V_t^0\varphi)(x^{-1}). \tag{3.78}$$

Dividing (3.76), (3.77), (3.78) by $\log x < 0$ and passing to the limit $x \to 0$, we obtain

$$\alpha(V_t^0 w) + \alpha(V_t^0\varphi) \leq \alpha(V_t^0\varphi w) \leq \min\Big\{\alpha(V_t^0 w) + \beta(V_t^0\varphi), \beta(V_t^0 w) + \alpha(V_t^0\varphi)\Big\},$$

which in conjunction with Lemma 3.16 gives the first chain of the asserted inequalities. The other inequalities can be shown similarly. □

Corollary 3.18. *If the hypotheses of Lemma 3.17 are satisfied and $\alpha(W_t\varphi) = \beta(W_t\varphi)$ or $\alpha(V_t^0 w) = \beta(V_t^0 w)$, then*

$$\alpha(V_t^0\varphi w) = \alpha(W_t\varphi) + \alpha(V_t^0 w), \quad \beta(V_t^0\varphi w) = \beta(W_t\varphi) + \beta(V_t^0 w).$$

Proof. Obvious consequence of Lemma 3.17. □

Corollary 3.19. *The function $V_t^0 \varphi_{t,\gamma} w$ is regular for every number $\gamma \in \mathbf{C}$ and every weight $w \in A_p(\Gamma)$, and*

$$\alpha(V_t^0 \varphi_{t,\gamma} w) = \operatorname{Re} \gamma + \alpha(V_t^0 \eta_t^{\operatorname{Im} \gamma} w), \tag{3.79}$$

$$\beta(V_t^0 \varphi_{t,\gamma} w) = \operatorname{Re} \gamma + \beta(V_t^0 \eta_t^{\operatorname{Im} \gamma} w). \tag{3.80}$$

Proof. Proposition 3.1 and Lemma 3.15 imply that $V_t^0 \varphi_{t,\gamma} w$ is regular. The equalities (3.79) and (3.80) so follow from (3.62) and Corollary 3.18. □

The previous corollary shows that $\alpha_t(x)$ and $\beta_t(x)$ are well-defined for all $x \in \mathbf{R}$. Moreover, (3.70) with $\varphi = \varphi_{t,ix} = \eta_t^x$ gives

$$\alpha_t^*(x) \le \alpha_t(x) \le \beta_t(x) \le \beta_t^*(x) \quad \text{for all} \ \ x \in \mathbf{R}. \tag{3.81}$$

The inequalities (3.81) are the reason for calling α_t^* and β_t^* the "outer" indicator functions.

Our next objective is to show that we may drop the $*$ in (3.68), i.e. that in (3.68) we may replace the outer indicator functions by the indicator functions. This will be done by Theorem 3.21. The proof of this theorem makes use of the following interesting fact.

Proposition 3.20. *The functions α_t^* and α_t are concave on $\mathbf{R}$, while the functions β_t^* and β_t are convex on $\mathbf{R}$. In particular, these four functions are continuous on all of $\mathbf{R}$.*

Remark. As usual, a function $f : \mathbf{R} \to \mathbf{R}$ is said to be *convex* if $f(\theta x + (1 - \theta)y) \le \theta f(x) + (1 - \theta)f(y)$ for all $x, y \in \mathbf{R}$ and all $\theta \in [0, 1]$. A function $f : \mathbf{R} \to \mathbf{R}$ is called *concave* if $-f$ is convex.

Proof. Since $\int f^\theta g^{1-\theta} |d\tau| \le (\int f|d\tau|)^\theta (\int g|d\tau|)^{1-\theta}$ for $\theta \in (0,1)$ by Hölder's inequality, we get

$$\left(\Delta_t(\eta_t^{p(x\theta + y(1-\theta))} w^p, \kappa\xi R, \xi R) \right)^{1/p}$$
$$\le \left(\Delta_t(\eta_t^{px} w^p, \kappa\xi R, \xi R) \right)^{\theta/p} \left(\Delta_t(\eta_t^{py} w^p, \kappa\xi R, \xi R) \right)^{(1-\theta)/p},$$
$$\left(\Delta_t(\eta_t^{-q(x\theta + y(1-\theta))} w^{-q}, \kappa R, R) \right)^{1/q}$$
$$\le \left(\Delta_t(\eta_t^{-qx} w^{-q}, \kappa R, R) \right)^{\theta/q} \left(\Delta_t(\eta_t^{-qy} w^{-q}, \kappa R, R) \right)^{(1-\theta)/q}.$$

Multiplying these two inequalities and taking $\sup_{R>0}$ we arrive at the inequality

$$(U_t \eta_t^{x\theta + y(1-\theta)} w)(\xi) \le \left((U_t \eta_t^x w)(\xi) \right)^\theta \left((U_t \eta_t^y w)(\xi) \right)^{1-\theta}$$

for $\xi \in (0,1)$. Consequently,

$$\alpha_t^*(x\theta + y(1-\theta)) = \lim_{\xi \to 0} \frac{1}{\log \xi} \log(U_t \eta_t^{x\theta + y(1-\theta)} w)(\xi)$$

$$\geq \lim_{\xi \to 0} \frac{\theta}{\log \xi} \log(U_t \eta_t^x w)(\xi) + \lim_{\xi \to 0} \frac{1-\theta}{\log \xi} \log(U_t \eta_t^y w)(\xi)$$

$$= \theta \alpha_t^*(x) + (1-\theta)\alpha_t^*(y),$$

which shows that α_t^* is concave. The convexity of β_t^* can be verified analogously. Further,

$$\frac{\exp\left(\Delta_t(\log \eta_t^{x\theta + y(1-\theta)} w, 0, \xi R)\right)}{\exp\left(\Delta_t(\log \eta_t^{x\theta + y(1-\theta)} w, 0, R)\right)}$$

$$= \left(\frac{\exp\left(\Delta_t(\log \eta_t^x w, 0, \xi R)\right)}{\exp\left(\Delta_t(\log \eta_t^x w, 0, R)\right)}\right)^\theta \left(\frac{\exp\left(\Delta_t(\log \eta_t^y w, 0, \xi R)\right)}{\exp\left(\Delta_t(\log \eta_t^y w, 0, R)\right)}\right)^{1-\theta},$$

and taking $\limsup_{R \to 0}$ we obtain

$$(V_t^0 \eta_t^{x\theta + y(1-\theta)} w)(\xi) \leq \left((V_t^0 \eta_t^x w)(\xi)\right)^\theta \left((V_t^0 \eta_t^y w)(\xi)\right)^{1-\theta}$$

for $\xi \in (0,1)$, which implies the concavity of α_t as above. In the same way one can prove that β_t is convex. Finally, it is well known that convex or concave functions are automatically continuous (see, e.g. [177, Corollary 10.1.1]). □

Theorem 3.21. *Let Γ be a bounded simple Carleson curve and $w \in A_p(\Gamma)$. Then for every $t \in \Gamma$ the indicator set $N_t(\Gamma, p, w) =: N_t$ is given by*

$$N_t = \left\{ y + ix \in \mathbf{C} : -\frac{1}{p} - \alpha_t(x) < y < \frac{1}{q} - \beta_t(x) \right\}. \tag{3.82}$$

Proof. We know that N_t is the set (3.68). Put $c(x) := \beta_t^*(x) - \alpha_t^*(x)$. The function c is convex due to Proposition 3.20, and $c(0) < 1/p + 1/q = 1$ since $0 \in N_t$. Consequently, either there exists a unique $x_t^+ > 0$ such that $c(x_t^+) = 1$ or we have $0 \leq c(x) < 1$ for all $x > 0$, in which case we put $x_t^+ := +\infty$. Analogously, either $c(x_t^-) = 1$ for some uniquely determined $x_t^- < 0$ or $0 \leq c(x) < 1$ for all $x < 0$, in which case we define $x_t^- := -\infty$.

Fix $x \in (x_t^-, x_t^+)$. Then $0 \leq \beta_t^*(x) - \alpha_t^*(x) < 1$ by the convexity of c. Hence, there is a real number μ such that

$$-1/p < \mu + \alpha_t^*(x) \leq \mu + \beta_t^*(x) < 1/q,$$

and from (3.68) we deduce that $\mu + ix \in N_t$, i.e. that $\varphi_{t,\mu+ix} \in A_p(\Gamma)$. Thus, by (3.29),

$$\alpha(V_t^0 \varphi_{t,\mu+ix} w) = \alpha(U_t \varphi_{t,\mu+ix} w),$$

and by virtue of (3.79) and (3.60), the latter equality may be written in the form

$$\mu + \alpha(V_t^0 \eta_t^x w) = \mu + \alpha(U_t \eta_t^x w),$$

which proves that $\alpha_t(x) = \alpha_t^*(x)$. Analogously one can show that $\beta_t(x) = \beta_t^*(x)$ for $x \in (x_t^-, x_t^+)$.

If $x_t^- = -\infty$ and $x_t^+ = +\infty$, the proof is complete. Suppose $x_t^+ < +\infty$. Denote the set on the right of (3.82) by M_t. Since $\alpha_t, \alpha_t^*, \beta_t, \beta_t^*$ are continuous and $\alpha_t = \alpha_t^*$, $\beta_t = \beta_t^*$ on (x_t^-, x_t^+), it follows that $\alpha_t = \alpha_t^*$, $\beta_t = \beta_t^*$ on $(x_t^-, x_t^+]$. Thus,

$$\beta_t(x_t^+) - \alpha_t(x_t^+) = \beta_t^*(x_t^+) - \alpha_t^*(x_t^+) = 1,$$

whence, by the convexity of $\beta_t^* - \alpha_t^*$ and $\beta_t - \alpha_t$,

$$\beta_t^*(x) - \alpha_t^*(x) \geq 1, \; \beta_t(x) - \alpha_t(x) \geq 1 \; \text{ for } \; x \geq x_t^+.$$

It follows that

$$N_t = \left\{ y + ix \in \mathbf{C} : x < x_t^+, -\frac{1}{p} - \alpha_t^*(x) < y < \frac{1}{q} - \beta_t^*(x) \right\},$$

$$M_t = \left\{ y + ix \in \mathbf{C} : x < x_t^+, -\frac{1}{p} - \alpha_t(x) < y < \frac{1}{q} - \beta_t(x) \right\}.$$

If $x_t^- = -\infty$, this shows that $N_t = M_t$ and completes the proof. In case $x_t^- > -\infty$, we analogously get

$$N_t = \left\{ y + ix \in \mathbf{C} : x_t^- < x < x_t^+, -\frac{1}{p} - \alpha_t^*(x) < y < \frac{1}{q} - \beta_t^*(x) \right\},$$

$$M_t = \left\{ y + ix \in \mathbf{C} : x_t^- < x < x_t^+, -\frac{1}{p} - \alpha_t(x) < y < \frac{1}{q} - \beta_t(x) \right\},$$

and as $\alpha_t = \alpha_t^*$, $\beta_t = \beta_t^*$ on (x_t^-, x_t^+), we get the desired equality $N_t = M_t$. $\square$

We wish to emphasize an observation we made in the proof of Theorem 3.21. If $\beta_t(x_0) - \alpha_t(x_0) \geq 1$ for some $x_0 \in \mathbf{R}$, then there is no $y \in \mathbf{R}$ satisfying $-1/p - \alpha_t(x_0) < y < 1/q - \beta_t(x_0)$ and hence the line $\{y + ix_0 : y \in \mathbf{R}\}$ and the set (3.82) are disjoint sets. Since $\beta_t - \alpha_t$ is convex, the set $\{x \in \mathbf{R} : \beta_t(x) - \alpha_t(x) < 1\}$ is connected and open. We denote this set by (x_t^-, x_t^+), where the cases $x_t^- = -\infty$ and $x_t^+ = +\infty$ are admitted. Thus, the whole information about N_t is contained in the values of the indicator functions on (x_t^-, x_t^+); their behavior outside this set does not contribute anything to the understanding of N_t and may therefore be neglected. We also note that (3.68) and (3.82) do not imply that $\alpha_t^*(x) = \alpha_t(x)$ and $\beta_t^*(x) = \beta_t(x)$ for all $x \in \mathbf{R}$. It only follows that coincidence happens for $x \in (x_t^-, x_t^+)$ and, by continuity, also at x_t^- and x_t^+ if $x_t^- > -\infty$ or $x_t^+ < +\infty$. Since $0 \in N_t$, we necessarily have $x_t^- < 0$ and $x_t^+ > 0$, implying that always $\alpha_t^*(x) = \alpha_t(x)$ and $\beta_t^*(x) = \beta_t(x)$ in some open neighborhood of $x = 0$.

Example 3.22. Let us consider the situation studied in the end of Section 3.1 once again. Thus, suppose Γ is a bounded simple Carleson curve and suppose $w(\tau) = 1$ for all $\tau \in \Gamma$. From Proposition 3.1 (or from (3.9)) we know that then

$$N_t = \left\{ y + ix \in \mathbf{C} : -\frac{1}{p} - a(x) < y < \frac{1}{q} - b(x) \right\}$$

where $a(x) := \min\{\delta_t^- x, \delta_t^+ x\}$ and $b(x) := \max\{\delta_t^- x, \delta_t^+ x\}$. If $\delta_t^- = \delta_t^+ =: \delta$, then $a(x) = b(x) = \delta x$, so $(x_t^-, x_t^+) = \mathbf{R}$, N_t is the stripe

$$\left\{ y + ix \in \mathbf{C} : -\frac{1}{p} - \delta x < y < \frac{1}{q} - \delta x \right\}$$

and

$$\alpha_t^*(x) = \alpha_t(x) = \beta_t(x) = \beta_t^*(x) = \delta x \quad \text{for all} \quad x \in \mathbf{R}.$$

However, if $\delta_t^- < \delta_t^+$ then the equation $b(x) - a(x) = 1$ has exactly two solutions,

$$x_t^- = 1/(\delta_t^- - \delta_t^+) \quad \text{and} \quad x_t^+ = 1/(\delta_t^+ - \delta_t^-),$$

N_t is the parallelogram

$$\left\{ y + ix \in \mathbf{C} : x_t^- < x < x_t^+, \ -\frac{1}{p} - a(x) < y < \frac{1}{q} - b(x) \right\},$$

and all we can say is that

$$\alpha_t^*(x) = \alpha_t(x) = a(x), \ \beta_t^*(x) = \beta_t(x) = b(x) \quad \text{for} \quad x \in [x_t^-, x_t^+]. \qquad \square$$

3.6 Indices of powerlikeness

Let Γ be a bounded simple Carleson curve and $t \in \Gamma$. Given a weight $\psi : \Gamma \to [0, \infty]$ for which $V_t^0 \psi$ is a regular function, we call the two numbers

$$\mu_t := \alpha(V_t^0 \psi) \quad \text{and} \quad \nu_t := \beta(V_t^0 \psi) \tag{3.83}$$

the *indices of powerlikeness of ψ at the point t*. By Theorem 3.3(a),(c), the indices of powerlikeness are in particular defined whenever $\log \psi \in BMO(\Gamma, t)$ and, all the more, if $\log \psi \in BMO(\Gamma)$ or even $\psi \in A_p(\Gamma)$. If $\psi = w \in A_p(\Gamma)$, then (3.69) implies that

$$\mu_t = \alpha_t(0) \quad \text{and} \quad \nu_t = \beta_t(0), \tag{3.84}$$

i.e. the indices of powerlikeness of a weight $w \in A_p(\Gamma)$ are the values of the indicator functions of the quadruple (Γ, p, w, t) at the origin. In that case we always have $-1/p < \mu_t \le \nu_t < 1/q$, since the origin belongs to $N_t(\Gamma, p, w)$.

The spirality indices of Γ at t are defined by $\delta_t^- := \alpha(W_t^0 \eta_t)$ and $\delta_t^+ := \beta(W_t^0 \eta_t)$ (recall (3.4)) and are indeed intrinsic characteristics of the curve Γ. In the definition (3.69) of the indicator functions, both the weight w and the curve Γ, represented by η_t^x, explicitly occur. The factor η_t^x disappears for $x = 0$, and hence, the curve Γ does not enter (3.83) or (3.84) explicitly. This is why we consider μ_t and ν_t as characteristics of the weight, although, the curve is implicitly present as the carrier of the weight.

The weight ψ is said to be *powerlike* at $t \in \Gamma$ if $V_t^0\psi$ is regular and $\mu_t = \nu_t$.

We now proceed to the computation of the indices of powerlikeness of some concrete weights.

Proposition 3.23. *Let Γ be a bounded simple Carleson curve and $t \in \Gamma$. Then for every $\gamma \in \mathbf{C}$, the indices of powerlikeness of $\varphi_{t,\gamma}$ at t are given by*

$$\mu_t := \alpha(V_t^0\varphi_{t,\gamma}) = \operatorname{Re}\gamma + \min\{\delta_t^-\operatorname{Im}\gamma, \delta_t^+\operatorname{Im}\gamma\},$$
$$\nu_t := \beta(V_t^0\varphi_{t,\gamma}) = \operatorname{Re}\gamma + \max\{\delta_t^-\operatorname{Im}\gamma, \delta_t^+\operatorname{Im}\gamma\},$$

where δ_t^- and δ_t^+ are the spirality indices of Γ at t.

Proof. Combine Proposition 3.1 and Lemma 3.16.   □

Consequently, $\varphi_{t,\gamma}$ is powerlike at t if and only if γ is real or $\delta_t^- = \delta_t^+$. In particular, if $w(\tau) = |\tau - t|^\lambda$ then the indices of powerlikeness at t are always both equal to λ. We also infer from Propositions 3.23 and 1.19 and from the observation made after Proposition 1.19 that for any numbers μ, ν satisfying $\mu \le \nu$ there exist a simple Carleson curve and a point $t \in \Gamma$ such that μ and ν are the indices of powerlikeness of the weight $\varphi_{t,i}(\tau) = e^{-\arg(\tau-t)}$ at the point t.

Example 3.24. The purpose of this example is to fix some notation for further reference. Let

$$\Gamma_1 = \{0\} \cup \{\tau \in \mathbf{C} : \tau = re^{i\varphi(r)}, \ 0 < r \le 1\},$$
$$\Gamma_2 = \{0\} \cup \{\tau \in \mathbf{C} : \tau = re^{i(\varphi(r)+b(r))}, \ 0 < r \le 1\}$$

be as in Example 1.11. Suppose $b \in C(0,1] \cap C^1(0,1)$, $0 < b(r) < 2\pi$ for $r \in (0,1)$, and $|rb'(r)|$ is bounded for $r \in (0,1)$. Suppose further that

$$\varphi(r) = h\big(\log(-\log r)\big)(-\log r)$$

with a function $h \in C^2(\mathbf{R})$ for which h, h', h'' are bounded on $\mathbf{R}$. Let Γ_0 be any of the (Carleson) curves $\Gamma_1, \Gamma_2, \Gamma_1 \cup \Gamma_2$ and put $\Gamma := t + \Gamma_0$. Define a weight w on Γ by $w(\tau) = e^{v(|\tau-t|)}$ where

$$v(r) = g\big(\log(-\log r)\big)\log r$$

with a function $g \in C^2(\mathbf{R})$ such that g, g', g'' are bounded on $\mathbf{R}$. The next proposition gives the indices of powerlikeness of v at t.   □

Proposition 3.25. *Let* Γ *and* w *be as in Example* 3.24. *Then* $V_t^0 w$ *is regular and the indices of powerlikeness of* w *at* t *are*

$$\mu_t := \alpha(V_t^0 w) = \liminf_{y \to +\infty} \big(g(y) + g'(y)\big),$$

$$\nu_t := \beta(V_t^0 w) = \limsup_{y \to +\infty} \big(g(y) + g'(y)\big).$$

The weight w *belongs to* $A_p(\Gamma)$ *if and only if* $-1/p < \mu_t \leq \nu_t < 1/q$.

Proof. Lemma 2.35 with $F(r) := v(r)$ implies that $W_t w$ and $W_t^0 w$ are regular. From Lemma 3.16 we deduce that $V_t^0 w$ is also regular and that $\alpha(V_t^0 w) = \alpha(W_t^0 w)$, $\beta(V_t^0 w) = \beta(W_t^0 w)$. The remaining assertions of the proposition now follow from Theorem 2.36 (with F and f replaced by v and g, respectively). □

Example 3.26. Choose Γ and w as in Example 3.24. Let $g \in C^2(\mathbf{R})$ be any function such that

$$g(y) = \lambda + (1/y) \sin y \quad \text{for } |y| > 1.$$

Then g, g', g'' are bounded on $\mathbf{R}$ and we have

$$\liminf_{y \to \infty} \big(g(y) + g'(y)\big) = \limsup_{y \to \infty} \big(g(y) + g'(y)\big) = \lambda.$$

Thus, by Proposition 3.25, the indices of powerlikeness of w at t are both equal to λ. Notice that

$$v(r) = \lambda \log r + \frac{\log r}{\log(-\log r)} \sin\big(\log(-\log r)\big) \quad \text{as } r \to 0$$

and that therefore $v(r)$ is not even of the form $\lambda \log r + O(1)$; we actually have

$$\liminf_{r \to 0} \big(v(r) - \lambda \log r\big) = -\infty, \ \limsup_{r \to 0} \big(v(r) - \lambda \log r\big) = +\infty$$

and hence

$$\liminf_{\tau \to t} w(\tau) = 0, \ \limsup_{\tau \to t} w(\tau) = \infty.$$

As will be seen in Chapter 7, the motivation for nevertheless calling this weight "powerlike" comes from the spectral theory of Toeplitz operators. □

Example 3.27. We know from the discussion after Proposition 3.23 that we may find weights with arbitrarily prescribed indices of powerlikeness due to the choice of a sufficiently complicated curve carrying the weight. Proposition 3.25 yields the existence of such weights on every curve as in Example 3.24 and, in particular, on line segments. Indeed, if we let

$$g(y) = \lambda + \varepsilon \sin(\eta y)$$

then $\mu_t = \lambda - |\varepsilon|\sqrt{\eta^2 + 1}$ and $\nu_t = \lambda + |\varepsilon|\sqrt{\eta^2 + 1}$. □

   The following proposition is, in a sense, a fusion of Propositions 3.23 and
3.25. It beautifully illustrates the interaction between curve and weight.

Proposition 3.28. *Let Γ and w be as in Example 3.24. Then for every $\gamma \in \mathbf{C}$, the
function $V_t^0 \varphi_{t,\gamma} w$ is regular, the indices of powerlikeness of $\varphi_{t,\gamma} w$ at t are*

$$\mu_t := \alpha(V_t^0 \varphi_{t,\gamma} w) = \operatorname{Re} \gamma + A(\operatorname{Im} \gamma),$$
$$\nu_t := \beta(V_t^0 \varphi_{t,\gamma} w) = \operatorname{Re} \gamma + B(\operatorname{Im} \gamma)$$

with

$$A(\operatorname{Im} \gamma) = \liminf_{y \to +\infty} \Big(g(y) + g'(y) + \operatorname{Im} \gamma \big(h(y) + h'(y)\big)\Big), \qquad (3.85)$$

$$B(\operatorname{Im} \gamma) = \limsup_{y \to +\infty} \Big(g(y) + g'(y) + \operatorname{Im} \gamma \big(h(y) + h'(y)\big)\Big). \qquad (3.86)$$

The weight $\varphi_{t,\gamma} w$ belongs to $A_p(\Gamma)$ if and only if $-1/p < \mu_t \le \nu_t < 1/q$.

Proof. We have

$$\varphi_{t,\gamma}(\tau) w(\tau) = \exp\Big(\operatorname{Re} \gamma \log |\tau - t| - \operatorname{Im} \gamma \arg(\tau - t) + v(|\tau - t|)\Big),$$

and since $|\arg(\tau - t) - \varphi(|\tau - t|)| < 2\pi$ for $\tau \in \Gamma \setminus \{t\}$, it follows that $\varphi_{t,\gamma} w = c\psi$
where

$$\psi(\tau) = e^{F(|\tau - t|)}, \quad F(r) = \operatorname{Re} \gamma \log r - \operatorname{Im} \gamma \varphi(r) + v(r),$$

and c is some function in $GL^\infty(\Gamma)$ such that $\| \log c \|_\infty \le 2\pi$. Lemma 2.35 and
Theorem 2.36 with w replaced by ψ and with

$$f(y) = \operatorname{Re} \gamma + g(y) + \operatorname{Im} \gamma \, h(y)$$

show that $W_t \psi$ and $W_t^0 \psi$ are regular, that

$$\alpha(W_t^0 \psi) = \operatorname{Re} \gamma + A(\operatorname{Im} \gamma), \ \ \beta(W_t^0 \psi) = \operatorname{Re} \gamma + B(\operatorname{Im} \gamma)$$

with $A(\operatorname{Im} \gamma)$ and $B(\operatorname{Im} \gamma)$ as in (3.85), (3.86), and that $\psi \in A_p(\Gamma)$ if and only if

$$-1/p < \alpha(W_t^0 \psi) \le \beta(W_t^0 \psi) < 1/q.$$

By Lemma 3.16, $V_t^0 \psi$ is also regular and $\alpha(V_t^0 \psi) = \alpha(W_t^0 \psi)$, $\beta(V_t^0 \psi) = \beta(W_t^0 \psi)$.
From the definition of V_t^0 it is easily seen that

$$e^{-4\pi} \big| (V_t^0 \psi)(x) \big| \le \big| (V_t^0 c\psi)(x) \big| \le e^{4\pi} \big| (V_t^0 \psi)(x) \big|$$

for all $x \in (0, \infty)$, which implies that $V_t^0 c\psi$ is regular and that $\alpha(V_t^0 c\psi) = \alpha(V_t^0 \psi)$,
$\beta(V_t^0 c\psi) = \beta(V_t^0 \psi)$. This gives all assertions of the proposition. $\qquad\square$

3.7 Shape of the indicator functions

Given a bounded simple Carleson curve Γ and a Muckenhoupt weight $w \in A_p(\Gamma)$ $(1 < p < \infty)$, we defined the indicator function $\alpha_t, \beta_t : \mathbf{R} \to \mathbf{R}$ by

$$\alpha_t(x) := \alpha(V_t^0 \eta_t^x w) = \alpha(V_t^0 \varphi_{t,ix} w), \quad \beta_t(x) := \beta(V_t^0 \eta_t^x w) = \beta(V_t^0 \varphi_{t,ix} w)$$

for each point $t \in \Gamma$. Recall that, by Corollary 3.19, $V_t^0 \eta_t^x w$ is always regular. Thus, in accordance with (3.84), the indices of powerlikeness of w may be given by $\mu_t = \alpha_t(0)$ and $\nu_t = \beta_t(0)$. Conversely, we may also regard the values $\alpha_t(x)$ and $\beta_t(x)$ of the indicator functions at $x \in \mathbf{R}$ as the indices of powerlikeness of the weight $\eta_t^x w$.

The following result provides full information about the indicator functions in case Γ is "nice" at t.

Theorem 3.29. *Let Γ be a bounded simple Carleson curve and $w \in A_p(\Gamma)$. If the spirality indices of Γ at $t \in \Gamma$ coincide, $\delta_t^- = \delta_t^+ =: \delta_t$, then*

$$\alpha_t(x) = \mu_t + \delta_t x \quad and \quad \beta_t(x) = \nu_t + \delta_t x$$

where μ_t and ν_t are the indices of powerlikeness of w at t.

Proof. Consider the indicator function $\alpha_t(x) := \alpha(V_t^0 \eta_t^x w) = \alpha(V_t^0 \varphi_{t,ix} w)$. Since $W_t \varphi_{t,ix}$ is regular and $\alpha(W_t \varphi_{t,ix}) = \beta(W_t \varphi_{t,ix}) = \delta_t x$ by virtue of Proposition 3.1, it follows from Corollary 3.18 that $\alpha_t(x) = \alpha_t(V_t^0 w) + \delta_t x$ for all $x \in \mathbf{R}$. Taking into account that $\alpha(V_t^0 w) =: \mu_t$, we so obtain that $\alpha_t(x) = \mu_t + \delta_t x$ for all $x \in \mathbf{R}$. Analogously one can show that $\beta_t(x) = \nu_t + \delta_t x$ for all $x \in \mathbf{R}$. $\square$

The next theorem allows us to describe the indicator functions in the case where the weight w is "nice" at t.

Theorem 3.30. *Let Γ be a bounded simple Carleson curve and $w \in A_p(\Gamma)$. If the weight w is powerlike at $t \in \Gamma$, $\mu_t = \nu_t =: \lambda_t$, then*

$$\alpha_t(x) = \lambda_t + \min\{\delta_t^- x, \delta_t^+ x\} \quad and \quad \beta_t(x) = \lambda_t + \max\{\delta_t^- x, \delta_t^+ x\}$$

where δ_t^- and δ_t^+ are the spirality indices of Γ at t.

Proof. The function $W_t \varphi_{t,ix}$ is regular due to Proposition 3.1. Since $\alpha(V_t^0 w) =: \mu_t =: \lambda_t$, $\beta(V_t^0 w) =: \nu_t =: \lambda_t$, we obtain from Corollary 3.18 that $\alpha_t(x) = \lambda_t + \alpha(W_t \varphi_{t,ix})$. Therefore $\alpha_t(x) = \lambda_t + \min\{\delta_t^- x, \delta_t^+ x\}$ due to Proposition 3.1. A similar argument gives the assertion for $\beta_t(x)$. $\square$

In the general case, we already know the first three statements of the following theorem. The fourth statement is in fact the key to understanding the shape of the indicator functions. It reveals in particular why we denoted the indices of powerlikeness by μ_t and ν_t rather than by λ_t^- and λ_t^+.

Theorem 3.31. *Let Γ be a bounded simple Carleson curve, $t \in \Gamma$, and $w \in A_p(\Gamma)$ $(1 < p < \infty)$. Then the indicator functions α_t and β_t enjoy the following properties:*

(a) $-\infty < \alpha_t(x) \le \beta_t(x) < +\infty$ *for all $x \in \mathbf{R}$;*

(b) $-1/p < \alpha_t(0) \le \beta_t(0) < 1/q$;

(c) α_t *is concave and β_t is convex;*

(d) $\alpha_t(x)$ *and $\beta_t(x)$ have asymptotes as $x \to \pm\infty$ and the convex regions*

$$\{x + iy \in \mathbf{C} : y < \alpha_t(x)\} \quad and \quad \{x + iy \in \mathbf{C} : y > \beta_t(x)\}$$

may be separated by parallels to each of these asymptotes; to be more precise, there exist real numbers $\mu_t^-, \mu_t^+, \nu_t^-, \nu_t^+$ such that

$$-1/p < \mu_t^- \le \nu_t^- < 1/q, \quad -1/p < \mu_t^+ \le \nu_t^+ < 1/q$$

and

$$\begin{aligned}
\beta_t(x) &= \nu_t^+ + \delta_t^+ x + o(1) \quad as \ x \to +\infty, \\
\beta_t(x) &= \nu_t^- + \delta_t^- x + o(1) \quad as \ x \to -\infty, \\
\alpha_t(x) &= \mu_t^+ + \delta_t^+ x + o(1) \quad as \ x \to -\infty, \\
\alpha_t(x) &= \mu_t^- + \delta_t^- x + o(1) \quad as \ x \to +\infty,
\end{aligned}$$

where δ_t^- and δ_t^+ are the spirality indices of Γ at t.

Proof. Property (a) results from Corollary 3.19 and Theorem 1.13, property (b) is a consequence of Theorem 3.21 and the fact that N_t contains the origin, and property (c) follows from Proposition 3.20. We are so left with proving property (d).

Fix $x > 0$. From Lemma 3.17 we infer that

$$\alpha(V_t^0 w) + \beta(W_t \eta_t^x) \le \beta(V_t^0 \eta_t^x w) \le \beta(V_t^0 w) + \beta(W_t \eta_t^x).$$

Since $\delta_t^- \le \delta_t^+$ and $x > 0$, equality (3.6) shows that $\beta(W_t \eta_t^x) = \delta_t^+ x$, and since $\beta_t(x) := \beta(V_t^0 \eta_t^x w)$, we obtain that

$$\alpha(V_t^0 w) + \delta_t^+ x \le \beta_t(x) \le \beta(V_t^0 w) + \delta_t^+ x \quad \text{for } x > 0. \tag{3.87}$$

Taking into account that β_t is convex, we conclude from (3.87) that $\beta_t(x)$ has an asymptote of slope δ_t^+ as $x \to +\infty$. Let $y = \nu_t^+ + \delta_t^+ x$ be the equation of this asymptote. Then

$$\nu_t^+ + \delta_t^+ x \le \beta_t(x) \le \nu_t^+ + \delta_t^+ x + o_1(x) \quad \text{for } x > 0 \tag{3.88}$$

where $o_1(x) \to 0$ as $x \to +\infty$. From (3.88) and property (b) we get $\nu_t^+ \le \beta_t(0) < 1/q$. In the same way we may derive from Lemma 3.17 and (3.5) that

$$\mu_t^+ + \delta_t^+ x - o_2(x) \le \alpha_t(x) \le \mu_t^+ + \delta_t^+ x \quad \text{for } x < 0 \tag{3.89}$$

with $\mu_t^+ \ge \alpha_t(0) > -1/p$ and $o_2(x) \to 0$ as $x \to -\infty$.

It remains to show that $\mu_t^+ \leq \nu_t^+$. Assume $\mu_t^+ > \nu_t^+ + 2\varepsilon$ with $\varepsilon > 0$. By (3.88) and (3.89),

$$\beta_t(x) \leq \nu_t^+ + \delta_t^+ x + o_1(x), \quad \alpha_t(-x) \geq \mu_t^+ - \delta_t^+ x - o_2(-x)$$

for every $x > 0$. Hence, there exists an $x > 0$ such that

$$\beta_t(x) \leq \nu_t^+ + \delta_t^+ x + \varepsilon, \quad \alpha_t(-x) \geq \mu_t^+ - \delta_t^+ x - \varepsilon,$$

whence

$$\beta_t(x) - \alpha_t(-x) \leq \nu_t^+ + \delta_t^+ x + \varepsilon - \mu_t^+ + \delta_t^+ x + \varepsilon < 2\delta_t^+ x$$

and thus

$$\beta(V_t^0 \eta_t^x w) - \alpha(V_t^0 \eta_t^{-x} w) < 2\delta_t^+ x. \tag{3.90}$$

Since $\eta_t^x w = \eta_t^{2x}(\eta_t^{-x} w)$, we obtain from Lemma 3.17 with η_t^{2x} and $\eta_t^{-x} w$ in place of φ and w, respectively, that

$$\alpha(V_t^0 \eta_t^{-x} w) + \beta(W_t \eta_t^{2x}) \leq \beta(V_t^0 \eta_t^x w). \tag{3.91}$$

The inequalities (3.90) and (3.91) imply that $\beta(W_t \eta_t^{2x}) < 2\delta_t^+ x$. This, however, is impossible, because $\beta(W_t \eta_t^{2x}) = \beta(W_t \varphi_{t,ix}) = 2\delta_t^+ x$ by Proposition 3.1. This contradiction shows that $\mu_t^+ \leq \nu_t^+$.

The remaining assertion of (d) can be proved analogously.   □

We remark that Theorem 3.31 remains valid with α_t^* and β_t^* in place of α_t and β_t, respectively.

Now recall that our primary goal is the description of the indicator set

$$N_t := N_t(\Gamma, p, w) = \left\{ y + ix \in \mathbf{C} : -\frac{1}{p} - \alpha_t(x) < y < \frac{1}{q} - \beta_t(x) \right\}.$$

This set is nonempty, and Theorem 3.31(c) tells us that N_t is convex and open. For a nonempty convex open subset N of $\mathbf{C}$ and a number $x \in \mathbf{R}$, we denote by $N(x)$ the intersection of N with the horizontal straight line $\{y + ix \in \mathbf{C} : y \in \mathbf{R}\}$ through the point ix. The set $N(x)$ is a (possibly empty) interval, and we denote by $|N(x)| \in [0, \infty]$ the length of this interval. The *width* of N is defined as $\sup_{x \in \mathbf{R}} |N(x)|$. The open set between two parallel straight lines of the complex plane is called an *open stripe*.

Theorem 3.32. *Let Γ be a bounded simple Carleson curve, $t \in \Gamma$, and $w \in A_p(\Gamma)$ $(1 < p < \infty)$. Denote by δ_t^-, δ_t^+ the spirality indices of Γ at t and by μ_t, ν_t the indices of powerlikeness of w at t.*

(a) If $\delta_t^- = \delta_t^+$, then $N_t(\Gamma, p, w)$ is an open stripe of width at most 1 containing the origin. Given $p \in (1, \infty)$ and any open stripe of width at most 1 which contains

the origin, there exists a Carleson Jordan curve Γ *spiralic at some point* $t \in \Gamma$ *and a weight* $w \in A_p(\Gamma)$ *such that* $N = N_t(\Gamma, p, w)$.

(b) *If* $\mu_t = \nu_t$, *then* $N_t(\Gamma, p, w)$ *is an open stripe of width equal to 1 containing the origin in case* $\delta_t^- = \delta_t^+$ *and an open parallelogram of width equal to 1 containing the origin in case* $\delta_t^- < \delta_t^+$. *Given* $p \in (1, \infty)$ *and any open stripe or parallelogram* N *of width equal to 1 containing the origin, there exists a Carleson Jordan curve* Γ *and a weight* $w \in A_p(\Gamma)$ *powerlike at some point* $t \in \Gamma$ *such that* $N = N_t(\Gamma, p, w)$.

Proof. (a) From Theorem 3.29 we know that

$$N_t = \{y + ix \in \mathbf{C} : -1/p - \mu_t - \delta_t x < y < 1/q - \nu_t - \delta_t x\},$$

and since $-1/p < \mu_t \le \nu_t < 1/q$, we get the first assertion. The second assertion follows from Example 3.27.

(b) This can be shown as part (a) using Theorem 3.30 as well as Propositions 1.19 and 3.23.   □

Thus, everything is clear if $\delta_t^- = \delta_t^+$ or $\mu_t = \nu_t$. So suppose $\delta_t^- < \delta_t^+$. Then Theorem 3.31(c) says that $\beta_t(x) - \alpha_t(x)$ is a convex function of x which, by Theorem 3.31(b), is less than 1 at $x = 0$ and, by Theorem 3.31(d), increases to infinity as $x \to \pm\infty$. Consequently, the equation

$$\beta_t(x) - \alpha_t(x) = 1 \tag{3.92}$$

has exactly two solutions, $x_t^- \in (-\infty, 0)$ and $x_t^+ \in (0, \infty)$. From the remark after Theorem 3.21 we know actually

$$N_t = \left\{y + ix \in \mathbf{C} : x_t^- < x < x_t^+, \ -\frac{1}{p} - \alpha_t(x) < y < \frac{1}{q} - \beta_t(x)\right\}$$

and that therefore we may restrict ourselves to looking at the values of $\alpha_t(x)$ and $\beta_t(x)$ for $x \in (x_t^-, x_t^+)$. Denote by $\Pi_t := \Pi_t(\Gamma, p, w)$ the closed parallelogram whose vertices are

$$x_t^- + i\left(\frac{1}{p} + \beta_t(x_t^-)\right), \ x_t^+ + i\left(\frac{1}{p} + \beta_t(x_t^+)\right), \tag{3.93}$$

$$x_t^- + i\left(\frac{1}{p} + \alpha_t(x_t^-)\right), \ x_t^+ + i\left(\frac{1}{p} + \alpha_t(x_t^+)\right). \tag{3.94}$$

By Theorem 3.31(c), the graphs

$$\left\{x + iy \in \mathbf{C} : x_t^- \le x \le x_t^+, \ y = \frac{1}{p} + \alpha_t(x)\right\},$$

$$\left\{x + iy \in \mathbf{C} : x_t^- \le x \le x_t^+, \ y = \frac{1}{p} + \beta_t(x)\right\}$$

are contained in the parallelogram Π_t.

Theorem 3.33. *If the equation* (3.92) *has exactly two solutions* $x_t^- \in (-\infty, 0)$ *and* $x_t^+ \in (0, \infty)$, *then the indicator functions* α_t *and* β_t *possess the following properties:*

(P$_1$) $\beta_t(x_t^-) - \alpha_t(x_t^-) = 1$, $\beta_t(x_t^+) - \alpha_t(x_t^+) = 1$;

(P$_2$) $-1/p < \alpha_t(0) \le \beta_t(0) < 1/q$;

(P$_3$) α_t *is concave and* β_t *is convex on* $[x_t^-, x_t^+]$;

(P$_4$) *each diagonal of the parallelogram* Π_t *spanned by the four points* (3.93), (3.94) *separates the convex regions*

$$\left\{ x + iy \in \mathbf{C} : x_t^- < x < x_t^+, \ y < 1/p + \alpha_t(x) \right\}$$

and

$$\left\{ x + iy \in \mathbf{C} : x_t^- < x < x_t^+, \ y > 1/p + \beta_t(x) \right\}.$$

Proof. Immediate from Theorem 3.31. $\square$

We say that a subset N of the plane $\mathbf{C}$ is *narrow* if there are two open stripes S_1 and S_2 of width at most 1 such that N is contained in $\Pi := S_1 \cap S_2$ and

$$\inf_{y+ix\in\Pi} x = \inf_{y+ix\in N} x, \quad \sup_{y+ix\in\Pi} x = \sup_{y+ix\in N} x.$$

Clearly, open stripes of width at most 1 are narrow, and these are the only unbounded open and narrow sets. A bounded narrow set has necessarily two "peaks" at opposite vertices of the parallelogram $\Pi = S_1 \cap S_2$. In particular, ellipses or any regions with smooth boundary are never narrow.

Corollary 3.34. *The indicator set* N_t *is always an open, convex, narrow set containing the origin.*

Proof. This is a simple consequence of Theorems 3.33 and 3.21. $\square$

3.8 Indicator functions of prescribed shape

The purpose of this section is to show that Theorem 3.33 and Corollary 3.34 are sharp, i.e. to show that any pair of functions with the properties (P$_1$) to (P$_4$) is the pair of indicator functions for some Carleson curve as in Example 3.24 and of some Muckenhoupt weight on this curve.

Example 3.35. Theorem 3.30 gives us the indicator functions for an arbitrary bounded simple Carleson curve and a powerlike weight. We know from the remark after Proposition 3.23 that $\varphi_{t,\gamma}$ is in general not powerlike. So let us try our luck with the indicator functions of $\varphi_{t,\gamma}$. By Proposition 3.23,

$$\alpha_t(x) \ := \ \alpha(V_t^0 \eta_t^x \varphi_{t,\gamma}) = \alpha(V_t^0 \varphi_{t,\gamma+ix})$$
$$= \ \operatorname{Re}\gamma + \min\left\{ \delta_t^-(\operatorname{Im}\gamma + x), \ \delta_t^+(\operatorname{Im}\gamma + x) \right\}$$

and analogously,

$$\beta_t(x) = \operatorname{Re}\gamma + \max\left\{\delta_t^-(\operatorname{Im}\gamma + x),\ \delta_t^+(\operatorname{Im}\gamma + x)\right\}.$$

Thus, the functions α_t and β_t differ from the functions in Theorem 3.30 only in a horizontal shift and do not supply us with indicator functions of a qualitatively new shape. □

We now show that curves and weights as in Example 3.24 produce arbitrary indicator functions. Thus, throughout the rest of this section let Γ and w be as in Example 3.24. Proposition 3.25 says that $w \in A_p(\Gamma)$ $(1 < p < \infty)$ if and only if

$$-1/p < \liminf_{y\to+\infty}\left(g(y) + g'(y)\right) \le \limsup_{y\to+\infty}\left(g(y) + g'(y)\right) < 1/q. \tag{3.95}$$

From Proposition 3.28 we obtain

$$
\begin{aligned}
\alpha_t(x) &:= \alpha(V_t^0\eta_t^x w) = \alpha(V_t^0\varphi_{t,ix}w) = A(x) \\
&= \liminf_{y\to+\infty}\left(g(y) + g'(y) + x\left(h(y) + h'(y)\right)\right), \tag{3.96}\\
\beta_t(x) &:= \beta(V_t^0\eta_t^x w) = \beta(V_t^0\varphi_{t,ix}w) = B(x) \\
&= \limsup_{y\to+\infty}\left(g(y) + g'(y) + x\left(h(y) + h'(y)\right)\right). \tag{3.97}
\end{aligned}
$$

Our aim is to choose g and h so that $\alpha_t(x)$ and $\beta_t(x)$ have prescribed values (satisfying the conditions $(P_1) - (P_4)$ of Theorem 3.33) on (x_t^-, x_t^+), where (x_t^-, x_t^+) is the maximal connected open subset on which $\beta_t(x) - \alpha_t(x) < 1$.

We first need a function $\chi(y)$ for which $\chi(y) + \chi'(y)$ has prescribed maximum and minimum and which, dictated by the construction that will follow, is subject to some constraints.

Lemma 3.36. *Given real numbers $c > 0, d > 0, a \in \mathbf{R}$, there exists a function $\chi \in C^2[0,1]$ with support in $(0,1)$ such that*

$$\max_{y\in[0,1]}\left(\chi(y) + \chi'(y)\right) = c, \quad \min_{y\in[0,1]}\left(\chi(y) + \chi'(y)\right) = -d \tag{3.98}$$

and

$$\|\chi\|_\infty \le 1, \ \|a\chi\|_\infty \le 1, \ \|\chi'\|_\infty \le 2\max\{c,d\}, \tag{3.99}$$

$$\|\chi''\|_\infty \le 100\max\{c^2, d^2, |a|c^2, |a|d^2, c, d, d^2/c, c^2/d\}. \tag{3.100}$$

Proof. It suffices to prove the lemma for $a \ne 0$. Choose any function $\psi \in C^2[0,\infty)$ such that

$$\psi(x) = 0 \text{ for } x \in [0, 1/6], \ \psi(x) = 1 \text{ for } x \in [2/6, \infty),$$

$$\psi(x) \text{ increases monotonically from } 0 \text{ to } 1 \text{ on } [1/6, 2/6],$$

and $\|\psi'\|_\infty = 7$. So $\psi'(x)$ goes from 0 to 7 and then from 7 to 0 on $[1/6, 2/6]$, and a little thought shows that we may choose ψ so that $\|\psi''\|_\infty < 85$. Put

$$\varepsilon := \min\left\{1, \frac{1}{|a|}, \frac{c}{14}, \frac{d}{14}\right\}. \tag{3.101}$$

For $\lambda \geq 1$, define $\psi_{\varepsilon,\lambda}(x) := \varepsilon\psi(\lambda x)$. The function

$$m(\lambda) := \max_{x\in[0,1/2]} \left(\psi_{\varepsilon,\lambda}(x) + \psi'_{\varepsilon,\lambda}(x)\right) = \varepsilon \max_{x\in[0,1/2]} \left(\psi(\lambda x) + \lambda\psi'(\lambda x)\right)$$

is continuous, we have $m(1) \leq 8\varepsilon < c$ and $m(\lambda) \geq 7\varepsilon\lambda$. Hence, there exists a $\lambda_1 > 1$ such that $m(\lambda_1) = c$. The inequality $m(\lambda) \geq 7\varepsilon\lambda$ implies that

$$c \geq 7\varepsilon\lambda_1. \tag{3.102}$$

From (3.101) and (3.102) we obtain that

$$\lambda_1 \leq \frac{c}{7\varepsilon} = \frac{1}{7}\max\left\{c, |a|c, 14, \frac{14c}{d}\right\}. \tag{3.103}$$

Put $\chi(x) := \psi_{\varepsilon,\lambda_1}(x)$ for $x \in [0, 1/2]$. Then

$$\min_{x\in[0,1/2]} \left(\chi(x) + \chi'(x)\right) = 0, \quad \max_{x\in[0,1/2]} \left(\chi(x) + \chi'(x)\right) = c. \tag{3.104}$$

By (3.101) – (3.103), the following estimates hold on $[0, 1/2]$:

$$\|\chi\|_\infty = \varepsilon \leq 1, \quad \|a\chi\|_\infty = |a|\varepsilon \leq 1,$$
$$\|\chi'\|_\infty = \varepsilon\lambda_1\|\psi'\|_\infty = 7\varepsilon\lambda_1 \leq c < 2c,$$
$$\|\chi''\|_\infty = \varepsilon\lambda_1^2\|\psi''\|_\infty < 85\,\varepsilon\lambda_1\lambda_1$$
$$\leq\ 85\,\frac{c}{7}\frac{1}{7}\max\left\{c, |a|c, 14, \frac{14c}{d}\right\} < 28\max\left\{c^2, |a|c^2, c, \frac{c^2}{d}\right\}.$$

Now define $\tilde\psi(x) := 1 - \psi(x)$ and $\tilde\psi_{\varepsilon,\lambda}(x) := \varepsilon\tilde\psi(\lambda x)$. Again

$$\tilde m(\lambda) := \min_{x\in[0,1/2]} \left(\tilde\psi_{\varepsilon,\lambda}(x) + \tilde\psi'_{\varepsilon,\lambda}(x)\right) = \varepsilon \min_{x\in[0,1/2]} \left(\tilde\psi(\lambda x) + \lambda\tilde\psi'(\lambda x)\right)$$

is continuous, $\tilde m(1) \geq -7\varepsilon > -d$ and $\tilde m(\lambda) \leq \varepsilon - 7\varepsilon\lambda$ for $\lambda \geq 1$, so that there is a $\lambda_2 > 1$ satisfying $\tilde m(\lambda_2) = -d$. We have $-d = \tilde m(\lambda_2) \leq \varepsilon - 7\varepsilon\lambda_2$ and thus,

$$7\varepsilon\lambda_2 \leq \varepsilon + d < 2d. \tag{3.105}$$

Combining (3.101) and (3.105) we get

$$\lambda_2 \leq \frac{2d}{7\varepsilon} = \frac{2}{7}\max\left\{d, |a|d, \frac{14d}{c}, 14\right\}. \tag{3.106}$$

Put $\chi(x) := \tilde{\psi}_{\varepsilon,\lambda_2}(x - 1/2)$ for $x \in [1/2, 1]$. Then

$$\min_{x \in [1/2,1]} \left(\chi(x) + \chi'(x)\right) = -d, \quad \max_{x \in [1/2,1]} \left(\chi(x) + \chi'(x)\right) = \varepsilon \leq c. \qquad (3.107)$$

From (3.104) and (3.107) we obtain (3.98). Taking into account (3.101), (3.105), (3.106) we see that on $[1/2, 1]$ the following estimates are valid:

$$
\begin{aligned}
\|\chi\|_\infty &= \varepsilon \leq 1, \ \|a\chi\|_\infty = |a|\varepsilon \leq 1, \\
\|\chi'\|_\infty &= \varepsilon\lambda_2\|\psi'\|_\infty = 7\varepsilon\lambda_2 < 2d, \\
\|\chi''\|_\infty &= \varepsilon\lambda_2^2|\psi''\|_\infty < 85\,\varepsilon\lambda_2\lambda_2 \\
&\leq 85\,\frac{2d}{7}\frac{2}{7}\max\left\{d, |a|d, \frac{14d}{c}, 14\right\} < 100\max\left\{d^2, |a|d^2, \frac{d^2}{c}, d\right\}.
\end{aligned}
$$

We so have defined χ on $[0, 1/2] \cup [1/2, 1] = [0, 1]$ and have shown that χ possesses all the properties required. $\qquad \square$

Put $I_n := [n, n+1)$, and let R_j $(j = 1, \ldots, N)$ be the union of the sets in the jth row of the matrix

$$
\begin{pmatrix}
I_0 & I_N & I_{2N} & \cdots \\
I_1 & I_{N+1} & I_{2N+1} & \cdots \\
\cdots & \cdots & \cdots & \cdots \\
I_{N-1} & I_{2N-1} & I_{3N-1} & \cdots
\end{pmatrix}.
$$

Clearly, $R_i \cap R_j = \emptyset$ for $i \neq j$ and $\bigcup_j R_j = [0, \infty)$. For each j, choose real numbers $c_j > 0$, $d_j > 0$, $a_j \in \mathbf{R}$, denote by $\chi_j \in C^2[0, 1]$ the corresponding function from Lemma 3.36 with $c = c_j$, $d = d_j$, $a = a_j$, and extend χ_j periodically (with period 1) to all of $\mathbf{R}$. Finally, pick numbers $\lambda \in \mathbf{R}$, $\delta \in \mathbf{R}$ and define

$$g(y) := \begin{cases} \lambda + a_j\chi_j(y) & \text{for } y \in R_j \\ \lambda & \text{for } y < 0 \end{cases}, \quad h(y) := \begin{cases} \delta + \chi_j(y) & \text{for } y \in R_j \\ \delta & \text{for } y < 0 \end{cases}. \qquad (3.108)$$

From Lemma 3.36 we infer that g and h are in $C^2(\mathbf{R})$ and that g, g', g'', h, h', h'' are bounded on $\mathbf{R}$. It is easy to compute (3.96) and (3.97):

$$
\begin{aligned}
\alpha_t(x) &:= \liminf_{y \to +\infty} \left(g(y) + g'(y) + x\big(h(y) + h'(y)\big)\right) \\
&= \inf_j \inf_{y \in R_j} \left(g(y) + g'(y) + x\big(h(y) + h'(y)\big)\right) \\
&= \inf_j \inf_{y \in [0,1]} \left(\lambda + a_j\chi_j(y) + a_j\chi_j'(y) + x\big(\delta + \chi_j(y) + \chi_j'(y)\big)\right) \\
&= \lambda + \delta x + \inf_j \min\{c_j a_j + c_j x, -d_j a_j - d_j x\} \qquad (3.109)
\end{aligned}
$$

and analogously,

$$\beta_t(x) = \lambda + \delta x + \sup_j \max\{c_j a_j + c_j x, -d_j a_j - d_j x\}. \qquad (3.110)$$

Thus, α_t and β_t are continuous and piecewise linear functions of a more general shape than in Theorems 3.29 and 3.30. One can indeed show that if α_t and β_t are any continuous and piecewise linear functions with a finite number of "corners", if $x_t^- \in (-\infty, 0)$, $x_t^+ \in (0, \infty)$ are any given numbers, and if α_t and β_t enjoy the properties $(P_1) - (P_4)$ of Theorem 3.33, then there are numbers $c_j > 0$, $d_j > 0$, $a_j \in \mathbf{R}$ $(j = 1, \ldots, N$; N sufficiently large) such that α_t and β_t may be represented by (3.109) and (3.110) for $x \in (x_t^-, x_t^+)$. Since the proof of this fact is not essentially simpler than the proof of the more general Theorem 3.37, which will follow below, we leave it aside.

Convex piecewise linear functions with only finitely many "corners" may be represented as the maximum of a finite number of affine linear functions. An arbitrary convex function is the maximum of at most countably many affine linear functions. Thus, let us modify the above construction as follows.

Let again $I_n := [n, n+1)$, but denote now by R_j $(j = 1, 2, 3, \ldots)$ the union of the sets in the jth row of the matrix

$$\begin{pmatrix} I_0 & I_1 & I_3 & I_6 & \cdots \\ I_2 & I_4 & I_7 & \cdots & \cdots \\ I_5 & I_8 & \cdots & \cdots & \cdots \\ I_9 & \cdots & \cdots & \cdots & \cdots \\ \cdots & \cdots & \cdots & \cdots & \cdots \end{pmatrix}.$$

Obviously, $R_i \cap R_j = \emptyset$ for $i \neq j$ and $\bigcup_{j=0}^{\infty} R_j = [0, \infty)$. Then choose $c_j > 0$, $d_j > 0$, $a_j \in \mathbf{R}$ for each $j \in \{1, 2, 3, \ldots\}$, construct the corresponding χ_j, and define g, h by (3.108). It is clear that $g, h \in C^2(\mathbf{R})$, and Lemma 3.36 implies that g, g', g'', h, h', h'' are bounded on $\mathbf{R}$ whenever

$$\sup_j \max \left\{ c_j, d_j, \frac{d_j}{c_j}, \frac{c_j}{d_j}, |a_j|c_j, |a_j|d_j \right\} < \infty. \tag{3.111}$$

Moreover, if (3.111) holds, then the equalities (3.109) and (3.110) remain literally true. Letting $x = 0$ in these equalities, we see that (3.95) is equivalent to the requirement

$$-1/p < \lambda + \inf_j \min\{c_j a_j, -d_j a_j\} \leq \lambda + \sup_j \max\{c_j a_j, -d_j a_j\} < 1/q. \tag{3.112}$$

Note that (3.95) and (3.112) are nothing but property (P_2) in Theorem 3.33.

Theorem 3.37. *Given any number $p \in (1, \infty)$, any numbers $x_t^- \in (-\infty, 0)$, $x_t^+ \in (0, \infty)$, and any functions α_t and β_t on $[x_t^-, x_t^+]$ with the properties (P_1) to (P_4) of Theorem 3.33, there exist a Carleson curve Γ and a weight $w \in A_p(\Gamma)$ as in Example 3.24 such that the equation $\beta(V_t^0 \eta_t^x w) - \alpha(V_t^0 \eta_t^x w) = 1$ has exactly the solutions x_t^- and x_t^+ and α_t, β_t are the restrictions to $[x_t^-, x_t^+]$ of the indicator functions of Γ, p, w at the point t.*

Proof. Consider the parallelogram **P** with the vertices

$$x_t^- + i\alpha_t(x_t^-), \ x_t^- + i\beta_t(x_t^-), \ x_t^+ + i\alpha_t(x_t^+), \ x_t^+ + i\beta_t(x_t^+)$$

and let the diagonals of this parallelogram have the equations $y = \mu_1 + \delta_1 x$ and $y = \mu_2 + \delta_2 x$. We have $\mathbf{P} = -i/p + \Pi_t$ where Π_t is the parallelogram whose vertices are (3.93), (3.94). Thus, by (P$_4$) and (P$_2$),

$$-1/p < \alpha_t(0) \le \mu_1 \le \beta_t(0) < 1/q, \ -1/p < \alpha_t(0) \le \mu_2 \le \beta_t(0) < 1/q. \quad (3.113)$$

Without loss of generality assume that $\delta_1 < \delta_2$. We extend α_t and β_t to functions α and β on all **R** as follows:

$$\alpha(x) := \begin{cases} \mu_2 + \delta_2 x, & x \in (-\infty, x_t^-) \\ \alpha_t(x), & x \in [x_t^-, x_t^+] \\ \mu_1 + \delta_1 x, & x \in (x_t^+, \infty) \end{cases} \quad \beta(x) := \begin{cases} \mu_1 + \delta_1 x, & x \in (-\infty, x_t^-) \\ \beta_t(x), & x \in [x_t^-, x_t^+] \\ \mu_2 + \delta_2 x, & x \in (x_t^+, \infty) \end{cases}.$$

By (P$_3$) and (P$_4$), α is concave, β is convex, and the straight lines given by $y = \mu_1 + \delta_1 x$ and $y = \mu_2 + \delta_2 x$ separate the convex regions $\{x + iy \in \mathbf{C} : y < \alpha(x)\}$ and $\{x + iy \in \mathbf{C} : y > \beta(x)\}$.

Suppose for a moment that we have numbers $c_j > 0$, $d_j > 0$, $a_j \in \mathbf{R}$ ($j = 1, 2, 3, \ldots$) and numbers $\lambda \in \mathbf{R}$, $\delta \in \mathbf{R}$ satisfying (3.111) and (3.112) and such that

$$\alpha(x) = \lambda + \delta x + A(x), \ \beta(x) = \lambda + \delta x + B(x) \quad (3.114)$$

for all $x \in \mathbf{R}$ where

$$A(x) := \inf_j \min\{c_j a_j + c_j x, -d_j a_j - d_j x\},$$
$$B(x) := \sup_j \max\{c_j a_j + c_j x, -d_j a_j - d_j x\}.$$

Define g, h and thus Γ, w as above. From (3.111) and (3.112) we deduce that Γ is a Carleson curve and $w \in A_p(\Gamma)$, and from (3.109), (3.110), (3.114) we conclude that α and β are the indicator functions of Γ, p, w at t, i.e. $\alpha(x) = \alpha(V_t^0 \eta_t^x w)$ and $\beta(x) = \beta(V_t^0 \eta_t^x w)$. Since, by (P$_1$)–(P$_3$), the equation

$$\beta(V_t^0 \eta_t^x w) - \alpha(V_t^0 \eta_t^x w) = \beta(x) - \alpha(x) = 1$$

has exactly the solutions x_t^- and x_t^+ and since $\alpha(x) = \alpha_t(x)$ and $\beta(x) = \beta_t(x)$ for $x \in [x_t^-, x_t^+]$, the proof is complete.

So let us construct the numbers $c_j, d_j, a_j, \lambda, \delta$. Denote by $x_0 + iy_0$ the intersection of the diagonals of the parallelogram **P**. From (P$_3$) and (P$_4$) we see that $\alpha(x_0) = y_0 = \beta(x_0)$ happens if and only if

$$\alpha(x) = \min\{\mu_1 + \delta_1 x, \mu_2 + \delta_2 x\}, \ \beta(x) = \max\{\mu_1 + \delta_1 x, \mu_2 + \delta_2 x\}. \quad (3.115)$$

Choose any $\delta \in (\delta_1, \delta_2)$, put

$$c = \delta_2 - \delta, \ d = \delta - \delta_1, \ a = \frac{\mu_2 - \mu_1}{c + d}, \ \lambda = \frac{c\mu_1 + d\mu_2}{c + d},$$

and let $c_j = c$, $d_j = d$, $a_j = a$ for all j. Then $c_j > 0$, $d_j > 0$, an easy computation using (3.115) shows that (3.114) holds, and (3.111) is obviously satisfied. Since $\lambda + ca = \mu_2$ and $\lambda - da = \mu_1$, we obtain from (3.113) that (3.112) is also satisfied. Thus, the case where the graphs of $y = \alpha(x)$ and $y = \beta(x)$ have the point $x_0 + iy_0$ in common is settled.

Suppose $\alpha(x_0) < \beta(x_0)$. Then there is a line $y = \lambda + \delta x$ which separates the two convex sets $\{x + iy \in \mathbf{C} : y \geq \beta(x)\}$ and $\{x + iy \in \mathbf{C} : y \leq \alpha(x)\}$. Thus, if we put

$$\alpha_0(x) := \alpha(x) - \delta x - \lambda, \ \beta_0(x) := \beta(x) - \delta x - \lambda,$$

then

$$\alpha^* := \sup_{x \in \mathbf{R}} \alpha_0(x) < 0 < \inf_{x \in \mathbf{R}} \beta_0(x) =: \beta^*. \tag{3.116}$$

Clearly, α_0 is concave and β_0 is convex. Moreover, we have $\delta_1 < \delta < \delta_2$ and hence, $c := \delta_2 - \delta > 0$ and $d := \delta - \delta_1 > 0$. Obviously,

$$\mu_2 - \lambda + cx \leq \beta_0(x) \leq \mu_2 - \lambda + cx + o_+(1),$$
$$\mu_1 - \lambda - dx \leq \beta_0(x) \leq \mu_1 - \lambda - dx + o_-(1),$$
$$\mu_1 - \lambda - dx \geq \alpha_0(x) \geq \mu_1 - \lambda - dx - o_+(1),$$
$$\mu_2 - \lambda + cx \geq \alpha_0(x) \geq \mu_2 - \lambda + cx - o_-(1),$$

where $o_\pm(1)$ denote nonnegative functions which vanish in a neighborhood of $\pm\infty$. Put

$$\xi_1 := -(\lambda - \mu_1)/d, \ \xi_2 := (\lambda - \mu_2)/c,$$

i.e. let ξ_1 and ξ_2 be the points at which the lines $y = \mu_1 - \lambda - dx$ and $y = \mu_2 - \lambda + cx$ meet the real axis.

The set $\mathbf{Q}$ of all rational numbers is countable. Let $\{a_j\}_{j=1}^\infty = \mathbf{Q}$. We construct the sequences $\{c_j\}_{j=1}^\infty$ and $\{d_j\}_{j=1}^\infty$ as follows:

for $-\infty < -a_j \leq \xi_2$ (resp. $\xi_2 < -a_j < +\infty$) let $c_j > 0$ be the number defined by requiring that $y \geq c_j a_j + c_j x$ (resp. $y \leq c_j a_j + c_j x$) is a supporting half-plane to $y \geq \beta_0(x)$ (resp. $y \leq \alpha_0(x)$);

for $-\infty < -a_j \leq \xi_1$ (resp. $\xi_1 < -a_j < +\infty$) define $d_j > 0$ as the number for which $y \leq -d_j a_j - d_j x$ (resp. $y \geq -d_j a_j - d_j x$) is a supporting half-plane to $y \leq \alpha_0(x)$ (resp. $y \geq \beta_0(x)$).

It is clear that $0 < c_j \leq c$ for all j and that $c_j \to c$ as $-a_j \to \xi_2$. This proves that $\sup_j c_j = c < \infty$. Analogously, $0 < d_j \leq d$ for all j, $d_j \to d$ as $-a_j \to \xi_j$, and thus $\sup_j d_j = d < \infty$. The lines $y = c_j a_j + c_j x$ always meet the imaginary axis between $i\alpha_0(0)$ and $i\beta_0(0)$, whence, by (3.113),

$$-1/p < \alpha_t(0) = \alpha(0) = \lambda + \alpha_0(0)$$
$$\leq \lambda + c_j a_j \leq \lambda + \beta_0(0) = \beta(0) = \beta_t(0) < 1/q. \tag{3.117}$$

Equally,
$$-1/p < \alpha_t(0) \leq \lambda - d_j a_j \leq \beta_t(0) < 1/q, \tag{3.118}$$

which shows that (3.112) is satisfied. From (3.117) and (3.118) we also obtain that $\sup_j |a_j| c_j < \infty$ and $\sup_j |a_j| d_j < \infty$. Thus, the proof of (3.111) will be complete once we have shown that $\sup_j (d_j/c_j) < \infty$ and $\sup_j (c_j/d_j) < \infty$. We now prove these estimates.

If $-a_j \to -\infty$, then $c_j a_j$, the imaginary part of the intersection of the line $y = c_j a_j + c_j x$ with the imaginary axis, goes to the minimal value $\beta^* := \inf_{x \in \mathbf{R}} \beta_0(x)$ of β_0. A similar argument shows that $-d_j a_j$ converges to $\alpha^* := \sup_{x \in \mathbf{R}} \alpha_0(x)$ as $-a_j \to -\infty$. It follows from (3.116) that

$$\frac{d_j}{c_j} = \frac{d_j a_j}{c_j a_j} \to \frac{-\alpha^*}{\beta^*} > 0 \text{ as } -a_j \to -\infty.$$

In the same manner one can show that

$$\frac{d_j}{c_j} = \frac{d_j a_j}{c_j a_j} \to \frac{\beta^*}{-\alpha^*} > 0 \text{ as } -a_j \to +\infty.$$

Hence, there exists a rational number $M := a_{j_1} > \max\{|\xi_1|, |\xi_2|\}$ such that

$$\frac{d_j}{c_j} < 2 \max\left\{\frac{-\alpha^*}{\beta^*}, \frac{\beta^*}{-\alpha^*}\right\} \text{ whenever } |a_j| \geq M. \tag{3.119}$$

Let $a_{j_2} := -M$. If $-a_j$ changes from $-a_j = -a_{j_1} = -M$ to $-a_j = -a_{j_2} = +M$, then c_j first increases from c_{j_1} to c and then decreases from c to c_{j_2}. Consequently, for $|a_j| < M$ we have $c_j \geq \min\{c_{j_1}, c_{j_2}\}$ and thus

$$\frac{d_j}{c_j} \leq d \max\left\{\frac{1}{c_{j_1}}, \frac{1}{c_{j_2}}\right\} \text{ whenever } |a_j| < M. \tag{3.120}$$

Combining (3.119) and (3.121) we get $\sup_j (d_j/c_j) < \infty$. In the same way one can show that $\sup_j (c_j/d_j) < \infty$. At this point the proof of (3.111) and (3.112) is complete.

We now show that $\beta_0(x) = B(x)$ for all $x \in \mathbf{R}$. Since $\beta_0(x) \geq c_j a_j + c_j x$ and $\beta_0(x) \geq -d_j a_j - d_j x$ for all j and all x, we have

$$\beta_0(x) \geq \sup_j \max\{c_j a_j + c_j x, -d_j a_j - d_j x\} = B(x) \tag{3.121}$$

for all $x \in \mathbf{R}$. Assume there is a $\xi_0 \in \mathbf{R}$ such that $\beta_0(\xi_0) > B(\xi_0)$. A convex function is not differentiable at at most countably many points (see e.g. Theorem 25.3 of [177]). Thus, there is a ξ in a neighborhood of ξ_0 such that

$$\beta_0(\xi) > B(\xi) \tag{3.122}$$

and both β_0 and B are differentiable at ξ. The supporting line to $y = \beta_0(x)$ through the point $\xi + i\beta_0(\xi)$ is the tangent

$$y = \beta_0(\xi) + \beta_0'(\xi)(x - \xi). \tag{3.123}$$

Suppose $\beta_0'(\xi) > 0$. If $\beta_0(\xi) + \beta_0'(\xi)(x - \xi) = 0$ has a rational solution x, then there exists a number j such that (3.123) is the line $y = c_j a_j + c_j x$ with $c_j = \beta_0'(\xi)$, $a_j \in \mathbf{Q}$, $-a_j \leq \xi_2$. It follows that

$$\beta_0(\xi) = c_j a_j + c_j \xi \leq \sup_j \max\{c_j a_j + c_j \xi, -d_j a_j - d_j \xi\} = B(\xi),$$

which contradicts (3.122). If the solution $x^* := \xi - \beta_0(\xi)/\beta_0'(\xi)$ of the equation $\beta_0(\xi) + \beta_0'(\xi)(x - \xi) = 0$ is irrational, we may choose

$$-a_{j_n} \in (-\infty, x^*) \cap \mathbf{Q} \subset (-\infty, \xi_2) \cap \mathbf{Q}$$

such that $-a_{j_n}$ approaches x^* monotonically. Since c_{j_n} is monotonically increasing and $c_{j_n} \leq \beta_0'(\xi)$, there is a $c^* \leq \beta_0'(\xi)$ such that $c_{j_n} \to c^*$. If $c^* < \beta_0'(\xi)$, then the line $y = c_{j_n} a_{j_n} + c_{j_n} x$ cannot be a supporting line to the curve $y = \beta_0(x)$ whenever $-a_{j_n}$ is close enough to x^* (recall (3.116)). Hence $c^* = \beta_0'(\xi)$ and consequently,

$$c_{j_n} \to \beta_0'(\xi), \quad c_{j_n} a_{j_n} \to \beta_0'(\xi)(-x^*) = \beta_0(\xi) - \xi\beta_0'(\xi).$$

It follows that $\beta_0(\xi) = \lim_{n \to \infty}(c_{j_n} a_{j_n} + c_{j_n}\xi)$ and thus, there is an n_0 such that

$$\beta_0(\xi) < c_{j_{n_0}} a_{j_{n_0}} + c_{j_{n_0}}\xi + \big(\beta_0(\xi) - B(\xi)\big)/2. \tag{3.124}$$

From (3.124) and (3.121) we obtain that

$$\beta_0(\xi) < B(\xi) + \big(\beta_0(\xi) - B(\xi)\big)/2 = \big(B(\xi) + \beta_0(\xi)\big)/2 \leq \beta_0(\xi), \tag{3.125}$$

which is a contradiction. Analogously one can dispose of the case $\beta_0'(\xi) < 0$. So let $\beta_0'(\xi) = 0$. Choosing $-a_{j_n} \in (-\infty, \xi_2)$ so that $-a_{j_n} \to -\infty$ and taking into account that, by convexity, the function β_0 assumes its minimum at ξ, we see as above that $c_{j_n} a_{j_n} + c_{j_n}\xi \to \beta_0(\xi)$, which again results in the contradiction (3.125).

We so have proved that $\beta_0(x) = B(x)$ for all $x \in \mathbf{R}$. In the same way one can show that $\alpha_0(x) = A(x)$ for all $x \in \mathbf{R}$. This gives (3.114) and completes the proof. $\qquad\square$

Corollary 3.38. *Given $p \in (1, \infty)$ and any open, convex, narrow set N containing the origin, there exist a Carleson curve Γ and a weight $w \in A_p(\Gamma)$ as in Example 3.24 such that N coincides with the indicator set $N_t(\Gamma, p, w)$.*

Proof. Combine Theorems 3.37 and 3.21. □

3.9 Notes and comments

All results of this chapter are from our papers [16] and [18]. We confine ourselves to a couple of additional remarks.

3.1–3.3. The U and V transforms, which may be considered as integral analogues of the W-transform, were introduced in [18] (also see [17]). While the W-transform works well for weights which are continuous on $\Gamma \setminus \{t\}$, the U and V transforms can be used to tackle arbitrary weights an appropriate power of which is locally integrable. These transforms assign a regular submultiplicative function to a weight whose indices tell us whether the weight belongs to $A_p(\Gamma)$ or not. As a rule, the U-transform is more convenient when proving something, whereas the V-transform is the more handy tool for computing the indices.

Theorem 3.7 is almost a criterion for the membership of a weight in the (local) class $A_p(\Gamma, t)$ in terms of the U-transform. The gap between the sufficient conditions (3.33) and the necessary conditions (3.34) in Theorem 3.7 is caused by the lack of a local version of Theorem 2.31. The estimates (3.44) and (3.45) were communicated to us by Alexei Yu. Karlovich.

In order to get a criterion for the membership of a weight in the (global) class $A_p(\Gamma)$ we have, in contrast to Theorem 2.33, to introduce the new function $U\psi := \sup_{t \in \Gamma} U_t \psi$. Theorem 3.8 is the desired criterion and may be viewed as the main result of Section 3.3. We have not been able to establish an $A_p(\Gamma)$ criterion in terms of the indices of only the functions $U_t \psi$ ($t \in \Gamma$). We also do not know an example of a weight ψ for which the most left and most right inequalities in (3.31) and (3.32) are strong inequalities.

3.4. The aim of this section is to prove Corollary 3.14, but the central result of the section is Theorem 3.13. Given a continuous function $\varphi : \Gamma \setminus \{t\} \to (0, \infty)$ and an arbitrary weight $w \in A_p(\Gamma)$, this theorem allows us to decide whether φw belongs to $A_p(\Gamma)$ by having recourse to the indices of only a single regular submultiplicative function, namely $U_t \varphi w$. The main ingredient of the proof of Theorem 3.13 is Lemma 3.9, which guarantees the regularity of the submultiplicative function $U_t \varphi w$ and yields two-sided estimates for the indices $\alpha(U_t \varphi w)$, $\beta(U_t \varphi w)$ via the indices $\alpha(U_t w)$, $\beta(U_t w)$, $\alpha(W_t \varphi)$, $\beta(W_t \varphi)$. Originally we had only a weaker version of Lemma 3.9 (see [18, Lemma 9.1] and [17, Lemma 4.4]); the lemma in its present form as well as the Corollaries 3.10 and 3.11 are due to Alexei Yu. Karlovich.

3.6. Spitkovsky [200] was probably the first to introduce a pair of numbers which measure the powerlikeness of a weight. Given a weight $w \in A_p(\Gamma)$ on a *piecewise*

Lyapunov curve Γ, he considered the set

$$I_t(\Gamma, p, w) := \{\lambda \in \mathbf{R} : |\tau - t|^\lambda w(\tau) \in A_p(\Gamma)\}$$

and showed that this is an open interval of length at most 1 containing the origin. Consequently, $I_t(\Gamma, p, w)$ can be written in the form

$$I_t(\Gamma, p, w) = (-\nu_t^-(\Gamma, p, w), 1 - \nu_t^+(\Gamma, p, w))$$

with $0 < \nu_t^-(\Gamma, p, w) \leq \nu_t^+(\Gamma, p, w) < 1$. Using the Fredholm criterion for Toeplitz operators with piecewise quasicontinuous symbols on $L_+^p(\Gamma, \varrho)$ (ϱ being a power weight) obtained in [25], Spitkovsky also showed that given any α, β such that $0 < \alpha \leq \beta < 1$, there exists a weight $w \in A_p(\Gamma)$ such that $\nu_t^-(\Gamma, p, w) = \alpha$ and $\nu_t^+(\Gamma, p, w) = \beta$. His numbers $\nu_t^\pm(\Gamma, p, w)$ are connected with our indices of powerlikeness (3.83) or (3.84) by the equalities

$$\nu_t^-(\Gamma, p, w) = 1/p + \nu_t, \quad \nu_t^+(\Gamma, p, w) = 1/p + \mu_t.$$

It should be noted that in [200] only the existence of the indices $\nu_t^\pm(\Gamma, p, w)$ is established, while we can give more or less constructive formulas for their computation.

Chapter 4
Boundedness of the Cauchy singular integral

In this chapter our leading actor, the Cauchy singular integral operator S, enters the scene. A very deep theorem, which should actually be Theorem 1 of this book, says that S is bounded on $L^p(\Gamma, w)$ $(1 < p < \infty)$ if and only if Γ is a Carleson curve and w is a Muckenhoupt weight in $A_p(\Gamma)$. The proof of this theorem is difficult and goes beyond the scope of this book. We nevertheless decided to write down a proof, but this proof will only be given in Chapter 5. The purpose of this chapter is to provide the reader with some facts and results that should suffice to understand Chapters 6 to 10 without browsing in Chapter 5.

This chapter contains a proof of the necessity portion of the afore-mentioned theorem, i.e. we show that if S is bounded on $L^p(\Gamma, w)$ then $w \in A_p(\Gamma)$. We also give an elementary proof of the sufficiency part for certain classes of oscillating curves Γ_0 and oscillating weights w_0. These curves Γ_0 and weights w_0 play a fundamental role in the spectral theory of singular integral and Toeplitz operators, because all possible phenomena, or, to be more precise, all possible indicator sets are already attained by these curves and weights: given any Carleson curve Γ and any weight $w \in A_p(\Gamma)$, there exist Γ_0 and w_0 such that $N_t(\Gamma, p, w) = N_t(\Gamma_0, p, w_0)$.

We conclude the chapter by some technical constructions which imply that in connection with boundedness questions everything can be reduced to the case of unbounded simple arcs.

4.1 The Cauchy singular integral

Let Γ be a simple locally rectifiable curve and equip Γ with length measure $|d\tau|$. Recall that local rectifiability means that $\Gamma \cap \{z \in \mathbf{C} : |z| < R\}$ is rectifiable for all $R > 0$. Bounded curves of infinite length, such as

$$\Gamma = \{(1 - e^{-r})e^{ir} : r \geq 0\},$$

are not locally rectifiable in our sense. Thus, a locally rectifiable curve is rectifiable if and only if it is bounded.

We denote by $L^p(\Gamma)$ $(1 \le p \le \infty)$ the usual Lebesgue spaces on Γ, and given a weight w on Γ, we let $L^p(\Gamma, w)$ $(1 \le p < \infty)$ stand for the weighted Lebesgue space with the norm

$$\|f\|_{p,w} := \left(\int_\Gamma |f(\tau)|^p w(\tau)^p |d\tau| \right)^{1/p}.$$

If $w \in L^p(\Gamma)$ and $w^{-1} \in L^q(\Gamma)$ $(1 < p < \infty, 1/p + 1/q = 1)$ then clearly

$$L^\infty(\Gamma) \subset L^p(\Gamma, w) \subset L^1(\Gamma).$$

Suppose henceforth that Γ is oriented. For $\varepsilon > 0$, the *truncated singular integral* $S_\varepsilon f$ of a function $f : \Gamma \to \mathbf{C}$ at a point $t \in \Gamma$ is given by

$$(S_\varepsilon f)(t) := \frac{1}{\pi i} \int_{\Gamma \backslash \Gamma(t,\varepsilon)} \frac{f(\tau)}{\tau - t} d\tau.$$

Obviously, $(S_\varepsilon f)(t)$ is a well-defined and finite number if $f \in L^1(\Gamma)$. The following proposition shows that $(S_\varepsilon f)(t)$ is well-defined and finite for much more functions.

Proposition 4.1. *Let Γ be a simple Carleson curve, $1 < p < \infty$, and $w \in A_p(\Gamma)$. Then $(S_\varepsilon f)(t)$ is well-defined and finite for every $t \in \Gamma$ and every $f \in L^p(\Gamma, w)$.*

Proof. This is trivial if Γ is bounded, since then $w \in L^p(\Gamma)$, $w^{-1} \in L^q(\Gamma)$ and thus $L^p(\Gamma, w) \subset L^1(\Gamma)$. So assume Γ is unbounded. For $f \in L^p(\Gamma, w)$, we have

$$\left| (S_\varepsilon f)(t) \right| \le \left(\int_{|\tau - t| \ge \varepsilon} |f(\tau)|^p w(\tau)^p |d\tau| \right)^{1/p} \left(\int_{|\tau - t| \ge \varepsilon} \frac{w(\tau)^{-q}}{|\tau - t|^q} |d\tau| \right)^{1/q},$$

where integration over $|\tau - t| \ge \varepsilon$ means integration over $\Gamma \backslash \Gamma(t, \varepsilon)$. The first factor on the right is at most $\|f\|_{p,w}$, and hence the assertion will follow as soon as we have shown that the second factor is finite.

Since $w \in A_p(\Gamma)$, it follows that $w^{-1} \in A_q(\Gamma)$. From Corollary 2.32 (with w^{-1} in place of w and q in place of p) we deduce that there is an $r \in (1, q)$ such that $w^{-q/r} \in A_r(\Gamma)$, which implies the existence of a constant $C < \infty$ such that

$$\left(\int_{|\tau - t| < y} w(\tau)^{-q} |d\tau| \right)^{1/r} \left(\int_{|\tau - t| < y} w(\tau)^{qs/r} |d\tau| \right)^{1/s} \le Cy$$

with $1/r + 1/s = 1$ for every $y > \varepsilon$. Thus, all the more,

$$\left(\int_{\varepsilon \le |\tau - t| < y} w(\tau)^{-q} |d\tau| \right)^{1/r} \left(\int_{|\tau - t| < \varepsilon} w(\tau)^{qs/r} |d\tau| \right)^{1/s} \le Cy. \qquad (4.1)$$

By Hölder's inequality

$$|\Gamma(t,\varepsilon)| = \int\limits_{|\tau-t|<\varepsilon} |d\tau| \leq \left(\int\limits_{|\tau-t|<\varepsilon} w(\tau)^{-q}|d\tau|\right)^{1/r} \left(\int\limits_{|\tau-t|<\varepsilon} w(\tau)^{qs/r}|d\tau|\right)^{1/s}. \quad (4.2)$$

From (4.1) and (4.2) we conclude that

$$\left(\int\limits_{\varepsilon\leq|\tau-t|<y} w(\tau)^{-q}|d\tau|\right)^{1/r} |\Gamma(t,\varepsilon)| \left(\int\limits_{|\tau-t|<\varepsilon} w(\tau)^{-q}|d\tau|\right)^{-1/r} \leq Cy,$$

whence

$$\int\limits_{\varepsilon\leq|\tau-t|<y} w(\tau)^{-q}|d\tau| \leq \frac{C^r}{|\Gamma(t,\varepsilon)|^r}\left(\int\limits_{|\tau-t|<\varepsilon} w(\tau)^{-q}|d\tau|\right) y^r =: Ny^r. \quad (4.3)$$

Using (4.3) with $y = 2^{k+1}\varepsilon$ we finally obtain

$$\int\limits_{|\tau-t|\geq\varepsilon} \frac{w(\tau)^{-q}}{|\tau-t|^q}|d\tau| = \sum_{k=0}^{\infty} \int\limits_{2^k\varepsilon\leq|\tau-t|<2^{k+1}\varepsilon} \frac{w(\tau)^{-q}}{|\tau-t|^q}|d\tau|$$

$$\leq \sum_{k=0}^{\infty} \frac{1}{2^{kq}\varepsilon^q} \int\limits_{\varepsilon\leq|\tau-t|<2^{k+1}\varepsilon} w(\tau)^{-q}|d\tau|$$

$$\leq \sum_{k=0}^{\infty} \frac{N}{2^{kq}\varepsilon^q}(2^{k+1}\varepsilon)^r = \frac{N2^r}{\varepsilon^{q-r}}\sum_{k=0}^{\infty}\left(\frac{1}{2^{q-r}}\right)^k < \infty. \quad (4.4)$$

$\square$

Now let $t \in \Gamma$ and suppose f is a function for which $(S_\varepsilon f)(t)$ is well-defined and finite for all $\varepsilon > 0$. If the limit

$$\lim_{\varepsilon\to 0}(S_\varepsilon f)(t) := \lim_{\varepsilon\to 0}\frac{1}{\pi i}\int\limits_{\Gamma\backslash\Gamma(t,\varepsilon)} \frac{f(\tau)}{\tau-t}\,d\tau \quad (4.5)$$

exists and is finite, it is denoted by $(Sf)(t)$ and referred to as the value of the *Cauchy singular integral* of f at the point t. In order to make explicit the dependence of S on Γ, we will occasionally write S_Γ in place of S. The following theorem concerns the case where Γ is bounded and f is constant.

Theorem 4.2. *Let Γ be a simple rectifiable curve and let $t \in \Gamma$ be a point at which the (two-sided) tangent to Γ exists. Then the limit*

$$(S1)(t) := \lim_{\varepsilon\to 0}\frac{1}{\pi i}\int\limits_{\Gamma\backslash\Gamma(t,\varepsilon)} \frac{d\tau}{\tau-t}$$

exists and is finite. If Γ is a Jordan curve, then $(S1)(t) = 1$. If Γ is an arc with the starting point A and the endpoint B, then

$$(S1)(t) = \frac{1}{\pi i} \log \frac{B - t}{A - t} - 1, \qquad (4.6)$$

where $\log\big((B - t)/(A - t)\big)$ is the boundary value of the branch of the function $\log\big((B - z)/(A - z)\big)$ which is analytic in $\mathbf{C} \setminus \Gamma$ and vanishes at infinity as z approaches t from the left.

Proof. First note that the right-hand side of (4.6) equals

$$\frac{1}{\pi i}\big(\log(B - t) - \log(A - t) - i\sigma(t)\big),$$

where $\log(z - t)$ is any branch of the logarithm continuous in $\mathbf{C} \setminus \Lambda$, Λ being any curve joining t to infinity such that $\Gamma \cap \Lambda = \{t\}$, and the number $\sigma(t) \in \{-\pi, \pi\}$ is defined by

$$\sigma(t) := \lim_{\tau \to t+0} \arg(\tau - t) - \lim_{\tau \to t-0} \arg(\tau - t).$$

Let $s \mapsto z(s)$ be the natural parametrization of Γ in a neighborhood of $t = z(s_0)$. Then

$$z(s) = z(s_0) + z'(s_0)(s - s_0) + o\big(|s - s_0|\big) \quad \text{as } s \to s_0.$$

Without loss of generality assume $z(s_0) = 0$ and $z'(s_0) = 1$, i.e. let

$$z(s) = s - s_0 + o\big(|s - s_0|\big). \qquad (4.7)$$

Given $n \in \{2, 3, \ldots\}$, there is a $\delta_n > 0$ such that

$$\big|o\big(|s - s_0|\big)\big| < (1/n)|s - s_0| \quad \text{for } |s - s_0| \le \delta_n. \qquad (4.8)$$

Clearly, we may assume that δ_n decreases monotoneously to zero. From (4.7), (4.8), and the inequality $\arcsin(1/n) < \pi/(2n)$ we see that the arc $\gamma_n := \{\tau \in \Gamma : \ell(\tau, t) \le \delta_n\}$ is contained in the double sector

$$\Delta_n := \{re^{i\theta} : r \in \mathbf{R}, -\pi/(2n) < \theta < \pi/(2n)\}.$$

Denote by a and b the endpoints of γ_2. The set $(\Gamma \setminus \gamma_2) \cup \{a, b\}$ is compact and the function $\tau \mapsto |\tau| \,(= |\tau - t|)$ is continuous and positive on this set. Let $\mu_0 > 0$ be the minimum of this function. Then $\Gamma(0, \mu_0)$ contains only points of γ_2:

$$\Gamma(0, \varepsilon) = \gamma_2(0, \varepsilon) \quad \text{for } \varepsilon \le \mu_0. \qquad (4.9)$$

Now fix an integer $n \ge 2$ such that $\delta_n \le \mu_0$ and let $0 < \varepsilon < \delta_n/2$. For $\tau_1, \tau_2 \in \gamma_2$, we denote by $\ell(\tau_1, \tau_2)$ the length of the subarc of γ_2 between τ_1 and τ_2.

Let $\tau = z(s) \in \gamma_2$ and $|\tau| \le \varepsilon$. Then (4.8) implies that

$$\varepsilon \ge |z(s)| = \left|s - s_0 + o(|s - s_0|)\right| > |s - s_0| - \frac{1}{2}|s - s_0| = \frac{1}{2}|s - s_0|,$$

whence $|s - s_0| \le 2\varepsilon < \delta_n$. Thus, again by (4.8),

$$\varepsilon \ge |z(s)| = \left|s - s_0 + o(|s - s_0|)\right| > |s - s_0| - \frac{1}{n}|s - s_0| = \frac{n - 1}{n}|s - s_0|.$$

In summary, we obtain that $\tau \in \gamma_n \subset \Delta_n$ and that

$$\ell(0, \tau) = |s - s_0| < \frac{n}{n - 1}\,\varepsilon. \tag{4.10}$$

The set $P(\varepsilon) := \gamma_2 \cap \{z \in \mathbf{C} : |z| = \varepsilon,\ \mathrm{Re}\, z \ge 0\}$ is compact and the function $z \mapsto s - s_0 := \ell(0, z)$ is continuous on $P(\varepsilon)$. Let

$$s_-(\varepsilon) := \min\{s : z(s) \in P(\varepsilon)\}, \quad s_+(\varepsilon) := \max\{s : z(s) \in P(\varepsilon)\},$$
$$\alpha(\varepsilon) := \varepsilon e^{i \arg z(s_-(\varepsilon))}, \qquad \beta(\varepsilon) := \varepsilon e^{i \arg z(s_+(\varepsilon))}.$$

The arc $\gamma_2 \cap \{z \in \mathbf{C} : \mathrm{Re}\, z \ge 0\}$ is the union of its subarc $\eta_1(\varepsilon)$ between 0 and $\alpha(\varepsilon)$, its subarc $\eta_2(\varepsilon)$ between $\alpha(\varepsilon)$ and $\beta(\varepsilon)$, and its subarc $\eta_3(\varepsilon)$ between $\beta(\varepsilon)$ and b. Put $\eta_2^+(\varepsilon) := \{\tau \in \eta_2(\varepsilon) : |\tau| \ge \varepsilon\}$. From (4.10) we infer that $\ell(0, \beta(\varepsilon)) < n\varepsilon/(n - 1)$. Since $\ell(0, \alpha(\varepsilon)) \ge |\alpha(\varepsilon)| = \varepsilon$, it follows that

$$\left|\eta_2^+(\varepsilon)\right| \le \left|\eta_2(\varepsilon)\right| = \ell\big(\alpha(\varepsilon), \beta(\varepsilon)\big) < \frac{n}{n - 1}\,\varepsilon - \varepsilon = \frac{\varepsilon}{n - 1}.$$

Consequently,

$$\left|\int_{\eta_2^+(\varepsilon)} \frac{d\tau}{\tau}\right| \le \int_{\eta_2^+(\varepsilon)} \frac{|d\tau|}{|\tau|} \le \frac{1}{\varepsilon} \int_{\eta_2^+(\varepsilon)} |d\tau| \le \frac{1}{n - 1}$$

and thus,

$$\left|\int_{(\gamma_2 \setminus \gamma_2(0, \varepsilon)) \cap \{z \in \mathbf{C} : \mathrm{Re}\, z \ge 0\}} d\tau/\tau\ - \log b + \log \varepsilon\right|$$

$$= \left|\int_{\eta_2^+(\varepsilon)} d\tau/\tau + \int_{\eta_3(\varepsilon)} d\tau/\tau - \log b + \log \varepsilon\right|$$

$$\le \frac{1}{n - 1} + \left|\int_{\eta_3(\varepsilon)} d\tau/\tau - \log b + \log \varepsilon\right|$$

$$= \frac{1}{n - 1} + \left|\log b - \log \beta(\varepsilon) - \log b + \log \varepsilon\right|$$

$$= \frac{1}{n - 1} + \left|\arg \beta(\varepsilon)\right| < \frac{1}{n - 1} + \frac{\pi}{2n}, \tag{4.11}$$

the latter estimate resulting from the fact that $\beta(\varepsilon) \in \gamma_n \subset \Delta_n$. Analogously one can prove that

$$\left| \int\limits_{(\gamma_2 \setminus \gamma_2(0,\varepsilon)) \cap \{z \in \mathbf{C} : \mathrm{Re}\, z \leq 0\}} d\tau/\tau \; + \log a - \log \varepsilon \right| < \frac{1}{n-1} + \frac{\pi}{2n}. \tag{4.12}$$

Adding (4.11) and (4.12) we obtain

$$\left| \int\limits_{\gamma_2 \setminus \gamma_2(0,\varepsilon)} d\tau/\tau \; - \log b + \log a \right| < \frac{2}{n-1} + \frac{\pi}{n}$$

whenever $n \geq 2$, $\delta_n \leq \mu_0$, and $0 < \varepsilon < \delta_n/2$. This shows that

$$\lim_{\varepsilon \to 0} \int\limits_{\gamma_2 \setminus \gamma_2(0,\varepsilon)} d\tau/\tau = \log b - \log a.$$

Taking into account (4.9) we easily get the remaining assertion of the theorem. $\square$

We remark that in almost all the classical books on singular integrals, $(Sf)(t)$ is not defined by (4.5) but rather by

$$(Sf)(t) := \lim_{\varepsilon \to 0} \frac{1}{\pi i} \int\limits_{\Gamma \setminus \Gamma[t,\varepsilon]} \frac{f(\tau)}{\tau - t} \, d\tau. \tag{4.13}$$

From the equality (4.9) we see that if Γ is Carleson and has a two-sided tangent at t (which happens for almost all $t \in \Gamma$), then $\Gamma[t,\varepsilon] \subset \Gamma(t,\varepsilon) \subset \Gamma[t, C_\Gamma \varepsilon]$ for all sufficiently small $\varepsilon > 0$. Thus, at these points t both definitions are equivalent.

To evaluate $(Sf)(t)$, we may write

$$(S_\varepsilon f)(t) = \frac{1}{\pi i} \int\limits_{\Gamma \setminus \Gamma(t,\varepsilon)} \frac{f(\tau) - f(t)}{\tau - t} \, d\tau + \frac{f(t)}{\pi i} \int\limits_{\Gamma \setminus \Gamma(t,\varepsilon)} \frac{d\tau}{\tau - t}. \tag{4.14}$$

Theorem 4.2 tells us that the second term on the right of (4.14) has a limit a.e. on every rectifiable curve Γ. It is clear that the first term on the right of (4.14) does not possess a singularity if only f is sufficiently nice.

We denote by $C_0^\infty(\mathbf{R}^2)$ the infinitely differentiable functions $g : \mathbf{R}^2 \to \mathbf{C}$ with compact support, and we let $C_0^\infty(\Gamma)$ stand for the set of the restrictions of functions in $C_0^\infty(\mathbf{R}^2)$ to Γ. Clearly $C_0^\infty(\Gamma) \subset L^1(\Gamma)$.

Theorem 4.3. *Let Γ be a simple locally rectifiable curve and $g \in C_0^\infty(\Gamma)$. If the (two-sided) tangent to Γ at $t \in \Gamma$ exists, then the limit $(Sg)(t)$ exists and is finite.*

Proof. If Γ is bounded, we see from Theorem 4.2 and formula (4.14) that we are left with proving that

$$\int_{\Gamma\backslash\Gamma(t,\varepsilon)} \frac{g(\tau) - g(t)}{\tau - t} \, d\tau \tag{4.15}$$

has a finite limit as $\varepsilon \to 0$. Because $g(\tau) = g(t) + g'(t)(\tau - t) + O(|\tau - t|^2)$ as $\tau \to t$, the integral (4.15) equals

$$\int_{\Gamma\backslash\Gamma(t,\varepsilon)} \frac{O(|\tau - t|^2)}{|\tau - t|} \, d\tau + g'(t) \int_{\Gamma\backslash\Gamma(t,\varepsilon)} d\tau. \tag{4.16}$$

The second term in (4.16) is obviously convergent. The function sending τ to $O(|\tau - t|^2)/|\tau - t|$ belongs to $L^1(\Gamma)$ and hence,

$$G := \int_{\Gamma} \frac{O(|\tau - t|^2)}{|\tau - t|} \, d\tau$$

exists and is finite. Since for all sufficiently small $\varepsilon > 0$

$$\left| \int_{\Gamma\backslash\Gamma(t,\varepsilon)} \frac{O(|\tau - t|^2)}{|\tau - t|} \, d\tau - G \right| \leq \int_{\Gamma(t,\varepsilon)} \frac{|O(|\tau - t|^2)|}{|\tau - t|} \, d\tau \leq C\varepsilon|\Gamma(t,\varepsilon)| \leq C\varepsilon|\Gamma|$$

with some constant $C < \infty$, it follows that the first term in (4.16) converges to G.

Now assume Γ is unbounded. We then have

$$\int_{\Gamma\backslash\Gamma(t,\varepsilon)} \frac{g(\tau)}{\tau - t} \, d\tau = \int_{\Gamma\backslash\Gamma(t,R)} \frac{g(\tau)}{\tau - t} \, d\tau + \int_{\Gamma(t,R)\backslash\Gamma(t,\varepsilon)} \frac{g(\tau)}{\tau - t} \, d\tau \tag{4.17}$$

for every $R > 0$. The first term on the right of (4.17) vanishes if $\operatorname{supp} g \subset \Gamma(t, R)$, and the convergence of the second term was proved in the preceding paragraph. $\square$

4.2   Necessary conditions for boundedness

Let Γ be a simple locally rectifiable curve, let $1 < p < \infty$, and let w be a weight on Γ. From Theorem 4.3 we know that if $g \in C_0^\infty(\Gamma)$, then $(Sg)(t)$ exists for almost all $t \in \Gamma$. The Cauchy singular integral S is said to *generate a bounded operator on $L^p(\Gamma, w)$* if

(i) $C_0^\infty(\Gamma)$ is dense in $L^p(\Gamma, w)$,

(ii) $\|Sg\|_{p,w} \leq M\|g\|_{p,w}$ for all $g \in C_0^\infty(\Gamma)$ with some constant $M < \infty$ independent of g.

Notice that (ii) includes the requirement that $Sg \in L^p(\Gamma, w)$ whenever g belongs to $C_0^\infty(\Gamma)$. If (i) and (ii) hold, then S extends to a bounded operator $\tilde{S}$ on all of $L^p(\Gamma, w)$ in a unique way. The purpose of this section is to prove that for S to generate a bounded operator on $L^p(\Gamma, w)$ ($1 < p < \infty$) it is necessary that $w \in A_p(\Gamma)$. Recall that if $w \in A_p(\Gamma)$, then Γ is necessarily a Carleson curve.

Lemma 4.4. *Let Γ be a simple rectifiable (and thus bounded) curve, let $1 < p < \infty$, and let w be a weight in $L^p(\Gamma)$. Then $C_0^\infty(\Gamma)$ is dense in $L^p(\Gamma, w)$.*

Proof. Suppose that Γ is an arc. Without loss of generality assume $|\Gamma| = 2\pi$. Let $I := [0, 2\pi]$ and let $\eta : I \to \Gamma$ be a homeomorphism such that $|\eta'(x)| = 1$ for almost all $x \in I$. Since $w \in L^p(\Gamma)$ and

$$\int_\Gamma w(\tau)^p |d\tau| = \int_I w(\eta(x))^p \, dx,$$

it follows that $\varrho := w \circ \eta \in L^p(I)$.

We claim that $C(I)$ is dense in $L^p(I, \varrho)$. To see this, assume $C(I)$ is not dense in $L^p(I, \varrho)$. Then, by the Hahn-Banach theorem, there is a nonzero function $h \in (L^p(I, \varrho))^* = L^q(I, \varrho^{-1})$ such that

$$\int_I h(x)\varphi(x) \, dx = 0 \quad \text{for all} \ \ \varphi \in C(I). \tag{4.18}$$

Hölder's inequality gives

$$\int_I |h(x)| \, dx \leq \left(\int_I |h(x)|^q \varrho(x)^{-q} \, dx \right)^{1/q} \left(\int_I \varrho(x)^p \, dx \right)^{1/p},$$

and as $h \in L^q(I, \varrho^{-1})$ and $\varrho \in L^p(I)$, we conclude that $h \in L^1(I)$. Taking $\varphi(x) = e^{inx}$ ($n \in \mathbf{Z}$) in (4.18), we obtain that all Fourier coefficients of h vanish, which implies that $h = 0$ a.e. on I (see, e.g., [52, Corollary 6.5]). This contradiction proves that $C(I)$ is dense in $L^p(I, \varrho)$.

The map $f \mapsto f \circ \eta$ is an isometric isomorphism of $L^p(\Gamma, w)$ onto $L^p(I, \varrho)$. Therefore, given $f \in L^p(\Gamma, w)$, we infer from the density of $C(I)$ in $L^p(I, \varrho)$ that there exist $\varphi_n \in C(I)$ such that $f \circ \eta - \varphi_n \to 0$ in $L^p(I, \varrho)$, whence $f - \varphi_n \circ \eta^{-1} \to 0$ in $L^p(\Gamma, w)$. As $\varphi_n \circ \eta^{-1} \in C(\Gamma)$, we deduce that $C(\Gamma)$ is dense in $L^p(\Gamma, w)$. By Mergelyan's theorem (see e.g. [70, p.119]), the set $R(\Gamma)$ of all rational functions without poles on Γ is uniformly dense in $C(\Gamma)$. Thus, given $\varepsilon > 0$ and $f \in L^p(\Gamma, w)$, we can find a function $\psi \in R(\Gamma)$ such that $\|f - \psi\|_{p,w} < \varepsilon$. Let $z_1, \ldots, z_n \in \mathbf{C} \setminus \Gamma$ be the poles of ψ in $\mathbf{C}$ and choose open neighborhoods $U(z_1), \ldots, U(z_n)$ of these poles such that $U(z_j) \cap \Gamma = \emptyset$ for all j. Obviously, there is a function $\chi \in C_0^\infty(\mathbf{R}^2)$ satisfying

$$\chi|U(z_1) \cup \ldots \cup U(z_n) = 0 \quad \text{and} \quad \chi|\Gamma = 1.$$

Then $g := \chi\psi \in C_0^\infty(\Gamma)$ and $\|f - g\|_{p,w} < \varepsilon$. This shows that $C_0^\infty(\Gamma)$ is dense in $L^p(\Gamma, w)$.

In the case where Γ is a Jordan curve, the above reasoning works with $\mathbf{T}$ in place of I. $\square$

Proposition 4.5. *Let Γ be a simple locally rectifiable curve, let $1 < p < \infty$, and let w be a weight on Γ. The set $C_0^\infty(\Gamma)$ is dense in $L^p(\Gamma, w)$ if and only if $w \in L_{loc}^p(\Gamma)$.*

Proof. Since $C_0^\infty(\mathbf{R}^2)$ contains functions which are identically 1 on arbitrarily large disks, the inclusion $C_0^\infty(\Gamma) \subset L^p(\Gamma, w)$ implies that $w \in L_{loc}^p(\Gamma)$, which gives the "only if" part.

If Γ is bounded, the "if" portion follows from Lemma 4.4. So suppose Γ is unbounded, and for the sake of definiteness, assume Γ is homeomorphic to $\mathbf{R}_+$. Denote by t the endpoint of Γ and put $\Gamma_N := \{\tau \in \Gamma : \ell(t, \tau) \le N\}$. Obviously, Γ_N is contained in the disk $\Delta_{2N} := \{z \in \mathbf{C} : |z - t| \le 2N\}$. As we trace out Γ from t to infinity, there is a last point $t_0(N) \in \Gamma$ lying in Δ_{2N} (recall that Γ is locally rectifiable). This point divides Γ into a bounded arc $\Gamma_1(N)$ and an unbounded arc $\Gamma_2(N)$. Given $f \in L^p(\Gamma, w)$ and $\varepsilon > 0$, there is an $N > 0$ such that

$$\int_{\Gamma \setminus \Gamma_N} |f(\tau)|^p w(\tau)^p |d\tau| < \varepsilon,$$

whence

$$\int_{\Gamma_1(N) \setminus \Gamma_N} |f(\tau)|^p w(\tau)^p |d\tau| < \varepsilon, \quad \int_{\Gamma_2(N)} |f(\tau)|^p w(\tau)^p |d\tau| < \varepsilon. \tag{4.19}$$

Let χ_N denote the characteristic function of Γ_N. From Lemma 4.4 we know that there is a function $\varphi \in C_0^\infty(\mathbf{R}^2)$ such that

$$\int_{\Gamma_1(N)} |\chi_N(\tau)f(\tau) - \varphi(\tau)|^p w(\tau)^p |d\tau| < \varepsilon.$$

It follows in particular that

$$\int_{\Gamma_N} |f(\tau) - \varphi(\tau)|^p w(\tau)^p |d\tau| < \varepsilon, \quad \int_{\Gamma_1(N) \setminus \Gamma_N} |\varphi(\tau)|^p w(\tau)^p |d\tau| < \varepsilon. \tag{4.20}$$

Let $\psi \in C_0^\infty(\mathbf{R}^2)$ be any function for which

$$0 \le \psi \le 1, \ \psi|\mathbf{R}^2 \setminus \Delta_{2N} = 0, \ \psi|\Gamma_N = 1. \tag{4.21}$$

Then $\psi\varphi \in C_0^\infty(\mathbf{R}^2)$. Taking into account (4.19) to (4.21) and using the inequality $(|a| + |b|)^p \leq 2^{p-1}(|a|^p + |b|^p)$, we get

$$\int_\Gamma |f(\tau) - \psi(\tau)\varphi(\tau)|^P w(\tau)^p |d\tau| = \int_{\Gamma_N} |f(\tau) - \varphi(\tau)|^P w(\tau)^p |d\tau|$$

$$+ \int_{\Gamma_1(N)\backslash\Gamma_N} |f(\tau) - \psi(\tau)\varphi(\tau)|^P w(\tau)^p |d\tau| + \int_{\Gamma_2(N)} |f(\tau)|^P w(\tau)^p |d\tau|$$

$$< 2\varepsilon + \int_{\Gamma_1(N)\backslash\Gamma_N} |f(\tau) - \psi(\tau)\varphi(\tau)|^P w(\tau)^p |d\tau|$$

$$\leq 2\varepsilon + 2^{p-1} \int_{\Gamma_1(N)\backslash\Gamma_N} |f(\tau)|^P w(\tau)^p |d\tau| + 2^{p-1} \int_{\Gamma_1(N)\backslash\Gamma_N} |\psi(\tau)\varphi(\tau)|^P w(\tau)^p |d\tau|$$

$$< (2 + 2^{p-1})\varepsilon + 2^{p-1} \int_{\Gamma_1(N)\backslash\Gamma_N} |\varphi(\tau)|^P w(\tau)^p |d\tau| < (2 + 2^p)\varepsilon.$$

This shows that $C_0^\infty(\Gamma)$ is dense in $L^p(\Gamma, w)$. The proof is analogous for unbounded curves homeomorphic to $\mathbf{R}$. $\qquad\square$

Lemma 4.6. *Let Γ be a simple locally rectifiable curve, let $1 < p < \infty$, and let w be a weight on Γ. If S generates a bounded operator on $L^p(\Gamma, w)$ then $w \in L_{\mathrm{loc}}^p(\Gamma)$ and $w^{-1} \in L_{\mathrm{loc}}^q(\Gamma)$.*

Proof. Suppose S extends to a bounded operator $\tilde{S}$ on $L^p(\Gamma, w)$. Then $C_0^\infty(\Gamma)$ is dense in $L^p(\Gamma, w)$, so that $w \in L_{\mathrm{loc}}^p(\Gamma)$ by Proposition 4.5. We show that w^{-1} belongs to $L_{\mathrm{loc}}^q(\Gamma)$.

Suppose first that Γ is bounded. Then the operator V defined by $(Vf)(\tau) := \tau f(\tau)$ is clearly bounded on $L^p(\Gamma, w)$, and V maps $C_0^\infty(\Gamma)$ into itself. Thus, if $g \in C_0^\infty(\Gamma)$, we deduce from Theorem 4.3 that

$$\big((\tilde{S}V - V\tilde{S})g\big)(t) = \big((SV - VS)g\big)(t) = \frac{1}{\pi i} \int_\Gamma g(\tau)\, d\tau$$

for almost all $t \in \Gamma$. It follows that

$$\|(\tilde{S}V - V\tilde{S})g\|_{p,w} = \frac{1}{\pi} \left| \int_\Gamma g(\tau)\, d\tau \right| \|w\|_p,$$

and since

$$\|(\tilde{S}V - V\tilde{S})g\|_{p,w} \leq 2\|\tilde{S}\|\, \|V\|\, \|g\|_{p,w},$$

we arrive at the estimate

$$\left| \int_{\Gamma} g(\tau)\, d\tau \right| \le \frac{2\|\tilde{S}\|\, \|V\|\pi}{\|w\|_p} \|g\|_{p,w} =: K\|g\|_{p,w} \tag{4.22}$$

for all $g \in C_0^{\infty}(\Gamma)$. We have $d\tau = e^{i\theta_{\Gamma}(\tau)}|d\tau|$, where $\theta_{\Gamma}(\tau)$ is the angle between the (almost everywhere existing) tangent to Γ at τ and the x-axis. From (4.22) and Lemma 4.4 we infer that the map

$$\mathbf{C}_0^{\infty}(\Gamma) \to \mathbf{C}, \; g \mapsto \int_{\Gamma} g(\tau)e^{i\theta_{\Gamma}(\tau)}|d\tau|$$

extends to a bounded linear functional on $L^p(\Gamma, w)$. This implies that $1 = |e^{i\theta_{\Gamma}(\tau)}|$ is a function in $L^q(\Gamma, w^{-1})$, whence $w^{-1} \in L^q(\Gamma)$.

Now suppose Γ is arbitrary. Then $\Gamma_R := \Gamma \cap \{z \in \mathbf{C} : |z| < R\}$ is contained in some bounded subarc $\gamma_R \subset \Gamma$. Let χ_R be the characteristic function of γ_R. The operator

$$L^p(\gamma_R, w) \to L^p(\gamma_R, w), \; f \mapsto \chi_R \tilde{S} f$$

is clearly bounded and hence $w^{-1} \in L^q(\gamma_R)$ by what was already proved. $\qquad\square$

Lemma 4.7. *Let Γ be a simple locally rectifiable curve, let $1 < p < \infty$, and let w be a weight on Γ. Suppose $\Gamma_0 \subset \Gamma$ is a bounded measurable subset, $f \in L^p(\Gamma, w)$, and supp $f \subset \Gamma_0$. If S extends to a bounded operator $\tilde{S}$ on $L^p(\Gamma, w)$, then*

$$(\tilde{S}f)(t) = \frac{1}{\pi i} \int_{\Gamma_0} \frac{f(\tau)}{\tau - t}\, d\tau$$

for almost all t in the set $E_\delta := \{t \in \Gamma : \operatorname{dist}(t, \Gamma_0) \ge \delta > 0\}$.

Proof. Put $V_n := \{z \in \mathbf{C} : \operatorname{dist}(z, \Gamma_0) < \delta/n\}$ and $\Gamma_n := \Gamma \cap V_n$. By Proposition 4.5, there are functions $h_n \in C_0^{\infty}(\Gamma)$ such that

$$\|f - h_n\|_{L^p(\Gamma, w)} \to 0.$$

Choose $\chi_n \in C_0^{\infty}(\Gamma)$ so that

$$0 \le \chi_n \le 1, \; \chi_n|V_{n+1} = 1, \; \chi_n|\mathbf{C} \setminus V_n = 0$$

and put $g_n := \chi_n h_n$. Then

$$\|f - g_n\|_{L^p(\Gamma,w)}^p = \|f - g_n\|_{L^p(\Gamma_0,w)}^p + \|f - g_n\|_{L^p(\Gamma \setminus \Gamma_0,w)}^p$$
$$= \|f - h_n\|_{L^p(\Gamma_0,w)}^p + \|g_n\|_{L^p(\Gamma \setminus \Gamma_0,w)}^p$$
$$\le \|f - h_n\|_{L^p(\Gamma_0,w)}^p + \|h_n\|_{L^p(\Gamma \setminus \Gamma_0,w)}^p = \|f - h_n\|_{L^p(\Gamma,w)}^p \to 0.$$

Since $\tilde{S}$ is bounded on $L^p(\Gamma, w)$ and coincides with S on $C_0^\infty(\Gamma)$, it follows that $\|Sg_n - \tilde{S}f\|_{p,w} \to 0$. Hence, there is a subsequence of $\{Sg_n\}$ converging a.e. to $\tilde{S}f$ (see, e.g., [126, Chap. VII, Sec. 2], [171, Corollary of Theorem I.12], or [180, Theorem 3.12]). Without loss of generality assume $\{Sg_n\}$ itself converges a.e. to $\tilde{S}f$. For $n \geq 2$ and all $t \in E_\delta$, we have

$$(Sg_n)(t) = \frac{1}{\pi i} \int\limits_{\Gamma_n} \frac{g_n(\tau)}{\tau - t}\, d\tau,$$

and thus we are left with proving that

$$\left| \int\limits_{\Gamma_n} \frac{g_n(\tau)}{\tau - t}\, d\tau - \int\limits_{\Gamma_0} \frac{f(\tau)}{\tau - t}\, d\tau \right| \tag{4.23}$$

converges to zero for almost all $t \in E_\delta$. Clearly, (4.23) is not greater than

$$\int\limits_{\Gamma_0} \frac{|g_n(\tau) - f(\tau)|}{|\tau - t|}\, |d\tau| + \int\limits_{\Gamma_n \backslash \Gamma_0} \frac{|g_n(\tau)|}{|\tau - t|}\, |d\tau|$$

$$\leq \frac{1}{\delta} \|g_n - f\|_{L^p(\Gamma_0, w)} \|w^{-1}\|_{L^q(\Gamma_0)}$$

$$+ \frac{2}{\delta} \|g_n\|_{L^p(\Gamma_n \backslash \Gamma_0, w)} \|w^{-1}\|_{L^q(\Gamma_n \backslash \Gamma_0)} \tag{4.24}$$

From Lemma 4.6 we deduce that $w^{-1} \in L^q(\Gamma_2)$. This implies that the first term in (4.24) goes to zero. The local rectifiability of Γ yields that $|\Gamma_n \backslash \Gamma_0| \to 0$, which shows that the second term in (4.24) also tends to zero. $\qquad \square$

Theorem 4.8. *Suppose Γ is a simple locally rectifiable curve, $1 < p < \infty$, and w is a weight on Γ. If S generates a bounded operator on $L^p(\Gamma, w)$, then $w \in A_p(\Gamma)$.*

Proof. Let $\tilde{S}$ be the continuous extension of S from $C_0^\infty(\Gamma)$ to all of $L^p(\Gamma, w)$. Lemma 4.6 says that $w \in L_{\mathrm{loc}}^p(\Gamma)$ and $w^{-1} \in L_{\mathrm{loc}}^q(\Gamma)$.

If Γ is bounded, then the function $\psi : \Gamma \to (0, \infty)$, $t \mapsto \max_{\tau \in \Gamma} |\tau - t|$ is continuous. We define $d := \min_{t \in \Gamma} \psi(t)$. Obviously, $d > 0$. We have

$$\sup_{t \in \Gamma} \sup_{\varepsilon \geq d/3} \frac{1}{\varepsilon} \left(\int\limits_{\Gamma(t,\varepsilon)} w(\tau)^p |d\tau| \right)^{1/p} \left(\int\limits_{\Gamma(t,\varepsilon)} w(\tau)^{-q} |d\tau| \right)^{1/q}$$

$$\leq \frac{3}{d} \|w\|_{L^p(\Gamma)} \|w^{-1}\|_{L^q(\Gamma)} < \infty.$$

Thus, in what follows we assume that either Γ is bounded and $\varepsilon \in (0, d/3)$ or that Γ is unbounded and $\varepsilon \in (0, \infty) =: (0, d/3)$.

Fix $t \in \Gamma$ and pick any $x \in \Gamma$ such that $|x - t| = 3\,\varepsilon$. Given a nonnegative function $f \in L^p(\Gamma, w)$, define a function $h \in L^p(\Gamma, w)$ by

$$
h(\tau) := \begin{cases} f(\tau)\frac{|d\tau|}{d\tau}e^{i\arg(t-x)} & \text{for} \quad \tau \in \Gamma(t, \varepsilon), \\ 0 & \text{for} \quad \tau \in \Gamma \setminus \Gamma(t, \varepsilon) \end{cases} .
$$

Lemma 4.7 shows that for almost all $z \in \Gamma(x, \varepsilon)$,

$$
(\tilde{S}h)(z) = \frac{1}{\pi i} \int\limits_{\Gamma(t,\varepsilon)} \frac{h(\tau)}{\tau - z}\, d\tau = \frac{1}{\pi i} \int\limits_{\Gamma(t,\varepsilon)} \frac{f(\tau)}{|\tau - z|}e^{i\alpha(\tau,z)}|d\tau|
$$

where $\alpha(\tau, z) := \arg(t - x) - \arg(\tau - z)$. Since $|\tau - z| \leq |t - x| + 2\,\varepsilon = 5\,\varepsilon$ and

$$
\big|\sin\alpha(\tau, z)\big| \leq \frac{\varepsilon}{(3/2)\varepsilon} = \frac{2}{3}, \quad \cos\alpha(\tau, z) \geq \left(1 - \left(\frac{2}{3}\right)^2\right)^{1/2} = \frac{\sqrt{5}}{3}
$$

for $\tau \in \Gamma(t, \varepsilon)$ and $z \in \Gamma(x, \varepsilon)$, we see that for these τ and z,

$$
|(\tilde{S}h)(z)| \geq \frac{1}{\pi} \int\limits_{\Gamma(t,\varepsilon)} \frac{f(\tau)}{|\tau - z|}\cos\alpha(\tau, z)|d\tau|
$$

$$
\geq \frac{\sqrt{5}}{3\pi} \int\limits_{\Gamma(t,\varepsilon)} \frac{f(\tau)}{|\tau - z|}|d\tau| \geq \frac{c}{\varepsilon} \int\limits_{\Gamma(t,\varepsilon)} f(\tau)|d\tau| \tag{4.25}
$$

with $c := \sqrt{5}/(15\pi)$. From (4.25) we infer that

$$
\left(\int\limits_{\Gamma(x,\varepsilon)} |(\tilde{S}h)(z)|^p w(z)^p|dz|\right)^{1/p} \geq \frac{c}{\varepsilon} \int\limits_{\Gamma(t,\varepsilon)} f(\tau)|d\tau| \left(\int\limits_{\Gamma(x,\varepsilon)} w(z)^p|dz|\right)^{1/p}.
$$

On the other hand, the boundedness of $\tilde{S}$ and the definition of h yield that

$$
\left(\int\limits_{\Gamma(x,\varepsilon)} |(\tilde{S}h)(z)|^p w(z)^p|dz|\right)^{1/p} \leq \|\tilde{S}h\|_{p,w} \leq M\|h\|_{p,w}
$$

$$
= M\left(\int\limits_{\Gamma(t,\varepsilon)} f(\tau)^p w(\tau)^p|d\tau|\right)^{1/p}
$$

with $M := \|\tilde{S}\|$. Combining these two estimates we get

$$
\frac{c}{\varepsilon} \int\limits_{\Gamma(t,\varepsilon)} f(\tau)\,|d\tau|\left(\int\limits_{\Gamma(x,\varepsilon)} w(z)^p|dz|\right)^{1/p} \leq M\left(\int\limits_{\Gamma(t,\varepsilon)} f(\tau)^p w(\tau)^p|d\tau|\right)^{1/p}, \tag{4.26}
$$

and letting $f = 1$, this implies that

$$c\frac{|\Gamma(t,\varepsilon)|}{\varepsilon}\left(\int\limits_{\Gamma(x,\varepsilon)} w(z)^p|dz|\right)^{1/p} \le M\left(\int\limits_{\Gamma(t,\varepsilon)} w(\tau)^p|d\tau|\right)^{1/p}.$$

Since $\varepsilon \le |\Gamma(t,\varepsilon)|$, it follows that

$$\left(\int\limits_{\Gamma(x,\varepsilon)} w(z)^p|dz|\right)^{1/p} \le \frac{M}{c}\left(\int\limits_{\Gamma(t,\varepsilon)} w(\tau)^p|d\tau|\right)^{1/p}.$$

Analogously one can show that

$$\left(\int\limits_{\Gamma(t,\varepsilon)} w(z)^p|d\tau|\right)^{1/p} \le \frac{M}{c}\left(\int\limits_{\Gamma(x,\varepsilon)} w(\tau)^p|d\tau|\right)^{1/p}. \tag{4.27}$$

From (4.26) and (4.27) we obtain

$$\frac{1}{\varepsilon}\int\limits_{\Gamma(t,\varepsilon)} f(\tau)|d\tau|\left(\int\limits_{\Gamma(t,\varepsilon)} w(\tau)^p|d\tau|\right)^{1/p} \le \frac{M^2}{c^2}\left(\int\limits_{\Gamma(t,\varepsilon)} f(\tau)^p w(\tau)^p|d\tau|\right)^{1/p}. \tag{4.28}$$

Let $X := \{f \in L^p(\Gamma, w) : \int\limits_{\Gamma(t,\varepsilon)} |f(\tau)|^p w(\tau)^p|d\tau| = 1\}$. From (4.28) we deduce that

$$\frac{1}{\varepsilon}\left(\sup\limits_{f\in X}\int\limits_{\Gamma(t,\varepsilon)} |f(\tau)|\,|d\tau|\right)\left(\int\limits_{\Gamma(t,\varepsilon)} w(\tau)^p|d\tau|\right)^{1/p} \le \frac{M^2}{c^2}.$$

Since

$$\sup\limits_{f\in X}\int\limits_{\Gamma(t,\varepsilon)} |f(\tau)|\,|d\tau| = \|\chi_{\Gamma(t,\varepsilon)}\|_{q,w^{-1}} = \left(\int\limits_{\Gamma(t,\varepsilon)} w(\tau)^{-q}|d\tau|\right)^{1/q}$$

(see e.g. [181, p. 94]), it results that

$$\sup\limits_{t\in\Gamma}\sup\limits_{\varepsilon\in(0,d/3)}\frac{1}{\varepsilon}\left(\int\limits_{\Gamma(t,\varepsilon)} w(\tau)^p|d\tau|\right)^{1/p}\left(\int\limits_{\Gamma(t,\varepsilon)} w(\tau)^{-q}|d\tau|\right)^{1/q} \le \frac{M^2}{c^2}. \qquad \square$$

4.3 Special curves and weights

The converse of Theorem 4.8 is also true but much harder to prove. We will present a proof in Chapter 5. The main topic of this book is the spectral theory of Toeplitz and singular integral operators. We will show that any possible spectral picture is attained for curves and weights as in Example 3.24. Therefore we will give a separate proof of the converse of Theorem 4.8 for such curves and weights now. To start somewhere, we take the following theorem for granted.

Theorem 4.9 (Hunt, Muckenhoupt, Wheeden). *If $\varrho \in A_p(\mathbf{R})$ $(1 < p < \infty)$ then $S_{\mathbf{R}}$ generates a bounded operator $\tilde{S}_{\mathbf{R}}$ on $L^p(\mathbf{R}, \varrho)$. Moreover, if $f \in L^p(\mathbf{R}, \varrho)$ then the limit*

$$(S_{\mathbf{R}}f)(x) := \lim_{\varepsilon \to 0} \frac{1}{\pi i} \int_{|y-x|>\varepsilon} \frac{f(y)}{y-x} \, dy \qquad (4.29)$$

exists and coincides with $(\tilde{S}_{\mathbf{R}}f)(x)$ for almost all $x \in \mathbf{R}$.

Notice that the integral on the right of (4.29) is well-defined for every f in $L^p(\mathbf{R}, \varrho)$ by virtue of Lemma 4.1. Clear and self-contained (though long) proofs of this result are in Hunt, Muckenhoupt, Wheeden's original paper [107] and Garnett's book [74]. $\qquad \square$

Suppose $\varrho \in A_p(\mathbf{R})$ is an even weight, i.e. $\varrho(-x) = \varrho(x)$ for all $x \in \mathbf{R}$. Then the operator

$$W : L^p(\mathbf{R}; \varrho) \to L^p(\mathbf{R}, \varrho), \ (Wf)(x) := f(-x)$$

is bounded. Put $I = [0, 1]$ and denote by χ_I the (bounded) operator of multiplication by χ_I, the characteristic function of I. Theorem 4.9 implies that the two operators

$$S_I := \chi_I S_{\mathbf{R}} \chi_I \ \text{ and } \ H_I := \chi_I W S_{\mathbf{R}} \chi_I$$

are bounded on $L^p(I, \varrho) := L^p(I, \varrho|I)$. Also by Theorem 4.9, for every $f \in L^p(I, \varrho)$,

$$(S_I f)(x) = \lim_{\varepsilon \to 0} \frac{1}{\pi i} \int_{|y-x|>\varepsilon} \frac{f(y)}{y-x} \, dy, \qquad (4.30)$$

$$(H_I f)(x) = \lim_{\varepsilon \to 0} \frac{1}{\pi i} \int_{|y+x|>\varepsilon} \frac{f(y)}{y+x} \, dy \qquad (4.31)$$

for almost all $x \in I$. Note that in (4.31) we may simply write

$$(H_I f)(x) = \frac{1}{\pi i} \int_I \frac{f(y)}{y+x} \, dx \qquad (4.32)$$

if only $x \in (0, 1]$.

Now let Γ_1 and w be as in Example 3.24 and abbreviate Γ_1 to Γ. For the sake of definiteness, assume

$$\Gamma = \{0\} \cup \{\tau \in \mathbf{C} : \tau = xe^{i\varphi(x)} : 0 < x < 1\}, \qquad (4.33)$$
$$w(\tau) = e^{v(|\tau|)} \quad (\tau \in \Gamma \setminus \{0\})$$

with

$$\varphi(x) = h\big(\log(-\log x)\big)(-\log x), \quad v(x) = g\big(\log(-\log x)\big)\log x$$

where h and g are real-valued functions in $C^2(\mathbf{R})$ such that $h|(-\infty, 0) = 0$ and $g|(-\infty, 0) = 0$, and h, h', h'', g, g', g'' are bounded on $\mathbf{R}$. We know from Proposition 1.4 that Γ is a bounded Carleson arc, and Theorem 2.36 tells us that $w \in A_p(\Gamma)$ if and only if

$$-\frac{1}{p} < \liminf_{x \to 0} \left(xv'(x)\right) \le \limsup_{x \to 0} \left(xv'(x)\right) < \frac{1}{q}. \tag{4.34}$$

The latter theorem also says that

$$\liminf_{x \to 0} \left(xv'(x)\right) = \liminf_{\xi \to +\infty} \left(g(\xi) + g'(\xi)\right),$$

$$\limsup_{x \to 0} \left(xv'(x)\right) = \limsup_{\xi \to +\infty} \left(g(\xi) + g'(\xi)\right).$$

Define a weight ϱ on I by $\varrho(x) := e^{v(x)}$. Theorem 2.36 with $\Gamma = I$ implies that $\varrho \in A_p(I)$ if and only if (4.34) holds. Extend ϱ by symmetric reproduction to $\mathbf{R} = \bigcup_{n \in \mathbf{Z}}(n + I)$. Clearly, the extended weight (which will also be denoted by ϱ) is even.

Put $\gamma(x) := xe^{i\varphi(x)}$. Formally substituting $\tau = \gamma(y)$ and $t = \gamma(x)$ in the integral

$$(S_\Gamma \psi)(t) = \frac{1}{\pi i} \int_\Gamma \frac{\psi(\tau)}{\tau - t} \, d\tau \quad (t \in \Gamma)$$

we arrive at the formula

$$(S_\Gamma \psi)(\gamma(x)) = \frac{1}{\pi i} \int_I \frac{\psi(\gamma(y))\gamma'(y) \, dy}{\gamma(y) - \gamma(x)} \quad (x \in I).$$

Lemma 4.10. *If $x, y \in (0, 1)$ and $x \ne y$, then*

$$\frac{\gamma'(y)}{\gamma(y) - \gamma(x)} = \frac{1}{y - x} + k(x, y)$$

where $|k(x, y)| \le M_1/(x + y)$ with some constant $M_1 < \infty$.

Proof. Clearly,

$$|k(x, y)| = \left| \frac{\gamma(x) - \gamma(y) - \gamma'(y)(x - y)}{(x - y)(\gamma(x) - \gamma(y))} \right|. \tag{4.35}$$

Since $\gamma'(x) = (1 + ix\varphi'(x))e^{i\varphi(x)}$ and $|x\varphi'(x)| \le M$ for all $x \in (0, 1)$, it follows that

$$1 \le |\gamma'(x)| = \sqrt{1 + (x\varphi'(x))^2} \le \sqrt{1 + M^2} \text{ for all } x \in (0, 1). \tag{4.36}$$

A straightforward computation gives

$$x^2\varphi''(x) = \Phi\big(\log(-\log x)\big) - \Phi'\big(\log(-\log x)\big)/\log x$$

where $\Phi(\xi) := h(\xi) + h'(\xi)$. This shows that $x^2\varphi''(x)$ is bounded on $(0,1)$. As

$$\gamma''(x) = \left(2i\varphi'(x) + ix\varphi''(x) - x(\varphi'(x))^2\right)e^{i\varphi(x)},$$

we therefore see that $|\gamma''(x)| \leq M_2/x$ with some constant $M_2 < \infty$. Write $\gamma(x) = \alpha(x) + i\beta(x)$ with real-valued functions α and β. From Taylor's formula we get

$$\gamma(x) - \gamma(y) - \gamma'(y)(x-y) = \frac{\alpha''(\xi)}{2}(x-y)^2 + i\frac{\beta''(\eta)}{2}(x-y)^2$$

with ξ and η between x and y. Since $|\alpha''(\xi)| \leq |\gamma''(\xi)|$ and $|\beta''(\eta)| \leq |\gamma''(\eta)|$, it results that

$$\left|\gamma(x) - \gamma(y) - \gamma'(y)(x-y)\right| \leq \frac{M_2}{2}\left(\frac{1}{\xi} + \frac{1}{\eta}\right)|x-y|^2. \tag{4.37}$$

The points $\gamma(x)$ and $\gamma(y)$ lie on concentric circles with the radii x and y. Hence,

$$|x-y| \leq |\gamma(x) - \gamma(y)|. \tag{4.38}$$

Combining (4.35), (4.37), (4.38) we obtain

$$|k(x,y)| \leq \frac{M_2}{2}\left(\frac{1}{\xi} + \frac{1}{\eta}\right)\frac{|x-y|}{|\gamma(x) - \gamma(y)|} \leq \frac{M_2}{2}\left(\frac{1}{\xi} + \frac{1}{\eta}\right). \tag{4.39}$$

From (4.36) and (4.38) we also get

$$|k(x,y)| \leq \frac{|\gamma'(y)|}{|\gamma(y) - \gamma(x)|} + \frac{1}{|y-x|} \leq \frac{\sqrt{1+M^2}}{|y-x|} + \frac{1}{|y-x|} =: \frac{M_3}{|y-x|}. \tag{4.40}$$

Suppose $y < x$. Then $\xi > y$ and $\eta > y$, whence, by (4.39),

$$|k(x,y)| \leq \frac{M_2}{2}\left(\frac{1}{y} + \frac{1}{y}\right) = \frac{M_2}{y}. \tag{4.41}$$

The two estimates (4.40) and (4.41) give

$$\begin{aligned}|k(x,y)| &\leq (M_2 + M_3)\min\left\{\frac{1}{x-y}, \frac{1}{y}\right\} \\ &\leq (M_2 + M_3)\frac{2}{x} = (M_2 + M_3)\frac{4}{2x} < (M_2 + M_3)\frac{4}{x+y}.\end{aligned}$$

For $y > x$, this inequality can be derived analogously. Thus, we arrive at the assertion with $M_1 := 4(M_2 + M_3)$. □

Lemma 4.11. *If $k(x,y)$ is measurable on $I \times I$ and $|k(x,y)| \leq 1/(x+y)$ for almost all $(x,y) \in I \times I$, then the operator K given for $f \in L^p(I, \varrho)$ by*

$$(Kf)(x) = \int\limits_I k(x,y)f(y)\,dy \quad (x \in (0,1))$$

is bounded on $L^p(I, \varrho)$.

Proof. We abbreviate $\|\cdot\|_{p,\varrho}$ to $\|\cdot\|$. We know that the operator H_I given by (4.32) is bounded on $L^p(I, \varrho)$. If $f \in L^p(I, \varrho)$ and $f \geq 0$ a.e., then

$$
\begin{aligned}
\|Kf\|^p &= \int_I \left| \int_I k(x,y) f(y)\, dy \right|^p \varrho(x)^p dx \\
&\leq \int_I \left(\int_I |k(x,y)| f(y)\, dy \right)^p \varrho(x)^p dx \\
&\leq \int_I \left(\int_I \frac{f(y)}{x+y}\, dy \right)^p \varrho(x)^p dx = \|H_I f\|^p \leq \|H_I\|^p \|f\|^p. \qquad (4.42)
\end{aligned}
$$

For a real-valued function $f \in L^p(I, \varrho)$, put

$$
f_1(x) = \begin{cases} f(x) & \text{if } f(x) \geq 0 \\ 0 & \text{if } f(x) < 0 \end{cases}, \quad f_2(x) = \begin{cases} 0 & \text{if } f(x) \geq 0 \\ -f(x) & \text{if } f(x) < 0 \end{cases}.
$$

Then $f = f_1 - f_2$ and

$$
\begin{aligned}
\|f\|^p &= \int_I |f_1(x) - f_2(x)|^p \varrho(x)^p dx \\
&= \int_{\text{supp } f_1} |f_1(x)|^p \varrho(x)^p dx + \int_{\text{supp } f_2} |f_2(x)|^p \varrho(x)^p dx \\
&= \|f_1\|^p + \|f_2\|^p.
\end{aligned}
$$

Since $f_1 \geq 0$ and $f_2 \geq 0$, we obtain from (4.42) that

$$
\|Kf\| = \|K(f_1 - f_2)\| \leq \|H_I\| (\|f_1\| + \|f_2\|) \leq 2\,\|H_I\|\,\|f\|.
$$

Finally, if $f \in L^p(I, \varrho)$ is complex-valued, we may write $f = f_1 + if_2$ with real-valued functions $f_1, f_2 \in L^p(I, \varrho)$. From what was already proved we get

$$
\begin{aligned}
\|Kf\| = \|K(f_1 + if_2)\| &\leq \|Kf_1\| + \|Kf_2\| \\
&\leq 2\,\|H_I\| (\|f_1\| + \|f_2\|) \leq 2\,\|H_I\| (\|f\| + \|f\|). \qquad \square
\end{aligned}
$$

Theorem 4.12. *If the arc $\Gamma_1 =: \Gamma$ and the weight $w \in A_p(\Gamma)$ are as in Example 3.24, then S_Γ generates a bounded operator $\tilde{S}_\Gamma$ on $L^p(\Gamma, w)$. For $\psi \in L^p(\Gamma, w)$, the limit*

$$
(S_\Gamma \psi)(t) := \lim_{\varepsilon \to 0} \frac{1}{\pi i} \int_{\Gamma \setminus \Gamma(t, \varepsilon)} \frac{\psi(\tau)}{\tau - t}\, d\tau
$$

exists and coincides with $(\tilde{S}_\Gamma \psi)(t)$ for almost all $t \in \Gamma$.

Proof. Since S_I and H_I are bounded on $L^p(I, \varrho)$ and given by (4.30) and (4.32), it follows from Lemmas 4.10 and 4.11 that the operator A given by

$$(Af)(x) := \lim_{\varepsilon \to 0} \int\limits_{|y-x|>\varepsilon} \frac{f(y)\gamma'(y)\,dy}{\gamma(y) - \gamma(x)}$$

$$= \lim_{\varepsilon \to 0} \int\limits_{|y-x|>\varepsilon} \frac{f(y)\,dy}{y - x} + \int\limits_I k(x,y)f(y)\,dy \quad \left(x \in (0,1)\right)$$

is well-defined and bounded on $L^p(I, \varrho)$. From (4.36) we infer that the operator

$$U : L^p(\Gamma, w) \to L^p(I, \varrho), \quad (U\psi)(x) := \psi\big(\gamma(x)\big)$$

is an isomorphism. A straightforward computation shows that if $f \in L^p(I, \varrho)$, then

$$(US_\Gamma U^{-1} f)(x) = \lim_{\varepsilon \to 0} \int\limits_{|\gamma(y)-\gamma(x)|>\varepsilon} \frac{f(y)\gamma'(y)}{\gamma(y) - \gamma(x)}\,dy,$$

$$= \lim_{\varepsilon \to 0} \int\limits_{|y-x|>\varepsilon} \frac{f(y)\gamma'(y)}{\gamma(y) - \gamma(x)}\,dy = (Af)(x),$$

the second equality resulting from the fact that

$$|x - y| \le \big|\gamma(x) - \gamma(y)\big| \le \|\gamma'\|_\infty |x - y|.$$

Thus, $S_\Gamma = U^{-1}AU$. As $U^{-1}AU$ is bounded, we get the assertion. $\qquad\square$

Things are more complicated if Γ is not an arc but a Jordan curve $\Gamma = \Gamma_1 \cup \Gamma_2$ as in Example 3.24. However, one case can be managed quite easily.

Theorem 4.13. *The conclusions of the previous theorem remain true if $\Gamma = \Gamma_1 \cup \Gamma_2$ is a Jordan curve and $w \in A_p(\Gamma)$ is a weight as in Example 3.24 and if, in addition, $b(r) = \beta \in (0, 2\pi)$ for all r sufficiently close to 0.*

Proof. Suppose $b(r) = \beta$ for $r < r_0 < 1$, put $\eta_j := \{\tau \in \Gamma_j : |\tau| \le r_0\}$ $(j = 1, 2)$, and $\eta := \eta_1 \cup \eta_2$. A little thought shows that it suffices to prove that S_η is bounded on $L^p(\eta, w|\eta)$. We may represent S_η as an operator matrix

$$\begin{pmatrix} S_{11} & S_{12} \\ S_{21} & S_{22} \end{pmatrix} : \begin{pmatrix} L^p(\eta_1, w|\eta_1) \\ L^p(\eta_2, w|\eta_2) \end{pmatrix} \to \begin{pmatrix} L^p(\eta_1, w|\eta_1) \\ L^p(\eta_2, w|\eta_2) \end{pmatrix}.$$

From Theorem 4.12 we deduce that S_{11} and S_{22} are bounded. Also as above, the boundedness of S_{12} may be reduced to the boundedness of the operator

$$(Bf)(x) := \int\limits_0^{r_0} \frac{f(y)\gamma_2'(y)}{\gamma_2(y) - \gamma_1(x)}\,dy \quad \left(x \in (0, r_0)\right)$$

on the space $L^p([0, r_0], e^{v(x)})$ where

$$\gamma_1(x) := xe^{i\varphi(x)}, \ \gamma_2(y) := ye^{i\beta}e^{i\varphi(y)}.$$

We show that there is a constant $M_4 < \infty$ such that

$$1/|\gamma_1(x) - \gamma_2(y)| \le M_4/(x+y) \ \text{ for all } \ x, y \in (0, r_0). \tag{4.43}$$

Once this is established, one can prove that B is bounded as above with the help of Lemma 4.11. For S_{21} one may proceed in the same way.

Suppose $y \le x$ and let $\delta \in (0, 1)$. We have

$$|\gamma_1(x) - \gamma_2(y)| = |xe^{i\varphi(x)} - e^{i\beta}ye^{i\varphi(y)}| \ge x - y,$$

and $x - y > \delta(x+y)$ if only $y < ((1-\delta)/(1+\delta))x$. So assume

$$\frac{1-\delta}{1+\delta}x \le y \le x.$$

Then

$$-\log\frac{1-\delta}{1+\delta} - \log x \ge -\log y \ge -\log x, \tag{4.44}$$

whence

$$\log\left(-\log x - \log\frac{1-\delta}{1+\delta}\right) \ge \log(-\log y) \ge \log(-\log x). \tag{4.45}$$

The most left term of (4.45) is

$$\log(-\log x) + \log\left(1 + \frac{1}{|\log x|}\left|\log\frac{1-\delta}{1+\delta}\right|\right) = \log(-\log x) + O\left(\frac{\delta}{|\log x|}\right). \tag{4.46}$$

From (4.44), (4.45), (4.46), and the boundedness of h' on $\mathbf{R}$ we obtain

$$
\begin{aligned}
\varphi(y) &= h\big(\log(-\log y)\big)(-\log y) \\
&= h\big(\log(-\log x) + O(\delta/|\log x|)\big)(-\log x + O(\delta)) \\
&= \Big(h\big(\log(-\log x)\big) + O(\delta/|\log x|)\Big)(-\log x + O(\delta)) \\
&= \varphi(x) + O(\delta) + O(\delta) + O\big(\delta^2/|\log x|\big) = \varphi(x) + O(\delta),
\end{aligned}
$$

the last equality following from the fact that $0 < x < r_0 < 1$. Thus,

$$
\begin{aligned}
|\gamma_1(x) - \gamma_2(y)| &= |xe^{i\varphi(x)} - e^{i\beta}ye^{i\varphi(y)}| \\
&= |xe^{i\varphi(x)} - e^{i\beta}ye^{i(\varphi(x)+O(\delta))}| = |x - e^{i\beta}ye^{iO(\delta)}| \\
&= |x - e^{i\beta}y - e^{i\beta}y(e^{iO(\delta)} - 1)| \\
&\ge |x - e^{i\beta}y| - y\,O(\delta).
\end{aligned}
\tag{4.47}
$$

If $\beta = \pi$, we have $|x - e^{i\beta}y| = x + y$. In case $\beta \neq \pi$, we get

$$|x - e^{i\beta}y| \geq \text{dist}\,(x, e^{i\beta}\mathbf{R}_+) = x|\sin\beta|,$$
$$|x - e^{i\beta}y| \geq \text{dist}\,(e^{i\beta}y, \mathbf{R}_+) = y|\sin\beta|,$$

whence $|x - e^{i\beta}y| \geq (x + y)|\sin\beta|/2$. In either case, there is a constant $C(\beta)$ depending only on β such that $|x - e^{i\beta}y| \geq C(\beta)(x + y)$. If $\delta > 0$ is sufficiently small, then the $O(\delta)$ in (4.47) is at most $C(\beta)/2$. For this δ,

$$\left|\gamma_1(x) - \gamma_2(y)\right| \geq C(\beta)(x + y) - C(\beta)y/2 > C(\beta)(x + y)/2.$$

In summary,

$$\left|\gamma_1(x) - \gamma_2(y)\right| \geq \min\{\delta, C(\beta)/2\}\,(x + y). \qquad \square$$

4.4 Brief survey of results on general curves and weights

The following two theorems contain (in a sense) final answers to the questions on the existence and boundedness of the Cauchy singular integral on composed locally rectifiable curves.

Theorem 4.14. *If Γ is a composed locally rectifiable curve and $f \in L^1(\Gamma)$, then the limit*

$$(Sf)(t) := \lim_{\varepsilon \to 0} \int_{\Gamma \backslash \Gamma(t, \varepsilon)} \frac{f(\tau)}{\tau - t}\, d\tau \qquad (4.48)$$

exists and is finite for almost all $t \in \Gamma$.

Theorem 4.15. *Let Γ be a composed locally rectifiable curve, let $1 < p < \infty$, and let w be a weight on Γ. The Cauchy singular integral S generates a bounded operator $\tilde{S}$ on $L^p(\Gamma, w)$ if and only if $w \in A_p(\Gamma)$. In that case $Sf = \tilde{S}f$ a.e. on Γ for every $f \in L^p(\Gamma, w)$.*

Notice that $L^p(\Gamma, w) \subset L^1(\Gamma)$ if Γ is bounded and $w \in A_p(\Gamma)$. Also recall that the integral in (4.48) is well-defined for all $f \in L^p(\Gamma, w)$ whenever $w \in A_p(\Gamma)$ (Lemma 4.1). For nice (piecewise Lyapunov) curves and nice (power) weights, these two theorems have been well known for a long time and we confine ourselves with referring the reader to Khvedelidze's 1956 papers [122], [123], his 1975 survey [124], and the standard texts by Privalov [162], Muskhelishvili [151], Gakhov [72], Danilyuk [45], Gohberg and Krupnik [89], Mikhlin and Prössdorf [147].

A new period started in 1977 with Carlderón's paper [32]. A curve Γ is said to be a *Lipschitz curve* if it may be represented in the form $\Gamma = \{x + i\varphi(x) : x \in \mathbf{R}\}$ where $\varphi : \mathbf{R} \to \mathbf{R}$ satisfies a Lipschitz condition,

$$\left|\varphi(x_1) - \varphi(x_2)\right| \leq K|x_1 - x_2| \quad \text{for all}\ \ x_1, x_2 \in \mathbf{R}. \qquad (4.49)$$

The best constant in (4.49) is referred to as the *Lipschitz constant* of Γ. One can show that φ satisfies a Lipschitz condition if and only if φ is differentiable almost everywhere on $\mathbf{R}$ and $\varphi' \in L^\infty(\mathbf{R})$. The Lipschitz constant is $K = \|\varphi'\|_\infty$.

Theorem 4.16. *If Γ is a Lipschitz curve then*

$$\|S\varphi\|_{L^2(\Gamma)} \le C(K)\|\varphi\|_{L^2(\Gamma)} \ \ for \ all \ \ \varphi \in C_0^\infty(\Gamma)$$

where $C(K)$ depends only on the Lipschitz constant K of Γ.

The point of this theorem is that the norm of S is estimated by a constant depending only on the Lipschitz constant K. Calderón [32] proved this theorem under the assumption that K is sufficiently small. This extra assumption was removed by Coifman, McIntosh, and Meyer [38] in 1982. The proof of [38] is difficult. In the sequel, other proofs were presented by several mathematicians, including Murai [149], David [47], Coifman, Jones, and Semmes [37]. The last three authors' proof will be given in Chapter 5.

As for Theorem 4.14, Dynkin [60] writes that "it was de facto established with Calderón's 1977 paper, although this was not realized at once. Namely, already in 1964, Havin [101] observed that Sf exists almost everywhere on Γ for arbitrary $f \in L^1(\Gamma)$ if only Sf exists almost everywhere for $f \in C_0(\Gamma)$. A slight modification of his argument allows reducing the case of an arbitrary locally rectifiable curve to the case of a Lipschitz curve with arbitrarily small Lipschitz constant." Dynkin then presents the modified reasoning of Havin; thus, a derivation of Theorem 4.14 from Theorem 4.16 is in [60, pp. 247–249].

The next milestone on the way to Theorem 4.15 is the following theorem.

Theorem 4.17. *Let Γ be a composed locally rectifiable curve. Then S generates a bounded operator on $L^p(\Gamma)$ $(1 < p < \infty)$ if and only if Γ is a Carleson curve.*

This theorem was established by David [46] in the early eighties. David proved it using Calderón's original result, i.e. Theorem 4.16 for small Lipschitz constants. Independently, the "only part" of Theorem 4.17 was obtained in 1982 by Paatashvili and Khuskivadze [155], who also raised the conjuncture that S generates a bounded operator on $L^p(\Gamma)$ whenever Γ is Carleson. We also wish to mention that it had been clear for many decades that the boundedness of S on $L^p(\Gamma)$ is solely a matter of the curve and not of the exponent $p \in (1, \infty)$. For example, already in Khvedelidze's survey [124] one may read that S is bounded on $L^p(\Gamma)$ for all $p \in (1, \infty)$ if S is bounded on $L^{p_0}(\Gamma)$ for some $p_0 \in (1, \infty)$.

The Muckenhoupt class $A_p(\Gamma)$ entered the scene in 1972 with Muckenhoupt's paper [148], in which it was shown that the maximal operator M (recall Section 2.6) is bounded on $L^p(\mathbf{R}, w)$ $(1 < p < \infty)$ if and only if $w \in A_p(\mathbf{R})$. For $\Gamma = \mathbf{R}$ and $\Gamma = \mathbf{T}$, Theorem 4.15 is a 1973 result of Hunt, Muckenhoupt, and Wheeden [107]. For these two cases, a detailed proof is also in [74]. Theorem 4.15 was proved in 1975 by Kokilashvili [125] under the assumption that Γ is a piecewise Lyapunov curve without cusps and in 1989 by Dynkin [61] under the sole assumption that Γ is a Carleson curve. Combining Dynkin's result with Theorem 4.8, we get Theorem 4.15 as it is stated here.

The way from Theorem 4.16 via Theorems 4.17 and 4.8 to Theorems 4.14 and 4.15 is spiralic, long, and hard. Researchers in the field consider many things as "elementary" or "standard", and therefore it is difficult for amateurs to read and understand the relevant papers. An attempt of presenting a more or less self-contained proof is made in Chapter 5. Readers who are satisfied by the above remarks may entirely skip Chapter 5 and continue with Chapter 6.

Before leaving this section, we want at least cite a remarkable counterpart of Theorem 4.15 in the case where $\Gamma = \mathbf{T}$ and $p = 2$: the Helson-Szegö theorem. Given a real-valued function $v \in L^2(\mathbf{T})$, denote by $\tilde{v} \in L^2(\mathbf{T})$ the *conjugate function*, i.e. the function determined uniquely by the requirement that the harmonic extension of $v + i\tilde{v}$ into $\mathbf{D}$ is analytic and vanishes at the origin. Equivalently, if v has the Fourier series

$$v(t) = \sum_{n=-\infty}^{\infty} v_n t^n \quad (t \in \mathbf{T}),$$

then

$$\tilde{v}(t) := -i\left(-\sum_{n=-\infty}^{-1} v_n t^n + \sum_{n=1}^{\infty} v_n t^n \right) \quad (t \in \mathbf{T}).$$

Theorem 4.18 (Helson and Szegö). *A weight w on $\mathbf{T}$ belongs to $A_2(\mathbf{T})$ if and only if $w = e^{u+\tilde{v}}$ where u and v are real-valued functions in $L^\infty(\mathbf{T})$ and $\|v\|_\infty < \pi/4$.*

This theorem was established in 1960 by Helson and Szegö in [104]. Beautiful presentations of the proof are also in the books [74] and [127]. Let us note that until now no "direct" proof of Theorem 4.18 is available: all proofs of this theorem are based on the observation that both the conditions of Theorem 4.18 and Theorem 4.15 (for $\Gamma = \mathbf{T}$ and $p = 2$) are equivalent to the boundedness of S on $L^2(\mathbf{T}, w)$.

It turns out that a Helson-Szegö type condition is also true for the simultaneous membership of a weight in $A_p(\mathbf{T})$ and $A_q(\mathbf{T})$.

Theorem 4.19. *Let $2 \le p < \infty$ and $1/p + 1/q = 1$. For a weight w on $\mathbf{T}$ to belong to $A_p(\mathbf{T}) \cap A_q(\mathbf{T})$ it is necessary and sufficient that $w = e^{u+\tilde{v}}$ where u and v are real-valued functions in $L^\infty(\mathbf{T})$ and $\|v\|_\infty < \pi/(2p)$.*

The sufficiency portion of this theorem is due to Simonenko [194], the necessity part is Krupnik's [132].

4.5 Composing curves and weights

Theorem 4.15 is usually only proved for unbounded arcs Γ. The reason is that bounded arcs cause difficulties in connection with the Calderón-Zygmund decomposition and thus with obtaining a weak estimate of the type (1,1) for the singular integral for all $\lambda > 0$. The purpose of this section is to show how the case of an arbitrary (bounded or unbounded) composed curve can be reduced to the case of an unbounded arc.

Lemma 4.20. *Let Γ_0 be a bounded composed Carleson curve and $w_0 \in A_p(\Gamma_0)$. Let $t_0 \in \Gamma_0$ and suppose Γ_0 has a (two-sided) tangent at t_0. Assume further that $\mathbf{C} \setminus \Gamma_0$ is connected and let $z_0 \in \mathbf{C} \setminus \Gamma_0$. Then there exists a Carleson arc B joining t_0 to z_0 and a weight $w \in A_p(\Gamma_0 \cup B)$ such that $B \cap \Gamma_0 = \{t_0\}$ and $w|\Gamma_0 = w_0$.*

Proof. Choose an open arc $\gamma \subset \Gamma_0$ such that $t_0 \in \gamma$, and all points of γ have multiplicity 2, i.e. are not endpoints of more than two subarcs of Γ_0. The set $\Gamma_0 \setminus \gamma$ is compact and the function $\tau \mapsto |\tau - t_0|$ is continuous on this set. Let μ_0 (> 0) be the minimum of this function. Then the portion $\Gamma_0(t_0, \mu_0)$ contains only points of the arc γ:

$$\Gamma_0(t_0, \mu_0) = \gamma(t_0, \mu_0). \tag{4.50}$$

Let $s \mapsto z(s)$ be the natural parametrization of γ and suppose $t_0 = z(s_0)$. Since the tangent to γ exists at t_0, we have

$$z(s) = z(s_0) + z'(s_0)(s - s_0) + o(|s - s_0|) \quad \text{as } s \mapsto s_0. \tag{4.51}$$

Without loss of generality assume $z'(s_0) = 1$. There is a $\delta_0 > 0$ such that

$$\big|o(|s - s_0|)\big| < (1/2)|s - s_0| \quad \text{for } |s - s_0| \le \delta_0. \tag{4.52}$$

From (4.51), (4.52), and the inequality $\arcsin(1/2) < \pi/4$ we see that the arc $\gamma_0 := \gamma[t_0, \delta_0] := \{t \in \gamma : \ell(t, t_0) < \delta_0\}$ is contained in the double sector

$$\Delta := \{t_0 + e^{i\theta}(s - s_0) : s \in \mathbf{R}, \ -\pi/4 < \theta < \pi/4\}.$$

Put $\varepsilon_0 := (1/3)\min\{\delta_0, \mu_0\}$. Then, by (4.50),

$$\Gamma_0(t, 3\varepsilon_0) \subset \gamma(t_0, \mu_0) \cap \gamma[t_0, \delta_0] \subset \Delta. \tag{4.53}$$

Denote by Γ_1 the segment $\Gamma_1 := \{t_0 + i(s - s_0) : 0 \le s - s_0 \le 2\,\varepsilon_0\}$ and define a weight w_1 on Γ_1 by

$$w_1\big(t_0 + i(s - s_0)\big) := w_0\big(z(s)\big).$$

From (4.53) we infer that $\Gamma_0 \cap \Gamma_1 = \{t_0\}$ and that therefore $\Gamma_0 \cup \Gamma_1$ is a composed Carleson curve. If $z_0 \in \Gamma_1$, we put $B := [t_0, z_0]$ and restrict w_1 to B. If $z_0 \notin \Gamma_1$, we may find an arc $\Gamma_2 \subset \mathbf{C} \setminus (\Gamma_0 \cup \Gamma_1)$ joining $t_1 = t_0 + 2i\varepsilon_0$ to z_0. We may assume that Γ_2 is a broken line composed of a finite number of horizontal or vertical line segments of equal length $\lambda_0 < \varepsilon_0/2$. Obviously, there is a $\nu_0 < \varepsilon_0$ such that

$$\{z \in \mathbf{C} : \text{dist}\,(z, \Gamma_2) < \nu_0\} \cap \Gamma = \emptyset. \tag{4.54}$$

Let $[t_1, t_2] = [t_1, t_1 + \varrho_1\lambda_0]$ ($\varrho_1 \in \{\pm 1, i\}$) be the first segment of Γ_2. We define a weight w_2 on $[t_1, t_2]$ by

$$w_2(t_1 + \varrho_1 x\lambda_0) := w_1(t_1 - ix\lambda_0) \quad \big(x \in [0, 1]\big).$$

Further, if $[t_2, t_3] = [t_2, t_2 + \varrho_2\lambda_0]$ $(\varrho_2 \in \{\pm1, \pm i\} \setminus \{-\varrho_1\})$ is the next segment of Γ_2, we define w_3 on $[t_2, t_3]$ by

$$w_3(t_2 + \varrho_2 x\lambda_0) := w_2(t_2 - \varrho_1 x\lambda_0) \quad (x \in [0, 1]).$$

Continuing this procedure of "symmetric reproduction" we obtain weights w_j $(j = 2, 3, \ldots, N)$ on the segments of Γ_2. Finally, define w on $\Gamma_0 \cup \Gamma_1 \cup \Gamma_2$ by $w|\Gamma_0 = w_0$, $w|\Gamma_1 = w_1$, $w|[t_{j-1}, t_j] = w_j$.

It is clear that $B := \Gamma_1 \cup \Gamma_2$ is the desired Carleson arc. We are so left with proving that $w \in A_p(\Gamma_0 \cup B)$. Put $\eta_0 := \min\{\lambda_0, \nu_0\}$. We first consider $\Gamma := \Gamma_0 \cup \Gamma_1$ and show that $w|\Gamma \in A_p(\Gamma)$. This will follow as soon as we have shown that $\sup_{t\in\Gamma} \sup_{\eta<\eta_0} M_\eta(t) < \infty$ where

$$M_\eta(t) := \frac{1}{\eta}\left(\int_{\Gamma(t,\eta)} w(\tau)^p|d\tau|\right)^{1/p}\left(\int_{\Gamma(t,\eta)} w(\tau)^{-q}|d\tau|\right)^{1/q}.$$

Let first $t \in \Gamma_0$ and $\eta < \eta_0$. If $\Gamma(t,\eta) \cap \Gamma_1 = \emptyset$, then $\Gamma(t,\eta) = \Gamma_0(t,\eta)$ and hence

$$M_\eta(t) = \frac{1}{\eta}\left(\int_{\Gamma_0(t,\eta)} w_0(\tau)^p|d\tau|\right)^{1/p}\left(\int_{\Gamma_0(t,\eta)} w_0(\tau)^{-q}|d\tau|\right)^{1/q} \leq M < \infty$$

with some M independent of t and η. So let $\Gamma(t,\eta) \cap \Gamma_1 \neq \emptyset$. Then $|\mathrm{Re}\,(t-t_0)| < \eta$ and $|t - t_0| < 3\varepsilon_0$, which implies that t lies in the set

$$\Delta \cap \{z \in \mathbf{C} : |\mathrm{Re}\,(z - t_0)| < \eta\}.$$

Thus, $|t - t_0| \leq \sqrt{\eta^2 + (\eta\tan(\pi/4))^2} = \sqrt{2}\eta$ and therefore,

$$\Gamma_0(t,\eta) \subset \Gamma_0(t_0, (1 + \sqrt{2})\eta) \subset \Gamma_0(t_0, 3\eta).$$

The set $\Gamma_1(t,\eta)$ is an interval of length at most 2η whose center is at a distance of at most η to t_0. Consequently, $\Gamma_1(t,\eta) \subset \Gamma_1(t_0, 3\eta)$. Let $\gamma_{t,\eta} \subset \Gamma_0$ be the preimage of the set $\Gamma_1(t,\eta)$ under the map $z(s) \mapsto t_0 + i(s - s_0)$. Since $\Gamma_1(t,\eta) \subset \Gamma_1(t_0, 3\eta)$, we see that $\gamma_{t,\eta} \subset \gamma_0[t_0, 3\eta] \subset \Gamma_0(t_0, 3\eta)$. Thus,

$$M_\eta(t) = \frac{1}{\eta}\left(\int_{\gamma_{t,\eta}} w_0(\tau)^p|d\tau| + \int_{\Gamma_0(t,\eta)} w_0(\tau)^p|d\tau|\right)^{1/p} \times$$

$$\times \left(\int_{\gamma_{t,\eta}} w_0(\tau)^{-q}|d\tau| + \int_{\Gamma_0(t,\eta)} w_0(\tau)^{-q}|d\tau|\right)^{1/q}$$

$$\leq \frac{1}{\eta}\left(2\int_{\Gamma_0(t_0,3\eta)} w_0(\tau)^p|d\tau|\right)^{1/p}\left(2\int_{\Gamma_0(t_0,3\eta)} w_0(\tau)^{-q}|d\tau|\right)^{1/q} \leq 6M < \infty \quad (4.55)$$

with M independent of t and η.

Now let $t \in \Gamma_1$ and $\eta < \eta_0$. Denote by $\gamma_{t,\eta}$ the preimage of $\Gamma_1(t, \eta)$ under the map $z(s) \mapsto t_0 + i(s - s_0)$. Since $|\gamma_{t,\eta}| = |\Gamma_1(t, \eta)| \leq 2\eta$, we have $\gamma_{t,\eta} \subset \Gamma_0(\tilde{t}, 2\eta)$ where $\tilde{t}$ is any point of $\gamma_{t,\eta}$. Thus, if $\Gamma(t, \eta) \cap \Gamma_0 = \emptyset$, then

$$
\begin{aligned}
M_\eta(t) &= \frac{1}{\eta} \left(\int\limits_{\gamma_{t,\eta}} w_0(\tau)^p |d\tau| \right)^{1/p} \left(\int\limits_{\gamma_{t,\eta}} w_0(\tau)^{-q} |d\tau| \right)^{1/q} \\
&\leq \frac{1}{\eta} \left(\int\limits_{\Gamma_0(\tilde{t}, 2\eta)} w_0(\tau)^p |d\tau| \right)^{1/p} \left(\int\limits_{\Gamma_0(\tilde{t}, 2\eta)} w_0(\tau)^{-q} |d\tau| \right)^{1/q} \leq 2M < \infty
\end{aligned}
$$

with M independent of t and η. So assume $\Gamma(t, \eta) \cap \Gamma_0 \neq \emptyset$. Then the distance between t and the double sector Δ is at most η, implying that $|t - t_0| \leq \sqrt{\eta^2 + \eta^2} = \sqrt{2}\eta$. Consequently, $\Gamma_1(t, \eta) \subset \Gamma_1(t_0, (1 + \sqrt{2})\eta) \subset \Gamma_1(t_0, 3\eta)$ and thus, $\gamma_{t,\eta} \subset \Gamma_0(t_0, 3\eta)$. If τ is any point in $\Gamma(t, \eta) \cap \Gamma_0 = \Gamma_0(t, \eta)$, then $|\tau - t_0| \leq |\tau - t| + |t - t_0| < \eta + \sqrt{2}\eta < 3\eta$, whence $\Gamma_0(t, \eta) \subset \Gamma_0(t_0, 3\eta)$. Thus, again (4.55) holds.

The proof of the fact that $w|\Gamma_0 \cup \Gamma_1 \subset A_p(\Gamma_0 \cup \Gamma_1)$ is complete. Repeating the above reasoning (with only minor modifications and using (4.54)) one can show that $w|\Gamma_0 \cup \Gamma_1 \cup [t_1, t_2] \in A_p(\Gamma_0 \cup \Gamma_1 \cup [t_1, t_2])$. Continuing in this way, one finally gets that $w \in A_p(\Gamma_0 \cup B)$. $\qquad \square$

Lemma 4.21. *Let Γ_0 be a bounded composed Carleson curve and $w \in A_p(\Gamma_0)$. Let $t_1, t_2 \in \Gamma_0$ be two points at which the (two-sided) tangent to Γ_0 exists. If $\mathbf{C} \setminus \Gamma_0$ is connected, then there exist two disjoint Carleson arcs U_1 and U_2 joining t_1 and t_2, respectively, to infinity and a weight $w \in A_p(\Gamma_0 \cup U_1 \cup U_2)$ such that $U_1 \cap \Gamma_0 = \{t_1\}$, $U_2 \cap \Gamma_0 = \{t_2\}$, and $w|\Gamma_0 = w_0$.*

Proof. Suppose Γ_0 is contained in the disk $D_R := \{z \in \mathbf{C} : |z| < R\}$. Choose two points z_1 and z_2 outside D_R. Twice using Lemma 4.20, we may construct two disjoint bounded Carleson arcs B_j ($j = 1, 2$) joining t_j to z_j and a weight $w \in A_p(\Gamma_0 \cup B_1 \cup B_2)$ such that $B_j \cap \Gamma_0 = \{t_j\}$ and $w|\Gamma_0 = w_0$. Moreover, from the proof of Lemma 4.20 we know that the arcs B_j can be chosen to be broken lines of horizontal and vertical segments. When tracing out B_j from t_j to z_j, there is a first segment $\sigma_{n_j} = [t_{n-1}^{(j)}, t_n^{(j)}] \subset B_j$ such that $(t_{n-1}^{(j)}, t_n^{(j)}]$ hits ∂D_R. We remove the segments of B_j following this segment and extend σ_{n_j} by a ray L_j to infinity. Thus, let $U_j = \sigma_1 \cup \ldots \cup \sigma_{n_j} \cup L_j$. We may clearly assume that $L_1 \cap L_2 = \emptyset$. Then U_1 and U_2 have the required properties. Divide the ray L_j into segments $\Delta_j^{(1)}, \Delta_j^{(2)}, \Delta_j^{(3)}, \ldots$ of equal length $|\sigma_{n_j}|$ and extend $w|\sigma_{n_j}$ to a weight on $\sigma_{n_j} \cup L_j$ by symmetric reproduction. Slightly modifying the proof of Theorem 2.7, one can show that the extended weight belongs to $A_p(\Gamma_0 \cup U_1 \cup U_2)$. $\qquad \square$

Lemma 4.22. *Suppose we know that S_Γ is a bounded operator on $L^p(\Gamma, w)$ whenever Γ is an unbounded Carleson arc and $w \in A_p(\Gamma)$. Then S_γ is a bounded operator on $L^p(\gamma, \varrho)$ for every bounded Carleson arc γ and every $\varrho \in A_p(\gamma)$.*

Proof. Let γ be a bounded Carleson arc and $\varrho \in A_p(\gamma)$. Denote the endpoints of γ by a and b. Choose two distinct points t_1 and t_2 on γ such that the (two-sided) tangent to γ exists at t_1 and t_2. We may assume that t_1 lies before t_2 as γ is traced out from a to b. So t_1 and t_2 divide γ into three consecutive arcs: $\gamma = \gamma_1 \cup \gamma_2 \cup \gamma_3$, where γ_1 is the arc between a and t_1, γ_2 is the arc between t_1 and t_2, and γ_3 is the arc between t_2 and b. Extend γ to $\eta := \gamma \cup U_1 \cup U_2$ and $\varrho \in A_p(\gamma)$ to a weight $v \in A_p(\eta)$ as in Lemma 4.21. Put $\gamma_4 := U_1$, $\gamma_5 = U_2$. The space $L^p(\eta, v)$ decomposes into the direct sum

$$L^p(\eta, v) = \bigoplus_{i=1}^{5} L^p(\gamma_i, v|\gamma_i).$$

Accordingly, S_η may be represented by a 5×5 operator matrix $(S_{jk})_{j,k=1}^{5}$ where

$$S_{jk} := \chi_j S_\eta \chi_k : L^p(\gamma_k, w|\gamma_k) \to L^p(\gamma_j, w|\gamma_j)$$

and χ_i stands for characteristic function of γ_i. We have to show that the 9 entries of the 3×3 operator matrix $(S_{jk})_{j,k=1}^{3}$ are bounded operators.

Since $\Gamma_1 := \gamma_1 \cup \gamma_2 \cup \gamma_5$ is an unbounded Carleson arc and $v|\Gamma_1 \in A_p(\Gamma_1)$, the operator S_{Γ_1} is bounded on $L^p(\Gamma_1, v|\Gamma_1)$ by assumption. This implies that

$$S_{11} = \chi_1 S_{\Gamma_1} \chi_1, \ S_{12} = \chi_1 S_{\Gamma_1} \chi_2, \ S_{21} = \chi_2 S_{\Gamma_1} \chi_1, \ S_{22} = \chi_2 S_{\Gamma_1} \chi_2$$

are bounded. Analogously on gets the boundedness of S_{23}, S_{32}, S_{33} from the boundedness of $S_{\gamma_4 \cup \gamma_2 \cup \gamma_3}$. Since dist $(\gamma_1, \gamma_3) > 0$, it is easily seen that S_{13} and S_{31} are bounded (even compact). $\qquad \square$

Theorem 4.23. *Suppose we know that S_Γ is bounded on $L^p(\Gamma, w)$ whenever Γ is an unbounded Carleson arc and $w \in A_p(\Gamma)$. Then S_γ is bounded on $L^p(\gamma, \varrho)$ for every composed Carleson curve γ and every weight $\varrho \in A_p(\gamma)$.*

Proof. Let γ be a composed Carleson curve and $\varrho \in A_p(\Gamma)$. Then $\gamma = \gamma_1 \cup \ldots \cup \gamma_n$ where $\gamma_1, \ldots, \gamma_n$ are (bounded or unbounded) Carleson arcs each pair of which have at most one endpoint in common. Decomposing $L^p(\gamma, \varrho)$ into the direct sum

$$L^p(\gamma, \varrho) = \bigoplus_{i=1}^{n} L^p(\gamma_i, \varrho|\gamma_i),$$

we may represent S_γ as an operator matrix $(S_{jk})_{j,k=1}^{n}$. From Lemma 4.22 we infer that the operators S_{jj} are bounded for each j. If γ_j and γ_k have a common endpoint, then $\gamma_j \cup \gamma_k$ is an arc, and hence

$$S_{jk} = \chi_j S_{\gamma_j \cup \gamma_k} \chi_k, \ \text{and} \ S_{kj} = \chi_k S_{\gamma_j \cup \gamma_k} \chi_j$$

are bounded due to Lemma 4.22. Finally, if γ_j and γ_k have no point in common, we may connect γ_j and γ_k by arcs in $\{\gamma_1, \ldots, \gamma_n\} \setminus \{\gamma_j, \gamma_k\}$ to an arc $\gamma_j \cup \eta \cup \gamma_k$. Since $S_{\gamma_j \cup \eta \cup \gamma_k}$ is bounded by Lemma 4.22, we deduce that

$$S_{jk} = \chi_j S_{\gamma_j \cup \eta \cup \gamma_k} \chi_k \ \text{ and } \ S_{kj} = \chi_k S_{\gamma_j \cup \eta \cup \gamma_k} \chi_j$$

are also bounded. This shows that each entry of the operator matrix $(S_{jk})_{j,k=1}^n$ is a bounded operator. $\qquad\square$

4.6 Notes and comments

4.1. These things are well known. We only remark that the proof of Theorem 4.2 given in the text differs from the proofs one can find in the standard books [45], [151], [72]; the latter proofs are based on the definition (4.13).

4.2. That Γ must be a Carleson curve whenever S_Γ is bounded on $L^p(\Gamma)$ for some $p \in (1, \infty)$ was first proved independently by David [46], [47] and Paatashvili and Khuskivadze [155]. Subsequently it was observed by several authors, including Dynkin, Osilenker [62] and Gohberg, Krupnik [89, Vol. 1, p. 50], that the Hunt, Muckenhoupt, Wheeden result [107] can be generalized as follows: *if Γ is a Carleson curve*, then S_Γ generates a bounded operator on $L^p(\Gamma, w)$ if and only if $w \in A_p(\Gamma)$. One can easily show that the boundedness of S_Γ on $L^p(\Gamma, w)$ implies its boundedness on $L^2(\Gamma)$ (see, e.g., [15, Theorem 3.1]), and hence the a-priori assumption that Γ be a Carleson curve is in fact superfluous.

The idea of the straightforward proof of Theorem 4.8 given in the text is due to Coifmann and Fefferman [36] (see also the paper by Kazarian [120]).

4.3. We vacillated a long time between including Chapter 5 into the book or not, and Section 4.3 was originally intended to replace Chapter 5 in order to make the book self-contained. Eventually we included both. Note that without Chapter 5 the situation is as follows. Fix $p \in (1, \infty)$. Given a curve Γ and a weight w on Γ, let us write $(\Gamma, w) \in \mathcal{B}_p$ if S_Γ is bounded on $L^p(\Gamma, w)$. Our aim is to determine the spectra of Toeplitz operators with piecewise continuous symbols on $L_+^p(\Gamma, w)$ under the assumption that $(\Gamma, w) \in \mathcal{B}_p$. Theorem 4.8 shows Γ is Carleson and that $w \in A_p(\Gamma)$ whenever $(\Gamma, w) \in \mathcal{B}_p$. Thus, we may use the fact that Γ and w satisfy the Carleson and Muckenhoupt conditions, respectively, and in this way we obtain Theorem 7.4 with the information about the leaves contained in Section 3.7. We are then left with proving that the results of Section 3.7 are "sharp", which is done in Section 3.8, and for this purpose we need only consider curves and weights as in Section 4.3. In other words, Theorems 4.8, 4.12, and 4.13 are actually all we need in order to develop a self-contained spectral theory of Toeplitz operators.

4.4. References are in the text.

4.5. These results are certainly well known, but we have not found them in the literature.

Chapter 5
Weighted norm inequalities

In this chapter we prove that if Γ is a Carleson curve and w is a weight in $A_p(\Gamma)$ $(1 < p < \infty)$, then the Cauchy singular integral operator S is bounded on $L^p(\Gamma, w)$. There are now various proofs of this deep result, and the proof given in the following is certainly not the most elegant proof. However, it is reasonably self-contained and it contains several details which are usually disposed of as "standard" and are therefore omitted in the advanced texts on this topic.

5.1 Again the maximal operator

Let X be a space with a measure $d\mu$. The *distribution function* $m(f, \lambda)$ of a measurable function $f : X \to \mathbf{C}$ is the function $\lambda \mapsto m(f, \lambda)$ defined by

$$m(f, \lambda) := \mu(\{\tau \in X : |f(\tau)| > \lambda\}) \quad (\lambda > 0). \tag{5.1}$$

The dependence of the distribution function on X and $d\mu$ is suppressed in the notation $m(f, \lambda)$. For $0 \leq p < \infty$, we clearly have

$$\lambda^p m(f, \lambda) \leq \int\limits_{\{\tau \in X : |f(\tau)| > \lambda\}} |f(\tau)|^p d\mu(\tau) \leq \int\limits_X |f(\tau)|^p d\mu(\tau),$$

that is,

$$m(f, \lambda) \leq \frac{1}{\lambda^p} \int\limits_X |f(\tau)|^p d\mu(\tau). \tag{5.2}$$

The inequality (5.2) is usually referred to as *Chebyshev's inequality*. It is also not difficult to see that

$$\int\limits_0^\infty p\lambda^{p-1} m(f, \lambda) \, d\lambda = \int\limits_X |f(\tau)|^p d\mu(\tau) \tag{5.3}$$

(see, e.g., [74, Lemma I.4.1]).

145

Now suppose $1 \le p \le \infty$, let X and Y be spaces with measures $d\mu$ and $d\nu$, respectively, and let T be a map of $L^p(X, d\mu)$ into the linear space of all complex-valued ν-measurable functions on Y. The map T is bounded from $L^p(X, d\mu)$ to $L^p(Y, d\nu)$ if and only if

$$\|Tf\|_{L^p(Y,d\nu)} \le C\|f\|_{L^p(X,d\mu)} \text{ for all } f \in L^p(X, d\mu)$$

with some constant $C < \infty$ independent of f. For $1 \le p < \infty$, this inequality may be rewritten in the form

$$\int_Y |(Tf)(t)|^p d\nu(t) \le C^p \int_X |f(\tau)|^p d\mu(\tau) \text{ for all } f \in L^p(X, d\mu). \qquad (5.4)$$

From (5.2) we see that (5.4) implies the inequality

$$\nu(\{t \in Y : |(Tf)(t)| > \lambda\}) \le \frac{C^p}{\lambda^p} \int_X |f(\tau)|^p d\mu(\tau) \text{ for all } f \in L^p(X, d\mu). \qquad (5.5)$$

Thus, (5.5) is weaker than (5.4). For $1 \le p < \infty$, one says that T is of the *weak type* (p, p) with respect to the measures $d\mu$ and $d\nu$ if (5.5) holds for all $\lambda > 0$. The map T is said to be of the *weak type* (∞, ∞) with respect to the measures $d\mu$ and $d\nu$ if T is bounded from $L^\infty(X, d\mu)$ into $L^\infty(Y, d\nu)$, i.e. if

$$\|Tf\|_{L^\infty(Y,d\nu)} \le C\|f\|_{L^\infty(X,d\mu)} \text{ for all } f \in L^\infty(X, d\mu).$$

In case $X = Y$ and $d\mu = d\nu$, we simply speak of the weak type (p, p) with respect to $d\nu$. Finally, the map T is called a *subadditive operator* if

$$\left|(T(f_1 + f_2))(t)\right| \le \left|(Tf_1)(t)\right| + \left|(Tf_2)(t)\right|$$

for all $f_1, f_2 \in L^p(X, d\mu)$ and ν-almost all $t \in Y$.

Theorem 5.1 (Marcinkiewicz). *Let X and Y be spaces with measures $d\mu$ and $d\nu$, respectively. Suppose $1 \le p_0 < p_1 \le \infty$ and T is a subadditive operator from $L^{p_0}(X, d\mu) + L^{p_1}(X, d\mu)$ into the ν-measurable complex-valued functions on Y. If T is of the weak types (p_0, p_0) and (p_1, p_1) with respect to the measures $d\mu$ and $d\nu$, then T is a bounded operator from $L^p(X, d\mu)$ to $L^p(Y, d\nu)$ for all $p \in (p_0, p_1)$.*

A proof may be found in many standard texts, e.g., in [73] (also see [74] and [202]).   □

We remark that if $p \in (p_0, p_1)$ and $f \in L^p(X, d\mu)$, then

$$f = \chi_{\{\tau \in X : |f(\tau)| \ge 1\}} f + \chi_{\{\tau \in X : |f(\tau)| < 1\}} f =: f_0 + f_1$$

and

$$\int_X |f_0|^{p_0} d\mu \le \int_X |f|^p d\mu < \infty, \quad \int_X |f_1|^{p_1} d\mu \le \int_X |f|^p d\mu < \infty.$$

Thus, $L^p(X, d\mu) \subset L^{p_0}(X, d\mu) + L^{p_1}(X, d\mu)$, and if T is as in Theorem 5.1, then Tf is automatically well-defined for all $f \in L^p(X, d\mu)$.

Also note that estimates for the norm of T are available. For example, if we have $p_0 = 1$ and $p_1 = \infty$ in Theorem 5.1, then

$$\|T\|_p^p \leq 2^p \frac{p}{p-1} \|T\|_\infty^{p-1} \|T\|_\sim \quad \text{for all} \quad p \in (1, \infty) \tag{5.6}$$

where $\|T\|_\sim$ is the best constant C in the weak $(1, 1)$ inequality

$$\nu\left(\{t \in Y : |(Tf)(t)| > \lambda\}\right) \leq \frac{C}{\lambda} \int_X |f(\tau)| d\mu(\tau) \quad (\lambda > 0).$$

Now let Γ be a simple Carleson curve, $1 < p < \infty$, and $w \in A_p(\Gamma)$. Define a measure $d\nu$ on Γ by $d\nu(\tau) = w(\tau)^p |d\tau|$. Since $w^p \in L^1_{loc}(\Gamma) := L^1_{loc}(\Gamma, |d\tau|)$ and $w^{-p} \in L^1_{loc}(\Gamma, d\nu)$, measurability with respect to length measure is equivalent to ν-measurability. In what follows we will therefore simply speak of measurability. As in Section 2.7, we write

$$|E| := \int_E |d\tau|, \quad |E|_\nu := \nu(E) := \int_E d\nu(\tau) := \int_E w(\tau)^p |d\tau|$$

for measurable sets $E \subset \Gamma$.

In Section 2.6, we introduced the maximal operator

$$(Mf)(t) := \sup_{\varepsilon > 0} \frac{1}{\varepsilon} \int_{\Gamma(t, \varepsilon)} |f(\tau)| \, |d\tau| \quad (t \in \Gamma)$$

and showed that M maps functions in $L^1_{loc}(\Gamma)$ to measurable functions on Γ. Obviously, M is subadditive. We know from Theorem 2.18 that M is of the weak type $(1, 1)$ with respect to $|d\tau|$, which corresponds to the case where $w = 1$. Consideration of weights $w \in A_p(\Gamma)$ motivates the following result.

Theorem 5.2. *Let Γ be a simple Carleson curve, $1 < p < \infty$, and $w \in A_p(\Gamma)$. Then the maximal operator M is of the weak type (p, p) with respect to the measure $d\nu(\tau) = w(\tau)^p |d\tau|$, i.e.*

$$\left|\{t \in \Gamma : (Mf)(t) > \lambda\}\right|_\nu \leq \frac{C^p}{\lambda^p} \int_\Gamma |f(\tau)|^p w(\tau)^p |d\tau| \tag{5.7}$$

for all $f \in L^p(\Gamma, w)$ and all $\lambda > 0$. Moreover, (5.7) holds with $C = 5 C_\Gamma^2 c_w^2$, where C_Γ is the Carleson constant and c_w is given by (2.4).

Proof. Suppose first that Γ is bounded. Fix $f \in L^p(\Gamma, w)$ and $\lambda > 0$. Since $L^p(\Gamma, w) \subset L^1(\Gamma)$ if $w \in A_p(\Gamma)$, we deduce from Lemma 2.16 that the set $A_\lambda := \{t \in \Gamma : (Mf)(t) > \lambda\}$ is open. If $A_\lambda = \emptyset$, then (5.7) is trivial. So assume $A_\lambda \neq \emptyset$. For every $t \in A_\lambda$, choose a portion $\Gamma(t, \varepsilon(t))$ such that

$$0 < \varepsilon(t) \leq d_t := \max_{\tau \in \Gamma} |\tau - t| \leq \operatorname{diam} \Gamma \quad \text{and} \quad \frac{1}{\varepsilon(t)} \int\limits_{\Gamma(t,\varepsilon(t))} |f(\tau)| \, |d\tau| > \lambda.$$

We then have

$$\lambda \big|\Gamma(t, \varepsilon(t))\big| \leq C_\Gamma \lambda \varepsilon(t) < C_\Gamma \int\limits_{\Gamma(t,\varepsilon(t))} |f(\tau)| \, |d\tau|$$

$$\leq C_\Gamma \left(\int\limits_{\Gamma(t,\varepsilon(t))} |f(\tau)|^p w(\tau)^p |d\tau| \right)^{1/p} \left(\int\limits_{\Gamma(t,\varepsilon(t))} w(\tau)^{-q} |d\tau| \right)^{1/q}$$

$$\leq C_\Gamma \left(\int\limits_{\Gamma(t,\varepsilon(t))} |f(\tau)|^p d\nu(\tau) \right)^{1/p} c_w \big|\Gamma(t, \varepsilon(t))\big| \, \big|\Gamma(t, \varepsilon(t))\big|_\nu^{-1/p},$$

where c_w is the constant of (2.4). Thus,

$$\lambda^p \big|\Gamma(t, \varepsilon(t))\big|_\nu \leq (C_\Gamma c_w)^p \int\limits_{\Gamma(t,\varepsilon(t))} |f(\tau)|^p d\nu(\tau). \qquad (5.8)$$

Let $\mathcal{F} := \{\Gamma(t, \varepsilon(t))\}_{t \in A_\lambda}$. By Lemma 2.23, there exists an at most countable set $\{\Gamma(t_i, \varepsilon(t_i))\}$ of pairwise disjoint portions from $\mathcal{F}$ such that

$$A_\lambda \subset \bigcup_{t \in A_\lambda} \Gamma(t, \varepsilon(t)) \subset \bigcup_i \Gamma(t_i, 5\varepsilon(t_i)). \qquad (5.9)$$

From Lemma 2.20 we infer that

$$\frac{|\Gamma(t_i, 5\varepsilon(t_i))|_\nu}{|\Gamma(t_i, \varepsilon(t_i))|_\nu} \leq c_w^p \left(\frac{|\Gamma(t_i, 5\varepsilon(t_i))|}{|\Gamma(t_i, \varepsilon(t_i))|} \right)^p \leq (5\, C_\Gamma c_w)^p. \qquad (5.10)$$

Since the portions $\Gamma(t_i, \varepsilon(t_i))$ are pairwise disjoint, we obtain from (5.8) to (5.10) that

$$|A_\lambda|_\nu \leq \sum_i \big|\Gamma(t_i, 5\varepsilon(t_i))\big|_\nu \leq (5\, C_\Gamma c_w)^p \sum_i \big|\Gamma(t_i, \varepsilon(t_i))\big|_\nu$$

$$\leq \frac{(5\, C_\Gamma c_w)^p}{\lambda^p} (C_\Gamma c_w)^p \sum_i \int\limits_{\Gamma(t_i,\varepsilon(t_i))} |f(\tau)|^p d\nu(\tau) \leq \frac{C^p}{\lambda^p} \int\limits_\Gamma |f(\tau)|^p d\nu(\tau)$$

with $C := 5\, C_\Gamma^2 c_w^2$.

The case where Γ is unbounded may be reduced to the case of bounded curves as in the proof of Theorem 2.18. $\qquad\square$

Corollary 5.3. *If Γ is a simple Carleson curve, $1 < p < \infty$, and $w \in A_p(\Gamma)$, then the maximal operator M is bounded on $L^p(\Gamma, w)$.*

Proof. If $w \in A_p(\Gamma)$ then (p, p) belongs to the set $\tilde{G}$ introduced in Section 2.8. Since $\tilde{G}$ is open due to Corollary 2.32, there is an $r \in (1, p)$ such that $(r, p) \in \tilde{G}$, i.e. $w^{p/r} \in A_p(\Gamma)$. Theorem 5.2 implies that M is of the weak type (r, r) with respect to the measure

$$d\nu(\tau) := \left(w^{p/r}(\tau)\right)^r |d\tau| = w(\tau)^p |d\tau|.$$

As $L^\infty(\Gamma) = L^\infty(\Gamma, d\nu)$ and

$$(Mf)(t) \leq \sup_{\varepsilon > 0} \frac{|\Gamma(t, \varepsilon)|}{\varepsilon} \|f\|_\infty \leq C_\Gamma \|f\|_\infty, \tag{5.11}$$

it follows that M is of the weak type (∞, ∞) with respect to the measure $d\nu$. Consequently, by Theorem 5.1, M is bounded on $L^p(\Gamma, d\nu) = L^p(\Gamma, w)$. $\qquad\square$

5.2 The Calderón-Zygmund decomposition

Lemma 5.4. *Let Γ be an unbounded Carleson arc and $f \in L^1(\Gamma)$. Then for every $\lambda > 0$ there exists an at most countable (and possibly empty) set of portions $\Gamma(t_k, \varepsilon_k)$ centered at points $t_k \in \Gamma$ such that*

(a) $\lambda < \frac{1}{|\Gamma(t_k, \varepsilon_k)|} \int\limits_{\Gamma(t_k, \varepsilon_k)} |f(\tau)| \, |d\tau| \leq 2\, C_\Gamma \lambda$ *for all k;*

(b) $|f(t)| \leq \lambda$ *for almost all $t \in \Gamma \setminus \bigcup\limits_k \Gamma(t_k, \varepsilon_k)$;*

(c) *every point $t \in \Gamma$ is contained in at most θ_2 of the portions $\Gamma(t_k, \varepsilon_k)$, where θ_2 is the constant from Theorem 2.17.*

Proof. Fix $\lambda > 0$ and put $A_\lambda := \{t \in \Gamma : |f(t)| > \lambda\}$. If $|A_\lambda| = 0$, then (a),(b),(c) hold with $\{\Gamma(t_k, \varepsilon_k)\} = \emptyset$. So assume $|A_\lambda| > 0$ and define

$$A_\lambda^* := \left\{ t \in A_\lambda : \sup_{\varepsilon > 0} \frac{1}{|\Gamma(t, \varepsilon)|} \int\limits_{\Gamma(t, \varepsilon)} |f(\tau)| \, |d\tau| \geq |f(t)| \right\}.$$

From Theorem 2.19 we know that the inequality defining A_λ^* holds for almost all $t \in A_\lambda$, which implies that $A_\lambda^* \neq \emptyset$. Given $t \in A_\lambda^*$, there is an $\varepsilon > 0$ such that

$$\frac{1}{|\Gamma(t, \varepsilon)|} \int\limits_{\Gamma(t, \varepsilon)} |f(\tau)| \, |d\tau| > \lambda.$$

On the other hand, if $\varepsilon > 0$ is sufficiently large then

$$\frac{1}{|\Gamma(t,\varepsilon)|} \int_{\Gamma(t,\varepsilon)} |f(\tau)|\,|d\tau| \leq \frac{\|f\|_1}{|\Gamma(t,\varepsilon)|} \leq \lambda.$$

(Note: this conclusion cannot be drawn if Γ is bounded.) Consequently, for every $t \in A_\lambda^*$ we have an $\varepsilon = \varepsilon(t) > 0$ such that

$$\frac{1}{|\Gamma(t,2\varepsilon)|} \int_{\Gamma(t,2\varepsilon)} |f(\tau)|\,|d\tau| \leq \lambda < \frac{1}{|\Gamma(t,\varepsilon)|} \int_{\Gamma(t,\varepsilon)} |f(\tau)|\,|d\tau|,$$

whence

$$\lambda < \frac{1}{|\Gamma(t,\varepsilon)|} \int_{\Gamma(t,\varepsilon)} |f(\tau)|\,|d\tau|$$

$$\leq \frac{|\Gamma(t,2\varepsilon)|}{|\Gamma(t,\varepsilon)|} \frac{1}{|\Gamma(t,2\varepsilon)|} \int_{\Gamma(t,2\varepsilon)} |f(\tau)|\,|d\tau| \leq 2C_\Gamma\lambda. \tag{5.12}$$

The set $A_\lambda \supset A_\lambda^*$ is bounded because $f \in L^1(\Gamma)$. Therefore we may employ Theorem 2.17 to deduce that there is an at most countable set of portions $\{\Gamma(t_k,\varepsilon_k)\} \subset \{\Gamma(t,\varepsilon(t))\}_{t\in A_\lambda^*}$ such that $A_\lambda^* \subset \bigcup_k \Gamma(t_k,\varepsilon_k)$ and (c) is satisfied. Property (a) is immediate from (5.12). Finally, as $\Gamma \setminus \bigcup_k \Gamma(t_k,\varepsilon_k) \subset (\Gamma \setminus A_\lambda) \cup (A_\lambda \setminus A_\lambda^*)$, $|A_\lambda \setminus A_\lambda^*| = 0$, and $|f(t)| \leq \lambda$ for $t \in \Gamma \setminus A_\lambda$, we see that (b) is also satisfied. $\square$

Theorem 5.5 (Calderón-Zygmund decomposition). *Let Γ, f, λ, and $\{\Gamma(t_k,\varepsilon_k)\}$ be as in the preceding lemma. Then there exist functions g and h_k in $L^1(\Gamma)$ such that*

$$f = g + \sum_k h_k, \tag{5.13}$$

$$|g(t)| \leq 2C_\Gamma\theta_2\lambda \text{ for almost all } t \in \Gamma, \tag{5.14}$$

$$\|g\|_1 \leq \|f\|_1, \tag{5.15}$$

$$h_k(t) = 0 \text{ for all } t \in \Gamma \setminus \Gamma(t_k,\varepsilon_k), \tag{5.16}$$

$$\int_\Gamma h_k(\tau)\,|d\tau| = 0 \text{ for all } k, \tag{5.17}$$

$$\sum_k \|h_k\|_1 \leq 2\|f\|_1. \tag{5.18}$$

Proof. Let χ_k be the characteristic function of the portion $\Gamma(t_k, \varepsilon_k)$ and put $\Gamma_\lambda := \bigcup_k \Gamma(t_k, \varepsilon_k)$. Further, define

$$
\eta_k(t) \quad := \quad \begin{cases} \chi_k(t) / \sum_k \chi_k(t) & \text{for } t \in \Gamma_\lambda, \\ 0 & \text{for } t \in \Gamma \setminus \Gamma_\lambda, \end{cases}
$$

$$
g(t) \quad := \quad \begin{cases} f(t) & \text{for } t \in \Gamma \setminus \Gamma_\lambda, \\ \sum_k \frac{\chi_k(t)}{|\Gamma(t_k, \varepsilon_k)|} \int\limits_{\Gamma(t_k, \varepsilon_k)} f(\tau) \eta_k(\tau) \, |d\tau| & \text{for } t \in \Gamma_\lambda, \end{cases}
$$

$$
h_k(t) \quad := \quad f(t) \eta_k(t) - \frac{\chi_k(t)}{|\Gamma(t_k, \varepsilon_k)|} \int\limits_{\Gamma(t_k, \varepsilon_k)} f(\tau) \eta_k(\tau) \, |d\tau| \text{ for } t \in \Gamma.
$$

Since $\sum_k \eta_k$ is the characteristic function of Γ_λ, it is clear that (5.13) holds. From properties (a) and (c) of Lemma 5.4 we infer that

$$
|g(t)| \le \sum_k \chi_k(t) \frac{1}{|\Gamma(t_k, \varepsilon_k)|} \int\limits_{\Gamma(t_k, \varepsilon_k)} |f(\tau)| \, |d\tau| \le \sum_k \chi_k(t) 2 C_\Gamma \lambda \le 2 C_\Gamma \lambda \theta_2
$$

for all $t \in \Gamma_\lambda$, and property (b) of Lemma 5.4 tells us that

$$
|g(t)| = |f(t)| \le \lambda < 2 C_\Gamma \theta_2 \lambda
$$

for almost all $t \in \Gamma \setminus \Gamma_\lambda$. This proves (5.14). Since $|\Gamma(t_k, \varepsilon_k)| = \int_\Gamma \chi_k(\tau) \, |d\tau|$, the estimate (5.15) results from the inequality

$$
\begin{aligned}
\|g\|_1 &\le \|f\|_{L^1(\Gamma \setminus \Gamma_\lambda)} + \sum_k \int\limits_{\Gamma(t_k, \varepsilon_k)} |f(\tau)| \eta_k(\tau) \, |d\tau| \\
&= \|f\|_{L^1(\Gamma \setminus \Gamma_\lambda)} + \int_\Gamma |f(\tau)| \sum_k \eta_k(\tau) \, |d\tau| \\
&= \|f\|_{L^1(\Gamma \setminus \Gamma_\lambda)} + \|f\|_{L^1(\Gamma_\lambda)} = \|f\|_1.
\end{aligned}
$$

The equalities (5.16) and (5.17) are easily verified. Finally, in the same way we proved (5.15) we get (5.18) from the estimate

$$
\sum_k \|h_k\|_1 \le \sum_k \int_\Gamma |f(\tau)| \eta_k(\tau) \, |d\tau| + \sum_k \int\limits_{\Gamma(t_k, \varepsilon_k)} |f(\tau)| \eta_k(\tau) \, |d\tau| \le 2\|f\|_1. \quad \square
$$

5.3 Cotlar's inequality

We will see that S extends to a bounded operator $\tilde{S}$ on $L^p(\Gamma)$ whenever $1 < p < \infty$ and Γ is a simple Carleson curve. As this result is not at our disposal at the present moment, we include it into the hypotheses of the following theorems. This is what is meant by saying that the proof of Theorem 4.15 is "spiralic".

Theorem 5.6. *Let* Γ *be an unbounded Carleson arc and* $1 < p < \infty$. *Suppose* S *generates a bounded operator* $\tilde{S}$ *on* $L^p(\Gamma)$. *Then there is a constant* $C < \infty$ *such that*

$$\left|\{t \in \Gamma : |(\tilde{S}\psi)(t)| > \lambda\}\right| \le \frac{C}{\lambda}\|\psi\|_1 \tag{5.19}$$

for all $\psi \in L^p(\Gamma) \cap L^1(\Gamma)$ *and all* $\lambda > 0$.

Note that (5.19) is a weak $(1,1)$ estimate for $\tilde{S}$.

Proof. Since $\tilde{S}$ maps $L^p(\Gamma)$ to $L^p(\Gamma)$, the set on the left of (5.19) is measurable for every $\psi \in L^p(\Gamma)$.

Put $a(t) := |dt|/dt$, denote by M_a the operator of multiplication by a, and define $R := \tilde{S}M_a$. Since $|a| = 1$ a.e. on Γ, we have $\|R\| = \|\tilde{S}\|$ and

$$\left|\{t \in \Gamma : |(\tilde{S}\psi)(t)| > \lambda\}\right| \le \frac{C}{\lambda}\|\psi\|_1 \ \forall \, \psi \in L^p(\Gamma) \cap L^1(\Gamma)$$

$$\Longleftrightarrow \left|\{t \in \Gamma : |(\tilde{S}M_a f)(t)| > \lambda\}\right| \le \frac{C}{\lambda}\|af\|_1 = \frac{C}{\lambda}\|f\|_1 \ \forall \, f \in L^p(\Gamma) \cap L^1(\Gamma)$$

$$\Longleftrightarrow \left|\{t \in \Gamma : |(Rf)(t)| > \lambda\}\right| \le \frac{C}{\lambda}\|f\|_1 \ \forall \, f \in L^p(\Gamma) \cap L^1(\Gamma). \tag{5.20}$$

Fix $\lambda > 0$ and let $f = g + h$, $h = \sum_k h_k$ be the Calderón-Zygmund decomposition of f in accordance with Theorem 5.5. We have $Rf = Rg + Rh$ and therefore

$$\left|\{t \in \Gamma : |(Rf)(t)| > \lambda\}\right|$$
$$\le \left|\left\{t \in \Gamma : |(Rg)(t)| > \frac{\lambda}{2}\right\}\right| + \left|\left\{t \in \Gamma : |(Rh)(t)| > \frac{\lambda}{2}\right\}\right|. \tag{5.21}$$

Taking into account (5.2), (5.14), (5.15) we get

$$\left|\left\{t \in \Gamma : |(Rg)(t)| > \frac{\lambda}{2}\right\}\right| \le \left(\frac{2}{\lambda}\right)^p \int_\Gamma |(Rg)(\tau)|^p |d\tau|$$

$$\le \left(\frac{2}{\lambda}\right)^p \|\tilde{S}\|^p \int_\Gamma |g(\tau)|^p |d\tau| \le \frac{2^p (2C_\Gamma \theta_2)^{p-1}}{\lambda} \|\tilde{S}\|^p \int_\Gamma |g(\tau)| |d\tau|$$

$$= \frac{2^p (2C_\Gamma \theta_2)^{p-1}}{\lambda} \|\tilde{S}\|^p \|g\|_1 \le \frac{2^p (2C_\Gamma \theta_2)^{p-1}}{\lambda} \|\tilde{S}\|^p \|f\|_1. \tag{5.22}$$

Put $Y := \Gamma \setminus \bigcup_k \Gamma(t_k, 2\varepsilon_k)$ and let χ_k be the characteristic function of the portion $\Gamma(t_k, \varepsilon_k)$. From properties (a) and (c) of Lemma 5.4 we obtain

$$\left|\bigcup_k \Gamma(t_k, 2\varepsilon_k)\right| \le \sum_k |\Gamma(t_k, 2\varepsilon_k)| \le 2C_\Gamma \sum_k |\Gamma(t_k, \varepsilon_k)|$$

$$< \frac{2C_\Gamma}{\lambda} \sum_k \int_{\Gamma(t_k,\varepsilon_k)} |f(\tau)| \, |d\tau| = \frac{2C_\Gamma}{\lambda} \int_\Gamma \sum_k \chi_k(\tau)|f(\tau)| \, |d\tau| \le \frac{2C_\Gamma \theta_2}{\lambda} \|f\|_1.$$

Consequently,

$$\left|\left\{t \in \Gamma : |(Rh)(t)| > \frac{\lambda}{2}\right\}\right| \leq \left|\left\{t \in Y : |(Rh)(t)| > \frac{\lambda}{2}\right\}\right| + \left|\bigcup_k \Gamma(t_k, 2\varepsilon_k)\right|$$

$$\leq \left|\left\{t \in Y : |(Rh)(t)| > \frac{\lambda}{2}\right\}\right| + \frac{2C_\Gamma \theta_2}{\lambda}\|f\|_1. \tag{5.23}$$

Since $\operatorname{supp} h_k \subset \Gamma(t_k, \varepsilon_k)$ by (5.16), we see from Lemma 4.7 that for almost all $t \in Y$,

$$(Rh_k)(t) = (\tilde{S}(ah_k))(t) = \frac{1}{\pi i} \int\limits_{\Gamma(t_k,\varepsilon_k)} \frac{a(\tau)h_k(\tau)}{\tau - t}\, d\tau = \frac{1}{\pi i} \int\limits_{\Gamma(t_k,\varepsilon_k)} \frac{h_k(\tau)}{\tau - t}\, |d\tau|.$$

This together with (5.16) and (5.17) gives

$$(Rh_k)(t) = \frac{1}{\pi i} \int\limits_{\Gamma(t_k,\varepsilon_k)} \left(\frac{1}{\tau - t} - \frac{1}{t_k - t}\right) h_k(\tau)\, |d\tau|. \tag{5.24}$$

If $\tau \in \Gamma(t_k, \varepsilon_k)$ and $t \in Y$, then

$$|\tau - t_k| \leq \varepsilon_k < \frac{1}{2}|t - t_k|, \quad |\tau - t| \geq |t - t_k| - |\tau - t_k| \geq \frac{1}{2}|t - t_k|,$$

whence

$$\left|\frac{1}{\tau - t} - \frac{1}{t_k - t}\right| = \frac{|\tau - t_k|}{|\tau - t||t - t_k|} \leq \frac{2\varepsilon_k}{|t - t_k|^2}. \tag{5.25}$$

Combining (5.24) and (5.25) shows that for almost all $t \in Y$,

$$|(Rh_k)(t)| \leq \frac{1}{\pi} \frac{2\varepsilon_k}{|t - t_k|^2} \int\limits_{\Gamma(t_k,\varepsilon_k)} |h_k(\tau)|\, |d\tau|$$

and thus,

$$\int\limits_Y |(Rh_k)(t)|\, |d\tau| \leq \frac{1}{\pi}\left(2\varepsilon_k \int\limits_{\Gamma \setminus \Gamma(t_k, 2\varepsilon_k)} \frac{|dt|}{|t - t_k|^2}\right) \int\limits_{\Gamma(t_k,\varepsilon_k)} |h_k(\tau)|\, |d\tau|. \tag{5.26}$$

Proceeding as in (4.4) we get

$$\varepsilon \int\limits_{\Gamma \setminus \Gamma(t,\varepsilon)} \frac{|d\tau|}{|\tau - t|^2} \leq \sum_{k=0}^{\infty} \frac{1}{2^{2k}\varepsilon} \int\limits_{\Gamma(t, 2^{k+1}\varepsilon)} |d\tau| \leq \sum_{k=0}^{\infty} \frac{C_\Gamma 2^{k+1}\varepsilon}{2^{2k}\varepsilon} = 4C_\Gamma. \tag{5.27}$$

Consequently, (5.26) implies that

$$\int\limits_Y |(Rh_k)(t)|\, |dt| \leq \frac{8C_\Gamma}{\pi}\|h_k\|_1.$$

This and (5.18) yield

$$\int\limits_Y |(Rh)(t)|\,|dt| \leq \sum_k \int\limits_Y |(Rh_k)(t)|\,|dt| \leq \frac{16C_\Gamma}{\pi}\|f\|_1$$

and thus, by (5.2),

$$\left|\left\{t \in Y : |(Rh)(t)| > \frac{\lambda}{2}\right\}\right| \leq \frac{2}{\lambda}\int\limits_Y |(Rh)(t)|\,|dt| \leq \frac{32C_\Gamma}{\pi\lambda}\|f\|_1. \qquad (5.28)$$

From (5.23) and (5.28) we obtain

$$\left|\left\{t \in \Gamma : |(Rh)(t)| > \frac{\lambda}{2}\right\}\right| \leq \left(\frac{32C_\Gamma}{\pi} + 2C_\Gamma\theta_2\right)\frac{1}{\lambda}\|f\|_1,$$

which together with (5.20), (5.21), (5.22) shows that (5.19) is valid with

$$C := 2^p(2C_\Gamma\theta_2)^{p-1}\|\tilde{S}\|^p + 32C_\Gamma/\pi + 2C_\Gamma\theta_2. \qquad (5.29)$$

$\square$

Let Γ be a simple Carleson curve. If $f \in L^p(\Gamma)$, then $(S_\varepsilon f)(t)$ is well-defined and finite for all $t \in \Gamma$ by virtue of Lemma 4.1. The *maximal singular integral operator* S_* is defined by

$$(S_* f)(t) := \sup_{\varepsilon>0}\left|(S_\varepsilon f)(t)\right| \quad (t \in \Gamma). \qquad (5.30)$$

Thus, $(S_* f)(t) \in [0,\infty]$ for every $f \in L^p(\Gamma)$ and every $t \in \Gamma$. The operator S_* is obviously sublinear. In what follows we will also need certain ("noncentered") modifications of the maximal operator M. Let $\delta \in (0,\infty)$ and suppose $|f|^\delta$ is in $L^1_{\mathrm{loc}}(\Gamma)$. Then the *modified maximal operator* $\mathcal{M}_\delta$ is defined by

$$(\mathcal{M}_\delta f)(t) := \sup_{\varepsilon>0}\ \sup_{x\in\Gamma(t,\varepsilon)}\left(\frac{1}{\varepsilon}\int\limits_{\Gamma(x,\varepsilon)} |f(\tau)|^\delta|d\tau|\right)^{1/\delta} \quad (t \in \Gamma). \qquad (5.31)$$

It is easily seen that

$$(Mf)(t) \leq (\mathcal{M}_1 f)(t) \leq 2(Mf)(t) \qquad (5.32)$$

for all $f \in L^1_{\mathrm{loc}}(\Gamma)$ and all $t \in \Gamma$.

Theorem 5.7 (Cotlar's inequality). *Let Γ be an unbounded Carleson arc, $1 < p < \infty$, and $0 < \delta \leq p$. Suppose S extends to a bounded operator $\tilde{S}$ on $L^p(\Gamma)$. If $f \in L^p(\Gamma)$, then*

$$(S_\varepsilon f)(t) \leq 3^{1/\delta}(\mathcal{M}_\delta\tilde{S}f)(t) + c_0(Mf)(t) \qquad (5.33)$$

for all $\varepsilon > 0$ and almost all $t \in \Gamma$, and

$$(S_* f)(t) \leq 3^{1/\delta}(\mathcal{M}_\delta\tilde{S}f)(t) + c_0(Mf)(t) \qquad (5.34)$$

for almost all $t \in \Gamma$. Here $c_0 < \infty$ is some constant depending only on Γ and p.

Proof. Let $t \in \Gamma$, $\varepsilon > 0$, $f \in L^p(\Gamma)$. Since $\tilde{S}f \in L^p(\Gamma)$, everything in (5.33) and (5.34) is well-defined. Put $Q := \Gamma(t, \varepsilon/2)$ and decompose f into the sum

$$f = f\chi_{\Gamma(t,\varepsilon)} + f\chi_{\Gamma\setminus\Gamma(t,\varepsilon)} =: f_1 + f_2.$$

Approximating f_2 by functions in $C_0^\infty(\Gamma)$ and taking into account Lemma 4.7 we get

$$(\tilde{S}f_2)(z) = \frac{1}{\pi i} \int\limits_{\Gamma\setminus\Gamma(t,\varepsilon)} \frac{f(\tau)}{\tau - z}\, d\tau \quad \text{for almost all } z \in Q. \tag{5.35}$$

Thus, for almost all $t \in \Gamma$ and almost all $x \in Q = \Gamma(t, \varepsilon/2)$ we have

$$\left| \frac{1}{\pi i} \int\limits_{\Gamma\setminus\Gamma(t,\varepsilon)} \frac{f(\tau)}{\tau - t}\, d\tau \right| = \left|(\tilde{S}f_2)(t)\right| \le \left|(\tilde{S}f_2)(x)\right| + \left|(\tilde{S}f_2)(t) - (\tilde{S}f_2)(x)\right|. \tag{5.36}$$

Proceeding as in the proof of (5.25) and (5.27) we obtain from (5.35) that

$$\pi|(\tilde{S}f_2)(t) - (\tilde{S}f_2)(x)| \le \int\limits_{\Gamma\setminus\Gamma(t,\varepsilon)} \left| \frac{1}{\tau - t} - \frac{1}{\tau - x} \right| |f(\tau)|\, |d\tau|$$

$$\le \varepsilon \int\limits_{\Gamma\setminus\Gamma(t,\varepsilon)} \frac{|f(\tau)|}{|\tau - t|^2}\, |d\tau| \le \sum_{k=0}^{\infty} \frac{1}{2^{2k}\varepsilon} \int\limits_{\Gamma(t,2^{k+1}\varepsilon)} |f(\tau)|\, |d\tau|$$

$$\le \sum_{k=0}^{\infty} \frac{1}{2^{k-1}}(Mf)(t) = 4(Mf)(t). \tag{5.37}$$

We have $(\tilde{S}f_2)(x) = (\tilde{S}f)(x) - (\tilde{S}f_1)(x)$, and we will estimate the two terms on the right separately. Since $f_1 \in L^1(\Gamma)$, we may employ Theorem 5.6 to conclude that for every $\lambda > 0$,

$$\left|\{x \in Q : |(\tilde{S}f_1)(x)| > \lambda\}\right| \le \frac{C}{\lambda}\|f_1\|_1 = \frac{C}{\lambda} \int\limits_{\Gamma(t,\varepsilon)} |f(\tau)|\, |d\tau|$$

$$\le \frac{C}{\lambda}\varepsilon(Mf)(t) \le \frac{C}{\lambda}2|\Gamma(t,\varepsilon/2)|\,(Mf)(t) = \frac{2C}{\lambda}|Q|\,(Mf)(t).$$

Taking $\lambda = 6C(Mf)(t)$ we get

$$\left|\{x \in Q : |(\tilde{S}f_1)(x)| > 6C(Mf)(t)\}\right| \le |Q|/3$$

and hence

$$|(\tilde{S}f_1)(x)| \le 6C(Mf)(t) \tag{5.38}$$

on a set of points $x \in Q$ whose measure is at least $2|Q|/3$.

Since $\tilde{S}f \in L^\delta_{\text{loc}}(\Gamma)$ for $0 < \delta \leq p$, we obtain from (5.2) that

$$\frac{1}{|Q|}|\{x \in Q : |(\tilde{S}f)(x)| > \lambda\}| \leq \frac{1}{\lambda^\delta}\frac{1}{|Q|}\int_Q |(\tilde{S}f)(\tau)|^\delta|d\tau|,$$

and letting

$$\lambda_0 = 3^{1/\delta}\left(\frac{1}{|Q|}\int_Q |(\tilde{S}f)(\tau)|^\delta|d\tau|\right)^{1/\delta}$$

we arrive at the estimate

$$|\{x \in Q : |(\tilde{S}f)(x)| > \lambda_0\}| \leq |Q|/3.$$

Consequently, on a set of points $x \in Q$ whose measure is at least $2|Q|/3$ we have

$$|(\tilde{S}f)(x)| \leq \lambda_0 \leq 3^{1/\delta}\left(\frac{1}{\varepsilon/2}\int_{\Gamma(t,\varepsilon/2)} |(\tilde{S}f)(\tau)|^\delta|d\tau|\right)^{1/\delta}$$
$$\leq 3^{1/\delta}(\mathcal{M}_\delta\tilde{S}f)(t). \tag{5.39}$$

Since $2/3 + 2/3 > 1$, there is a point $x \in Q$ for which both (5.38) and (5.39) hold. For this x,

$$|(\tilde{S}f_2)(x)| \leq |(\tilde{S}f)(x)| + |(\tilde{S}f_1)(x)| \leq 3^{1/\delta}(\mathcal{M}_\delta\tilde{S}f)(t) + 6C(Mf)(t),$$

and (5.36), (5.37), (5.38), (5.29) then imply (5.33) with

$$c_0 = 6C + 4/\pi = 6 \cdot 2^p(2C_\Gamma\theta_2)^{p-1}\|\tilde{S}\|^p + 192C_\Gamma/\pi + 12C_\Gamma\theta_2 + 4/\pi. \tag{5.40}$$

The inequality (5.34) is an immediate consequence of (5.33). $\qquad\square$

Corollary 5.8. *If Γ is an unbounded Carleson arc, $1 < p < \infty$, and S generates a bounded operator on $L^p(\Gamma)$, then $S_*f \in L^p(\Gamma)$ and $\|S_*f\|_p \leq C_p\|f\|_p$ for all $f \in L^p(\Gamma)$ with some constant $C_p < \infty$ independent of f.*

Proof. From (5.34) with $\delta = 1$ and from (5.32) we get

$$(S_*f)(t) \leq 6(M\tilde{S}f)(t) + c_0(Mf)(t).$$

Since M is bounded on $L^p(\Gamma)$ by Corollary 5.3 and $\tilde{S}$ is bounded on $L^p(\Gamma)$ by assumption, it follows that

$$\|S_*f\|_p \leq (6\|M\|\,\|\tilde{S}\| + c_0\|M\|)\,\|f\|_p. \qquad\square$$

Lemma 5.9 (Fatou's theorem). *Let X be a space with a σ-finite measure $d\mu$. If $f_n \in L^p(X, d\mu)$, $f_n \geq 0$, $\|f_n\|_{L^p(X,d\mu)} \leq K < \infty$ for all n, and $f_n \to f$ a.e. on X, then $f \in L^p(X, d\mu)$ and $\|f\|_{L^p(X,d\mu)} \leq K$.*

This very useful result is proved in many standard texts (see, e.g., [126, Chapter V, Section 5, Theorem 8]). $\qquad\square$

Corollary 5.10. *Suppose Γ is an unbounded Carleson arc, $1 < p < \infty$, and S extends to a bounded operator $\tilde{S}$ on $L^p(\Gamma)$. Then for every function $f \in L^p(\Gamma)$ the limit*

$$(Sf)(t) := \lim_{\varepsilon \to 0} (S_\varepsilon f)(t) \tag{5.41}$$

exists and coincides with $(\tilde{S}f)(t)$ for almost all $t \in \Gamma$.

Proof. For $t \in \Gamma$, define

$$(\Lambda f)(t) := \limsup_{\varepsilon \to 0} \operatorname{Re}(S_\varepsilon f)(t) - \liminf_{\varepsilon \to 0} \operatorname{Re}(S_\varepsilon f)(t).$$

Choose $g_n \in C_0^\infty(\Gamma)$ so that $\|f - g_n\|_p \to 0$. By Theorem 4.3 we have for almost all $t \in \Gamma$,

$$\limsup_{\varepsilon \to 0} \Big(\operatorname{Re}(S_\varepsilon f)(t) - \operatorname{Re}(S_\varepsilon g_n)(t) \Big) = \limsup_{\varepsilon \to 0} \operatorname{Re}(S_\varepsilon f)(t) - \operatorname{Re}(Sg_n)(t),$$

$$\liminf_{\varepsilon \to 0} \Big(\operatorname{Re}(S_\varepsilon f)(t) - \operatorname{Re}(S_\varepsilon g_n)(t) \Big) = \liminf_{\varepsilon \to 0} \operatorname{Re}(S_\varepsilon f)(t) - \operatorname{Re}(Sg_n)(t),$$

whence $\Lambda f = \Lambda(f - g_n)$ a.e. on Γ. Thus

$$0 \le \Lambda f = \Lambda(f - g_n) \le 2S_*(f - g_n) \quad \text{a.e. on } \Gamma. \tag{5.42}$$

Corollary 5.8 implies that $\|S_*(f - g_n)\|_p \to 0$. Since every sequence converging in the norm contains a subsequence converging almost everywhere (see, e.g., [126, Chap. VII, Sec. 2] or [180, Theorem 3.12]), there is a subsequence $\{S_*(f - g_{n_k})\}$ converging to zero almost everywhere on Γ. Hence, by (5.42), $\Lambda f = 0$ a.e. on Γ, which means that $\lim_{\varepsilon \to 0} \operatorname{Re}(S_\varepsilon f)(t)$ exists for almost all $t \in \Gamma$. Analogously one can consider $\operatorname{Im}(S_\varepsilon f)$. In summary, it results that the limit (5.41) exists for almost all $t \in \Gamma$. Since the functions $S_\varepsilon f$ are measurable, so also is Sf.

As $\|S_\varepsilon f\|_p \le \|S_* f\|_p \le C_p \|f\|_p$ for all $\varepsilon > 0$ due to Corollary 5.8 and $|S_\varepsilon f| \to |Sf|$ a.e. by what was just proved, Fatou's theorem shows that $|Sf| \in L^p(\Gamma)$ and that $\|Sf\|_p \le \|S_* f\|_p \le C_p \|f\|_p$. Replacing in this inequality f by $f - g_n$, we obtain that $\|Sf - Sg_n\|_p \to 0$. On the other hand, $\|\tilde{S}f - \tilde{S}g_n\|_p \to 0$. Because $Sg_n = \tilde{S}g_n$ by Theorem 4.3, we see that $Sf = \tilde{S}f$. $\qquad\square$

5.4 Good λ inequalities

Let Γ be a simple Carleson curve and let $d\mu$ be a Borel measure on Γ. Two μ-measurable functions $f, g : \Gamma \to \mathbf{C}$ are said to satisfy a *good λ inequality* if there are $r \in (0, 1)$, $\gamma \in (0, \infty)$, and $A \in (1, \infty)$ such that

$$\mu\big(\{t \in \Gamma : |f(t)| > A\lambda, \ |g(t)| \le \gamma\lambda\}\big) \le r\mu\big(\{t \in \Gamma : |f(t)| > \lambda\}\big) \tag{5.43}$$

for all $\lambda > 0$.

Theorem 5.11. *Let Γ be a simple Carleson curve, let $d\mu$ be a Borel measure on Γ, and let $0 < p < \infty$. Suppose f and g are μ-measurable functions on Γ satisfying* (5.43) *for certain $r \in (0,1)$, $\gamma \in (0,\infty)$, $A \in (1,\infty)$. Also suppose that $rA^p < 1$. If $\mu(\Gamma) < \infty$ or if $f \in L^{p_0}(\Gamma \setminus \Gamma_0, d\mu)$ for some compact subset $\Gamma_0 \subset \Gamma$ and some $p_0 \in (0,p]$, then*

$$\int_\Gamma |f(\tau)|^p d\mu(\tau) \le \frac{A^p}{\gamma^p(1 - rA^p)} \int_\Gamma |g(\tau)|^p d\mu(\tau). \tag{5.44}$$

Proof. From (5.43) and (5.1) we infer that

$$\begin{aligned}
m(f, A\lambda) &= \mu\big(\{t \in \Gamma : |f(t)| > A\lambda\}\big) \\
&\le \mu\big(\{t \in \Gamma : |f(t)| > A\lambda, \ |g(t)| \le \gamma\lambda\}\big) + \mu\big(\{t \in \Gamma : |g(t)| > \gamma\lambda\}\big) \\
&\le r\mu\big(\{t \in \Gamma : |f(t)| > \lambda\}\big) + \mu\big(\{t \in \Gamma : |g(t)| > \gamma\lambda\}\big) \\
&= rm(f, \lambda) + m(g, \gamma\lambda).
\end{aligned}$$

On multiplying this by $p\lambda^{p-1}$ and integrating from 0 to $\Lambda < \infty$ we get

$$p \int_0^\Lambda \lambda^{p-1} m(f, A\lambda)\, d\lambda \le rp \int_0^\Lambda \lambda^{p-1} m(f, \lambda)\, d\lambda + p \int_0^\Lambda \lambda^{p-1} m(g, \gamma\lambda)\, d\lambda.$$

Substitution of variables yields

$$\frac{p}{A^p} \int_0^{A\Lambda} \lambda^{p-1} m(f, \lambda)\, d\lambda \le rp \int_0^\Lambda \lambda^{p-1} m(f, \lambda)\, d\lambda + \frac{p}{\gamma^p} \int_0^{\gamma\Lambda} \lambda^{p-1} m(g, \lambda)\, d\lambda.$$

Since $A > 1$, it follows that

$$\frac{p}{A^p} \int_0^\Lambda \lambda^{p-1} m(f, \lambda)\, d\lambda \le rp \int_0^\Lambda \lambda^{p-1} m(f, \lambda)\, d\lambda + \frac{p}{\gamma^p} \int_0^\infty \lambda^{p-1} m(g, \lambda)\, d\lambda. \tag{5.45}$$

Both sides of (5.45) may be infinite. However, if $\mu(\Gamma) < \infty$ then $m(f, \lambda) \le \mu(\Gamma) < \infty$ and thus

$$\int_0^\Lambda \lambda^{p-1} m(f, \lambda)\, d\lambda < \infty. \tag{5.46}$$

On the other hand, if $f \in L^{p_0}(\Gamma \setminus \Gamma_0, d\mu)$ for some compact subset $\Gamma_0 \subset \Gamma$ and some $p_0 \in (0,p]$, we put

$$m_0(f, \lambda) := \mu\big(\{t \in \Gamma_0 : |f(t)| > \lambda\}\big), \quad m_1(f, \lambda) := \mu\big(\{t \in \Gamma \setminus \Gamma_0 : |f(t)| > \lambda\}\big).$$

Then

$$p\int_0^\Lambda \lambda^{p-1} m(f,\lambda)\,d\lambda = p\int_0^\Lambda \lambda^{p-1} m_0(f,\lambda)\,d\lambda + p\int_0^\Lambda \lambda^{p-1} m_1(f,\lambda)\,d\lambda$$

$$\leq \mu(\Gamma_0)\Lambda^p + p\int_0^\Lambda \lambda^{p-1} m_1(f,\lambda)\,d\lambda.$$

Chebyshev's inequality gives

$$\lambda^{p_0} m_1(f,\lambda) \leq \int_{\Gamma\setminus\Gamma_0} |f(\tau)|^{p_0}\,d\mu(\tau) < \infty.$$

Consequently, if $p > p_0$ then

$$\int_0^\Lambda \lambda^{p-1} m_1(f,\lambda)\,d\lambda = \int_0^\Lambda \lambda^{p-p_0-1}\lambda^{p_0} m_1(f,\lambda)\,d\lambda$$

$$\leq \left(\int_{\Gamma\setminus\Gamma_0} |f(\tau)|^{p_0}\,d\mu(\tau)\right)\int_0^\Lambda \lambda^{p-p_0-1}\,d\lambda = \frac{1}{p-p_0}\Lambda^{p-p_0}\int_{\Gamma\setminus\Gamma_0} |f(\tau)|^{p_0}\,d\mu(\tau) < \infty.$$

On the other hand, if $p = p_0$, we have

$$\int_0^\Lambda \lambda^{p-1} m_1(f,\lambda)\,d\lambda = \frac{1}{p_0}\int_{\Gamma\setminus\Gamma_0} |f(\tau)|^{p_0}\,d\mu(\tau) < \infty$$

due to (5.3). Thus, the assumptions of the theorem ensure that (5.46) is always valid. From (5.45) we therefore obtain that

$$(1 - rA^p)p\int_0^\Lambda \lambda^{p-1} m(f,\lambda)\,d\lambda \leq \left(\frac{A}{\gamma}\right)^p p\int_0^\infty \lambda^{p-1} m(g,\lambda)\,d\lambda,$$

whence $\qquad (1 - rA^p)p\int_0^\infty \lambda^{p-1} m(f,\lambda)\,d\lambda \leq \left(\frac{A}{\gamma}\right)^p p\int_0^\infty \lambda^{p-1} m(g,\lambda)\,d\lambda.$

This in conjunction with (5.3) gives (5.44). $\qquad\qquad\square$

Our aim is to establish a good λ inequality for the functions $f = S_*\psi$ and $g = M\psi$. Corollary 5.3 tells us that M is bounded, and Theorem 5.11 may therefore be used to prove the boundedness of S_*. Standard arguments finally show that S generates a bounded operator whenever S_* is bounded. Unfortunately, this is still a long way ...

5.5   Modified maximal operators

In this section, we prove a weak $(1, 1)$ estimate for the maximal singular integral operator on bounded curves.

Recall the definition (5.31) of the modified maximal operator $\mathcal{M}_\delta$. It can be easily seen (and is really simpler than the proof of Lemma 2.16) that the sets $\{t \in \Gamma : (\mathcal{M}_\delta f)(t) > \lambda\}$ are open and therefore measurable for every $\delta \in (0, \infty)$ and every $\lambda > 0$.

Lemma 5.12. *If Γ is a bounded simple Carleson curve and $f \in L^1(\Gamma)$ then*

$$\left|\{t \in \Gamma : (\mathcal{M}_1 f)(t) > \lambda\}\right| \le \frac{C}{\lambda} \int\limits_{\{t \in \Gamma : (\mathcal{M}_1 f)(t) > \lambda\}} |f(\tau)| \, |d\tau|$$

with $C := 2C_\Gamma \theta_2$.

Proof. Put $E := \{t \in \Gamma : (\mathcal{M}_1 f)(t) > \lambda\}$ and let χ_E be the characteristic function of E. Pick $t \in E$. Then there is a portion $\Gamma(x, \varepsilon)$ containing t such that

$$\frac{1}{\varepsilon} \int\limits_{\Gamma(x,\varepsilon)} |f(\tau)| \, |d\tau| > \lambda. \tag{5.47}$$

If there were a point y in $\Gamma(x, \varepsilon) \setminus E$, then $x \in \Gamma(y, \varepsilon)$ and

$$\lambda \ge (\mathcal{M}_1 f)(y) = \sup_{\varepsilon > 0} \sup_{z \in \Gamma(y,\varepsilon)} \frac{1}{\varepsilon} \int\limits_{\Gamma(z,\varepsilon)} |f(\tau)| \, |d\tau| \ge \frac{1}{\varepsilon} \int\limits_{\Gamma(x,\varepsilon)} |f(\tau)| \, |d\tau|,$$

which contradicts (5.47). Consequently, $\Gamma(x, \varepsilon) \subset E$. Put $f_E := f\chi_E$. Since $\Gamma(x, \varepsilon) \subset E$, the functions f_E and f coincide on $\Gamma(x, \varepsilon)$. Thus, by (5.47),

$$(\mathcal{M}_1 f_E)(t) \ge \frac{1}{\varepsilon} \int\limits_{\Gamma(x,\varepsilon)} |f_E(\tau)| \, |d\tau| = \frac{1}{\varepsilon} \int\limits_{\Gamma(x,\varepsilon)} |f(\tau)| \, |d\tau| > \lambda.$$

In summary, we have shown that $E \subset \{t \in \Gamma : (\mathcal{M}_1 f_E)(t) > \lambda\}$. Applying Theorem 2.18 to the function f_E and taking into account (5.32), we get

$$
\begin{aligned}
|E| & \le \left|\{t \in \Gamma : (\mathcal{M}_1 f_E)(t) > \lambda\}\right| \le \left|\{t \in \Gamma : (M f_E)(t) > \lambda/2\}\right| \\
& \le \frac{2C}{\lambda} \int\limits_{\Gamma} |f_E(\tau)| \, |d\tau| = \frac{2C}{\lambda} \int\limits_{E} |f(\tau)| \, |d\tau|
\end{aligned}
$$

with $C = C_\Gamma \theta_2$. $\qquad\square$

We define the weak L^1 space $L^\sim(\Gamma)$ and $\| \cdot \|_\sim$ as in Section 2.6.

Lemma 5.13. *If* Γ *is a bounded simple Carleson curve and* $0 < \delta < 1$, *then* $\mathcal{M}_\delta$ *is bounded on* $L^\sim(\Gamma)$:

$$\|\mathcal{M}_\delta f\|_\sim \leq C\|f\|_\sim \text{ for all } f \in L^\sim(\Gamma)$$

with $C := (2C_\Gamma \theta_2/(1-\delta))^{1/\delta}$.

Proof. Let $f \in L^\sim(\Gamma)$ and let E be a measurable subset of Γ. For $\lambda > 0$, put $E_\lambda := \{t \in E : |f(t)| > \lambda\}$. Clearly, $|E_\lambda| \leq |E|$. By the definition of $\|\cdot\|_\sim$, we have $|E_\lambda| \leq (1/\lambda)\|f\|_\sim$. This and (5.3) show that for every $\Lambda > 0$,

$$\int_E |f(\tau)|^\delta |d\tau| = \delta \int_0^\infty \lambda^{\delta-1} |E_\lambda|\, d\lambda$$

$$\leq \delta \int_0^\Lambda \lambda^{\delta-1} |E|\, d\lambda + \delta \int_\Lambda^\infty \lambda^{\delta-2}\|f\|_\sim d\lambda = \Lambda^\delta |E| + \frac{\delta}{1-\delta}\Lambda^{\delta-1}\|f\|_\sim.$$

Letting $\Lambda := \|f\|_\sim/|E|$ we get

$$\int_E |f(\tau)|^\delta |d\tau| \leq \frac{1}{1-\delta}|E|^{1-\delta}\|f\|_\sim^\delta, \qquad (5.48)$$

and taking $E = \Gamma$ in (5.48) we see that $|f|^\delta \in L^1(\Gamma)$. Now let

$$E := \{t \in \Gamma : (\mathcal{M}_\delta f)(t) > \lambda\} = \{t \in \Gamma : (\mathcal{M}_1(|f|^\delta))(t) > \lambda^\delta\}.$$

From Lemma 5.12 and (5.48) we obtain

$$|E| \leq \frac{2C_\Gamma \theta_2}{\lambda^\delta} \int_E |f(\tau)|^\delta |d\tau| \leq \frac{2C_\Gamma \theta_2}{1-\delta}\frac{1}{\lambda^\delta}|E|^{1-\delta}\|f\|_\sim^\delta$$

and thus $\lambda|E| \leq C\|f\|_\sim^\delta$ with $C = (2C_\Gamma \theta_2/(1-\delta))^{1/\delta}$. $\qquad\square$

The following two propositions are certain analogues of Theorems 5.6 and 5.7 for bounded curves.

Proposition 5.14. *Let* Γ *be an unbounded Carleson arc and let* $\Gamma_0 \subset \Gamma$ *be a bounded arc. Suppose* S_Γ *generates a bounded operator on* $L^p(\Gamma)$ *for some* $p \in (1, \infty)$. *If* $f \in L^p(\Gamma_0)$, *then*

$$(S_{\Gamma_0} f)(t) := \lim_{\varepsilon \to 0} (S_\varepsilon^{\Gamma_0} f)(t) := \lim_{\varepsilon \to 0} \frac{1}{\pi i} \int_{\Gamma_0 \backslash \Gamma_0(t,\varepsilon)} \frac{f(\tau)}{\tau - t}\, d\tau$$

exists for almost all $t \in \Gamma$ *and*

$$\left|\{t \in \Gamma_0 : (S_{\Gamma_0} f)(t) > \lambda\}\right| \leq \frac{C}{\lambda}\|f\|_{L^1(\Gamma_0)}.$$

Here C *is the constant* (5.29).

Proof. Let $f \in L^p(\Gamma_0)$ and let ψ be the extension of f by zero to all of Γ. Then $\psi \in L^p(\Gamma) \cap L^1(\Gamma)$. Corollary 5.10 implies that $(S_\Gamma \psi)(t)$ exists and $(S_\Gamma \psi)(t) = (\tilde{S}_\Gamma \psi)(t)$ for almost all $t \in \Gamma_0$. For these t we have $(S_{\Gamma_0} f)(t) = (S_\Gamma \psi)(t)$. This shows that $(S_{\Gamma_0} f)(t)$ exists and equals $(\tilde{S}_\Gamma \psi)(t)$ for almost all $t \in \Gamma_0$. It follows that

$$\left| \{ t \in \Gamma_0 : |(S_{\Gamma_0} f)(t)| > \lambda \} \right| = \left| \{ t \in \Gamma_0 : |(\tilde{S}_\Gamma \psi)(t)| > \lambda \} \right|$$
$$\leq \left| \{ t \in \Gamma : |(\tilde{S}_\Gamma \psi)(t)| > \lambda \} \right|,$$

and Theorem 5.6 says that this does not exceed

$$(C/\lambda) \| \psi \|_{L^1(\Gamma)} = (C/\lambda) \| f \|_{L^1(\Gamma_0)}$$

with C as in (5.29). $\qquad \square$

Proposition 5.15. *Let Γ be an unbounded Carleson arc, let $\Gamma_0 \subset \Gamma$ be a bounded arc, let $1 < p < \infty$, and suppose S_Γ generates a bounded operator on $L^p(\Gamma)$. If $f \in L^p(\Gamma_0)$ and $0 < \delta \leq p$, then*

$$(S_*^{\Gamma_0} f)(t) \leq 3^{1/\delta} (\mathcal{M}_\delta^{\Gamma_0} S_{\Gamma_0} f)(t) + c_0 (M^{\Gamma_0} f)(t)$$

for all $t \in \Gamma$ with c_0 given by (5.40).

Proof. Note that $S_{\Gamma_0} f$ is well-defined by virtue of Proposition 5.14. Fix $t \in \Gamma_0$, $\varepsilon > 0$, $f \in L^p(\Gamma_0)$. Put $Q := \Gamma_0(t, \varepsilon/2)$ and decompose f into the sum

$$f = f \chi_{\Gamma_0(t,\varepsilon)} + f \chi_{\Gamma_0 \setminus \Gamma_0(t,\varepsilon)} =: f_1 + f_2.$$

By Proposition 5.14,

$$(S_\varepsilon^{\Gamma_0} f)(t) = \frac{1}{\pi i} \int\limits_{\Gamma_0 \setminus \Gamma_0(t,\varepsilon)} \frac{f(\tau)}{\tau - t} \, d\tau = (S_{\Gamma_0} f_2)(t)$$

for almost all $t \in \Gamma$. Therefore

$$\left| (S_\varepsilon^{\Gamma_0} f)(t) \right| \leq \left| (S_{\Gamma_0} f_2)(x) \right| + \left| (S_{\Gamma_0} f_2)(t) - (S_{\Gamma_0} f_2)(x) \right|$$

for almost all $x \in Q$. As in the proof of Theorem 5.7 we get

$$\pi \left| (S_{\Gamma_0} f_2)(t) - (S_{\Gamma_0} f_2)(x) \right| \leq 4 (M^{\Gamma_0} f)(t).$$

Further, $(S_{\Gamma_0} f_2)(x) = (S_{\Gamma_0} f)(x) - (S_{\Gamma_0} f_1)(x)$. Letting $d_t := \max_{\tau \in \Gamma_0} |\tau - t|$, we obtain from Proposition 5.14 that

$$\left| \{ x \in Q : |(S_{\Gamma_0} f_1)(x)| > \lambda \} \right| \leq (C/\lambda) \| f_1 \|_{L^1(\Gamma_0)}$$
$$= \frac{C}{\lambda} \int\limits_{\Gamma_0(t,\varepsilon)} |f(\tau)| \, |d\tau| = \frac{C}{\lambda} \min\{\varepsilon, d_t\} \frac{1}{\min\{\varepsilon, d_t\}} \int\limits_{\Gamma_0(t,\min\{\varepsilon,d_t\})} |f(\tau)| \, |d\tau|$$
$$\leq \frac{C}{\lambda} \min\{\varepsilon, d_t\} (M^{\Gamma_0} f)(t) \leq \frac{2C}{\lambda} |Q| (M^{\Gamma_0} f)(t).$$

Taking $\lambda = 6C(M^{\Gamma_0}f)(t)$ we see as in the proof of Theorem 5.7 that

$$\left|(S_{\Gamma_0}f_1)(x)\right| \leq 6C(M^{\Gamma_0}f)(t)$$

for all $x \in Q$ belonging to a set of measure at least $(2/3)|Q|$. The reasoning of the proof of Theorem 5.7 also shows that

$$\left|(S_{\Gamma_0}f)(x)\right| \leq 3^{1/\delta}\left(\frac{1}{|\Gamma_0(t,\varepsilon/2)|} \int_{\Gamma_0(t,\varepsilon/2)} \left|(S_{\Gamma_0}f)(\tau)\right|^\delta |d\tau|\right)^{1/\delta}$$

for all x in a subset of Q with measure at least $(2/3)|Q|$. The term on the right of the last estimate does not exceed

$$3^{1/\delta}\left(\frac{1}{\min\{\varepsilon/2, d_t\}} \int_{\Gamma_0(t,\min\{\varepsilon/2, d_t\})} \left|(S_{\Gamma_0}f)(\tau)\right|^\delta |d\tau|\right)^{1/\delta} \leq 3^{1/\delta}(\mathcal{M}_\delta^{\Gamma_0}S_{\Gamma_0}f)(t).$$

The rest is as in the proof of Theorem 5.7. $\qquad\square$

Lemma 5.16. *Let Γ be a simple Carleson curve and $f \in L^1(\Gamma)$. Then S_*f is lower semi-continuous on Γ. In particular, the set $\{t \in \Gamma : (S_*f)(t) > \lambda\}$ is open and therefore measurable for every $\lambda > 0$.*

Proof. Fix $t_0 \in \Gamma$ and suppose $(S_*f)(t_0) > \lambda$. Then there are $\varepsilon_0 > 0$ and $\mu > 0$ such that

$$\left|(S_{\varepsilon_0}f)(t_0)\right| = \frac{1}{\pi}\left|\int_{\Gamma\setminus\Gamma(t_0,\varepsilon_0)} \frac{f(\tau)}{\tau - t_0}\, d\tau\right| \geq \lambda + 2\mu > \lambda. \tag{5.49}$$

The function $\varepsilon \mapsto |\Gamma(t_0,\varepsilon)|$ is continuous from the left (recall the proof of Lemma 2.22). Hence,

$$\lim_{\delta\to 0+0} \left|\Gamma(t_0,\varepsilon_0) \setminus \Gamma(t_0,\varepsilon_0 - 2\delta)\right| = 0.$$

Consequently, there is a $\delta \in (0, \varepsilon_0/2)$ such that

$$\frac{2}{\pi\varepsilon_0} \int_{\Gamma(t_0,\varepsilon_0)\setminus\Gamma(t_0,\varepsilon_0-2\delta)} |f(\tau)|\,|d\tau| < \mu. \tag{5.50}$$

Moreover, we may also assume that

$$\left(2/(\pi\varepsilon_0^2)\right)\delta\|f\|_{L^1(\Gamma)} < \mu. \tag{5.51}$$

For $t \in \Gamma(t_0,\varepsilon_0/2)$, define

$$(A_{\varepsilon_0})(t) := \frac{1}{\pi} \int_{\Gamma\setminus\Gamma(t_0,\varepsilon_0)} \frac{f(\tau)}{\tau - t}\, d\tau.$$

Now pick $t \in \Gamma(t_0, \delta) \subset \Gamma(t_0, \varepsilon_0/2)$. We have

$$\left|(A_{\varepsilon_0}f)(t) - (S_{\varepsilon_0}f)(t_0)\right| \leq \frac{|t - t_0|}{\pi} \int\limits_{\Gamma \setminus \Gamma(t_0, \varepsilon_0)} \frac{|f(\tau)|}{|\tau - t|\,|\tau - t_0|}\,|d\tau|,$$

and since

$$|\tau - t_0| \geq \varepsilon_0, \quad |\tau - t| \geq |\tau - t_0| - |t_0 - t| \geq \varepsilon_0 - \delta > \varepsilon_0/2 \quad \text{for} \quad \tau \in \Gamma \setminus \Gamma(t_0, \varepsilon_0),$$

we obtain from (5.51) that

$$\left|(A_{\varepsilon_0}f)(t) - (S_{\varepsilon_0}f)(t_0)\right| \leq \frac{|t - t_0|}{\pi} \frac{2}{\varepsilon_0^2} \int\limits_{\Gamma} |f(\tau)|\,|d\tau| \leq \frac{2\delta}{\pi\varepsilon_0^2} \|f\|_{L^1(\Gamma)} < \mu. \quad (5.52)$$

Further, if $|\tau - t| < \varepsilon_0 - \delta$ then

$$|\tau - t_0| \leq |\tau - t| + |t - t_0| < \varepsilon_0 - \delta + \delta = \varepsilon_0$$

and thus, $\Gamma(t, \varepsilon_0 - \delta) \subset \Gamma(t_0, \varepsilon_0)$. It follows that $\Gamma \setminus \Gamma(t_0, \varepsilon_0) \supset \Gamma \setminus \Gamma(t, \varepsilon_0 - \delta)$ and therefore,

$$\frac{1}{\pi} \int\limits_{\Gamma \setminus \Gamma(t, \varepsilon_0 - \delta)} \frac{f(\tau)}{\tau - t}\,d\tau = \frac{1}{\pi} \int\limits_{\Gamma \setminus \Gamma(t_0, \varepsilon_0)} \frac{f(\tau)}{\tau - t}\,d\tau + \frac{1}{\pi} \int\limits_{\Gamma(t_0, \varepsilon_0) \setminus \Gamma(t, \varepsilon_0 - \delta)} \frac{f(\tau)}{\tau - t}\,d\tau. \quad (5.53)$$

For $\tau \in \Gamma(t_0, \varepsilon_0) \setminus \Gamma(t, \varepsilon_0 - \delta)$ we have

$$|\tau - t_0| \geq |\tau - t| - |t_0 - t| \geq \varepsilon_0 - \delta - \delta = \varepsilon_0 - 2\delta,$$

which implies that

$$\Gamma(t_0, \varepsilon_0) \setminus \Gamma(t, \varepsilon_0 - \delta) \subset \Gamma(t_0, \varepsilon_0) \setminus \Gamma(t_0, \varepsilon_0 - 2\delta).$$

This in conjunction with (5.50) gives

$$\frac{1}{\pi} \int\limits_{\Gamma(t_0, \varepsilon_0) \setminus \Gamma(t, \varepsilon_0 - \delta)} \frac{|f(\tau)|}{|\tau - t|}\,|d\tau| \leq \frac{1}{\pi(\varepsilon_0 - \delta)} \int\limits_{\Gamma(t_0, \varepsilon_0) \setminus \Gamma(t_0, \varepsilon_0 - 2\delta)} |f(\tau)|\,|d\tau|$$

$$\leq \frac{2}{\pi\varepsilon_0} \int\limits_{\Gamma(t_0, \varepsilon_0) \setminus \Gamma(t_0, \varepsilon_0 - 2\delta)} |f(\tau)|\,|d\tau| < \mu. \quad (5.54)$$

Combining (5.53), (5.54), (5.52), (5.49), we get

$$\left|(S_{\varepsilon_0 - \delta}f)(t)\right| \geq \left|(A_{\varepsilon_0}f)(t)\right| - \frac{1}{\pi}\left|\int\limits_{\Gamma(t_0, \varepsilon_0) \setminus \Gamma(t, \varepsilon_0 - \delta)} \frac{f(\tau)}{\tau - t}\,d\tau\right|$$

$$> \left|(A_{\varepsilon_0}f)(t)\right| - \mu \geq \left|(S_{\varepsilon_0}f)(t_0)\right| - \left|(A_{\varepsilon_0}f)(t) - (S_{\varepsilon_0}f)(t_0)\right| - \mu$$

$$> \left|(S_{\varepsilon_0}f)(t_0)\right| - 2\mu > \lambda + 2\mu - 2\mu = \lambda$$

for all $t \in \Gamma(t_0, \delta)$. Thus, $(S_*f)(t) \geq \left|(S_{\varepsilon_0 - \delta}f)(t)\right| > \lambda$ for these t. $\qquad\square$

Corollary 5.17. *Let Γ be an unbounded Carleson arc and let $\Gamma_0 \subset \Gamma$ be a bounded subarc. Suppose $1 < p < \infty$ and S_Γ generates a bounded operator on $L^p(\Gamma)$. If $f \in L^p(\Gamma_0)$ then*

$$\left|\{t \in \Gamma_0 : (S_*^{\Gamma_0} f)(t) > \lambda\}\right| \leq (C_0/\lambda)\|f\|_{L^1(\Gamma_0)}$$

for all $\lambda > 0$ with some constant C_0 independent of f.

Proof. By Lemma 5.16 and Proposition 5.15,

$$\left|\{t \in \Gamma_0 : (S_*^{\Gamma_0} f)(t) > \lambda\}\right|$$
$$\leq \left|\left\{t \in \Gamma_0 : 3^{1/\delta}(\mathcal{M}_\delta^{\Gamma_0} S_{\Gamma_0} f)(t) > \frac{\lambda}{2}\right\}\right| + \left|\left\{t \in \Gamma_0 : c_0(M^{\Gamma_0} f)(t) > \frac{\lambda}{2}\right\}\right| \quad (5.55)$$

for every $\delta \in (0,1)$. Lemma 5.13 shows that

$$\frac{\lambda}{2}\left|\left\{t \in \Gamma_0 : (\mathcal{M}_\delta^{\Gamma_0} S_{\Gamma_0} f)(t) > \frac{\lambda}{2}\right\}\right| \leq C_1 \sup_{\lambda > 0}\left(\frac{\lambda}{2}\left|\left\{t \in \Gamma_0 : |(S_{\Gamma_0} f)(t)|\right\} > \frac{\lambda}{2}\right\}\right|$$

with $C_1 := (2C_\Gamma \theta_2/(1-\delta))^{1/\delta}$. From Proposition 5.14 we infer that

$$\frac{\lambda}{2}\left|\left\{t \in \Gamma_0 : |(S_{\Gamma_0} f)(t)| > \frac{\lambda}{2}\right\}\right| \leq C_2\|f\|_{L^1(\Gamma_0)}$$

with C_2 given by the right-hand side of (5.29). Finally, Theorem 2.18 tells us that

$$\left|\left\{t \in \Gamma_0 : c_0(M^{\Gamma_0} f)(t) > \frac{\lambda}{2}\right\}\right| \leq (C_3/\lambda)\|f\|_{L^1(\Gamma_0)}$$

with $C_3 := 2c_0 C_\Gamma \theta_2$. Thus, the right-hand side of (5.55) is not greater than

$$(2 \cdot 3^{1/\delta} C_1 C_2 + C_3)(1/\lambda)\|f\|_{L^1(\Gamma_0)}. \qquad \square$$

5.6 The maximal singular integral operator

Here now is a good λ inequality for $S_* g$ and Mg.

Theorem 5.18. *Let Γ be an unbounded Carleson arc, $1 < p < \infty$, and $w \in A_p(\Gamma)$. Suppose S generates a bounded operator on $L^2(\Gamma)$. Put $d\mu(\tau) := w(\tau)^p|d\tau|$. Then for every $r \in (0,1)$ there exists a $\gamma > 0$ such that*

$$\mu(\{t \in \Gamma : (S_* g)(t) > 2\lambda, \, (Mg)(t) \leq \gamma\lambda\}) \leq r\mu(\{t \in \Gamma : (S_* g)(t) > \lambda\})$$

for all $g \in C_0^\infty(\Gamma)$ and all $\lambda > 0$.

Proof. Fix $g \in C_0^\infty(\Gamma)$ and $\lambda > 0$, and consider the set

$$U_\lambda := \{t \in \Gamma : (S_*g)(t) > \lambda\}. \tag{5.56}$$

The set U_λ is an open subset of Γ due to Lemma 5.16. If $t \in \Gamma \setminus \operatorname{supp} g$, then

$$\left|(S_\varepsilon g)(t)\right| \leq \int\limits_{\operatorname{supp} g \setminus \Gamma(t,\varepsilon)} \frac{|g(\tau)|}{|\tau - t|}\,|d\tau| \leq \frac{1}{\operatorname{dist}(t, \operatorname{supp} g)}\|g\|_1 \leq \lambda \tag{5.57}$$

if only $|t|$ is sufficiently large. Consequently, U_λ is bounded. For $t \in U_\lambda$, let $\varepsilon(t) := \operatorname{dist}(t, \Gamma \setminus U_\lambda)$. Further, define $\mathcal{F} := \{\Gamma(t, \varepsilon(t))\}_{t \in U_\lambda}$. From Theorem 2.17 we deduce that there is an at most countable subfamily $\tilde{\mathcal{F}} = \{Q_k\}$ of portions $Q_k = \Gamma(t_k, \varepsilon(t_k))$ such that $U_\lambda \subset \bigcup_k Q_k$ and $\sum_k \chi_k \leq \theta_2$, where χ_k is the characteristic function of Q_k. Since $Q_k \subset U_\lambda$, it is clear that actually $U_\lambda = \bigcup_k Q_k$.

We show that for every $\beta \in (0,1)$ there is a $\gamma > 0$ such that

$$\left|\{t \in Q_k : (S_*g)(t) > 2\lambda,\ (Mg)(t) \leq \gamma\lambda\}\right| \leq \beta|Q_k| \tag{5.58}$$

for all $Q_k \in \tilde{\mathcal{F}}$. Let first $\gamma > 0$ be an arbitrary number. If Q_k does not contain a point ξ for which $(Mg)(\xi) \leq \gamma\lambda$, then (5.58) trivially holds for Q_k. So assume there is a point $\xi \in Q_k = \Gamma(t_k, \varepsilon(t_k))$ such that $(Mg)(\xi) \leq \gamma\lambda$. Let $a \in \Gamma \setminus U_\lambda$ be any point such that $|a - t_k| = \varepsilon(t_k)$. From (5.56) we know that

$$(S_*g)(a) \leq \lambda. \tag{5.59}$$

Put $R := 2\varepsilon(t_k)$ and write g in the form

$$g = g\chi_{\Gamma(a,2R)} + g\chi_{\Gamma \setminus \Gamma(a,2R)} =: f_1 + f_2.$$

Since $\xi \in Q_k \subset \Gamma(a, 2R) \subset \Gamma(\xi, 3R)$, we get

$$\frac{1}{2R}\|f_1\|_{L^1(\Gamma)} = \frac{1}{2R} \int\limits_{\Gamma(a,2R)} |g(\tau)|\,|d\tau| \leq \frac{3/2}{3R} \int\limits_{\Gamma(\xi,3R)} |g(\tau)|\,|d\tau| \leq \frac{3}{2}(Mg)(\xi) \leq \frac{3}{2}\gamma\lambda, \tag{5.60}$$

whence

$$\|f_1\|_{L^1(\Gamma)} \leq 3R\gamma\lambda = 6\varepsilon_k(t)\gamma\lambda \leq 6|Q_k|\gamma\lambda. \tag{5.61}$$

If $t \in Q_k$ and $\varepsilon > 0$, then

$$\pi\left|(S_\varepsilon f_2)(t)\right| = \left|\int\limits_{\Gamma \setminus \Gamma(t,\varepsilon)} \frac{f_2(\tau)}{\tau - t}\,d\tau\right| \leq \left|\int\limits_{\Gamma \setminus \Gamma(a,\varepsilon)} \frac{f_2(\tau)}{\tau - t}\,d\tau\right| + \int\limits_{\Gamma(t,\varepsilon)\triangle\Gamma(a,\varepsilon)} \frac{|f_2(\tau)|}{|\tau - t|}\,|d\tau|$$

$$\leq \left|\int\limits_{\Gamma \setminus \Gamma(a,\varepsilon)} \frac{f_2(\tau)}{\tau - a}\,d\tau\right| + \int\limits_{\Gamma \setminus \Gamma(a,\varepsilon)} \left|\frac{1}{\tau - t} - \frac{1}{\tau - a}\right| |f_2(\tau)|\,|d\tau|$$

$$+ \int\limits_{\Gamma(t,\varepsilon)\triangle\Gamma(a,\varepsilon)} \frac{|f_2(\tau)|}{|\tau - t|}\,|d\tau|. \tag{5.62}$$

Put $\varepsilon' := \max\{\varepsilon, 2R\}$. Then, by (5.59),

$$\left| \int\limits_{\Gamma\setminus\Gamma(a,\varepsilon)} \frac{f_2(\tau)}{\tau - a}\, d\tau \right| = \left| \int\limits_{\Gamma\setminus\Gamma(a,\varepsilon')} \frac{g(\tau)}{\tau - a}\, d\tau \right| \le \pi(S_* g)(a) \le \pi\lambda. \tag{5.63}$$

Now let $t \in Q_k$ and $\tau \in \Gamma \setminus \Gamma(a, 2R)$. Then $|t - a| \le R$, and since $\xi \in Q_k$, we have

$$|\tau - \xi| \le |\tau - a| + |\xi - a| \le 2|\tau - a|, \quad |\tau - \xi| \le |\tau - t| + |t - \xi| \le 2|\tau - t|,$$

whence

$$\left| \frac{1}{\tau - t} - \frac{1}{\tau - a} \right| = \frac{|t - a|}{|\tau - t|\,|\tau - a|} \le \frac{4R}{|\tau - \xi|^2}$$

and thus, as in (4.4),

$$\int\limits_{\Gamma\setminus\Gamma(a,\varepsilon)} \left| \frac{1}{\tau - t} - \frac{1}{\tau - a} \right| |f_2(\tau)|\,|d\tau| \le \int\limits_{\Gamma\setminus\Gamma(a,2R)} \left| \frac{1}{\tau - t} - \frac{1}{\tau - a} \right| |g(\tau)|\,|d\tau|$$

$$\le 4R \int\limits_{\Gamma\setminus\Gamma(a,2R)} \frac{|g(\tau)|}{|\tau - \xi|^2}\,|d\tau| \le 4R \int\limits_{\Gamma\setminus\Gamma(\xi,R)} \frac{|g(\tau)|}{|\tau - \xi|^2}\,|d\tau|$$

$$\le 4R \sum_{k=0}^{\infty} \frac{1}{2^{2k} R^2} \int\limits_{\Gamma(\xi,2^{k+1}R)} |g(\tau)|\,|d\tau|$$

$$= 4 \sum_{k=0}^{\infty} \frac{1}{2^{k-1}} \frac{1}{2^{k+1}R} \int\limits_{\Gamma(\xi,2^{k+1}R)} |g(\tau)|\,|d\tau|$$

$$\le 4 \sum_{k=0}^{\infty} \frac{1}{2^{k-1}} (Mg)(\xi) = 16(Mg)(\xi) \le 16\gamma\lambda. \tag{5.64}$$

Finally, if $t \in Q_k$ and $\tau \in (\Gamma(t,\varepsilon)\Delta\Gamma(a,\varepsilon)) \setminus \Gamma(a, 2R) =: Y$, then $|\tau - t| \ge R$ for $0 < \varepsilon \le 2R$ and $|\tau - t| \ge \varepsilon - R$ for $\varepsilon > 2R$. Consequently,

$$\int\limits_{\Gamma(t,\varepsilon)\Delta\Gamma(a,\varepsilon)} \frac{|f_2(\tau)|}{|\tau - t|}\,|d\tau| = \int\limits_{Y} \frac{|g(\tau)|}{|\tau - t|}\,|d\tau|$$

$$\le \begin{cases} \frac{1}{R}\int\limits_{Y}|g(\tau)|\,|d\tau| \le \frac{3}{3R} \int\limits_{\Gamma(\xi,3R)} |g(\tau)|\,|d\tau| \le 3(Mg)(\xi) & \text{for } 0 < \varepsilon \le 2R, \\[2ex] \frac{1}{\varepsilon-R}\int\limits_{Y}|g(\tau)|\,|d\tau| \le \frac{\varepsilon+R}{\varepsilon-R}\frac{1}{\varepsilon+R} \int\limits_{\Gamma(\xi,\varepsilon+R)} |g(\tau)|\,|d\tau| \le 3(Mg)(\xi) & \text{for } \varepsilon > 2R. \end{cases}$$

In either case,

$$\int\limits_{\Gamma(t,\varepsilon)\Delta\Gamma(a,\varepsilon)} \frac{|f_2(\tau)|}{|\tau - t|}\,|d\tau| \le 3(Mg)(\xi) \le 3\gamma\lambda. \tag{5.65}$$

Putting (5.62), (5.63), (5.64), (5.65) together, we arrive at the estimate

$$\left|(S_\varepsilon f_2)(t)\right| \le \lambda + (19/\pi)\gamma\lambda =: \lambda + C\gamma\lambda$$

for all $t \in Q_k$ and all $\varepsilon > 0$. Thus,

$$(S_* f_2)(t) \le \lambda + C\gamma\lambda \quad \text{for all } t \in Q_k.$$

Now choose $\gamma \le 1/(2C) = \pi/38$. Then

$$(S_* f_2)(t) \le \lambda + \lambda/2 = (3/2)\lambda \quad \text{for all } t \in Q_k.$$

Therefore, if $t \in Q_k$ and $(S_* g)(t) > 2\lambda$, then

$$(S_* f_1)(t) \ge (S_* g)(t) - (S_* f_2)(t) > 2\lambda - (3/2)\lambda = \lambda/2.$$

What results is that

$$\left\{t \in Q_k : (S_* g)(t) > 2\lambda\right\} \subset \left\{t \in Q_k : (S_* f_1)(t) > \lambda/2\right\} =: E_k^0.$$

Let $\Gamma_0 \subset \Gamma$ denote a bounded arc containing $\Gamma(a, 2R)$. From Corollary 5.17 and (5.61) we infer that

$$
\begin{aligned}
|E_k^0| &= \left|\left\{t \in Q_k : (S_* f_1)(t) > \lambda/2\right\}\right| \\
&\le \left|\left\{t \in \Gamma_0 : (S_* f_1)(t) > \lambda/2\right\}\right| = \left|\left\{t \in \Gamma_0 : (S_*^{\Gamma_0} f_1)(t) > \lambda/2\right\}\right| \\
&\le (2C_0/\lambda)\|f_1\|_{L^1(\Gamma_0)} = (2C_0/\lambda)\|f_1\|_{L^1(\Gamma)} \le 12C_0\gamma|Q_k|.
\end{aligned}
$$

Thus, if we require that $\gamma = \beta/(12C_0)$, then $|E_k^0| \le \beta|Q_k|$.

Now let $0 < \gamma \le \pi/38$ and $0 < \gamma \le \beta/(12C_0)$. Put

$$E_k := \left\{t \in Q_k : (S_* g)(t) > 2\lambda,\ (Mg)(t) \le \gamma\lambda\right\}.$$

Then $E_k \subset E_k^0$, and we have proved that $|E_k| \le |E_k^0| \le \beta|Q_k|$ for all $Q_k \in \tilde{\mathcal{F}}$.

Since $w \in A_p(\Gamma)$, we deduce from Theorem 2.29 that there exist positive constants N and δ such that

$$\left(\frac{1}{|Q|} \int_Q w(\tau)^{p(1+\delta)}|d\tau|\right)^{\frac{1}{1+\delta}} \le N\frac{1}{|Q|}\mu(Q)$$

for every portion $Q = \Gamma(t, \varepsilon)$. This and Hölder's inequality show that if E is a measurable subset of Q, then

$$
\begin{aligned}
\mu(E) &= \int_Q \chi_E(\tau)w(\tau)^p|d\tau| \le \left(\int_Q w(\tau)^{p(1+\delta)}|d\tau|\right)^{\frac{1}{1+\delta}} |E|^{\frac{\delta}{1+\delta}} \\
&\le N|Q|^{\frac{1}{1+\delta}}|Q|^{-1}\mu(Q)|E|^{\frac{\delta}{1+\delta}} = N|Q|^{-\frac{\delta}{1+\delta}}\mu(Q)|E|^{\frac{\delta}{1+\delta}},
\end{aligned}
$$

whence

$$\frac{\mu(E)}{\mu(Q)} \le N\left(\frac{|E|}{|Q|}\right)^\eta \quad \text{with } \eta := \frac{\delta}{1+\delta}.$$

In particular,
$$\frac{\mu(E_k)}{\mu(Q_k)} \leq N\left(\frac{|E_k|}{|Q_k|}\right)^{\eta} \leq N\beta^{\eta} \text{ for all } Q_k \in \tilde{\mathcal{F}}.$$

Since $U_\lambda = \bigcup_k Q_k$ and $\sum_k \chi_k \leq \theta_2 \chi_{U_\lambda}$, we obtain

$$\mu\big(\{t \in \Gamma : (S_* g)(t) > 2\lambda, \ (Mg)(t) \leq \gamma\lambda\}\big) \leq \sum_k \mu(E_k)$$

$$\leq N\beta^{\eta} \sum_k \mu(Q_k) = N\beta^{\eta} \int_\Gamma \sum_k \chi_k(\tau)\, d\mu(\tau)$$

$$\leq N\beta^{\eta}\theta_2 \int_\Gamma \chi_{U_\lambda}(\tau)\, d\mu(\tau) = N\beta^{\eta}\theta_2\mu(U_\lambda)$$

$$= N\beta^{\eta}\theta_2\mu\big(\{t \in \Gamma : (S_* g)(t) > \lambda\}\big).$$

Finally, choose $\beta \in (0,1)$ so that $N\beta^{\eta}\theta_2 \leq r$. Then the assertion of the theorem follows for

$$0 < \gamma \leq \min\left\{\pi/38, \ \beta/(12C_0)\right\}. \qquad \square$$

Lemma 5.19. *Let Γ be an unbounded Carleson curve and $g \in C_0^\infty(\Gamma)$. Then there exists a compact subset $\Gamma_0 \subset \Gamma$ such that $\operatorname{supp} g \subset \Gamma_0$ and*

$$S_* g \in L^p(\Gamma \setminus \Gamma_0, w) \text{ for all } p \in (1,\infty) \text{ and all } w \in A_p(\Gamma).$$

Proof. There are $a \in \Gamma$ and $R > 0$ such that $\operatorname{supp} g \subset \Gamma(a, R)$. If $\tau \in \Gamma(a, R)$ and $t \in \Gamma \setminus \Gamma(a, 2R)$, then $|\tau - t| \geq |t - a| - R$ and $\Gamma(a, R) \subset \Gamma(t, |t - a| + R)$, and since $|t - a| \geq 2R$, we have $(|t - a| + R)/(|t - a| - R) \leq 3$. Consequently,

$$\pi\big|(S_\varepsilon g)(t)\big| \leq \int_{\Gamma(a,R)\setminus\Gamma(t,\varepsilon)} \frac{|g(\tau)|}{|\tau - t|}\, |d\tau| \leq \frac{1}{|t - a| - R} \int_{\Gamma(t,|t-a|+R)} |g(\tau)|\, |d\tau|$$

$$\leq \frac{|t - a| + R}{|t - a| - R} \frac{1}{|t - a| + R} \int_{\Gamma(t,|t-a|+R)} |g(\tau)|\, |d\tau| \leq 3(Mg)(t),$$

whence $(S_* g)(t) \leq (3/\pi)(Mg)(t)$ for all $t \in \Gamma \setminus \Gamma(a, 2R)$. Let Γ_0 be the closure of $\Gamma(a, 2R)$. Then

$$\|S_* g\|_{L^p(\Gamma \setminus \Gamma_0, w)} \leq (3/\pi)\|Mg\|_{L^p(\Gamma \setminus \Gamma_0, w)} \leq (3/\pi)\|Mg\|_{L^p(\Gamma, w)}$$

and since M is a bounded operator on $L^p(\Gamma, w)$ due to Corollary 5.3, we arrive at the assertion. $\qquad \square$

Corollary 5.20. *Let Γ be an unbounded Carleson curve and suppose S generates a bounded operator on $L^2(\Gamma)$. If $1 < p < \infty$ and $w \in A_p(\Gamma)$, then*

$$\|S_* g\|_{L^p(\Gamma, w)} \leq C(\Gamma, p, w)\|g\|_{L^p(\Gamma, w)}$$

for all $g \in C_0^\infty(\Gamma)$ with some constant $C(\Gamma, p, w)$ depending only on Γ, p, w.

Proof. Let $g \in C_0^\infty(\Gamma)$. Then, by Lemma 5.19, $S_* g \in L^p(\Gamma \setminus \Gamma_0, w)$ where Γ_0 is some compact subset of Γ. Choose $r \in (0, 1/2^p)$ and let γ be as in Theorem 5.18. From this theorem and Theorem 5.11 with $A = 2$ we deduce that

$$\|S_* g\|_{L^p(\Gamma,w)}^p \leq \frac{2^p}{\gamma^p(1 - r2^p)} \|Mg\|_{L^p(\Gamma,w)}^p.$$

Corollary 5.3 completes the proof. $\qquad\square$

Corollary 5.21. *Let Γ be an unbounded Carleson curve and suppose S generates a bounded operator on $L^2(\Gamma)$. If $1 < p < \infty$ and $w \in A_p(\Gamma)$, then S extends to a bounded operator on $L^p(\Gamma, w)$, i.e. $C_0^\infty(\Gamma)$ is dense in $L^p(\Gamma, w)$ and*

$$\|Sg\|_{L^p(\Gamma,w)} \leq C(\Gamma, p, w)\|g\|_{L^p(\Gamma,w)}$$

for all $g \in C_0^\infty(\Gamma)$.

Proof. Proposition 4.5 implies that $C_0^\infty(\Gamma)$ is dense in $L^p(\Gamma, w)$. Fix $g \in C_0^\infty(\Gamma)$. Theorem 4.3 shows that $S_\varepsilon g \to Sg$ a.e. on Γ. In particular, Sg is measurable. Since $|(S_\varepsilon g)(t)| \leq (S_* g)(t)$ for almost all $t \in \Gamma$, it follows that $\|S_\varepsilon g\|_{p,w} \leq \|S_*\|_{p,w}$ for all $\varepsilon > 0$. From Corollary 5.20 we therefore deduce that $\|S_\varepsilon g\|_{p,w} \leq C(\Gamma, p, w)\|g\|_{g,w}$ for all $\varepsilon > 0$. Now Fatou's theorem (Lemma 5.9) gives that $|Sg| \in L^p(\Gamma, w)$ and that $\|Sg\|_{p,w} \leq C(\Gamma, p, w)\|g\|_{p,w}$. $\qquad\square$

Corollary 5.22. *Let Γ be an unbounded Carleson curve and suppose S generates a bounded operator on $L^2(\Gamma)$. Let $1 < p < \infty$ and $w \in A_p(\Gamma)$. If $f \in L^p(\Gamma, w)$ and the limit*

$$(Sf)(t) := \lim_{\varepsilon \to 0} (S_\varepsilon f)(t) \tag{5.66}$$

exists and is finite for almost all $t \in \Gamma$, then $Sf \in L^p(\Gamma, w)$ and $Sf = \tilde{S}f$ a.e. on Γ where $\tilde{S}$ is the bounded extension of S to $L^p(\Gamma, w)$ guaranteed by the preceding corollary.

Proof. Let $f \in L^p(\Gamma, w)$ and choose $g_n \in C_0^\infty(\Gamma)$ such that $\|f - g_n\|_{p,w} \to 0$. By Hölder's inequality,

$$\pi \left|(S_\varepsilon f)(t) - (S_\varepsilon g_n)(t)\right| \leq \int\limits_{|\tau - t| \geq \varepsilon} \frac{|f(\tau) - g_n(\tau)|}{|\tau - t|} \, |d\tau| \leq \|f - g_n\|_{p,w} \int\limits_{\Gamma \setminus \Gamma(t,\varepsilon)} \frac{w(\tau)^{-q}}{|\tau - t|^q} \, |d\tau|.$$
$$\tag{5.67}$$

The proof of Proposition 4.1 shows that the second factor in (5.67) is finite. Consequently, $S_\varepsilon g_n \to S_\varepsilon f$ a.e. on Γ for each $\varepsilon > 0$. Let $\delta > 0$ be arbitrarily fixed. Since, by Corollary 5.20,

$$\|S_\varepsilon g_n\|_{p,w} \leq \|S_* g_n\|_{p,w} \leq C(\Gamma, p, w)\|g_n\|_{p,w}$$

and because $\|g_n\|_{p,w} < (1 + \delta)\|f\|_{p,w}$ for all sufficiently large n, it follows that

$$\|S_\varepsilon g_n\|_{p,w} \leq C(\Gamma, p, w)(1 + \delta)\|f\|_{p,w}$$

for all n large enough. Fatou's theorem now implies that $|S_\varepsilon f| \in L^p(\Gamma, w)$ and

$$\|S_\varepsilon f\|_{p,w} \leq C(\Gamma, p, w)(1 + \delta)\|f\|_{p,w}.$$

As $\delta > 0$ was arbitrary, we even have

$$\|S_\varepsilon f\|_{p,w} \leq C(\Gamma, p, w)\|f\|_{p,w}. \tag{5.68}$$

By assumption, $S_\varepsilon f \to Sf$ a.e. on Γ. Hence, Sf is measurable, and again having recourse to Fatou's theorem, we infer from (5.68) that $|Sf| \in L^p(\Gamma, w)$ and

$$\|Sf\|_{p,w} \leq C(\Gamma, p, w)\|f\|_{p,w}. \tag{5.69}$$

We have shown that (5.69) holds for all $f \in L^p(\Gamma, w)$. Since $Sg_n = \tilde{S}g_n$, we finally get

$$\|Sf - \tilde{S}f\|_{p,w} \leq \|Sf - Sg_n\|_{p,w} + \|\tilde{S}g_n - \tilde{S}f\|_{p,w}$$
$$\leq \left(C(\Gamma, p, w) + \|\tilde{S}\|\right)\|f - g_n\|_{p,w} = o(1),$$

which proves that $Sf = \tilde{S}f$ (a.e.). $\qquad\square$

Remark 5.23. Combining the previous corollary with Theorem 4.14, we see that $Sf = \tilde{S}f$ for all $f \in L^p(\Gamma, w) \cap L^1(\Gamma)$. One can prove that actually (5.66) exists and is finite a.e. on Γ for all $f \in L^p(\Gamma, w)$. A proof may be based on the fact that Theorems 5.6 and 5.7 have weighted analogues. In particular, (5.34) can be shown to hold for all $f \in L^p(\Gamma, w)$ provided S extends to a bounded operator on $L^p(\Gamma, w)$ (which is known from Corollary 5.21). The rest of the proof follows the reasoning of the proofs of Corollaries 5.8 and 5.10. $\qquad\square$

Notice that at this point we have proved Theorem 4.9, since the boundedness of S on $L^2(\mathbf{R})$ is a classical result. We are now going to remove the superfluous assumption in the last three corollaries requiring that S generates a bounded operator on $L^2(\Gamma)$. In other words, the purpose of what follows is to show that S always generates a bounded operator on $L^2(\Gamma)$ if only Γ is an unbounded Carleson curve.

5.7 Lipschitz curves

A function $\varphi : \mathbf{R} \to \mathbf{R}$ is said to *satisfy a Lipschitz condition* or to be a *Lipschitz function* if

$$|\varphi(x_1) - \varphi(x_2)| \leq K|x_1 - x_2| \quad \text{for all } x_1, x_2 \in \mathbf{R} \tag{5.70}$$

with some constant $K < \infty$ independent of x_1, x_2. One can show that if φ satisfies a Lipschitz condition then φ differentiable almost everywhere on $\mathbf{R}$ and $\varphi' \in L^\infty(\mathbf{R})$. The best constant in (5.70) is $K = \|\varphi'\|_\infty$ and is referred to as the *Lipschitz constant* of φ.

A curve Γ is called a *Lipschitz graph* if $\Gamma = \{x + i\varphi(x) : x \in \mathbf{R}\}$ with some Lipschitz function φ. A *Lipschitz curve* is a curve which results by rotation from some Lipschitz graph. We denote by $\Lambda(K)$ the set of all Lipschitz curves corresponding to Lipschitz graphs $\Gamma = \{x + i\varphi(x) : x \in \mathbf{R}\}$ with $\|\varphi'\|_\infty \leq K$.

Let Γ be a Lipschitz graph. We orient Γ so that the positive direction is induced by the change of x from $-\infty$ to $+\infty$. Put

$$\Omega_\pm := \{x + i(\varphi(x) \pm y) : x \in \mathbf{R}, y > 0\}.$$

For $g \in C_0^\infty(\Gamma)$, we define the *Cauchy integral*

$$(Cg)(z) := \frac{1}{2\pi i} \int_\Gamma \frac{g(\tau)\, d\tau}{\tau - z} \quad (z \in \mathbf{C} \setminus \Gamma). \tag{5.71}$$

One can show that the non-tangential limits

$$(Cg)^\pm(t) := \lim_{z \to t, z \in \Omega^\pm} (Cg)(z)$$

exist and are finite for almost all $t \in \Gamma$. Moreover, it turns out that

$$(Cg)^\pm(t) = \frac{1}{2}\big(\pm g(t) + (Sg)(t)\big) \tag{5.72}$$

and thus

$$(Sg)(t) = (Cg)^+(t) + (Cg)^-(t) \tag{5.73}$$

for almost all $t \in \Gamma$. Here Sg is the Cauchy singular integral of g. Formulas (5.72) are usually referred to as the *Sokhotski-Plemelj formulas*.

Let $\mathbf{C}_+ := \{z = x + iy : x \in \mathbf{R}, y > 0\}$ be the upper half-plane and put $\overline{\mathbf{C}_+} = \mathbf{C}_+ \cup \mathbf{R} \cup \{\infty\}$. We denote by $H^2 := H^2(\mathbf{C}_+)$ and $H^\infty := H^\infty(\mathbf{C}_+)$ the usual Hardy spaces of $\mathbf{C}_+$. Further, we let $E^2(\Omega_\pm)$ stand for the Smirnov spaces of $\Omega_\pm$. For precise definitions and information about these spaces we refer the reader to the standard texts, e.g., to Duren's book [59].

Lemma 5.24. *If $f \in H^2$, $\psi \in H^\infty$, $\|\psi\|_{H^\infty} \leq 1$, then*

$$\iint_{\mathbf{C}_+} |f(z)\psi'(z)|^2 y\, dx\, dy \leq \int_{\mathbf{R}} |f(x)|^2 dx = 4 \iint_{\mathbf{C}_+} |f'(z)|^2 y\, dx\, dy.$$

Proof. Since $f \in H^2$, we have

$$f(x+iy) = \int_0^\infty g(t) e^{i(x+iy)t}\, dt = \int_0^\infty g(t) e^{-yt} e^{ixt}\, dt$$

for all $x + iy \in \mathbf{C}_+$ with some function $g \in L^2(\mathbf{R}_+)$. It follows that

$$f'(x+iy) = \frac{1}{2}\Big[(\partial_x f)(x+iy) - i(\partial_y f)(x+iy)\Big]$$

$$= \frac{1}{2}\int_0^\infty g(t)e^{-yt}ite^{ixt}dt + \frac{1}{2}\int_0^\infty g(t)e^{-yt}ite^{ixt}dt = i\int_0^\infty g(t)e^{-yt}te^{ixt}dt,$$

whence

$$4\iint_{\mathbf{C}_+} |f'(z)|^2 y\, dx\, dy = 4\int_0^\infty \left(\int_{\mathbf{R}} |f'(x+iy)|^2 dx\right) y\, dy$$

$$= 4\int_0^\infty 2\pi\left(\int_0^\infty |g(t)|^2 e^{-2yt}t^2 dt\right) y\, dy \quad \text{(Parseval)}$$

$$= 2\pi\int_0^\infty 4\left(\int_0^\infty e^{-2yt}y\, dy\right) t^2|g(t)|^2 dt \quad \text{(Fubini)}$$

$$= 2\pi\int_0^\infty |g(t)|^2 dt = \int_{\mathbf{R}} |f(x)|^2 dx \quad \text{(again Parseval)}.$$

This proves the asserted equality.

Since $\|\psi\|_\infty \leq 1$ and $f\psi \in H^2$, we obtain from what was already proved that

$$\int_{\mathbf{R}} |f(x)|^2 dx \geq \int_{\mathbf{R}} |f(x)\psi(x)|^2 dx = 4\iint_{\mathbf{C}_+} |f'(z)\psi(z) + f(z)\psi'(z)|^2 y\, dx\, dy$$

and thus,

$$\left(\iint_{\mathbf{C}_+} |f(z)\psi'(z)|^2 y\, dx\, dy\right)^{1/2}$$

$$\leq \left(\iint_{\mathbf{C}_+} |f'(z)\psi(z) + f(z)\psi'(z)|^2 y\, dx\, dy\right)^{1/2} + \left(\iint_{\mathbf{C}_+} |f'(z)\psi(z)|^2 y\, dx\, dy\right)^{1/2}$$

$$\leq \frac{1}{2}\left(\int_{\mathbf{R}} |f(x)|^2 dx\right)^{1/2} + \left(\iint_{\mathbf{C}_+} |f'(z)|^2 y\, dx\, dy\right)^{1/2}$$

$$= \frac{1}{2}\left(\int_{\mathbf{R}} |f(x)|^2 dx\right)^{1/2} + \frac{1}{2}\left(\int_{\mathbf{R}} |f(x)|^2 dx\right)^{1/2} = \left(\int_{\mathbf{R}} |f(x)|^2 dx\right)^{1/2}. \qquad \square$$

We denote by $\mathcal{H}_{\pm} := L^2(\Omega_{\pm}, d)$ the Hilbert space of all measurable functions $f : \Omega_{\pm} \to \mathbf{C}$ such that

$$\|f\|_{\mathcal{H}_{\pm}} := \left(\iint\limits_{\Omega_{\pm}} |f(z)|^2 d(z) \, dx \, dy \right)^{1/2} < \infty$$

where $d(z) := \operatorname{dist}(z, \Gamma)$.

Lemma 5.25 (Kenig). *Let $\Gamma = \{x + i\varphi(x) : x \in \mathbf{R}\}$ be a Lipschitz graph and suppose $\varphi' \in C_0^{\infty}(\mathbf{R})$. If $F \in E^2(\Omega_{\pm})$, then*

$$c_1(K)\|F'\|_{\mathcal{H}_{\pm}} \leq \|F\|_{L^2(\Gamma)} \leq c_2(K)\|F'\|_{\mathcal{H}_{\pm}}$$

with constants $c_1(K), c_2(K)$ depending only on the Lipschitz constant $K = \|\varphi'\|_{\infty}$ of the function φ.

Proof. We only prove the $+$ version. Let $\Phi : \mathbf{C}_+ \to \Omega_+$ be a Riemann map such that $\Phi(\mathbf{R}) = \Gamma$ and $\Phi(\infty) = \infty$. The smoothness of Γ implies that Φ' and $1/\Phi'$ are continuous on $\overline{\mathbf{C}_+} := \mathbf{C} \cup \mathbf{R} \cup \{\infty\}$, where a neighborhood base of ∞ is given by $\{z \in \mathbf{C}_+ : |z| > R\}$ $(R > 0)$. In particular,

$$(\Phi')^{1/2} \in H^{\infty} \cap C(\overline{\mathbf{C}_+}) \quad \text{and} \quad \log \Phi' \in H^{\infty} \cap C(\overline{\mathbf{C}_+}).$$

Put $G := F \circ \Phi$. Since $F \in E^2(\Omega_+)$, we have $G(\Phi')^{1/2} \in H^2$. By Koebe's distortion theorem (see Theorem 9.3),

$$\gamma_1 |\Phi'(z)| y \leq d(\Phi(z)) \leq \gamma_2 |\Phi'(z)| y$$

with universal constants $\gamma_1, \gamma_2 \in (0, \infty)$. Since

$$\|F'\|_{\mathcal{H}_+}^2 = \iint\limits_{\mathbf{C}_+} |F'(\Phi(z))|^2 d(\Phi(z)) |\Phi'(z)|^2 dx \, dy = \iint\limits_{\mathbf{C}_+} |G'(z)|^2 d(\Phi(z)) \, dx \, dy,$$

we therefore see that the assertion is equivalent to the existence of constants $c_3(K), c_4(K) \in (0, \infty)$ depending only on K such that

$$c_3(K)B \leq A \leq c_4(K)B \tag{5.74}$$

where

$$B := \iint\limits_{\mathbf{C}_+} |G'(z)|^2 |\Phi'(z)| y \, dx \, dy, \quad A := \int\limits_{\mathbf{R}} |G(x)|^2 |\Phi'(x)| \, dx.$$

Applying Lemma 5.24 to $f = G(\Phi')^{1/2}$, we get

$$
\begin{aligned}
A^{1/2} &= 2\left(\iint_{\mathbf{C}_+} \left| G'(z)(\Phi'(z))^{1/2} + \frac{1}{2}G(z)(\Phi'(z))^{-1/2}\Phi''(z) \right|^2 y\,dx\,dy\right)^{1/2} \\
&\geq 2B^{1/2} - \left(\iint_{\mathbf{C}_+} |G(z)|^2 |\Phi'(z)|^{-1} |\Phi''(z)|^2 y\,dx\,dy\right)^{1/2}.
\end{aligned}
\tag{5.75}
$$

Since Γ is a Lipschitz graph, we have

$$
\|\arg \Phi'\|_{L^\infty(\mathbf{R})} = \|\arctan \varphi'\|_{L^\infty(\mathbf{R})} = \arctan K < \pi/2.
$$

As $\log \Phi' \in H^\infty \cap C(\overline{\mathbf{C}_+})$, the function $\arg \Phi'$ is harmonic in $\mathbf{C}_+$ and continuous on $\overline{\mathbf{C}_+}$. Thus, the maximum modulus principle implies that

$$
\left| \arg \Phi'(z) \right| \leq \arctan K < \pi/2 \quad \text{for all } z \in \overline{\mathbf{C}_+}.
\tag{5.76}
$$

Consider the H^∞ function $\Psi := e^{-\pi/2}e^{i \log \Phi'}$. From (5.76) we see that

$$
\|\Psi\|_{H^\infty} \leq 1, \quad \|1/\Psi\|_{H^\infty} \leq e^\pi.
\tag{5.77}
$$

Clearly,

$$
\Psi' = i\Psi\Phi''/\Phi'.
\tag{5.78}
$$

Combining (5.77), (5.78), and Lemma 5.24 for $f = G(\Phi')^{1/2} \in H^2$ and $\psi = \Psi \in H^\infty$ we obtain

$$
\begin{aligned}
&\iint_{\mathbf{C}_+} |G(z)|^2 |\Phi'(z)|^{-1} |\Phi''(z)|^2 y\,dx\,dy \\
&= \iint_{\mathbf{C}_+} \left| G(z)(\Phi'(z))^{1/2} \right|^2 \left| \Psi'(z)/\Psi(z) \right|^2 y\,dx\,dy \\
&\leq e^{2\pi} \iint_{\mathbf{C}_+} \left| G(z)(\Phi'(z))^{1/2}\Psi'(z) \right|^2 y\,dx\,dy \\
&\leq e^{2\pi} \int_{\mathbf{R}} \left| G(x)(\Phi'(x))^{1/2} \right|^2 dx = e^{2\pi} A.
\end{aligned}
\tag{5.79}
$$

From (5.75) and (5.79) we get the left inequality of (5.74) with $c_3(K) = 4(1+e^\pi)^{-2}$.

The function Φ' is continuous on $\overline{\mathbf{C}_+}$ and has no zeros on $\overline{\mathbf{C}_+}$. Hence, $G(\Phi')^{1/2} \in H^2$ if and only if $G \in H^2$. Therefore it suffices to prove the right inequality in (5.74) for all G in some dense subset of H^2. According to [74, Chapter II, Corollary 3.3] and [73, Chapter 3, Theorem 1.6], a dense subset of H^2 is given by

$$
\mathcal{A}_2 := \{G \in H^2 : G \in C^\infty(\mathbf{C}_+ \cup \mathbf{R}), |G(z)| = O(|z|^{-2}) \text{ as } |z| \to \infty\}.
$$

Furthermore, it suffices to prove the right inequality of (5.74) with z and x replaced by $z + i\varepsilon$ and $x + i\varepsilon$, respectively. Indeed, assuming we have proved that

$$\int_{\mathbf{R}} |G(x+i\varepsilon)|^2 |\Phi'(x+i\varepsilon)|\, dx \leq c_4(K) \iint_{\mathbf{C}_+} |G(z+i\varepsilon)|^2 |\Phi'(z+i\varepsilon)|\, y\, dx\, dy$$

for all $\varepsilon > 0$ and taking into account that

$$\int_{\mathbf{R}} |G(x)|^2 |\Phi'(x)|\, dx = \lim_{\varepsilon \to 0} \int_{\mathbf{R}} |G(x+i\varepsilon)|^2 |\Phi'(x+i\varepsilon)|\, dx$$

and

$$\iint_{\mathbf{C}_+} |G(z+i\varepsilon)|^2 |\Phi'(z+i\varepsilon)|\, y\, dx\, dy \leq \iint_{\mathbf{C}_+} |G(z+i\varepsilon)|^2 |\Phi'(z+i\varepsilon)|\, (y+\varepsilon)\, dx\, dy$$

$$= \iint_{\mathrm{Im}\, z > \varepsilon} |G(z)|^2 |\Phi'(z)|^2 y\, dx\, dy \leq \iint_{\mathbf{C}_+} |G(z)|^2 |\Phi'(z)|\, y\, dx\, dy,$$

we get the right inequality of (5.74). Let $\Phi_\varepsilon(z) := \Phi(z+i\varepsilon)$. Clearly, Φ'_ε belongs to $C^\infty(\mathbf{C}_+ \cup \mathbf{R})$. The estimate (5.76) holds with Φ'_ε in place of Φ, and Φ'_ε is bounded away from zero.

Thus, suppose $G \in \mathcal{A}_2 \,(\subset H^\infty)$ and that $\Phi' \in C^\infty(\mathbf{C}_+ \cup \mathbf{R})$. By (5.76),

$$\left| \int_{\mathbf{R}} |G(x)|^2 \Phi'(x)\, dx \right| = \left| \int_{\mathbf{R}} |G(x)|^2 |\Phi'(x)| e^{i\,\arg\,\Phi'(x)} dx \right|$$

$$\geq \int_{\mathbf{R}} |G(x)|^2 |\Phi'(x)| \cos(\arg \Phi'(x))\, dx \geq \frac{1}{\sqrt{1+K^2}} \int_{\mathbf{R}} |G(x)|^2 |\Phi'(x)|\, dx.$$

Consequently,

$$A \leq \sqrt{1+K^2} \left| \int_{\mathbf{R}} |G(x)|^2 \Phi'(x)\, dx \right|. \tag{5.80}$$

We now apply Green's formula

$$\int_{\partial D} \left(u \frac{\partial v}{\partial n} - v \frac{\partial u}{\partial n} \right) ds = \iint_D (u\Delta v - v\Delta u)\, dx\, dy, \tag{5.81}$$

where n is the outer normal, with $u(x,y) := y$ and $v(x,y) := |G(x+iy)|^2 \Phi'(x+iy)$ on $D := \{z \in \mathbf{C}_+ : |z| < R\}$. Since

$$\Delta = 4\partial\bar{\partial} \quad \text{with} \quad \partial := \frac{1}{2}(\partial_x - i\partial_y), \quad \bar{\partial} := \frac{1}{2}(\partial_x + i\partial_y)$$

and since for every analytic function f we have

$$\partial f = f', \ \overline{\partial} f = 0, \ \partial \overline{f} = 0, \ \overline{\partial} \, \overline{f} = \overline{f'},$$

it follows that

$$\Delta\big(|G|^2\Phi'\big) = 4\partial\overline{\partial}(G\overline{G}\Phi') = 4\partial(G\overline{G'}\Phi') = 4\big(|G'|^2\Phi' + G\overline{G'}\Phi''\big).$$

Hence, after passing to the limit $R \to \infty$, the equality (5.81) takes the form

$$\int_{\mathbf{R}} |G(x)|^2 \Phi'(x)\,dx = 4 \iint_{\mathbf{C}_+} \Big(|G'(z)|^2\Phi'(z) + G(z)\overline{G'(z)}\Phi''(z)\Big) y\,dx\,dy \qquad (5.82)$$

provided we can show that

$$\int_{\gamma_R} \left| u\frac{\partial v}{\partial n} - v\frac{\partial u}{\partial n}\right| ds \to 0 \ \text{ as } \ R \to \infty \qquad (5.83)$$

where $\gamma_R := \{z \in \mathbf{C}_+ : |z| = R\}$. Let us show (5.83). Denote by θ the angle between n and the x-axis. Then

$$\frac{\partial}{\partial n} = \frac{\partial}{\partial x}\cos\theta + \frac{\partial}{\partial y}\sin\theta = e^{i\theta}\partial + e^{-i\theta}\overline{\partial}$$

and hence,

$$\begin{aligned}
\frac{\partial v}{\partial n} &= e^{i\theta}\partial(G\overline{G}\Phi') + e^{-i\theta}\overline{\partial}(G\overline{G}\phi') \\
&= e^{i\theta}(G'\overline{G}\Phi' + |G|^2\Phi'') + e^{-i\theta}G\overline{G'}\Phi'), \\
u\frac{\partial v}{\partial n} - v\frac{\partial u}{\partial n} &= y\big(e^{i\theta}G'\overline{G}\phi' + e^{i\theta}|G|^2\Phi'' + e^{-i\theta}G\overline{G'}\Phi'\big) - |G|^2\Phi'\sin\theta.
\end{aligned}$$

It results that

$$\int_{\gamma_R} \left| u\frac{\partial v}{\partial n} - v\frac{\partial u}{\partial n}\right| ds \le \int_{\gamma_R} \Big(2y|G|\,|G'|\,|\Phi'| + y|G|^2|\Phi''| + |G|^2|\Phi'|\Big)\,ds. \qquad (5.84)$$

For every function $\psi \in H^\infty$ we have

$$|\psi'(z)| \le \frac{1}{2\pi} \int_{|\tau-z|=R} \frac{|\psi(\tau)|}{|\tau - z|^2}\,|d\tau| \le \frac{\|\psi\|_\infty}{R} \ \text{ if } \ 0 < R < \operatorname{Im} z,$$

implying that $|\psi'(x + iy)| \le \|\psi\|_\infty/y$ for all $y > 0$. Thus, for $G \in \mathcal{A}_2 \subset H^\infty$ and $\Phi' \in H^\infty$ we get the estimates

$$y|G'(x + iy)| \le \|G\|_\infty, \quad y|\Phi''(x + iy)| \le \|\Phi'\|_\infty.$$

These estimates and (5.84) yield

$$\int_{\gamma_R} \left| u \frac{\partial v}{\partial n} - v \frac{\partial u}{\partial n} \right| ds \leq 2\|\Phi'\|_\infty \int_{\gamma_R} \left(\|G\|_\infty |G| + |G|^2 \right) ds.$$

Finally, since $|G(z)| = O(|z|^{-2})$ as $|z| \to \infty$, the latter integral goes to zero as $R \to \infty$. This completes the proof of (5.83) and thus gives (5.82).

Letting $C_0 := \sqrt{1 + K^2}$, we obtain from (5.80) and (5.82) that

$$
\begin{aligned}
A &\leq 4C_0 \left| \iint_{\mathbf{C}_+} \left(|G'(z)|^2 \Phi'(z) + G(z)\overline{G'(z)}\Phi''(z) \right) y\, dx\, dy \right| \\
&\leq 4C_0 \left(\iint_{\mathbf{C}_+} |G'(z)|^2 \Phi'(z) |y\, dx\, dy + \iint_{\mathbf{C}_+} |G(z)\overline{G'(z)}\Phi''(z)| y\, dx\, dy \right) \\
&= 4C_0 \left(B + \iint_{\mathbf{C}_+} |G(z)G'(z)\Phi''(z)| y\, dx\, dy \right).
\end{aligned}
\tag{5.85}
$$

Cauchy-Schwarz and (5.79) show that

$$
\begin{aligned}
&\iint_{\mathbf{C}_+} |G(z)G'(z)\Phi''(z)| y\, dx\, dy \\
&\leq \left(\iint_{\mathbf{C}_+} |G'(z)|^2 |\Phi'(z)| y\, dx\, dy \right)^{1/2} \left(\iint_{\mathbf{C}_+} |G(z)|^2 |\Phi'(z)|^{-1} |\Phi''(z)|^2 y\, dx\, dy \right)^{1/2} \\
&\leq B^{1/2} e^\pi A^{1/2}.
\end{aligned}
\tag{5.86}
$$

Combining (5.85) and (5.86) we get

$$A \leq 4C_0 e^\pi (B + A^{1/2}B^{1/2}) =: C(B + A^{1/2}B^{1/2}).$$

Thus, for $x = (A/B)^{1/2}$ we have the inequality $x^2 \leq C(1 + x)$. This implies that

$$0 \leq x \leq (C + \sqrt{C^2 + 4C})/2 < C + 1.$$

Consequently, $A \leq (C + 1)^2 B$, which yields the right inequality of (5.74) with

$$c_4(K) = (4C_0 e^\pi + 1)^2 = (4e^\pi \sqrt{1 + K^2} + 1)^2. \qquad \square$$

For $f \in C_0^\infty(\Omega_+)$ define

$$(Tf)(w) := \iint_{\Omega_+} \frac{f(z)d(z)}{(z - w)^2} dx\, dy \quad (w \in \mathbf{C} \setminus \operatorname{supp} f),$$

where, as above, $d(z) = \operatorname{dist}(z, \Gamma)$. The function $d(z)f(z)$ is continuous on the set $\operatorname{supp} f \subset \Omega_+$. Since

$$(Tf)(w) = \iint\limits_{\operatorname{supp} f} \frac{f(z)d(z)}{(z-w)^2} \, dx \, dy \quad (w \in \mathbf{C} \setminus \operatorname{supp} f),$$

it follows that Tf is analytic in $\mathbf{C} \setminus \operatorname{supp} f \supset \Gamma \cup \Omega_-$ and has a zero at infinity. Therefore $Tf \in E^2(\Omega_-) \cap C(\overline{\Omega_-})$ and

$$(Tf)(t) = (Tf)^-(t) := \lim_{w \to t, w \in \Omega_-} \iint\limits_{\Omega_+} \frac{f(z)d(z)}{(z-w)^2} \, dx \, dy \tag{5.87}$$

for all $t \in \Gamma$.

Lemma 5.26. *If $\varphi' \in C_0^\infty(\mathbf{R})$ and $f \in C_0^\infty(\Omega_+)$ then*

$$\|(Tf)^-\|_{L^2(\Gamma)} \le c_5(K)\|f\|_{\mathcal{H}_+} \tag{5.88}$$

where $c_5(K)$ depends only on $K := \|\varphi'\|_\infty$.

Proof. Since $Tf \in E^2(\Omega_-)$, we deduce from Lemma 5.25 that

$$\|(Tf)^-\|_{L^2(\Gamma)} \le c_2(K)\|(Tf)'\|_{\mathcal{H}_-}. \tag{5.89}$$

Further, for $w \in \Omega_-$,

$$|(Tf)'(w)| = \left|(-2) \iint\limits_{\Omega_+} \frac{f(z)d(z)}{(z-w)^3} \, dx \, dy\right| \le 2 \iint\limits_{\Omega_+} \frac{|f(z)|d(z)}{|z-w|^3} \, dx \, dy. \tag{5.90}$$

As $\|g\|_{\mathcal{H}_\pm} = \|\psi\|_{L^2(\Omega_\pm)}$ with $\psi(z) := |g(z)|(d(z))^{1/2}$, we see from the estimates (5.89) and (5.90) that (5.88) will follow as soon as we have shown that the integral operator $N : L^2(\Omega_+) \to L^2(\Omega_-)$ given by

$$(N\psi)(w) := (d(w))^{1/2} \iint\limits_{\Omega_+} \frac{\psi(z)(d(z))^{1/2}}{|z-w|^3} \, dx \, dy \quad (w \in \Omega_-)$$

is bounded. The kernel of the integral operator N is

$$N(z,w) := (d(z))^{1/2}(d(w))^{1/2}|z-w|^{-3} \quad (z \in \Omega_+, w \in \Omega_-).$$

Fix $w \in \Omega_-$ and put $\delta := d(w)$. If $z \in \Omega_+$, then $d(z) \le |z-w|$ and $|z-w| > \delta$, whence

$$\iint\limits_{\Omega_+} N(z,w) \, dx \, dy \le \delta^{1/2} \iint\limits_{|z-w|>\delta} \frac{|z-w|^{1/2}}{|z-w|^3} \, dx \, dy = \delta^{1/2} \int\limits_0^{2\pi} d\theta \int\limits_\delta^\infty \varrho^{-5/2} \varrho \, d\varrho = 4\pi.$$

Analogously,

$$\iint\limits_{\Omega_-} N(z,w)\,du\,dv \le 4\pi \quad (w = u + iv).$$

Thus, we may have recourse to Schur's criterion (see, e.g., [210, Theorem 3.2.2.]) to conclude that N is bounded and $\|N\| \le 4\pi$. In summary, we have (5.88) with $c_5(K) = 8\pi c_2(K)$. $\square$

The boundedness of the operators $g \mapsto (Cg)^\pm$ on $L^2(\Gamma)$ over nice curves Γ, and in particular over Lipschitz graphs with $\varphi' \in C_0^\infty(\mathbf{R})$, is a classical result (see, e.g., [72], [151], [206]). The point of the following theorem is that the norms of these operators are estimated by constants depending only on the Lipschitz constant $K = \|\varphi'\|_\infty$.

Theorem 5.27. *Let $\Gamma = \{x + i\varphi(x) : x \in \mathbf{R}\}$ be a Lipschitz graph and suppose $\varphi' \in C_0^\infty(\mathbf{R})$. Then for every $g \in C_0^\infty(\Gamma)$,*

$$\|(Cg)^\pm\|_{L^2(\Gamma)} \le \tilde{c}(K)\|g\|_{L^2(\Gamma)}$$

where $\tilde{c}(K)$ depends only on $K = \|\varphi'\|_\infty$.

Proof. Since Cg is analytic in $\Omega_- \cup \Omega_+ \cup \{\infty\}$ and has a zero at infinity, it follows that the boundary functions $(Cg)^\pm$ are continuous on $\dot{\Gamma} := \Gamma \cup \{\infty\}$ and that $Cg|\Omega_\pm \in E^2(\Omega_\pm)$. Hence, by Lemma 5.25,

$$\|(Cg)^+\|_{L^2(\Gamma)} \le c_2(K) \|(Cg)'\|_{\mathcal{H}_+}.$$

Put $B := \{f \in \mathcal{H}_+ : f \in C_0^\infty(\Omega_+), \|f\|_{\mathcal{H}_+} \le 1\}$. The set $C_0^\infty(\Omega_+)$ is easily seen to be dense in $\mathcal{H}_+$. Therefore,

$$\|(Cg)'\|_{\mathcal{H}_+} = \sup_{f \in B} \left|((Cg)', f)\right|$$

where $(h, f) := \iint_{\Omega_+} h(z)\overline{f(z)}d(z)\,dx\,dy$ is the scalar product in $\mathcal{H}_+$. For $f \in B$, we have by Fubini's theorem and (5.87),

$$\left|((Cg)', f)\right| = \left| \iint\limits_{\Omega_+} \left\{ -\int_\Gamma \frac{g(\tau)\,d\tau}{(\tau - z)^2} \right\} \overline{f(z)}d(z)\,dx\,dy \right|$$

$$= \left| \int_\Gamma g(\tau) \left\{ \iint\limits_{\Omega_+} \frac{\overline{f(z)}d(z)}{(\tau - z)^2}\,dx\,dy \right\} d\tau \right| = \left| \int_\Gamma g(\tau)(T\overline{f})^-(\tau)\,d\tau \right|. \qquad (5.91)$$

From the Cauchy-Schwarz inequality and Lemma 5.26 we infer that (5.91) is at most

$$\|g\|_{L^2(\Gamma)}\|(T\overline{f})^-\|_{L^2(\Gamma)} \le c_5(K)\|g\|_{L^2(\Gamma)}\|f\|_{\mathcal{H}_+}.$$

In summary,

$$\|(Cg)^+\|_{L^2(\Gamma)} \le c_2(K)c_5(K)\|g\|_{L^2(\Gamma)} \sup_{f\in B} \|f\|_{\mathcal{H}_+} = c_2(K)c_5(K)\|g\|_{L^2(\Gamma)},$$

which is the desired estimate with $\tilde{c}(K) = c_2(K)c_5(K)$. The proof is analogous for the function $(Cg)^-$. $\qquad\Box$

Combining Theorem 5.27 and formula (5.73) we arrive at the norm estimate

$$\|Sg\|_{L^2(\Gamma)} \le 2\tilde{c}(K)\|g\|_{L^2(\Gamma)} \tag{5.92}$$

for all $g \in C_0^\infty(\Gamma)$ provided $\varphi' \in C_0^\infty(\mathbf{R})$. We now remove the restriction to φ' in $C_0^\infty(\mathbf{R})$.

Lemma 5.28. *Given a Lipschitz function* $\varphi : \mathbf{R} \to \mathbf{R}$, *a finite segment* $[a,b] \subset \mathbf{R}$, *and* $\varepsilon > 0$, *there exists a Lipschitz function* $\psi : \mathbf{R} \to \mathbf{R}$ *such that*

$$\psi' \in C_0^\infty(\mathbf{R}), \quad \|\psi'\|_\infty \le \|\varphi'\|_\infty, \quad \|\varphi - \psi\|_{C[a,b]} < \varepsilon.$$

Proof. Choose $\delta > 0$ so that $|\varphi(x) - \varphi(y)| < \varepsilon/3$ whenever $x, y \in [a,b]$ and $|x - y| < \delta$. Then divide $[a,b]$ by points $a = x_0 < x_1 < \ldots < x_j < x_{j+1} = b$ into segments of length less than δ and consider the piecewise linear and continuous function φ_1 given by

$$\varphi_1(x) = \varphi(x_n)\frac{x_{n+1} - x}{x_{n+1} - x_n} + \varphi(x_{n+1})\frac{x - x_n}{x_{n+1} - x_n}$$

for $x \in [x_n, x_{n+1}]$ $(n = 0, 1, \ldots, j)$. Clearly, if $x \in [x_n, x_{n+1}]$, then

$$\varphi(x) - \varphi_1(x) = \big(\varphi(x) - \varphi(x_n)\big)\frac{x_{n+1} - x}{x_{n+1} - x_n} + \big(\varphi(x) - \varphi(x_{n+1})\big)\frac{x - x_n}{x_{n+1} - x_n},$$

whence

$$\|\varphi - \varphi_1\|_{C[a,b]} \le \max_n \max_{x\in[x_n,x_{n+1}]} \frac{\varepsilon}{3}\left(\frac{x_{n+1} - x}{x_{n+1} - x_n} + \frac{x - x_n}{x_{n+1} - x_n}\right) = \frac{\varepsilon}{3}.$$

Lipschitz functions are absolutely continuous, and therefore we may write

$$\varphi(x) = \varphi(a) + \int_a^x \varphi'(t)\, dt,$$

which implies that

$$\|\varphi_1'\|_\infty = \max_n \frac{|\varphi(x_{n+1}) - \varphi(x_n)|}{x_{n+1} - x_n} \le \max_n \left(\frac{1}{x_{n+1} - x_n}\int_{x_n}^{x_{n+1}} |\varphi'(t)|\, dt\right) \le \|\varphi'\|_\infty.$$

The function φ_1' is piecewise constant. There is a function $\varphi_2 \in C^1[a,b]$ such that φ_2' is piecewise linear,

$$\varphi_2'(a) = \varphi_2'(b) = 0, \quad \int_a^b |\varphi_1'(t) - \varphi_2'(t)|\, dt < \frac{\varepsilon}{3}, \quad \|\varphi_2'\|_{C[a,b]} = \|\varphi_1'\|_\infty.$$

Finally, we can find a function $\varphi_3 \in C^\infty[a,b]$ such that

$$\operatorname{supp}\varphi_3' \subset (a,b), \quad \|\varphi_2' - \varphi_3'\|_{C[a,b]} < \frac{\varepsilon}{3(b-a)}, \quad \|\varphi_3'\|_{C[a,b]} \doteq \|\varphi_2'\|_{C[a,b]}.$$

Extend φ_3' by zero to $\mathbf{R} \setminus [a,b]$ and put

$$\psi(x) := \varphi(a) + \int_a^x \varphi_3'(t)\, dt.$$

Then $\psi' = \varphi_3' \in C_0^\infty(\mathbf{R})$, and from the construction of ψ we immediately see that $\|\psi'\|_\infty \le \|\varphi'\|_\infty$. Obviously, ψ is a Lipschitz function. We have

$$\|\varphi - \psi\|_{C[a,b]} \le \|\varphi - \varphi_1\|_{C[a,b]} + \|\varphi_1 - \psi\|_{C[a,b]} < \varepsilon/3 + \|\varphi_1 - \psi\|_{C[a,b]}$$

and

$$\|\varphi_1 - \psi\|_{C[a,b]} \le \max_{x \in [a,b]} \left| \int_a^x (\varphi_1'(t) - \varphi_3'(t))\, dt \right|$$

$$\le \int_a^b |\varphi_1'(t) - \varphi_3'(t)|\, dt \le \int_a^b |\varphi_1'(t) - \varphi_2'(t)|\, dt + \int_a^b |\varphi_2'(t) - \varphi_3'(t)|\, dt.$$

Since each term in the last sum is at most $\varepsilon/3$, it follows that $\|\varphi - \psi\|_{C[a,b]} < \varepsilon$. $\square$

Theorem 5.29. If $\Gamma = \{x + i\varphi(x) : x \in \mathbf{R}\}$ is a Lipschitz graph, then

$$\|Sg\|_{L^2(\Gamma)} \le c(K)\|g\|_{L^2(\Gamma)} \text{ for all } g \in C_0^\infty(\Gamma)$$

with some finite constant $c(K)$ depending only on the Lipschitz constant $K = \|\varphi'\|_{L^\infty(\mathbf{R})}$.

Proof. Suppose first that $\varphi' \in C_0^\infty(\mathbf{R})$. Then (5.92) holds and hence we may employ Theorem 5.7 (Cotlar's inequality) with $p = 2$, $\delta = 1$ and (5.32) to get

$$|(S_\varepsilon g)(t)| \le 6(M\tilde{S}g)(t) + c_0(Mg)(t) \tag{5.93}$$

for all $t \in \Gamma$. Here, by (5.40) and (5.92),

$$c_0 \le 6 \cdot 2^3 C_\Gamma \theta_2 (2\tilde{c}(K))^2 + 96 C_\Gamma/\pi + 12 C_\Gamma \theta_2 + 4/\pi.$$

For a Lipschitz graph Γ we have

$$C_\Gamma = \sup_{t \in \Gamma} \sup_{\varepsilon>0} \frac{|\Gamma(t,\varepsilon)|}{\varepsilon} \leq \sup_{t \in \Gamma} \sup_{\varepsilon>0} \frac{1}{\varepsilon} \int\limits_{\operatorname{Re}t-\varepsilon}^{\operatorname{Re}t+\varepsilon} \sqrt{1+(\varphi'(x))^2}\, dx \leq 2\sqrt{1+K^2}.$$

Thus, $c_0 \leq c_0(K)$ with some constant $c_0(K) < \infty$ depending only on K. From (5.93) and Corollary 5.3 we infer that

$$\|S_\varepsilon g\|_2 \leq \left(6\|M\|_2\|\tilde{S}\|_2 + c_0(K)\|M\|_2\right)\|g\|_2.$$

By (5.6), Theorem 2.18, and (5.11),

$$\|M\|_2^2 \leq 8\|M\|_\infty\|M\|_\sim \leq 8\,C_\Gamma(C_\Gamma\theta_2) \leq 32\,\theta_2(1+K^2),$$

whence $\|S_\varepsilon g\|_2 \leq c_6(K)\|g\|_2$ with some constant $c_6(K) < \infty$ depending only on K. It follows that

$$\left|(S_\varepsilon g, f)\right| \leq c_6(K)\|g\|_2\|f\|_2 \quad \text{for all } g, f \in C_0^\infty(\Gamma). \tag{5.94}$$

Let now $\varphi : \mathbf{R} \to \mathbf{R}$ be an arbitrary Lipschitz function and fix $g, f \in C_0^\infty(\Gamma)$. Define $\zeta : \mathbf{R} \to \Gamma$ by $\zeta(x) := x + i\varphi(x)$. Fix a segment $[a,b] \subset \mathbf{R}$ containing the compact set $\operatorname{supp}(g \circ \zeta) \cup \operatorname{supp}(f \circ \zeta)$. By Lemma 5.28, there exists a sequence of Lipschitz functions $\varphi_n : \mathbf{R} \to \mathbf{R}$ such that $\|\varphi - \varphi_n\|_{C[a,b]} \to 0$ and $\|\varphi_n'\|_\infty \leq \|\varphi'\|_\infty = K$. Put $\zeta_n(x) := x + i\varphi_n(x)$ and consider the Lipschitz graphs $\Gamma_n := \{\zeta_n(x) : x \in \mathbf{R}\}$. Let $S_\varepsilon^{(n)}$ denote the corresponding truncated singular integral operators. Since $\varphi_n' \in C_0^\infty(\mathbf{R})$, we obtain from (5.94) that

$$\left|(S_\varepsilon^{(n)}h, k)\right| \leq c_6(K)\|h\|_{L^2(\Gamma_n)}\|k\|_{L^2(\Gamma_n)} \quad \text{for all } h, k \in C_0^\infty(\Gamma_n). \tag{5.95}$$

Put $\tilde{g}(x) := g(\zeta(x))\zeta'(x)$ and $\tilde{f}(x) := \overline{f(\zeta(x))}|\zeta'(x)|$. These are functions in $L^2(\mathbf{R})$ with finite support, we have

$$\|\tilde{g}\|_2^2 = \int\limits_{\mathbf{R}} \left|g(\zeta(x))\right|^2|\zeta'(x)|^2 dx$$

$$\leq \sqrt{1+K^2} \int\limits_{\mathbf{R}} \left|g(\zeta(x))\right|^2|\zeta'(x)|\, dx = \sqrt{1+K^2}\,\|g\|_2^2$$

and, analogously, $\|\tilde{f}\|_2^2 \leq \sqrt{1+K^2}\,\|f\|_2^2$. Clearly,

$$\begin{aligned}
(S_\varepsilon g, f) &= \frac{1}{\pi i} \int\limits_\Gamma \int\limits_{\Gamma \setminus \Gamma(t,\varepsilon)} \frac{g(\tau)}{\tau - t}\, d\tau\, \overline{f}(t)\, |dt| \\
&= \frac{1}{\pi i} \int\limits_a^b \int\limits_{[a,b] \setminus \{x : |\zeta(x) - \zeta(y)| < \varepsilon\}} \frac{\tilde{g}(x)}{\zeta(x) - \zeta(y)}\, dx\, \tilde{f}(y)\, dy. \tag{5.96}
\end{aligned}$$

Define $g_n, f_n \in L^2(\Gamma_n)$ by

$$\tilde{g}(x) = \big(g_n(\zeta_n(x))\big)\zeta_n'(x), \ \ \tilde{f}(x) = \overline{f_n(\zeta_n(x))}|\zeta_n'(x)|.$$

Then

$$
\begin{aligned}
\|g_n\|_{L^2(\Gamma_n)}^2 &= \int_{\mathbf{R}} |g_n(\zeta_n(x))|^2 |\zeta_n'(x)|\, dx = \int_{\mathbf{R}} |\tilde{g}(x)|^2 |\zeta_n'(x)|^{-1} dx \\
&\leq \int_{\mathbf{R}} |\tilde{g}(x)|^2 dx = \|\tilde{g}\|_2^2
\end{aligned}
$$

and, equally, $\|f_n\|_{L^2(\Gamma_n)}^2 \leq \|\tilde{f}\|_2^2$. Thus,

$$\|g_n\|_{L^2(\Gamma_n)}^2 \leq \sqrt{1+K^2}\,\|g\|_2^2, \quad \|f_n\|_{L^2(\Gamma_n)}^2 \leq \sqrt{1+K^2}\,\|f\|_2^2. \tag{5.97}$$

Further,

$$(S_\varepsilon^{(n)} g_n, f_n) = \frac{1}{\pi i} \int_a^b \int_{[a,b]\setminus\{x:|\zeta_n(x)-\zeta_n(y)|<\varepsilon\}} \frac{\tilde{g}(x)}{\zeta_n(x)-\zeta_n(y)}\, dx\, \tilde{f}(y)\, dy.$$

This together with (5.96) gives

$$\lim_{n\to\infty} (S_\varepsilon^{(n)} g_n, f_n) = (S_\varepsilon g, f)$$

whence, by (5.95) and (5.97),

$$
\begin{aligned}
\big|(S_\varepsilon g, f)\big| &\leq \sup_n \big|(S_\varepsilon^{(n)} g_n, f_n)\big| \leq c_6(K)\|g_n\|_{L^2(\Gamma_n)}\|f_n\|_{L^2(\Gamma_n)} \\
&\leq c_6(K)\sqrt{1+K^2}\,\|g\|_2\|f\|_2.
\end{aligned}
$$

Consequently, $\|S_\varepsilon g\|_2 \leq c_6(K)\sqrt{1+K^2}\,\|g\|_2$ for all $\varepsilon > 0$ and all $g \in C_0^\infty(\Gamma)$. Since $S_\varepsilon g \to Sg$ a.e. on Γ due to Theorem 4.3, we infer from Fatou's theorem (Lemma 5.9) that $\|Sg\|_2 \leq c_6(K)\sqrt{1+K^2}\,\|g\|_2$ for all $g \in C_0^\infty(\Gamma)$. $\qquad\square$

5.8 Measures in the plane

Using Theorem 5.29, we now show that S generates a bounded operator on $L^2(\Gamma)$ where Γ is any unbounded Carleson curve. Note that "one-dimensional beings" living in a curve Γ cannot find out whether their curve is Carleson by measuring only lengths along Γ. The Carleson condition is rather a condition on the embedding of the curve in the (two-dimensional) plane.

A non-negative Borel measure μ on $\mathbf{C}$ is called a *Radon measure* if μ is finite on compact sets and $\mu(A) = \sup \mu(K) = \inf \mu(U)$ for every Borel set A, where the supremum is taken over all compact sets $K \subset A$ and the infimum is over all open sets $U \supset A$. We refer to a Borel measure μ on $\mathbf{C}$ as a *Carleson measure* if μ is a Radon measure and there is a constant $C := C(\mu) \geq 0$ such that $\mu(D(z, \varepsilon)) \leq C\varepsilon$ for all disks $D(z, \varepsilon) := \{\zeta \in \mathbf{C} : |\zeta - z| < \varepsilon\}$. The set of Carleson measures will be denoted by Δ.

Let Γ be a simple locally rectifiable curve in the plane and consider the measure σ given by

$$\sigma(A) := |\Gamma \cap A| := \text{ length of } \Gamma \cap A. \tag{5.98}$$

One can show that σ is a Radon measure. Clearly, σ is a Carleson measure if and only if Γ is a Carleson curve.

Given a Radon measure μ and a function $f \in L^1_{\text{loc}}(d\mu) := L^1_{\text{loc}}(\mathbf{C}, d\mu)$, we denote by $M_\mu f : \mathbf{C} \to [0, \infty]$ the function defined by

$$(M_\mu f)(z) := \sup_{\varepsilon > 0} \frac{1}{\varepsilon} \int\limits_{D(z,\varepsilon)} |f(w)| \, d\mu(w).$$

Thus, M_μ is again some kind of maximal operator. It can be shown as in the proof of Lemma 2.16 that $M_\mu f$ is lower semi-continuous, which implies that the set $\{z \in \mathbf{C} : (M_\mu f)(z) > \lambda\}$ is open and therefore measurable for every $\lambda > 0$ with respect to every Borel measure on $\mathbf{C}$. In what follows we abbreviate $L^p(\mathbf{C}, d\mu)$ to $L^p(d\mu)$.

Theorem 5.30. *Let $\mu_1, \mu_2 \in \Delta$. Then for every $p \in (1, \infty]$ there exists a constant $C = C(\mu_1, \mu_2, p)$ such that*

$$\|M_{\mu_1} f\|_{L^p(d\mu_2)} \leq C\|f\|_{L^p(d\mu_1)} \text{ for all } f \in L^p(d\mu_1).$$

Proof. Let first $f \in L^\infty(d\mu_1)$. Then

$$(M_{\mu_1} f)(z) \leq \sup_{\varepsilon > 0} \frac{\mu_1(D(z, \varepsilon))}{\varepsilon} \|f\|_{L^\infty(d\mu_1)} \leq C(\mu_1)\|f\|_{L^\infty(d\mu_1)}$$

and consequently, $M_{\mu_1} : L^\infty(d\mu_1) \to L^\infty(d\mu_2)$ is bounded.

We now show that $M_{\mu_1} : L^1(d\mu_1) \to L^\sim(d\mu_2)$ is of the weak type $(1,1)$ with respect to the measures $d\mu_1, d\mu_2$, i.e. that there exists a constant $C < \infty$ such that

$$\mu_2(\{z \in \mathbf{C} : (M_{\mu_1} f)(z) > \lambda\}) \leq (C/\lambda)\|f\|_{L^1(d\mu_1)} \tag{5.99}$$

for all $f \in L^1(d\mu_1)$ and all $\lambda > 0$. Fix $f \in L^1(d\mu_1)$ and $\lambda > 0$. For $n \in \{1, 2, 3, \ldots\}$, put

$$A_{\lambda,n} := \{z \in \mathbf{C} : (M_{\mu_1} f)(z) > \lambda, \ |z| < n\}.$$

Also fix n. For each $z \in A_{\lambda,n}$ we pick a disk $D(z, \varepsilon(z))$ such that

$$\frac{1}{\varepsilon(z)} \int\limits_{D(z,\varepsilon(z))} |f(w)| \, d\mu_1(w) > \lambda.$$

By Theorem 2.17, the family $\{D(z, \varepsilon(z))\}_{z \in A_{\lambda,n}}$ contains an at most countable subfamily $\{D_k\} := \{D(z_k, \varepsilon(z_k))\}$ such that $A_{\lambda,n} \subset \bigcup_k D_k$ and $\sum_k \chi_k \leq \theta_2$, where χ_k is the characteristic function of D_k. Taking into account that $\mu_2 \in \Delta$, we get

$$\mu_2(A_{\lambda,n}) \leq \sum_k \mu_2(D_k) \leq \sum_k C(\mu_2)\varepsilon(z_k)$$

$$\leq C(\mu_2) \sum_k \frac{1}{\lambda} \int\limits_{D_k} |f(w)| \, d\mu_1(w)$$

$$= C(\mu_2) \frac{1}{\lambda} \int\limits_{\mathbf{C}} \sum_k \chi_k(w) |f(w)| \, d\mu_1(w)$$

$$\leq C(\mu_2)\theta_2 \frac{1}{\lambda} \int\limits_{\mathbf{C}} |f(w)| \, d\mu_1(w) = \frac{C(\mu_2)\theta_2}{\lambda} \|f\|_{L^1(d\mu_1)}.$$

Since $\mu_2(\{z \in \mathbf{C} : (M_{\mu_1}f)(z) > \lambda\}) = \lim_{n \to \infty} \mu_2(A_{\lambda,n})$, we arrive at (5.99).

The assertion of the theorem now follows from Theorem 5.1. $\qquad\qquad\square$

Given a Carleson measure $\mu \in \Delta$ and $\varepsilon > 0$, we define the *truncated singular integral operator* S_μ^ε and the *maximal singular integral operator* S_μ^* by

$$(S_\mu^\varepsilon f)(z) := \frac{1}{\pi i} \int\limits_{|w-z| \geq \varepsilon} \frac{f(w)}{w - z} \, d\mu(w) \quad (z \in \mathbf{C}),$$

$$(S_\mu^* f)(z) := \sup_{\varepsilon > 0} |(S_\mu^\varepsilon f)(z)| \quad (z \in \mathbf{C}).$$

Clearly, $(S_\mu^\varepsilon f)(z)$ and thus $(S_\mu^* f)(z)$ are well-defined for every $z \in \mathbf{C}$ in case $f \in L^1(d\mu)$. If $f \in L^p(d\mu)$ $(1 < p < \infty)$, then

$$\int\limits_{|w-z| \geq \varepsilon} \frac{|f(w)|}{|w - z|} \, d\mu(w)$$

$$\leq \left(\int\limits_{\mathbf{C}} |f(w)|^p d\mu(w) \right)^{1/p} \left(\int\limits_{|w-z| \geq \varepsilon} \frac{d\mu(w)}{|w - z|^q} \right)^{1/q} \qquad (5.100)$$

and

$$\int\limits_{|w-z|\geq\varepsilon} \frac{d\mu(w)}{|w-z|^q} = \sum_{k=0}^{\infty} \int\limits_{2^k\varepsilon\leq|w-z|<2^{k+1}\varepsilon} \frac{d\mu(w)}{|w-z|^q}$$

$$\leq \sum_{k=0}^{\infty} \frac{1}{(2^k\varepsilon)^q} \int\limits_{|w-z|<2^{k+1}\varepsilon} d\mu(w)$$

$$\leq \sum_{k=0}^{\infty} \frac{C(\mu)2^{k+1}\varepsilon}{(2^k\varepsilon)^q} = \frac{2C(\mu)}{\varepsilon^{q-1}} \sum_{k=0}^{\infty}(2^{1-q})^k < \infty. \tag{5.101}$$

This shows that $(S_\mu^\varepsilon f)(z)$ and thus $(S_\mu^* f)(z)$ are well-defined for every $z \in \mathbf{C}$ whenever $f \in L^p(d\mu)$ and $1 < p < \infty$.

Lemma 5.31. *If $f \in L^p(d\mu)$ ($1 \leq p < \infty$) then $S_\mu^* f$ is lower semi-continuous on $\mathbf{C}$. In particular, for every $\lambda > 0$ the set $\{z \in \mathbf{C} : (S_\mu^* f)(z) > \lambda\}$ is open and thus measurable with respect to every Borel measure.*

Proof. Using (5.100) and (5.101) for $p > 1$, this can be shown by slightly modifying the proof of Lemma 5.16.   □

5.9   Cotlar's inequality in the plane

The following Lemma 5.32 and Theorem 5.33 are analogues of Cotlar's inequality.

Lemma 5.32. *Let $\mu \in \Delta$ and $f \in L^p(d\mu)$ for some $1 < p < \infty$. Then*

$$(S_\mu^\varepsilon f)(z_0) \leq (S_\mu^* f)(z) + \frac{7}{\pi}(M_\mu f)(z_0)$$

for all $\varepsilon > 0$, $z_0 \in \mathbf{C}$, and $z \in D(z_0, \varepsilon/2)$.

Proof. Fix $\varepsilon > 0$, $z_0 \in \mathbf{C}$, and $z \in D(z_0, \varepsilon/2)$. If $w \in \mathbf{C} \setminus D(z_0, \varepsilon)$, then

$$|z - z_0| < \varepsilon/2 \leq |w - z_0|/2, \quad |w - z| \geq |w - z_0| - |z - z_0| \geq |w - z_0|/2,$$

whence

$$\left| \int\limits_{\mathbf{C}\setminus D(z_0,\varepsilon)} \left(\frac{1}{w - z} - \frac{1}{w - z_0}\right) f(w) \, d\mu(w) \right|$$

$$\leq |z - z_0| \int\limits_{\mathbf{C}\setminus D(z_0,\varepsilon)} \frac{|f(w)|}{|w - z||w - z_0|} \, d\mu(w)$$

$$\leq 2|z - z_0| \int\limits_{\mathbf{C}\setminus D(z_0,\varepsilon)} \frac{|f(w)|}{|w - z_0|^2} \, d\mu(w). \tag{5.102}$$

Consequently,

$$\pi\big|(S_\mu^\varepsilon f)(z) - (S_\mu^\varepsilon f)(z_0)\big|$$

$$\leq \left| \int\limits_{\mathbf{C}\backslash D(z_0,\varepsilon)} \left(\frac{1}{w-z} - \frac{1}{w-z_0}\right) f(w)\, d\mu(w) \right|$$

$$+ \left| \int\limits_{\mathbf{C}\backslash D(z,\varepsilon)} \frac{f(w)}{w-z}\, d\mu(w) - \int\limits_{\mathbf{C}\backslash D(z_0,\varepsilon)} \frac{f(w)}{w-z}\, d\mu(w) \right|$$

$$\leq 2|z-z_0| \int\limits_{\mathbf{C}\backslash D(z_0,\varepsilon)} \frac{|f(w)|}{|w-z_0|^2}\, d\mu(w) + \int\limits_{D(z,\varepsilon)\Delta D(z_0,\varepsilon)} \frac{|f(w)|}{|w-z|}\, d\mu(w),$$

where $A\Delta B := (A \setminus B) \cup (B \setminus A)$. Since $|z - z_0| < \varepsilon/2$, we have

$$\frac{2}{\pi}|z-z_0| \int\limits_{\mathbf{C}\backslash D(z_0,\varepsilon)} \frac{|f(w)|}{|w-z_0|^2}\, d\mu(w)$$

$$\leq \frac{\varepsilon}{\pi} \sum_{k=0}^{\infty} \int\limits_{2^k\varepsilon \leq |w-z_0| < 2^{k+1}\varepsilon} \frac{|f(w)|}{|w-z_0|^2}\, d\mu(w)$$

$$\leq \frac{1}{\pi} \sum_{k=0}^{\infty} \frac{1}{2^{2k}\varepsilon} \int\limits_{D(z_0,2^{k+1}\varepsilon)} |f(w)|\, d\mu(w)$$

$$\leq \frac{1}{\pi} \sum_{k=0}^{\infty} 2^{1-k} (M_\mu f)(z_0) = \frac{4}{\pi}(M_\mu f)(z_0). \qquad (5.103)$$

Further, since

$$|w - z| \geq \varepsilon/2 \ \text{ for } \ z \in D(z_0,\varepsilon/2) \ \text{ and } \ w \in D(z,\varepsilon)\Delta D(z_0,\varepsilon)$$

and since $D(z,\varepsilon)\Delta D(z_0,\varepsilon) \subset D(z_0,3\varepsilon/2)$, we get

$$\frac{1}{\pi} \int\limits_{D(z,\varepsilon)\Delta(z_0,\varepsilon)} \frac{|f(w)|}{|w-z|}\, d\mu(w) \leq \frac{2}{\pi\varepsilon} \int\limits_{D(z_0,3\varepsilon/2)} |f(w)|\, d\mu(w) \leq \frac{3}{\pi}(M_\mu f)(z_0).$$

In summary, $|(S_\mu^\varepsilon f)(z) - (S_\mu^\varepsilon f)(z_0)| \leq (7/\pi)(M_\mu f)(z_0)$, whence

$$\big|(S_\mu^\varepsilon f)(z_0)\big| \leq \big|(S_\mu^\varepsilon f)(z)\big| + \big|(S_\mu^\varepsilon f)(z) - (S_\mu^\varepsilon f)(z_0)\big|$$

$$\leq (S_\mu^* f)(z) + \frac{7}{\pi}(M_\mu f)(z_0). \qquad \square$$

Let Σ denote the set of all measures $\mu \in \Delta$ for which there exists a constant $\gamma := \gamma(\mu) > 0$ such that

$$\mu\big(D(z,\varepsilon)\big) \geq \gamma\varepsilon \ \text{ for all } \ \varepsilon > 0 \ \text{ and all } \ z \in \operatorname{supp}\mu. \qquad (5.104)$$

If Γ is an unbounded Carleson curve and σ is defined by (5.98), then (5.104) holds with $\gamma = 1$ and thus, $\sigma \in \Sigma$. Note that if Γ is bounded, then the measure (5.98) does not belong to Σ.

Theorem 5.33. *If $\mu \in \Sigma$, then there is a constant $C = C(\gamma) < \infty$ depending only on the γ in (5.104) such that*

$$(S_\mu^* f)(z) \leq C(M_\mu S_\mu^* f)(z) + C(M_\mu f)(z) \tag{5.105}$$

for all $z \in \mathbf{C}$ and all $f \in L^p(d\mu)$ $(1 < p < \infty)$.

Proof. Fix $z_0 \in \mathbf{C}$ and $\varepsilon > 0$. By Lemma 5.32,

$$\left|(S_\mu^\varepsilon f)(z_0)\right| \leq \inf_{z \in D(z_0, \varepsilon/2)} (S_\mu^* f)(z) + (7/\pi)(M_\mu f)(z_0). \tag{5.106}$$

Put $d := \operatorname{dist}(z_0, \operatorname{supp}\mu)$.

Assume first that $\varepsilon \geq 4d$. Letting $\delta := \varepsilon/2$ we get

$$\frac{1}{\delta} \int_{D(z_0,\delta)} (S_\mu^* f)(w)\, d\mu(w) \geq \left(\inf_{z \in D(z_0,\delta)} (S_\mu^* f)(z) \right) \frac{\mu(D(z_0,\delta))}{\delta}. \tag{5.107}$$

Since $\delta \geq 2d$, we can find a $z_1 \in \operatorname{supp}\mu$ such that $D(z_1, \delta/2) \subset D(z_0, \delta)$. Because $\mu \in \Sigma$, we obtain from (5.104) that

$$\mu\big(D(z_0, \delta)\big) \geq \mu\big(D(z_1, \delta/2)\big) \geq \gamma\delta/2. \tag{5.108}$$

Thus, (5.107) and (5.108) show that

$$\inf_{z \in D(z_0,\delta)} (S_\mu^* f)(z) \leq \frac{2}{\gamma\delta} \int_{D(z_0,\delta)} (S_\mu^* f)(w)\, d\mu(w) \leq \frac{2}{\gamma}(M_\mu S_\mu^* f)(z_0).$$

This and (5.106) with $\varepsilon = 2\delta$ imply that

$$\left|(S_\mu^\varepsilon f)(z_0)\right| \leq \frac{2}{\gamma}(M_\mu S_\mu^* f)(z_0) + \frac{7}{\pi}(M_\mu f)(z_0).$$

If $d/2 \leq \varepsilon < 4d$, then

$$\pi\left|(S_\mu^\varepsilon f)(z_0) - (S_\mu^{4d} f)(z_0)\right| \leq \int_{D(z_0,4d) \setminus D(z_0,d/2)} \frac{|f(w)|}{|w - z_0|}\, d\mu(w)$$

$$\leq \frac{2}{d} \int_{D(z_0,4d)} |f(w)|\, d\mu(w) \leq 8(M_\mu f)(z_0)$$

and hence, by what was proved in the previous paragraph,

$$
\begin{aligned}
\left|(S_\mu^\varepsilon f)(z_0)\right| &\leq \left|(S_\mu^{4d} f)(z_0)\right| + \frac{8}{\pi}(M_\mu f)(z_0) \\
&\leq \frac{2}{\gamma}(M_\mu S_\mu^* f)(z_0) + \frac{15}{\pi}(M_\mu f)(z_0).
\end{aligned}
$$

Finally, if $0 < \varepsilon < d/2$ then $(S_\mu^\varepsilon f)(z_0) = (S_\mu^{d/2} f)(z_0)$, which leads us to the case $\varepsilon = d/2$. In summary, we have shown that (5.105) is true with $C = \max\{2/\gamma, 15/\pi\}$.

□

5.10 Maximal singular integrals in the plane

Theorem 5.34. *Let $\mu \in \Delta$ and $\sigma \in \Sigma$. Suppose that $S_\sigma^* : L^p(d\sigma) \to L^p(d\sigma)$ is bounded for all $p \in (1,\infty)$. Then the operator $S_\sigma^* : L^p(d\sigma) \to L^p(d\mu)$ is bounded for all $p \in (1,\infty)$.*

Proof. Theorem 5.33 with μ replaced by σ gives

$$
(S_\sigma^* f)(z) \leq C(M_\sigma S_\sigma^* f)(z) + C(M_\sigma f)(z)
$$

for all $z \in \mathbf{C}$ and all $f \in L^p(d\sigma)$. Theorem 5.30 implies that $\|S_\sigma^* f\|_{L^p(d\mu)}$ is not greater than

$$
C\left(\|M_\sigma\|_{L^p(d\sigma)\to L^p(d\mu)}\|S_\sigma^*\|_{L^p(d\sigma)\to L^p(d\sigma)} + \|M_\sigma\|_{L^p(d\sigma)\to L^p(d\mu)}\right)\|f\|_{L^p(d\sigma)}. \quad □
$$

Under the hypothesis of Theorem 5.34, the operators $S_\sigma^\varepsilon : L^p(d\sigma) \to L^p(d\mu)$ are uniformly bounded with respect to $\varepsilon > 0$:

$$
\|S_\sigma^\varepsilon\|_{L^p(d\sigma)\to L^p(d\mu)} \leq \|S_\sigma^*\|_{L^p(d\sigma)\to L^p(d\mu)} =: C_p < \infty. \tag{5.109}
$$

For a Borel measure ν, for $F \in L^p(d\nu)$, and for $G \in L^q(d\nu)$ $(1/p + 1/q = 1)$, define

$$
(F,G)_\nu := \int_{\mathbf{C}} F(\zeta)G(\zeta)\, d\nu(\zeta).
$$

Clearly, $|(F,G)_\nu| \leq \|F\|_{L^p(d\nu)}\|G\|_{L^q(d\nu)}$. From (5.109) we infer that

$$
|(S_\sigma^\varepsilon f, g)_\mu| \leq C_p \|f\|_{L^p(d\sigma)}\|g\|_{L^q(d\mu)}.
$$

As $(S_\sigma^\varepsilon f, g)_\mu = -(f, S_\mu^\varepsilon g)_\sigma$, it follows that

$$
|(f, S_\mu^\varepsilon g)_\sigma| \leq C_p \|f\|_{L^p(d\sigma)}\|g\|_{L^q(d\mu)}, \tag{5.110}
$$

implying that $S_\mu^\varepsilon : L^q(d\mu) \to L^q(d\sigma)$ is uniformly bounded with $\|S_\mu^\varepsilon\| \leq C_p$ for all $\varepsilon > 0$.

Now choose countable subsets X_σ and Y_μ of $C_0^\infty(\mathbf{C})$ which are dense in all $L^p(d\sigma)$ $(1 < p < \infty)$ and all $L^q(d\mu)$ $(1 < q < \infty)$, respectively. From (5.110) we deduce that for each pair $(f, g) \in X_\sigma \times Y_\mu$ the set $\{(f, S_\mu^\varepsilon g)_\sigma\}_{\varepsilon > 0}$ contains a subsequence $\{(f, S_\mu^{\varepsilon_k} g)_\sigma\}$ converging as $\varepsilon_k \to 0$. A standard diagonalizing process (see, e.g., [171, Theorem I.24]) yields a sequence $\varepsilon_j \to 0$ such that

$$[f, g] := \lim_{\varepsilon_j \to 0} (f, S_\mu^{\varepsilon_j} g)_\sigma \text{ for all } (f, g) \in X_\sigma \times Y_\mu.$$

Again invoking (5.110), we see that the bilinear form $(f, g) \mapsto [f, g]$ extends to a bounded bilinear form on $L^p(d\sigma) \times L^q(d\mu)$. Thus, there is a bounded operator $S_\mu^{(q)} : L^q(d\mu) \to L^q(d\sigma)$ such that

$$[f, g] = (f, S_\mu^{(q)} g)_\sigma \text{ for all } (f, g) \in L^p(d\sigma) \times L^q(d\mu) \tag{5.111}$$

and $\|S_\mu^{(q)}\| \le C_p$. Note that $S_\mu^{(q)}$ depends both on q and the choice of the sequence $\{\varepsilon_j\}$. Any bounded operator $S_\mu^{(q)} : L^q(d\mu) \to L^q(d\sigma)$ satisfying (5.111) will be called a *weak limit* of the operators $S_\mu^\varepsilon : L^q(d\mu) \to L^q(d\sigma)$. It is clear that

$$S_\mu^{(q_1)} g = S_\mu^{(q_2)} g \text{ if } g \in L^{q_1}(d\mu) \cap L^{q_2}(d\mu). \tag{5.112}$$

We therefore abbreviate $S_\mu^{(q)}$ simply to S_μ.

Lemma 5.35. *Let $1 < r \le p < \infty$, $\mu \in \Delta$, $\sigma \in \Sigma$, and suppose $S_\sigma^* : L^s(d\sigma) \to L^s(d\sigma)$ is bounded for all $s \in (1, \infty)$. Then there exist constants $c, c_r > 0$ such that*

$$(S_\mu^* f)(z_0) \le c(M_\sigma S_\mu f)(z_0) + c(M_\mu f)(z_0) + c_r \left[(M_\mu(|f|^r))(z_0) \right]^{1/r}$$

for every $f \in L^p(d\mu)$, every $z_0 \in \operatorname{supp} \sigma$, and every weak limit S_μ of the operators $S_\mu^\varepsilon : L^s(d\mu) \to L^s(d\sigma)$ $(1 < s < \infty)$.

Proof. Fix a weak limit $S_\mu : L^p(d\mu) \to L^p(d\sigma)$, $f \in L^p(d\mu)$, $z_0 \in \operatorname{supp} \sigma$, and $\varepsilon > 0$. Put

$$f = f\chi_{D(z_0,\varepsilon)} + f\chi_{\mathbf{C} \setminus D(z_0,\varepsilon)} =: f_1 + f_2.$$

Then for $z \in D(z_0, \varepsilon/2)$,

$$(S_\mu f_2)(z) = \frac{1}{\pi i} \int\limits_{\mathbf{C} \setminus D(z_0,\varepsilon)} \frac{f(w)}{w - z} \, d\mu(w),$$

and consequently, by (5.102) and (5.103),

$$\left| (S_\mu^\varepsilon f)(z_0) - (S_\mu f_2)(z)) \right| = \frac{1}{\pi} \left| \int\limits_{\mathbf{C} \setminus D(z_0,\varepsilon)} \left(\frac{1}{w - z_0} - \frac{1}{w - z} \right) f(w) \, d\mu(w) \right|$$

$$\le \frac{2}{\pi} |z - z_0| \int\limits_{\mathbf{C} \setminus D(z_0,\varepsilon)} \frac{|f(w)|}{|w - z_0|^2} \, d\mu(w) \le \frac{4}{\pi} (M_\mu f)(z_0).$$

Thus, since $f_2 = f - f_1$, we get

$$\left|(S_\mu^\varepsilon f)(z_0)\right| \leq \left|(S_\mu^\varepsilon f)(z_0) - (S_\mu f_2)(z)\right| + \left|(S_\mu f_2)(z)\right|$$
$$\leq \frac{4}{\pi}(M_\mu f)(z_0) + \left|(S_\mu f)(z)\right| + \left|(S_\mu f_1)(z)\right|. \tag{5.113}$$

Put

$$\Lambda := \frac{1}{\sigma(D(z_0, \varepsilon/2))} \int\limits_{D(z_0,\varepsilon/2)} \left|(S_\mu f)(w)\right| d\sigma(w) \tag{5.114}$$

and $E := \{z \in D(z_0, \varepsilon/2) : |(S_\mu f)(z)| > 3\Lambda\}$. If $\sigma(E) > (1/3)\sigma(D(z_0, \varepsilon/2))$, then (5.114) is at least

$$\frac{3\Lambda}{\sigma(D(z_0, \varepsilon/2))} \int\limits_E d\sigma(w) = 3\Lambda \frac{\sigma(E)}{\sigma(D(z_0, \varepsilon/2))} > \Lambda,$$

which is impossible. Consequently, $\sigma(E) \leq (1/3)\sigma(D(z_0, \varepsilon/2))$, which implies that $(S_\mu f)(z) \leq 3\Lambda$ on $D(z_0, \varepsilon/2) \setminus E$ and

$$\sigma(D(z_0, \varepsilon/2) \setminus E) > (2/3)\sigma(D(z_0, \varepsilon/2)).$$

Analogously, letting

$$\Lambda_1 := \frac{1}{\sigma(D(z_0, \varepsilon/2))} \int\limits_{D(z_0,\varepsilon/2)} \left|(S_\mu f_1)(w)\right| d\sigma(w)$$

and $E_1 := \{z \in D(z_0, \varepsilon/2) : |(S_\mu f_1)(z)| > 3\Lambda\}$, we see that $(S_\mu f_1)(z) \leq 3\Lambda$ on $D(z_0, \varepsilon/2) \setminus E_1$ with $\sigma(D(z_0, \varepsilon/2)) \setminus E_1) > (2/3)\sigma(D(z_0, \varepsilon/2))$. As $2/3 + 2/3 > 1$, it follows that there is a point $z \in D(z_0, \varepsilon/2)$ such that

$$(S_\mu f)(z) \leq 3\Lambda \quad \text{and} \quad (S_\mu f_1)(z) \leq 3\Lambda_1.$$

Since $\sigma(D(z_0, \varepsilon/2)) \geq \gamma\varepsilon/2$, we obtain

$$(S_\mu f)(z) \leq 3\Lambda \leq \frac{3}{\gamma}(M_\sigma S_\mu f)(z_0),$$

$$(S_\mu f_1)(z) \leq 3\left(\frac{1}{\sigma(D(z_0, \varepsilon/2))} \int\limits_{D(z_0,\varepsilon/2)} \left|(S_\mu f_1)(w)\right|^r d\sigma(w)\right)^{1/r}$$

$$\leq 3\left(\frac{2}{\gamma\varepsilon}\right)^{1/r} \|S_\mu f_1\|_{L^r(d\sigma)} \quad (1 < r < \infty).$$

If $1 < r \leq p$, then $f_1 \in L^p(d\mu) \cap L^r(d\mu)$ and so (5.112) and the boundedness of $S_\mu : L^r(d\mu) \to L^r(d\sigma)$ imply that

$$\|S_\mu f_1\|_{L^r(d\sigma)} \leq N\|f_1\|_{L^r(d\mu)} = N \int\limits_{D(z_0,\varepsilon)} |f(w)|d\mu(w) \leq N\left[\varepsilon\left(M_\mu(|f|^r)\right)(z_0)\right]^{1/r}$$

where $N := \|S_\mu\|_{L^r(d\mu)\to L^r(d\sigma)} \leq C_{r/(r-1)}$. Thus, from (5.113) we see that $|(S_\mu^\varepsilon f)(z_0)|$ is at most

$$\frac{4}{\pi}(M_\mu f)(z_0) + \frac{3}{\gamma}(M_\sigma S_\mu f)(z_0) + 3N\left(\frac{2}{\gamma}\right)^{1/r}\left[(M_\mu(|f|^r))(z_0)\right]^{1/r},$$

which gives the assertion with $c := \max\{4/\pi, 3/\gamma\}$ and $c_r := 3N(2/\gamma)^{1/r}$. $\qquad\square$

Theorem 5.36. *Let $\mu \in \Delta$, $\sigma \in \Sigma$, and suppose $S_\sigma^* : L^p(d\sigma) \to L^p(d\sigma)$ is bounded for all $p \in (1,\infty)$. Then $S_\mu^* : L^p(d\mu) \to L^p(d\sigma)$ is bounded for all $p \in (1,\infty)$.*

Proof. Let $f \in L^p(d\mu)$. Since $|f|^r \in L^{p/r}(d\mu)$ for $1 < r < p$ and

$$\|M_\mu g\|_{L^{p/r}(d\mu)} \leq c_0\|g\|_{L^{p/r}(d\mu)} \quad \text{for all } g \in L^{p/r}(d\mu)$$

by Theorem 5.30, we see that

$$\left\|\left[M_\mu(|f|)^r\right]^{1/r}\right\|_{L^p(d\sigma)} = \left\|M_\mu(|f|^r)\right\|_{L^{p/r}(d\sigma)}^{1/r}$$
$$\leq c_0^{1/r}\||f|^r\|_{L^{p/r}(d\mu)}^{1/r} = c_0^{1/r}\|f\|_{L^p(d\mu)}. \tag{5.115}$$

The construction of S_μ and Theorem 5.30 give the inequalities

$$\|S_\mu f\|_{L^p(d\sigma)} \leq c_1\|f\|_{L^p(d\mu)},$$
$$\|M_\sigma g\|_{L^p(d\sigma)} \leq c_2\|g\|_{L^p(d\sigma)},$$
$$\|M_\mu f\|_{L^p(d\sigma)} \leq c_3\|f\|_{L^p(d\mu)}.$$

Thus, by Lemma 5.35,

$$\|S_\mu^* f\|_{L^p(d\sigma)} \leq (cc_2 c_1 + cc_3 + c_r c_0^{1/r})\|f\|_{L^p(d\mu)}. \qquad\square$$

Corollary 5.37. *Let Γ be a simple Carleson curve and let $\tilde{\Gamma} \in \Lambda(K)$ be a curve resulting from a Lipschitz graph by rotation. Put*

$$(S_\Gamma^* f)(t) := \sup_{\varepsilon > 0}\left|\frac{1}{\pi i}\int\limits_{\Gamma\backslash\Gamma(t,\varepsilon)} \frac{f(\tau)}{\tau - t}\, d\tau\right| \quad (t \in \tilde{\Gamma}).$$

Then $S_\Gamma^* : L^p(\Gamma) \to L^p(\tilde{\Gamma})$ is bounded for all $p \in (1,\infty)$.

Proof. Define $\mu \in \Delta$ and $\sigma \in \Sigma$ by

$$\mu(A) = |A \cap \Gamma|, \quad \sigma(A) = |A \cap \tilde{\Gamma}|.$$

Given a Carleson curve γ, we let

$$(S_\gamma^0 f)(t) := \sup_{\varepsilon > 0}\left|\frac{1}{\pi i}\int\limits_{\gamma\backslash\gamma(t,\varepsilon)} \frac{f(\tau)}{\tau - t}\, |d\tau|\right|.$$

Clearly, $S^0_\gamma f = S^*_\gamma(f\psi)$ where $\psi(\tau) := |d\tau|/d\tau$. Since $|\psi| = 1$ a.e. on γ, S^0_γ is bounded if and only if S^*_γ is bounded. From Theorem 5.29 and Corollary 5.21 we know that $S_{\tilde\Gamma}$ generates a bounded operator on $L^p(\tilde\Gamma)$ for all $p \in (1,\infty)$. So Corollary 5.8 shows that $S^*_{\tilde\Gamma} : L^p(\tilde\Gamma) \to L^p(\tilde\Gamma)$ and thus $S^0_{\tilde\Gamma} : L^p(\tilde\Gamma) \to L^p(\tilde\Gamma)$ is bounded for all $p \in (1,\infty)$. Therefore $S^*_\sigma : L^p(d\sigma) \to L^p(d\sigma)$ is bounded for all $p \in (1,\infty)$. Now Theorem 5.36 implies that $S^*_\mu : L^p(d\mu) \to L^p(d\sigma)$ is bounded for all $p \in (1,\infty)$, which gives the boundedness of $S^0_\Gamma : L^p(\Gamma) \to L^p(\tilde\Gamma)$ and thus also of $S^*_\Gamma : L^p(\Gamma) \to L^p(\tilde\Gamma)$ for all $p \in (1,\infty)$. $\qquad\square$

5.11 Approximation by Lipschitz curves

In this section we prove a result on the approximation of Carleson curves by Lipschitz curves, which will be used to establish a good λ inequality. Theorem 5.11 then quickly gives the boundedness of the maximal singular integral operator on unbounded Carleson arcs.

Lemma 5.38 (Rising sun lemma of F. Riesz). *Let $g : [a,b] \to \mathbf{R}$ be continuous, and for $x \in [a,b]$, put $h(x) := \max\{g(t) : a \le t \le x\}$. Let*

$$E := \big\{x \in [a,b] : g(x) = h(x)\big\}, \quad \Omega := [a,b] \setminus E,$$

and let (a_k, b_k) be an open connected component of the set Ω. Then h is constant on $[a_k, b_k]$, $h(a_k) = h(b_k) = g(a_k)$, and $g(x) < g(a_k)$ for all $x \in (a_k, b_k)$. If $b \in \Omega$ and $(\beta, b]$ is a connected component of Ω, then $g(x) < g(\beta) = h(\beta) = h(x)$ for all $x \in (\beta, b]$.

For a *proof* see [213, p. 31], for example. $\qquad\square$

Lemma 5.39. *Let Γ be an unbounded Carleson arc and let $I \subset \Gamma$ be a bounded arc. Let $\nu \in (0,1]$ be any number such that $\operatorname{diam} I \ge \nu|I|$. Then there exists a Lipschitz curve $\tilde\Gamma \in \Lambda(2/\nu)$ such that*

$$|\tilde\Gamma \cap I| \ge \frac{\nu}{2}|I|. \tag{5.116}$$

Proof. Let $l := |I|$ and let $z(s) = x(s) + iy(s)$ be the natural parametrization of Γ for which $I = z([0,l])$.

Let $s_1 < s_2$ be points on $[0,l]$ such that $|z(s_2) - z(s_1)| = \operatorname{diam} I$. Then, by assumption, $|z(s_2) - z(s_1)| \ge \nu l$. Put $\nu' := \nu l/(s_2 - s_1)$. So

$$|z(s_2) - z(s_1)| \ge \nu l = \nu'(s_2 - s_1)$$

and thus, for the arc $I' := z([s_1, s_2])$ we have

$$\operatorname{diam} I' = |z(s_2) - z(s_1)| \ge \nu'(s_2 - s_1) = \nu'|I'|,$$

whence $\nu' \in (0,1]$. Assume we have shown the lemma for the arc I'. Then there exists a $\tilde{\Gamma}$ with Lipschitz constant

$$K \le \frac{2}{\nu'} = \frac{2}{\nu}\frac{s_2 - s_1}{l} \le \frac{2}{\nu}$$

such that

$$|\tilde{\Gamma} \cap I| \ge |\tilde{\Gamma} \cap I'| \ge \frac{\nu'}{2}|I'| = \frac{\nu}{2}|I|.$$

Thus, it suffices to prove the lemma for the case where $|z(l) - z(0)| \ge \nu l$. Also, we may without loss of generality assume that $z(0) = 0$ and $z(l) = b \in (0,\infty)$. Clearly, $b \ge \nu l$.

The functions

$$g(s) := x(s) - \frac{b}{2l}s, \quad h(s) := \max_{0 \le t \le s} g(t), \quad q(s) := h(s) + \frac{b}{2l}s \qquad (5.117)$$

are continuous on $[0,l]$. Since h does not decrease, the function q must be monotoneously increasing. Moreover, $q'(s) \ge b/(2l) \ge \nu/2$ for almost all $s \in [0,l]$. Let $\varphi : [0, q(l)] \to [0,l]$ be the inverse function of q and extend φ to all of $\mathbf{R}$ by defining

$$\varphi(x) = 0 \text{ for } x < 0, \quad \varphi(x) = l \text{ for } x > q(l).$$

Then φ is continuous on $\mathbf{R}$ and

$$\|\varphi'\|_{L^\infty(\mathbf{R})} = \|\varphi'\|_{L^\infty[0,q(l)]} = \|1/q'\|_{L^\infty[0,l]} \le \frac{2}{\nu}.$$

Thus, $\varphi : \mathbf{R} \to \mathbf{R}$ is a Lipschitz function with Lipschitz constant $\|\varphi'\|_\infty \le 2/\nu$.

Put $\tilde{\Gamma} := \{\tilde{z}(x) : x \in \mathbf{R}\}$ where

$$\tilde{z}(x) := x + i\psi(x), \quad \psi(x) := \operatorname{Im} z(\varphi(x)).$$

Since $|z'(s)| = 1$ for almost all $s \in \mathbf{R}$, it is clear that $\psi : \mathbf{R} \to \mathbf{R}$ is a Lipschitz function with Lipschitz constant

$$K := \|\psi'\|_{L^\infty(\mathbf{R})} \le \|\varphi'\|_{L^\infty(\mathbf{R})} \le 2/\nu.$$

We are left with proving (5.116). Let

$$E := \{s \in [0,l] : h(s) = g(s)\} \qquad (5.118)$$

By Lemma 5.38, h is constant on the connected components of $\Omega := [0,l] \setminus E$ and

$$h(s) > g(s) \text{ for } s \in \Omega. \qquad (5.119)$$

From (5.117) we infer that $q'(s) = b/(2l)$ for $s \in \Omega$, which implies that

$$|q(\Omega)| = \frac{b}{2l}|\Omega| \le \frac{bl}{2l} = \frac{b}{2}. \qquad (5.120)$$

Putting (5.117), (5.118), (5.119) together we get

$$q(s) = x(s) \text{ for } s \in E, \quad q(s) > x(s) \text{ for } s \in \Omega. \tag{5.121}$$

Since $q(0) = 0$ and $q(l) \geq x(l) = b$, we deduce from (5.120) and the inequality $b \geq \nu l$ that

$$|q(E)| = |q([0,l])| - |q(\Omega)| \geq b - \frac{b}{2} = \frac{b}{2} \geq \frac{\nu l}{2} = \frac{\nu}{2}|I|. \tag{5.122}$$

Further, by (5.121), $\varphi(x(s)) = \varphi(q(s)) = s$ for $s \in E$, implying that

$$\tilde{z}(x(s)) = x(s) + iy(\varphi(x(s))) = x(s) + iy(s) = z(s) \text{ for } s \in E.$$

Thus, $\tilde{\Gamma} \cap I \supset z(E)$, whence

$$|\tilde{\Gamma} \cap I| \geq |z(E)| \geq |x(E)| = |q(E)|.$$

This and (5.122) complete the proof. □

   The following theorem provides us with a good λ inequality for the functions $S_* g$ and $Mg + [M(|g^r|)]^{1/r}$.

Theorem 5.40. *Let Γ be an unbounded Carleson arc and let $q := 1 - 1/(4C_\Gamma)$. Then for every $\varepsilon > 0$ and every $r \in (1, \infty)$ there exists a $\gamma = \gamma(\Gamma, \varepsilon, r) > 0$ such that*

$$\left| \left\{ t \in \Gamma : (S_* g)(t) > (1 + \varepsilon)\lambda, (Mg)(t) + \left[(M(|g|^r))(t) \right]^{1/r} \leq \gamma\lambda \right\} \right|$$
$$\leq q \left| \left\{ t \in \Gamma : (S_* g)(t) > \lambda \right\} \right|. \tag{5.123}$$

for all $g \in C_0^\infty(\Gamma)$ and all $\lambda > 0$.

Proof. Fix $g \in C_0^\infty(\Gamma)$ and numbers $\lambda > 0, \varepsilon > 0, r \in (1, \infty)$. Put

$$U_\lambda := \left\{ t \in \Gamma : (S_* g)(t) > \lambda \right\}.$$

Lemma 5.16 implies that $U_\lambda \subset \Gamma$ is open, and from (5.57) we infer that U_λ is bounded. Let I be any of the open arcs constituting U_λ and suppose there is a point $\xi \in I$ such that

$$(Mg)(\xi) + \left[(M(|g|^r))(\xi) \right]^{1/r} \leq \gamma\lambda. \tag{5.124}$$

We show that if $\gamma > 0$ is sufficiently small, then

$$\left| \left\{ t \in I : (S_* g)(t) > (1 + \varepsilon)\lambda \right\} \right| \leq q|I|. \tag{5.125}$$

This gives the assertion, since arcs I which do not contain a point satisfying (5.124) make no contribution to the left-hand side of (5.123).

Let a and b be the endpoints of I and suppose I is oriented from a to b. Since I is an open arc, the point a belongs to $\Gamma \setminus U_\lambda$, whence

$$(S_* g)(a) \leq \lambda. \qquad (5.126)$$

Let $R := |I|$. Clearly, $R < \infty$. Write g in the form

$$g = g\chi_{\Gamma(a,2R)} + g\chi_{\Gamma \setminus \Gamma(a,2R)} =: f_1 + f_2.$$

Since $\xi \in I \subset \Gamma(a, 2R) \subset \Gamma(\xi, 3R)$, we obtain from (5.60) and (5.124) that

$$\|f_1\|_{L^1(\Gamma)} \leq 3R(Mg)(\xi) \leq 3R\gamma\lambda = 3|I|\gamma\lambda. \qquad (5.127)$$

With $Q_k = I$, we get from (5.62), (5.63), (5.64), (5.65) that if $t \in I$, then

$$\pi \left| (S_\varepsilon f_2)(t) \right| \leq \left| \int_{\Gamma \setminus \Gamma(a,\varepsilon)} \frac{f_2(\tau)}{\tau - a} \, d\tau \right| + \int_{\Gamma \setminus \Gamma(a,\varepsilon)} \left| \frac{1}{\tau - t} - \frac{1}{\tau - a} \right| |f_2(\tau)| \, |d\tau|$$

$$+ \int_{\Gamma(t,\varepsilon) \Delta \Gamma(a,\varepsilon)} \frac{|f_2(\tau)|}{|\tau - t|} \, |d\tau| \leq \pi (S_* g)(a) + 16(Mg)(\xi) + 3(Mg)(\xi).$$

Thus, (5.126) and (5.124) imply that

$$\left| (S_\varepsilon f_2)(t) \right| \leq \lambda + (19/\pi)\gamma\lambda,$$

whence

$$(S_* f_2)(t) = \sup_{\varepsilon > 0} \left| (S_\varepsilon f_2)(t) \right| \leq \lambda + (19/\pi)\gamma\lambda$$

for all $t \in I$. Letting $\gamma \leq \pi\varepsilon/38$, we so get $(S_* f_2)(t) \leq (1 + \varepsilon/2)\lambda$ for $t \in I$. Consequently, if $t \in I$ and $(S_* g)(t) > (1 + \varepsilon)\lambda$, then

$$(S_* f_1)(t) \geq (S_* g)(t) - (S_* f_2)(t) > (1 + \varepsilon)\lambda - (1 + \varepsilon/2)\lambda = \varepsilon\lambda/2$$

and therefore,

$$\{t \in I : (S_* g)(t) > (1 + \varepsilon)\lambda\} \subset \{t \in I : (S_* f_1)(t) > \varepsilon\lambda/2\} =: E. \qquad (5.128)$$

In summary, (5.125) will follow as soon as we have shown that

$$|E| \leq q|I|. \qquad (5.129)$$

Obviously, $\operatorname{diam} I \geq |I|/C_\Gamma$. Thus, we may use Lemma 5.39 with $\nu = 1/C_\Gamma$ to see that there is a Lipschitz curve $\tilde{\Gamma} \in \Lambda(2C_\Gamma)$ such that $|\tilde{\Gamma} \cap I| \geq |I|/(2C_\Gamma)$. Then

$$|E \setminus \tilde{\Gamma}| \leq |I \setminus \tilde{\Gamma}| \leq \left(1 - 1/(2C_\Gamma)\right)|I|. \qquad (5.130)$$

On the other hand, Corollary 5.37 tells us that the operator $S_\Gamma^* : L^r(\Gamma) \to L^r(\tilde{\Gamma})$ is bounded for all $r \in (1, \infty)$. In particular,

$$\|S_\Gamma^* f_1\|_{L^r(\tilde{\Gamma})}^r \leq N_r \|f_1\|_{L^r(\Gamma)}^r, \tag{5.131}$$

where $N_r^{1/r}$ is the norm of $S_\Gamma^* : L^r(\Gamma) \to L^r(\tilde{\Gamma})$. We have

$$\|f_1\|_{L^r(\Gamma)}^r = \int\limits_{\Gamma(a,2R)} |g(\tau)|^r |d\tau| \leq 3R \frac{1}{3R} \int\limits_{\Gamma(\xi,3R)} |g(\tau)|^r |d\tau|$$
$$\leq 3R\big(M(|g|^r)\big)(\xi) = 3|I|\big(M(|g|^r)\big)(\xi). \tag{5.132}$$

Combining (5.131), (5.132), (5.124) we obtain

$$\|S_\Gamma^* f_1\|_{L^r(\tilde{\Gamma})}^r \leq 3N_r |I| \gamma^r \lambda^r$$

and thus,

$$\|S_* f_1\|_{L^r(E \cap \tilde{\Gamma})}^r = \|S_\Gamma^* f_1\|_{L^r(E \cap \tilde{\Gamma})}^r \leq \|S_\Gamma^* f_1\|_{L^r(\tilde{\Gamma})}^r \leq 3N_r |I| \gamma^r \lambda^r.$$

By (5.128),

$$|E \cap \tilde{\Gamma}| \left(\frac{\varepsilon\lambda}{2}\right)^r \leq \int\limits_{E \cap \tilde{\Gamma}} \big|(S_* f_1)(\tau)\big|^r |d\tau| = \|S_* f_1\|_{L^r(E \cap \tilde{\Gamma})}^r.$$

Thus, we arrive at the estimate

$$|E \cap \tilde{\Gamma}| \leq \left(\frac{2}{\varepsilon\lambda}\right)^r 3N_r |I| \gamma^r \lambda^r = 3N_r \left(\frac{2}{\varepsilon}\right)^r \gamma^r |I|.$$

Imposing on γ not only the restriction $0 < \gamma \leq \pi\varepsilon/38$ but also requiring that $\gamma \leq (\varepsilon/2)(12C_\Gamma N_r)^{-1/r}$, it follows that

$$|E \cap \tilde{\Gamma}| \leq |I|/(4C_\Gamma). \tag{5.133}$$

Now (5.130) and (5.133) give (5.129) with $q = 1 - 1/(4C_\Gamma)$. $\qquad\square$

5.12 Completing the puzzle

We know from Corollary 5.3 that the maximal operator M is bounded on $L^p(\Gamma)$ for every $p \in (1, \infty)$ and every Carleson curve Γ. Let $\|M\|_p$ denote the norm of M on $L^p(\Gamma)$.

Corollary 5.41. *If Γ is an unbounded Carleson arc and $1 < r < 2$, then*

$$\|S_* g\|_{L^2(\Gamma)} \leq C\big(\|M\|_2 + \|M\|_{2/r}^{1/r}\big)\|g\|_{L^2(\Gamma)}$$

for all $g \in C_0^\infty(\Gamma)$ with some costant $C < \infty$ independent of g.

Proof. Fix $g \in C_0^\infty(\Gamma)$ and $r \in (1,2)$. Put $q := 1 - 1/(4C_\Gamma)$, choose $\varepsilon > 0$ so that $q(1 + \varepsilon)^2 < 1$, and let $\gamma = \gamma(\Gamma, \varepsilon, r) > 0$ be as in Theorem 5.40. By Lemma 5.19, there is a compact set $\Gamma_0 \subset \Gamma$ such that $S_*g \in L^2(\Gamma \setminus \Gamma_0)$. Thus, combining Theorems 5.40 and 5.11 we obtain

$$
\begin{aligned}
\|S_*g\|_{L^2(\Gamma)} &\leq \frac{1+\varepsilon}{\gamma(1 - q(1+\varepsilon)^2)^{1/2}} \left\| Mg + [M(|g|^r)]^{1/r} \right\|_{L^2(\Gamma)} \\
&\leq \frac{1+\varepsilon}{\gamma(1 - q(1+\varepsilon)^2)^{1/2}} \left(\|Mg\|_2 + \left\| [M(|g|^r)]^{1/r} \right\|_{L^2(\Gamma)} \right).
\end{aligned}
$$

Since $\|Mg\|_2 \leq \|M\|_2 \|g\|_2$ and

$$
\left\| [M(|g|^r)]^{1/r} \right\|_{L^2(\Gamma)} = \left\| M(|g|^r) \right\|_{L^{2/r}(\Gamma)}^{1/r} \leq \|M\|_{2/r}^{1/r} \|g\|_2,
$$

we arrive at the assertion. $\square$

Corollary 5.42. *If Γ is an unbounded Carleson arc then S generates a bounded operator on $L^2(\Gamma)$.*

Proof. We can proceed as in the proof of Corollary 5.21. From Theorem 4.3 we infer that $S_\varepsilon g \to Sg$ a.e. on Γ and that therefore Sg is measurable for every $g \in C_0^\infty(\Gamma)$. Since $|S_\varepsilon g| \leq S_*g$ a.e. on Γ, Corollary 5.41 shows that $\|S_\varepsilon g\|_2 \leq B\|g\|_2$ for all $\varepsilon > 0$, whence, by Fatou's theorem, $Sg \in L^2(\Gamma)$ and $\|Sg\|_2 \leq B\|g\|_2$; here $B := C\big(\|M\|_2 + \|M\|_{2/r}^{1/r}\big)$. $\square$

From Corollaries 5.21 and 5.42 we now deduce that S extends to a bounded operator $\tilde{S}$ on $L^p(\Gamma, w)$ whenever $1 < p < \infty$, Γ is an unbounded Carleson arc, and $w \in A_p(\Gamma)$. Remark 5.23 tells us that in this case Sf exists and coincides with $\tilde{S}f$ almost everywhere for every $f \in L^p(\Gamma, w)$. Taking into account the results of Section 4.5, we finally get the sufficiency part of Theorem 4.15.

5.13 Notes and comments

5.1. In Section 5.1 it is shown that the maximal operator M is bounded on $L^p(\Gamma, w)$ whenever Γ is Carleson and $w \in A_p(\Gamma)$ ($1 < p < \infty$). Such a result was first established by Muckenhoupt [148] in the case where $\Gamma = \mathbf{R}^n$. To be more precise, Muckenhoupt proved that if $1 < p < \infty$ and w is a weight on $\mathbf{R}^n$, then the following are equivalent:

  (i) $w \in A_p(\mathbf{R}^n)$;
 (ii) M is an operator of the weak type (p,p) with respect to the measure $d\nu(x) = w(x)^p dx$, where dx denotes Lebesgue measure on $\mathbf{R}^n$;
(iii) M is bounded on $L^p(\mathbf{R}^n, w)$.

A self-contained proof of this fundamental result can also be found in [73].

Subsequently, Muckenhoupt's proof was simplified by Coifman and Fefferman [36], who employed a reverse Hölder inequality for this purpose. Calderón
[31] proved the boundedness of M on $L^p(\Gamma, w)$ in case Γ is a so-called space of
homogeneous type; he also used a reverse Hölder inequality. A different approach
to weighted norm inequalities for the maximal operator was developed by Sawyer
[183], [184]: this approach does not have recourse to the fact that $A_p(\Gamma) \subset A_{p-\varepsilon}(\Gamma)$
for all sufficiently small $\varepsilon > 0$, which in the one or other manner prevailed in all
the afore-mentioned proofs. Our proof of Corollary 5.3 follows Calderón [31]. We
remark that the converse of Corollary 5.3 is also true: if Γ is a simple locally rectifiable curve and $1 < p < \infty$, then the boundedness of M on $L^p(\Gamma, w)$ implies that
Γ is Carleson and $w \in A_p(\Gamma)$.

The Marcinkiewicz interpolation theorem is discussed and proved in the
books by Garcia-Cuerva and Rubio de Francia [73], Stein [202], Garnett [74], for
example.

5.2. For more on the Calderón-Zygmund decomposition we refer to [202], [60],
[39]. We remark that in Lemma 5.4, which is a key result for getting the Calderón-
Zygmund decomposition, the restriction to *unbounded* curves is essential; if Γ is
bounded, one cannot construct the decomposition for all $\lambda > 0$ but only for $\lambda >
\|f\|_1/|\Gamma|$. The construction presented in Section 5.2 is due to Coifman and Weiss
[39].

5.3. The way of reasoning of Section 5.3 is as follows. With the help of the Calderón-
Zygmund decomposition, we show that if there is a bounded extension $\tilde{S}$ of S to
$L^p(\Gamma)$, then $\tilde{S}$ is of the weak type $(1, 1)$ (Theorem 5.6). This is then used to prove
Cotlar's inequality (Theorem 5.7), which gives an estimate for the values of the
maximal singular integral S_* in terms of the values of the maximal operators M
and $\mathcal{M}_\delta$ and the extension $\tilde{S}$ of the Cauchy singular integral operator S. The Cotlar
inequality in turn is employed in order to deduce the boundedness of S_* on $L^p(\Gamma)$
from the boundedness of $\tilde{S}$ on $L^p(\Gamma)$ (Corollary 5.8). In the end and after having
recourse to Fatou's theorem, we arrive at the conclusion that if $\tilde{S}$ is bounded on
$L^p(\Gamma)$, then the Cauchy singular integral Sf exists and coincides with $\tilde{S}f$ for all
$f \in L^p(\Gamma)$ (Corollary 5.10). This way of reasoning is nowadays "standard" and
can be found in the work by Dynkin [60], Garnett [74], Journé [109], Stein [202],
and others. We emphasize once more that all assertions of Section 5.3 are stated
under the assumption that S admits extension to a bounded linear operator $\tilde{S}$ on
$L^p(\Gamma)$.

5.4. Good λ inequalities were introduced by Burkholder and Gundy [30] and by
Coifman [35] (see also Dynkin [60] and Garnett [74]). The good λ inequality of
Theorem 5.11 rests on excluding a compact subset $\Gamma_0 \subset \Gamma$, which is an idea we
took from Lemma 12 of David's paper [47].

5.5. We learned Lemmas 5.12 and 5.13 from Dynkin [60, Chapter 1], who attributes
them to A.N. Kolmogorov.

5.6. This section is devoted to L^p spaces with weights. The proof of the good λ inequality contained in Theorem 5.18 follows the plan of the proof of [74, Theorem 6.12] and makes heavy use of the so-called A_∞ condition. More about A_∞ weights is in Coifman, Fefferman [36], Dynkin, Osilenker [62], Dynkin [60], Garcia-Cuerva, Rubio de Francia [73], Garnett [74], Strömberg, Torchinsky [203]. The rather puzzling conclusion of the section is as follows: if M is bounded on $L^p(\Gamma, w)$ and S is bounded on $L^2(\Gamma)$, then S_* is bounded on $L^p(\Gamma, w)$ (Corollary 5.20) and thus S itself is a bounded operator on $L^p(\Gamma, w)$ (Corollary 5.21).

5.7. Calderón's paper [32] was the starting point for proving that the Cauchy singular integral operator S is bounded on L^2 over Lipschitz curves. The proof in the text is due to Coifman, Jones, and Semmes [37]. Other proofs were given by R.R. Coifman, A. McIntosh, Y. Meyer [38], David [47], David, Journé [48], David, Journé, Semmes [49], Jones [108], Murai [150], Semmes [188]. The very recent papers by Melnikov, Verdera [145] and Mattila, Melnikov, Verdera [143] contain a "geometric" proof of the boundedness of S in $L^2(\Gamma)$; this proof is based on appropriately measuring the curvature of the curve Γ or, more general, the curvature of a measure (see Melnikov's paper [144]). Lemma 5.25 is a version of a result by Kenig [121]. The proof given in the text again follows Coifman, Jones, and Semmes [37] and Dynkin [61, Chapter 4].

5.8–5.12. There is little to say: all the material of these sections is taken from David's pioneering papers [46], [47]. We are very grateful to Sergei M. Grudsky for providing us with a Russian translation of [47] done by one of his daughters some years ago when learning French.

Chapter 6
General properties of Toeplitz operators

Let $1 < p < \infty$, let Γ be a Carleson Jordan curve, and let w be a weight in $A_p(\Gamma)$. We know that then the operator S is bounded on $L^p(\Gamma, w)$. It follows easily that $S^2 = I$, and hence $P := (I + S)/2$ is a bounded projection on $L^p(\Gamma, w)$. The image of P, i.e. the space $L^p_+(\Gamma, w) := PL^p(\Gamma, w)$, is therefore a closed subspace of $L^p(\Gamma, w)$, which is called the pth Hardy space of Γ and w. If $a \in L^\infty(\Gamma)$, then the operator of multiplication by a is obviously bounded on $L^p(\Gamma, w)$. The compression of this operator to $L^p_+(\Gamma, w)$ is referred to as the Toeplitz operator on $L^p_+(\Gamma, w)$ with the symbol a and is denoted by $T(a)$. In other words, $T(a)$ is the bounded operator which sends $g \in L^p_+(\Gamma, w)$ to $P(ag) \in L^p_+(\Gamma, w)$. A central problem in the spectral theory of singular integral operators is the determination of the essential spectrum of Toeplitz operators with piecewise continuous symbols. This problem will be completely solved in Chapter 7.

The purpose of the present chapter is to exhibit some basic properties of Toeplitz operators. These include the Coburn-Simonenko theorem, which says that the spectrum of $T(a)$ is the union of the essential spectrum of $T(a)$ and all complex numbers λ for which $T(a) - \lambda I = T(a - \lambda)$ has nonzero index. Another basic result, the Hartman-Wintner-Simonenko theorem, tells us that the essential range of the symbol a is always a subset of the essential spectrum of $T(a)$. These two theorems in conjunction with the fact that Hankel operators with continuous symbols are compact allow us to identify the essential spectrum of Toeplitz operators with continuous symbols as the range of the symbol.

Piecewise continuous symbols require more powerful tools. Due to some happy circumstances, one can relatively quickly dispose of the problem in case $p = 2$, $\Gamma = \mathbf{T}$, and $w = 1$, that is, for Toeplitz operators given by the classical Toeplitz matrices on l^2. In the last three sections of this chapter we prove certain separation, localization, and factorization theorems for Toeplitz operators. These theorems are of independent interest on the one hand and are just the sort of machinery we need to tackle piecewise continuous symbols in the general case on the other.

6.1   Smirnov classes

Let Γ be a rectifiable Jordan curve in the complex plane $\mathbf{C}$. We denote by D_+ and D_- the bounded and unbounded component of $\mathbf{C} \setminus \Gamma$, respectively. We orient Γ counter-clockwise, i.e. we suppose that D_+ stays on the left of Γ as the curve is traced out in the positive direction. Without loss of generality we will always assume that the origin belongs to D_+.

A function f analytic in D_+ is said to be in the *Smirnov class* $E^p(D_+)$ ($1 \leq p < \infty$) if there exists a sequence of rectifiable Jordan curves $\Gamma_1, \Gamma_2, \dots$ in D_+ tending to the boundary Γ in the sense that Γ_n eventually surrounds each compact subset of D_+ such that

$$\sup_{n \geq 1} \int_{\Gamma_n} |f(z)|^p |dz| < \infty. \tag{6.1}$$

Denote by $\mathbf{D} := \{z \in \mathbf{C} : |z| < 1\}$ the complex unit disk and let $\varphi : \mathbf{D} \to D_+$ be any Riemann map of $\mathbf{D}$ onto D_+. One can show that an analytic function $f : D_+ \to \mathbf{C}$ belongs to $E^p(D_+)$ if and only if

$$\sup_{0 < r < 1} \int_{\Gamma_r} |f(z)|^p |dz| < \infty$$

where $\Gamma_r := \varphi(\{z \in \mathbf{C} : |z| = r\})$. Choose any analytic branch of $(\varphi'(z))^{1/p}$ for $z \in \mathbf{D}$. It turns out that $f \in E^p(D_+)$ if and only if $F \in E^p(\mathbf{D})$ where $F : \mathbf{D} \to \mathbf{C}$ is defined by $F(z) := f(\varphi(z))(\varphi'(z))^{1/p}$.

Since Γ is rectifiable, the tangent to Γ exists at almost all points of Γ. Therefore, it makes sense to speak of non-tangential limits a.e. on Γ for functions defined in D_+. One can prove that every function $f \in E^p(D_+)$ has a non-tangential limit a.e. on Γ and that the boundary function belongs to $L^p(\Gamma)$. We denote by $E_+^p(\Gamma)$ the collection of all the boundary functions of functions in $E^p(D_+)$ and by

$$B_+ : E^p(D_+) \to E_+^p(\Gamma)$$

the linear operator which sends $f \in E^p(D_+)$ to its boundary function in $E_+^p(\Gamma)$. It can be shown that $E_+^p(\Gamma)$ is a closed subspace of $L^p(\Gamma)$ for all $p \in [1, \infty)$. Clearly, $E^1(D_+) \supset E^p(D_+)$ and thus $E_+^1(\Gamma) \supset E_+^p(\Gamma)$ for all $p \in [1, \infty)$.

Theorem 6.1 (Lusin-Privalov). *Let Γ be a rectifiable Jordan curve. A function in $E_+^1(\Gamma)$ vanishes either almost everywhere or almost nowhere on Γ. Moreover, if $f \in E^1(D_+)$ and $B_+ f$ vanishes on a set of positive measure, the f vanishes identically in D_+.*

For a *proof* see e.g. [59, Theorem 10.3]. □

The second part of Theorem 6.1 implies that B_+ is bijective. The inverse map can be shown to be given by the *Cauchy integral*: if we define

$$(C_{\pm}g)(z) := \frac{1}{2\pi i} \int_{\Gamma} \frac{g(\tau)}{\tau - z} \, d\tau \quad (z \in D_{\pm}),$$ (6.2)

then C_+ maps $E_+^p(\Gamma)$ bijectively onto $E^p(D_+)$ and $B_+C_+ = I$, $C_+B_+ = I$.

Theorem 6.2. *For every rectifiable Jordan curve Γ,*

$$E_+^1(\Gamma) = \left\{ g \in L^1(\Gamma) : \int_{\Gamma} g(\tau)\tau^n d\tau = 0 \text{ for } n = 0, 1, 2, \dots \right\}$$ (6.3)

$$= \left\{ g \in L^1(\Gamma) : (C_-g)(z) = 0 \text{ for } z \in D_- \right\}$$ (6.4)

A *proof* is in [59, Theorem 10.4].   □

Since $E_+^p(\Gamma) \subset E_+^1(\Gamma)$ and $E_+^p(\Gamma) \subset L^p(\Gamma)$, it follows from Theorem 6.2 that

$$E_+^p(\Gamma) \subset E_+^1(\Gamma) \cap L^p(\Gamma)$$ (6.5)

$$= \left\{ g \in L^p(\Gamma) : \int_{\Gamma} g(\tau)\tau^n d\tau = 0 \text{ for } n = 0, 1, 2, \dots \right\}$$ (6.6)

$$= \left\{ g \in L^p(\Gamma) : (C_-g)(z) = 0 \text{ for } z \in D_- \right\}.$$ (6.7)

For $p = 1$, we have equality in (6.5). However, if $1 < p < \infty$, then in (6.5) equality need not hold. It is well known that one has equality in (6.5) if and only if Γ is a so-called Smirnov curve (see, e.g., [137, p. 39]). In the next section we show that equality in (6.5) holds whenever Γ is a Carleson curve.

The above definitions and results have analogues in the case where D_+ is replaced by D_-, since we may consider everything on the Riemann sphere $\dot{\mathbf{C}} := \mathbf{C} \cup \{\infty\}$, where D_+ and D_- are sets of the same kind. The point ∞ nevertheless causes a few minor technical complications.

We define the *Smirnov class* $E^p(D_-)$ $(1 \le p < \infty)$ as the set of all analytic functions $f : D_- \cup \{\infty\} \to \mathbf{C}$ for which (6.1) holds with some sequence of curves $\Gamma_1, \Gamma_2, \dots \subset D_-$ tending to the boundary Γ in the sense that every compact subset of $D_- \cup \{\infty\}$ eventually lies outside Γ_n.

Many results on $E^p(D_-)$ can be almost immediately derived from their $E^p(D_+)$ counterparts with the help of the following observation. Put $\tilde{\Gamma} := \{1/\tau : \tau \in \Gamma\}$ and $\tilde{D}_- := \{1/z : z \in D_- \cup \{\infty\}\}$. Hence, $\tilde{\Gamma}$ is a rectifiable Jordan curve whose interior domain is $\tilde{D}_-$. For a function f analytic in $D_- \cup \{\infty\}$, define $\tilde{f}$ in $\tilde{D}_-$ by $\tilde{f}(z) := f(1/z)$. Then $f \in E^p(D_-)$ if and only if $\tilde{f} \in E^p(\tilde{D}_-)$.

Functions in $E^p(D_-)$ have non-tangential limits a.e. on Γ. The set of all their boundary functions will be denoted by $E^p_-(\Gamma)$. Again $E^p_-(\Gamma)$ is a closed subspace of $L^p(\Gamma)$. The Lusin-Privalov theorem remains true with "+" replaced by "−". Thus, the map

$$B_- : E^p(D_-) \to E^p_-(\Gamma)$$

sending a function on D_- to its boundary function on Γ is bijective. However, the Cauchy integral

$$C_- : E^p_-(\Gamma) \to E^p(D_-)$$

is only surjective; the set $\{g \in E^-_p(\Gamma) : C_-g = 0\}$ turns out to be the set $\mathbf{C}$ of all constant functions on Γ. If $g \in E^p_-(\Gamma)$ then

$$(B_-C_-g)(\tau) = (C_-g)(\infty) - g(\tau) \ \text{ for almost all } \ \tau \in \Gamma,$$

and if $f \in E^p(D_-)$ then

$$(C_-B_-f)(z) = f(\infty) - f(z) \ \text{ for all } \ z \in D_- \cup \{\infty\}.$$

We denote by $\dot{E}^p(D_-)$ the functions in $E^p(D_-)$ which vanish at infinity and by $\dot{E}^p_-(\Gamma)$ the boundary functions on Γ of the functions in $\dot{E}^p(D_-)$. Then

$$B_- : \dot{E}^p(D_-) \to \dot{E}^p_-(\Gamma), \ \ C_- : \dot{E}^p_-(\Gamma) \to \dot{E}^p(D_-)$$

are bijective and $B_-C_- = -I, C_-B_- = -I$, that is,

$$B_-C_-g = -g \ \text{ for } \ g \in \dot{E}^p_-(\Gamma) \ \text{ and } \ C_-B_-f = -f \ \text{ for } \ f \in \dot{E}^p(D_-).$$

Here is the analogue of Theorem 6.2.

Theorem 6.3. *For every rectifiable Jordan curve Γ,*

$$E^1_-(\Gamma) = \left\{ g \in L^1(\Gamma) : \int_\Gamma g(\tau)\tau^{-n} d\tau = 0 \ for \ n = 2, 3, 4, \dots \right\}$$

$$= \left\{ g \in L^1(\Gamma) : (C_+g)(z) = (C_-g)(\infty) \ for \ z \in D_+ \right\},$$

$$\dot{E}^1_-(\Gamma) = \left\{ g \in L^1(\Gamma) : \int_\Gamma g(\tau)\tau^{-n} d\tau = 0 \ for \ n = 1, 2, 3, \dots \right\}$$

$$= \left\{ g \in L^1(\Gamma) : (C_+g)(z) = 0 \ for \ z \in D_+ \right\}. \qquad \square$$

We remark that both the minus signs we encountered above and the "$n \geq 2$" appearing in Theorem 6.3 results from the equality $(1/z)' = -1/z^2$.

Finally, for further referencing we fix the following fact.

Theorem 6.4. *If Γ is a rectifiable Jordan curve then*

$$E_+^1(\Gamma) \cap E_-^1(\Gamma) = \mathbf{C} \ \text{and} \ E_+^1(\Gamma) \cap \dot{E}_-^1(\Gamma) = \{0\}.$$

Proof. If $g \in E_+^1(\Gamma) \cap E_-^1(\Gamma)$, then $(C_+g)(z) = (C_-g)(\infty)$ for all $z \in D_+$ by Theorem 6.3. Since $B_+C_+ = I$ on $E_+^1(\Gamma)$, it follows that

$$g(\tau) = (B_+C_+g)(\tau) = (C_-g)(\infty)$$

for all $\tau \in \Gamma$. $\qquad\square$

6.2 Weighted Hardy spaces

Let now Γ be a Carleson Jordan curve, $1 < p < \infty$, and $w \in A_p(\Gamma)$. Since Γ is a bounded curve, we have $L^p(\Gamma, w) \subset L^1(\Gamma)$. By Theorem 4.15, the Cauchy singular integral operator $S := S_\Gamma$ is bounded on $L^p(\Gamma, w)$.

We denote by $R(\Gamma)$ the rational functions without poles on Γ. Further, we let $R_+(\Gamma)$ stand for the functions in $R(\Gamma)$ which have no poles in D_+, and we define $R_-(\Gamma)$ as the set of all functions in $R(\Gamma)$ which have no poles in D_- and vanish at infinity. Clearly, every function $g \in R(\Gamma)$ may be written in the form $g = g_+ + g_-$ with $g_+ \in R_+(\Gamma)$ and $g_- \in R_-(\Gamma)$. In the proof of Lemma 4.4 we showed that $R(\Gamma)$ is dense in $L^p(\Gamma, w)$.

Lemma 6.5. *If $g_+ \in R_+(\Gamma)$ and $g_- \in R_-(\Gamma)$, then $Sg_+ = g_+$ and $Sg_- = -g_-$.*

Proof. Decompose $g_\pm$ into partial fractions and use formula (4.14) together with Theorem 4.2. $\qquad\square$

Corollary 6.6. *If Γ is a Carleson Jordan curve, $1 < p < \infty$, and $w \in A_p(\Gamma)$, then*

$$S^2g = g \ \text{for} \ g \in L^p(\Gamma, w). \tag{6.8}$$

Proof. If $g \in R(\Gamma)$ and $g = g_+ + g_-$ with $g_\pm \in R_\pm(\Gamma)$, then $Sg_+ = g_+$ and $Sg_- = -g_-$ by Lemma 6.5. Invoking this lemma once more we see that $S^2g_+ = g_+$ and $S^2g_- = g_-$, whence $S^2g = g$. Since $R(\Gamma)$ is dense in $L^p(\Gamma, w)$, we arrive at (6.8). $\qquad\square$

The previous theorem shows that the bounded operators P and Q defined by

$$P := P_\Gamma := \frac{1}{2}(I + S) \ \text{and} \ Q := Q_\Gamma = \frac{1}{2}(I - S) \tag{6.9}$$

are complementary projections $P^2 = P$, $Q^2 = Q$, $P + Q = I$. Consequently, their images (= ranges) on $L^p(\Gamma, w)$,

$$L_+^p(\Gamma, w) := PL^p(\Gamma, w), \quad \dot{L}_-^p(\Gamma, w) := QL^p(\Gamma, w),$$

are closed subspaces of $L^p(\Gamma, w)$ and, moreover, $L^p(\Gamma, w)$ decomposes into the direct sum of these two subspaces:

$$L^p(\Gamma, w) = \dot{L}^p_+(\Gamma, w) \oplus \dot{L}^p_-(\Gamma, w).$$

Also define $L^p_-(\Gamma, w) := \dot{L}^-_p(\Gamma, w) + \mathbf{C}$. The spaces $L^p_+(\Gamma, w)$ and $L^p_-(\Gamma, w)$ are called the pth *Hardy spaces* on Γ with the weight w or simply the Hardy spaces of $L^p(\Gamma, w)$.

Recall that $C_\pm$ stand for the Cauchy operators (6.2) and let $B_\pm$ be the "boundary" operators introduced in Section 6.1.

Theorem 6.7 (Sokhotski-Plemelj formulas). *Let Γ be a Carleson Jordan curve, $1 < p < \infty$, and $w \in A_p(\Gamma)$. Then for every $g \in L^p(\Gamma, w)$,*

$$Pg = B_+C_+g \quad and \quad Qg = -B_-C_-g, \tag{6.10}$$

i.e. (5.72) holds.

Proof. Suppose first that $g \in R(\Gamma)$ and let $g = g_+ + g_-$ with $g_+ \in R_+(\Gamma)$, $g_- \in R_-(\Gamma)$. By Cauchy's formula, $B_+C_+g = g_+$ and $B_-C_-g = -g_-$. On the other hand, Lemma 6.5 shows that $Sg = g_+ - g_-$. Thus,

$$Pg = \frac{1}{2}\left(g_+ + g_- + g_+ - g_-\right) = g_+ = B_+C_+g,$$

and similarly, $Qg = -B_-C_-g$.

We now prove that $L^p_+(\Gamma, w) \subset E^1_+(\Gamma)$. So let $f \in L^p_+(\Gamma, w)$. Then, by definition, $f = Ph$ with $h \in L^p(\Gamma, w)$. Because $R(\Gamma)$ is dense in $L^p(\Gamma, w)$, there is a sequence $g_n \in R(\Gamma)$ such that $\|h - g_n\|_{p,w} \to 0$. The boundedness of P on $L^p(\Gamma, w)$ implies that $\|f - Pg_n\|_{p,w} \to 0$, and since

$$\|f - Pg_n\|_1 \leq \|f - Pg_n\|_{p,w}\|w^{-1}\|_q,$$

it follows that $\|f - Pg_n\|_1 \to 0$. It is clear that $C_+g_n \in E^1(D_+)$ and from what was proved in the preceding paragraph, we deduce that $Pg_n = B_+C_+g_n \in E^1_+(\Gamma)$. As $E^1_+(\Gamma)$ is a closed subspace of $L^1(\Gamma)$, we obtain that $f \in E^1_+(\Gamma)$. This proves the desired inclusion. It can be shown analogously that $\dot{L}^-_p(\Gamma, w) \subset \dot{E}^-_1(\Gamma)$.

Now let g be any function in $L^p(\Gamma, w)$. Then $g = g_+ + g_-$ with $g_+ \in L^p_+(\Gamma, w)$ and $g_- \in \dot{L}^p_-(\Gamma, w)$. Since $g_+ \in E^1_+(\Gamma)$, we have $B_+C_+g_+ = g_+$, and because $g_- \in \dot{E}^1_-(\Gamma)$, we infer from Theorem 6.3 that $B_+C_+g_- = B_+0 = 0$. Hence, $B_+C_+g = g_+$. The definition of $L^p_+(\Gamma, w)$ implies that $Pg = g_+$. Consequently, $Pg = B_+C_+g$. In the same way one can show that $Qg = -B_-C_-g$. $\qquad\square$

In case w is identically 1, we abbreviate $L^p_\pm(\Gamma, w)$ and $\dot{L}^p_-(\Gamma, w)$ to $L^p_\pm(\Gamma)$ and $\dot{L}^p_-(\Gamma)$, respectively.

Corollary 6.8. *If Γ is a Carleson Jordan curve and $1 < p < \infty$, then*

$$L_{\pm}^p(\Gamma) = E_{\pm}^1(\Gamma) \cap L^p(\Gamma) = E_{\pm}^p(\Gamma), \tag{6.11}$$

$$\dot{L}_{-}^p(\Gamma) = \dot{E}_{-}^1(\Gamma) \cap L^p(\Gamma) = \dot{E}_{-}^p(\Gamma). \tag{6.12}$$

If, in addition, w is a weight in $A_p(\Gamma)$, then

$$L_{\pm}^p(\Gamma, w) = E_{\pm}^1(\Gamma) \cap L^p(\Gamma, w), \tag{6.13}$$

$$\dot{L}_{-}^p(\Gamma, w) = \dot{E}_{-}^1(\Gamma) \cap L^p(\Gamma, w). \tag{6.14}$$

Proof. From the proof of Theorem 6.7 we know that $L_{+}^p(\Gamma, w) \subset E_{+}^1(\Gamma) \cap L^p(\Gamma, w)$. To get the reverse inclusion, let $g \in E_{+}^1(\Gamma) \cap L^p(\Gamma, w)$. Since $B_+ C_+$ is the identity operator on $E_{+}^1(\Gamma)$, we have $g = B_+ C_+ g$ and hence, by Theorem 6.7, $g = Pg$. This shows that $g \in L_{+}^p(\Gamma, w)$ and gives the inclusion $E_{+}^1(\Gamma) \cap L^p(\Gamma, w) \subset L_{+}^p(\Gamma, w)$.

Thus, we have proved the "+" equality of (6.13). The "−" equality of (6.13) and equality (6.14) can be derived similarly.

We now show that $L_{+}^p(\Gamma) \subset E_{+}^p(\Gamma)$. Pick $g \in L_{+}^p(\Gamma)$. Then $g = Ph$ with $h \in L^p(\Gamma)$. Choose $h_n \in R(\Gamma)$ so that $\|h - h_n\|_p \to 0$, whence, by the continuity of P, $\|g - Ph_n\|_p \to 0$. Since $h_n \in R(\Gamma)$ and hence $B_+ C_+ h_n \in E^p(D_+)$ for all $p \geq 1$, we conclude from the equality $Ph_n = B_+ C_+ h_n$ that $Ph_n \in E_{+}^p(\Gamma)$. The closedness of $E_{+}^p(\Gamma)$ therefore yields that $g \in E_{+}^p(\Gamma)$. Thus, we have proved that

$$E_{+}^1(\Gamma) \cap L^p(\Gamma) = L_{+}^p(\Gamma) \subset E_{+}^p(\Gamma).$$

Taking into account (6.5) we arrive at the right "+" equality of (6.11). The right "−" equality of (6.11) and (6.12) follow analogously. $\qquad\square$

In general it is no easy task to decide whether a function belongs to $L_{\pm}^p(\Gamma, w)$, although the preceding theorem reduces this problem to finding out whether the function is in $E_{\pm}^1(\Gamma)$. Clearly, functions in $L_{\pm}^p(\Gamma, w)$ are in $L^p(\Gamma, w)$ and admit analytic extension into D_+. The delicacy of the matter is that the converse is in general not true.

Example 6.9. Let $f(z) = \exp(-(z+1)/(z-1))$. This function is analytic in $\mathbf{C} \setminus \{1\}$, $|f(\tau)| = 1$ for $\tau \in \mathbf{T} \setminus \{1\}$, and $|f(z)| \leq 1$ for $|z| > 1$. Consequently, $f|\mathbf{T} \in E_{-}^p(\mathbf{T})$ for all $p \in [1, \infty)$. If $f|\mathbf{T}$ were in $L_{+}^p(\mathbf{T}) = E_{+}^p(\mathbf{T})$ for some $p \geq 1$, then Theorem 6.4 would imply that f were constant on $\mathbf{T}$. This shows that $f|\mathbf{T} \notin L_{+}^p(\Gamma)$ for every $p \in [1, \infty)$. $\qquad\square$

To check whether a function belongs to $L_{\pm}^p(\Gamma, w)$, the following result is often useful.

Lemma 6.10. *Let Γ be a Carleson Jordan curve, $1 < p < \infty$, and $w \in A_p(\Gamma)$. Suppose $f_\pm$ is analytic in $D_\pm$ (where $\infty \in D_-$), continuous on $D_\pm \cup \Gamma$ with the possible exception of finitely many points $t_1, \ldots, t_n \in \Gamma$, and that $f_\pm$ admits the estimate*

$$|f_\pm(z)| \leq M/|z - t_k|^\mu \quad (k = 1, \ldots, n)$$

with some $\mu > 0$ for all $z \in D_\pm$ sufficiently close to t_k. Then $f_\pm|\Gamma \in L^p_\pm(\Gamma, w)$ whenever $f_\pm|\Gamma \in L^p(\Gamma, w)$.

Proof. We consider only the $L^p_-(\Gamma, w)$ case, which is slightly more complicated than the $L^p_+(\Gamma, w)$ case.

The functions $\varphi_+ := C_+ f_-$ and $\varphi_- := -C_- f_-$ are analytic in D_+ and D_-, respectively. The polynomial g_1 given by

$$g_1(z) := \prod_{k=1}^n (z - t_k)^{[\mu]+1},$$

where $[\mu]$ stands for the integral part of μ, is analytic in $D_- \setminus \{\infty\}$ and has a pole of order $\deg g_1$ at infinity. Hence, $g_1(f_- - \varphi_-)$ is analytic in $D_- \setminus \{\infty\}$ and has at most a pole of order $\deg g_1$ at infinity. Choose any polynomial g_2 such that $\deg g_2 \leq \deg g_1$ and

$$d_- := g_1(f_- - \varphi_-) - g_2 \tag{6.15}$$

is analytic in D_-. Since $g_1 f_-$ has at most a pole of order $\deg g_1$ at infinity, we can also find a polynomial h_1 such that $\deg h_1 \leq \deg g_1$ and $g_1 f_- - h_1$ is analytic in D_-. Put $h_2 := g_2 - h_1$. Then

$$d_- = (g_1 f_- - h_1) - (g_1 \varphi_- + h_2) =: \psi_1 - \psi_2,$$

and since d_- and $\psi_1 = g_1 f_- - h_1$ are analytic in D_-, so also is $\psi_2 = g_1 \varphi_- + h_2$. Because $\varphi_- = -C_- f_-$ vanishes at infinity, we have $\deg h_2 \leq \deg g_1 - 1$.

The function ψ_1 is analytic in D_- and continuous on $D_- \cup \Gamma$. Therefore $B_- \psi_1 \in E^1_-(\Gamma) \cap L^\infty(\Gamma) \subset L^p_-(\Gamma, w)$ due to (6.13).

By Theorem 6.7, $B_- \varphi_- = -B_- C_- f_- = Q f_-$ and hence, $B_- \varphi_- \in L^p_-(\Gamma, w)$. Since g_1 is a polynomial of degree g_1, it follows that

$$
\begin{aligned}
B_- \psi_2 &= g_1(B_- \varphi_-) + h_2 = g_1 Q f_- + h_2 \\
&\in L^p_-(\Gamma, w) + \mathrm{span}\{\chi_1, \ldots, \chi_{\deg g_1 - 1}\}
\end{aligned}
$$

where $\chi_j(\tau) := \tau^j$. As ψ_2 has no pole at infinity, we even see that $B_- \psi_2$ belongs to $L^p_-(\Gamma, w)$.

In summary, we have shown that $B_- d_- = B_- \psi_1 - B_- \psi_2 \in L^p_-(\Gamma, w)$. On the other hand, from (6.15) and Theorem 6.7 we get

$$
\begin{aligned}
B_- d_- &= g_1(f_- - B_- \varphi_-) - g_2 = g_1(f_- + B_- C_- f_-) - g_2 \\
&= g_1(f_- - Q f_-) - g_2 = g_1(P f_-) - g_2. \tag{6.16}
\end{aligned}
$$

Hence, $B_-d_- \in L_+^p(\Gamma, w)$ (recall that g_1 and g_2 are polynomials). Consequently,

$$B_-d_- \in L_+^p(\Gamma, w) \cap L_-^p(\Gamma, w) = \mathbf{C},$$

i.e. there is a constant c such that $(B_-d_-)(\tau) = c$ for almost all $\tau \in \Gamma$.

By virtue of (6.16), the rational function $(g_2 + c)/g_1$ equals $C_+f_- = \varphi_+$ on D_+ and is equal to $f_- + C_-f_- = f_- - \varphi_-$ on D_-. Thus, $(g_2 + c)/g_1$ is analytic in $D_+ \cup D_-$. Since $(g_2 + c)/g_1 = Pf_- \in L^p(\Gamma, w) \subset L^1(\Gamma)$, the rational function $(g_2 + c)/g_1$ cannot have poles on Γ. It results that $(g_2 + c)/g_1$ is analytic in $D_+ \cup \Gamma \cup D_- = \mathbf{C} \cup \{\infty\}$ and hence, by Liouville's theorem, it is constant. Therefore φ_+ is constant on D_+, $\varphi_+(z) = c_0$ for $z \in D_+$. We finally obtain from Theorem 6.7 that

$$f_- = Qf_- + Pf_- = Qf_- + B_+C_+f_- = Qf_- + B_+\varphi_+$$
$$= Qf_- + c_0 \in L_-^p(\Gamma, w). \qquad \Box$$

The following lemma will find repeated application in the sequel.

Lemma 6.11 *Let Γ be a Carleson Jordan curve, $1 < p < \infty$, $1/p + 1/q = 1$, and $w \in A_p(\Gamma)$. Then*

$$f \in L_\pm^p(\Gamma, w), \ g \in L_\pm^q(\Gamma, w^{-1}) \Longrightarrow fg \in E_\pm^1(\Gamma).$$

Proof. Suppose $f \in L_+^p(\Gamma, w)$ and $g \in L_+^q(\Gamma, w^{-1})$. We have $f = Ph$ and $g = Pk$ with $h \in L^p(\Gamma, w)$ and $k \in L^q(\Gamma, w^{-1})$. Choose functions $h_n, k_n \in R(\Gamma)$ so that $\|h - h_n\|_{p,w} \to 0$ and $\|k - k_n\|_{q,w^{-1}} \to 0$. The boundedness of P implies that $\|f - f_n\|_{p,w} \to 0$ and $\|g - g_n\|_{q,w^{-1}} \to 0$ where $f_n := Ph_n$ and $g_n := Pk_n$. Since f_n and g_n are rational functions without poles on $D_+ \cup \Gamma$ (Lemma 6.5), the product f_ng_n belongs to $E_+^1(\Gamma)$. Because

$$\|fg - f_ng_n\|_1 \le \|f - f_n\|_{p,w}\|g\|_{q,w^{-1}} + \|f_n\|_{p,w}\|g - g_n\|_{q,w^{-1}} = o(1)$$

and $E_+^1(\Gamma)$ is closed, we conclude that $fg \in E_+^1(\Gamma)$. Analogously one can show that $fg \in E_-^1(\Gamma)$ if $f \in L_-^p(\Gamma, w)$ and $g \in L_-^q(\Gamma, w^{-1})$. $\qquad \Box$

6.3 Fredholm operators

Let X be a (complex) Banach space. We denote by $\mathcal{B}(X)$ the Banach algebra of all bounded linear operators on X and by $\mathcal{K}(X)$ the collection of all compact linear operators on X. It is well known that $\mathcal{K}(X)$ is a closed two-sided ideal of $\mathcal{B}(X)$. The quotient algebra $\mathcal{B}(X)/\mathcal{K}(X)$ is often called the *Calkin algebra* of X.

The *kernel* and the *image* (=range) of an operator $A \in \mathcal{B}(X)$ are

$$\operatorname{Ker} A := \{x \in X : Ax = 0\}, \quad \operatorname{Im} A := \{Ax : x \in X\}.$$

Clearly, $\operatorname{Ker} A$ is always a closed subspace of X. The operator A is called *normally solvable* if $\operatorname{Im} A$ is a closed subspace of X.

An operator $A \in \mathcal{B}(X)$ is said to be *semi-Fredholm* if it is normally solvable and at least one of the numbers

$$\alpha(A) := \dim \operatorname{Ker} A, \quad \beta(A) := \dim(X/\operatorname{Im} A)$$

is finite.* In that case

$$\operatorname{Ind} A := \alpha(A) - \beta(A)$$

is referred to as the *index* of A. Thus, the index of a semi-Fredholm operator is either an integer or $+\infty$ or $-\infty$. If A is normally solvable and both $\alpha(A)$ and $\beta(A)$ are finite, then A is called a *Fredholm operator*. The index of a Fredholm operator is always an integer.

An operator $A \in \mathcal{B}(X)$ is said to be *invertible*(*invertible from the left* or *invertible from the right*) if there is an operator $B \in \mathcal{B}(X)$ such that $AB = BA = I$ ($BA = I$ or $AB = I$, respectively). Clearly, invertible operators are Fredholm of index zero.

If $A \in \mathcal{B}(X)$ is invertible from the left (from the right) then necessarily $\operatorname{Ker} A = \{0\}$ and thus $\alpha(A) = 0$ ($\operatorname{Im} A = X$ and thus $\beta(A) = 0$). The inverse is not true. We rather have the following:

A is invertible from the left if and only if $\operatorname{Ker} A = \{0\}$ *and* $\operatorname{Im} A$ *is a closed subspace of X possessing a direct complement in X;*

A is invertible from the right if and only if $\operatorname{Im} A = X$ *and* $\operatorname{Ker} A$ *has a direct complement in X.*

We also remark that a Fredholm or semi-Fredholm operator need not be invertible from any side.

An operator $A \in \mathcal{B}(X)$ turns out to be Fredholm if and only if the coset $A + \mathcal{K}(X)$ is invertible in the Calkin algebra $\mathcal{B}(X)/\mathcal{K}(X)$. In words, A is Fredholm if and only if it is invertible modulo compact operators. Some care is in order with one-sided invertibility in the Calkin algebra:

the coset $A + \mathcal{K}(X)$ is invertible from the left in $\mathcal{B}(X)/\mathcal{K}(X)$ if and only if A is semi-Fredholm, $\alpha(A) < \infty$, and $\operatorname{Im} A$ has a direct complement in X;

the coset $A + \mathcal{K}(X)$ is invertible from the right in $\mathcal{B}(X)/\mathcal{K}(X)$ if and only if A is semi-Fredholm, $\beta(A) < \infty$, and $\operatorname{Ker} A$ has a direct complement in X.

Any operator $R \in \mathcal{B}(X)$ for which $RA - I \in \mathcal{K}(X)$ is called a *left regularizer* of A, while a *right regularizer* of A is an operator $R \in \mathcal{B}(X)$ such that $AR - I \in \mathcal{K}(X)$. Thus, A is Fredholm if and only if it has a left and a right regularizer. The existence of a left (right) regularizer implies that A is semi-Fredholm with $\alpha(A) < \infty$ (with $\beta(A) < \infty$), but semi-Fredholmness does in general not imply the existence of one-sided regularizers.

---

*The letters α, β are in general use for both the indices of submultiplicative functions and the kernel/cokernel dimensions of an operator. This will not cause any confusion in the following.

Theorem 6.12. *If $A, B \in \mathcal{B}(X)$ are Fredholm, then the product AB is also Fredholm and*

$$\operatorname{Ind} AB = \operatorname{Ind} A + \operatorname{Ind} B.$$

Theorem 6.13. *Suppose $A \in \mathcal{B}(X)$ is semi-Fredholm. Then there is a number $\varepsilon = \varepsilon(A) > 0$ such that $A + K + D$ is semi-Fredholm and*

$$\operatorname{Ind}(A + K + D) = \operatorname{Ind}(A + K) = \operatorname{Ind} A,$$

$$\alpha(A + K + D) \le \alpha(A + K), \ \ \beta(A + K + D) \le \beta(A + K)$$

whenever $K \in \mathcal{K}(X)$, $D \in \mathcal{B}(X)$, $\|D\| < \varepsilon$.

Proofs of these two theorems are e.g. in [89, pp. 166–169 and 208–209]. □

For $A \in \mathcal{B}(X)$, we define the *spectrum* in the usual way,

$$\operatorname{sp} A := \big\{ \lambda \in \mathbf{C} : A - \lambda I \text{ is not invertible} \big\},$$

and we refer to the set

$$\operatorname{sp}_{\mathrm{ess}} A := \big\{ \lambda \in \mathbf{C} : A - \lambda I \text{ is not Fredholm} \big\}$$

as the *essential spectrum* of A. Since $\operatorname{sp} A$ is the spectrum of A as an element of the Banach algebra $\mathcal{B}(X)$ and $\operatorname{sp}_{\mathrm{ess}} A$ is the spectrum of $A + \mathcal{K}(X)$ as an element of the Banach algebra $\mathcal{B}(X)/\mathcal{K}(X)$, both spectra are compact nonempty subsets of $\mathbf{C}$. Clearly,

$$\operatorname{sp}_{\mathrm{ess}} A \subset \operatorname{sp} A \subset \big\{ \lambda \in \mathbf{C} : |\lambda| \le \|A\| \big\}.$$

Given $A \in \mathcal{B}(X)$, we denote by $A^* \in \mathcal{B}(X^*)$ the adjoint operator on the dual Banach space X^*. One can show that A is normally solvable, Fredholm, or invertible if and only if A^* has the corresponding property. If A is Fredholm, then

$$\alpha(A^*) = \beta(A), \ \ \beta(A^*) = \alpha(A), \ \ \operatorname{Ind} A^* = -\operatorname{Ind} A.$$

By virtue of the first of these equalities, $X/\operatorname{Im} A$ is also referred to as the *cokernel* of A and is denoted by Coker A.

6.4 Toeplitz operators

Let Γ be a Carleson Jordan curve, $1 < p < \infty$, and $w \in A_p(\Gamma)$. The operator of multiplication by a function $a \in L^\infty(\Gamma)$ is obviously bounded on $L^p(\Gamma, w)$ and is, by tradition, denoted by aI. In case aI follows another operator, B say, one usually abbreviates aIB to aB.

The *Toeplitz operator* generated by a function $a \in L^\infty(\Gamma)$ is the operator

$$T(a) : L^p_+(\Gamma, w) \to L^p_+(\Gamma, w), \ g \mapsto P(ag)$$

where $P := (I + S)/2$. Since P is bounded, so is $T(a)$ for every $a \in L^\infty(\Gamma)$. The function a is in this context referred to as the *symbol* of the operator $T(a)$.

If $\Gamma = \mathbf{T}$ is the complex unit circle, $p = 2$, and $w \equiv 1$, then the functions $\{(1/\sqrt{2\pi})e^{in\theta}\}_{n=0}^{\infty}$ constitute an orthonormal basis in the Hardy space $L_+^p(\Gamma, w) = L_+^2(\mathbf{T})$. The matrix representation of $T(a)$ with respect to this basis is the *infinite Toeplitz matrix*

$$\begin{pmatrix} a_0 & a_{-1} & a_{-2} & \cdots \\ a_1 & a_0 & a_{-1} & \cdots \\ a_2 & a_1 & a_0 & \cdots \\ \cdots & \cdots & \cdots & \cdots \end{pmatrix} \qquad (6.17)$$

composed by the Fourier coefficients of a,

$$a_n := \frac{1}{2\pi} \int_0^{2\pi} a(e^{i\theta})e^{-in\theta}\, d\theta \quad (n \in \mathbf{Z}).$$

Clearly, in the general case we have neither a nice basis $L_+^p(\Gamma, w)$ nor a nice matrix representation of Toeplitz operators.

Our main subject is invertibility and Fredholm criteria for Toeplitz operators on $L_+^p(\Gamma, w)$. Staying all the time in Hardy spaces is sometimes inconvenient, especially when passing to adjoint operators. We therefore also consider two operators closely related to $T(a)$, the operators

$$PaP + Q \text{ and } aP + Q \text{ on } L^p(\Gamma, w).$$

The following simple result shows that, in a sense, all these operators are one and the same thing.

Lemma 6.14. *Let $a \in L^\infty(\Gamma)$. If one of the operators $T(a)$, $PaP + Q$, $aP + Q$ is normally solvable, semi-Fredholm, Fredholm, injective, surjective, an operator with a dense image, invertible from the left or from the right, or invertible, then the other two operators also have the correponding property. If these operators are semi-Fredholm, then*

$$\alpha(T(a)) = \alpha(PaP + Q) = \alpha(aP + Q), \quad \beta(T(a)) = \beta(PaP + Q) = \beta(aP + Q),$$

and hence they all have the same index.

Proof. The space $L^p(\Gamma, w)$ decomposes into the direct sum

$$L^p(\Gamma, w) = L_+^p(\Gamma, w) \oplus \dot{L}_-^p(\Gamma, w)$$

and accordingly, $PaP + Q$ may be written as an operator matrix:

$$\begin{pmatrix} T(a) & 0 \\ 0 & I \end{pmatrix} : \begin{pmatrix} L_+^p(\Gamma, w) \\ \dot{L}_-^p(\Gamma, w) \end{pmatrix} \rightarrow \begin{pmatrix} L_+^p(\Gamma, w) \\ \dot{L}_-^p(\Gamma, w) \end{pmatrix}.$$

Hence,

$$\mathrm{Im}\,(PaP+Q) = \mathrm{Im}\,T(a) \oplus \dot{L}^p_-(\Gamma,w), \quad \mathrm{Ker}\,(PaP+Q) = \mathrm{Ker}\,T(a),$$

which gives all assertions for the pair $T(a)$ and $PaP+Q$. Because

$$aP+Q = (PaP+Q)(I+QaP)$$

(note that $PQ = 0$) and $I+QaP$ is always invertible (the inverse being $I-QaP$), we get the assertions for the pair $PaP+Q$ and $aP+Q$. $\qquad\square$

6.5 Adjoints

The following fact is well known and merely quoted in order to emphasize that it has nothing to do with the Carleson and Muckenhoupt conditions.

Proposition 6.15. *Let Γ be a rectifiable composed curve, let $1 < p < \infty$, and let $w : \Gamma \to [0,\infty]$ be a weight. Then*

$$\left(L^p(\Gamma,w)\right)^* = L^q(\Gamma,w^{-1}) \quad (1/p+1/q=1)$$

in the following sense: a map $\varphi : L^p(\Gamma,w) \to \mathbf{C}$ is a bounded linear functional if and only if there is a function $g \in L^q(\Gamma,w^{-1})$ such that

$$\varphi(f) = (f,g) := \int_\Gamma f(\tau)\overline{g(\tau)}\,|d\tau| \quad \text{for all } f \in L^p(\Gamma,w). \tag{6.18}$$

Proof. Clearly, φ is a bounded linear functional on $L^p(\Gamma,w)$ if and only if $\varphi w^{-1}I : L^p(\Gamma) \to \mathbf{C}$ is bounded. Since $(L^p(\Gamma))^* = L^q(\Gamma)$, this happens if and only if there exists a function $g_0 \in L^q(\Gamma)$ such that

$$\varphi(w^{-1}f_0) = \int_\Gamma f_0(\tau)\,\overline{g_0(\tau)}\,|d\tau| \quad \text{for all } f_0 \in L^p(\Gamma),$$

i.e. such that

$$\varphi(f) = \int_\Gamma f(\tau)w(\tau)\,\overline{g_0(\tau)}\,|d\tau| \quad \text{for all } f \in L^p(\Gamma,w).$$

Since $g_0 \in L^q(\Gamma)$ if and only if $g := wg_0 \in L^q(\Gamma,w^{-1})$, we get the assertion. $\qquad\square$

On a rectifiable simple oriented curve Γ we have

$$d\tau = e^{i\theta_\Gamma(\tau)}|d\tau| \tag{6.19}$$

where $\theta_\Gamma(\tau)$ is the angle made by the positively oriented real axis and the naturally oriented tangent of Γ at τ (which exists almost everywhere). Given any $r \in (1, \infty)$ and any weight $\psi : \Gamma \to [0, \infty]$, we define the operator H_Γ by

$$H_\Gamma : L^r(\Gamma, \psi) \to L^r(\Gamma, \psi), \quad (H_\Gamma g)(\tau) := e^{-i\theta_\Gamma(\tau)}\overline{g(\tau)}.$$

Notice that H_Γ is additive but that $H_\Gamma(\alpha g) = \overline{\alpha}(H_\Gamma g)$ for $\alpha \in \mathbf{C}$. Clearly, $\|H_\Gamma g\|_{r,\psi} = \|g\|_{r,\psi}$ and $H_\Gamma^2 = I$.

Proposition 6.16. *Let Γ be a Carleson Jordan curve, $1 < p < \infty$, and $w \in A_p(\Gamma)$. Then the adjoint operator of $S \in \mathcal{B}(L^p(\Gamma, w))$ is the operator $S^* = -H_\Gamma S H_\Gamma \in \mathcal{B}(L^q(\Gamma, w^{-1}))$ and consequently,*

$$P^* = H_\Gamma Q H_\Gamma, \quad Q^* = H_\Gamma P H_\Gamma.$$

Proof. For $f \in L^p(\Gamma, w)$ and $g \in L^q(\Gamma, w^{-1})$, put

$$(f, g)_0 := \int_\Gamma f(\tau)g(\tau)\,d\tau.$$

Let first f and g be functions in $R(\Gamma)$ and write $f = f_+ + f_-$, $g = g_+ + g_-$ with $f_\pm \in R_\pm(\Gamma)$, $g_\pm \in R_\pm(\Gamma)$ (recall Section 6.2). By Lemma 6.5,

$$(Pf, g)_0 = (f_+, g_+ + g_-)_0 = (f_+, g_+)_0 + (f_+, g_-)_0,$$

and since $f_+ g_+$ is analytic in D_+, Cauchy's theorem implies that $(f_+, g_+)_0 = 0$. On the other hand, Lemma 6.5 shows that

$$(f, Qg)_0 = (f_+ + f_-, g_-)_0 = (f_+, g_-)_0 + (f_-, g_-)_0,$$

and again we deduce from Cauchy's theorem that $(f_-, g_-)_0 = 0$. Thus,

$$(Pf, g)_0 = (f_+, g_-)_0 = (f, Qg)_0$$

for all $f, g \in R(\Gamma)$. Since $R(\Gamma)$ is dense in $L^p(\Gamma, w)$ and $L^q(\Gamma, w^{-1})$ and P, Q are bounded on these spaces, it follows that

$$(Pf, g)_0 = (f, Qg)_0 \text{ for all } f \in L^p(\Gamma, w), \ g \in L^q(\Gamma, w^{-1}). \tag{6.20}$$

From (6.18) and (6.19) we obtain

$$(f, g) = \int_\Gamma f(\tau)\overline{g(\tau)}e^{-i\theta_\Gamma(\tau)}d\tau = (f, H_\Gamma g)_0.$$

Combining this and (6.20) we get

$$(Pf, g) = (Pf, H_\Gamma g)_0 = (f, QH_\Gamma g)_0 = (f, H_\Gamma Q H_\Gamma g)$$

for all $f \in L^p(\Gamma, w)$ and all $g \in L^q(\Gamma, w^{-1})$. Thus, $P^* = H_\Gamma Q H_\Gamma$, whence $Q^* = I - P^* = H_\Gamma(I - Q)H_\Gamma = H_\Gamma P H_\Gamma$ and $S^* = -H_\Gamma S H_\Gamma$. $\qquad\square$

We remark that the adjoint of the multiplication operator aI is the multiplication operator $\bar{a}I$. Obviously, $\bar{a}I = H_\Gamma a H_\Gamma$ and therefore Proposition 6.16 gives

$$\begin{aligned}(aP + Q)^* &= P^*\bar{a}I + Q^* = H_\Gamma Q H_\Gamma H_\Gamma a H_\Gamma + H_\Gamma P H_\Gamma \\ &= H_\Gamma(QaI + P)H_\Gamma. \end{aligned} \tag{6.21}$$

6.6 Two basic theorems

If a is the zero function then $T(a)$ is the zero operator and hence normally solvable but not semi-Fredholm. The following theorem implies that every nonzero and normally solvable Toeplitz operator is automatically semi-Fredholm.

Theorem 6.17 (Coburn-Simonenko). *Let Γ be a Carleson Jordan curve and $w \in A_p(\Gamma)$ $(1 < p < \infty)$. If $a \in L^\infty(\Gamma) \setminus \{0\}$ then $T(a)$ has a trivial kernel or a dense image on $L^p_+(\Gamma, w)$.*

Proof. For brevity, we write $L^p := L^p(\Gamma, w)$, $L^q = L^q(\Gamma, w^{-1})$, $L^p_\pm := L^p_\pm(\Gamma, w)$, etc. We first show that $aP + Q$ is injective on L^p or $QaI + P$ is injective on L^q.

Assume the contrary, i.e. assume there are nonzero $f \in L^p$ and $g \in L^q$ such that

$$(aP + Q)f = 0 \text{ and } (QaI + P)g = 0.$$

The first of these equalities implies that $f_- := Qf \in \dot{L}^p_-$, $f_+ := Pf \in L^p_+$, and $af_+ + f_- = 0$. From the second equality we infer that $Qag = 0$ and $Pg = 0$, whence $g_+ := ag \in L^q_+$ and $g_- := g \in \dot{L}^q_-$. Multiplying the equality $af_+ = -f_-$ by g we get $agf_+ = -f_-g$, and hence,

$$g_+f_+ = -f_-g_-. \tag{6.22}$$

By Lemma 6.11, the left-hand side of (6.22) lies in $\dot{E}^1_+(\Gamma)$, while its right-hand side belongs to $\dot{E}^1_-(\Gamma)$. Consequently, by Theorem 6.4, $g_+f_+ = 0$ and $f_-g_- = 0$. From the Lusin-Privalov Theorem 6.1 we deduce that $g_- \neq 0$ a.e. on Γ, implying that $f_- = 0$. It follows that $f_+ \neq 0$ a.e. on Γ since otherwise, again by Lusin and Privalov's theorem, $f_+ = 0$ and thus $f = f_+ + f_- = 0$. Now the equality $g_+f_+ = 0$ shows that $0 = g_+ = ag$. Because $a \in L^\infty(\Gamma) \setminus \{0\}$, we arrive at the conclusion that $g = g_-$ must vanish on a set of positive measure. Once more invoking Theorem 6.1, we obtain that $g = 0$. This contradiction proves our claim.

If $aP + Q$ is injective on L^p, then so is $T(a)$ on L^p_+ by virtue of Lemma 6.14. On the other hand, if $QaI + P$ is injective on L^q, then $H_\Gamma(QaI + P)H_\Gamma$ is also injective on L^q. From (6.21) we infer that $aP + Q$ has a dense range on L^p, and Lemma 6.14 so yields that $T(a)$ has a dense image on L^p_+. $\qquad\square$

Corollary 6.18. *If $a \in L^\infty(\Gamma) \setminus \{0\}$ and $T(a)$ is normally solvable, then $T(a)$ is an injective or surjective semi-Fredholm operator.* $\qquad\square$

Corollary 6.19. *Let $a \in L^\infty(\Gamma)$. Then $T(a)$ is invertible if and only if it is Fredholm of index zero.*

Proof. The "only if" part is trivial. To show the "if" portion, suppose $T(a)$ is Fredholm and $\alpha(T(a)) = \beta(T(a))$. The previous corollary implies that then $\alpha(T(a)) = 0$ or $\beta(T(a)) = 0$. Thus, $\alpha(T(a)) = \beta(T(a)) = 0$, and since $\operatorname{Im} T(a)$ is closed, it follows that $T(a)$ is bijective and therefore invertible. $\qquad\square$

Corollary 6.19 is of fundamental importance. It divides the problem of deciding whether a Toeplitz operator is invertible into two subproblems: find out whether the operator is Fredholm and if yes, compute its index.

Taking into account that $T(a) - \lambda I = T(a - \lambda)$ and applying Corollary 6.19 to the operator $T(a - \lambda)$ we see that

$$\operatorname{sp} T(a) = \operatorname{sp}_{\mathrm{ess}} T(a) \cup \big\{\lambda \in \mathbf{C} : T(a - \lambda) \text{ is Fredholm with nonzero index}\big\},$$

which provides us with useful information about the parts of the spectrum. The following theorem gives another important piece of information about the spectrum of a Toeplitz operator.

Recall that $GL^\infty(\Gamma)$ stands for the functions in $L^\infty(\Gamma)$ which are invertible in $L^\infty(\Gamma)$. The spectrum of a function $a \in L^\infty(\Gamma)$ as an element of the Banach algebra $L^\infty(\Gamma)$ is its *essential range*

$$\mathcal{R}(a) := \Big\{\lambda \in \mathbf{C} : \big|\{t \in \Gamma : |a(t) - \lambda| < \varepsilon\}\big| > 0 \;\; \forall \varepsilon > 0\Big\}.$$

Theorem 6.20 (Hartman-Wintner-Simonenko). *Let Γ be a Carleson Jordan curve, $1 < p < \infty$, and $w \in A_p(\Gamma)$. If $a \in L^\infty(\Gamma) \setminus \{0\}$ and $T(a)$ is normally solvable, then $a \in GL^\infty(\Gamma)$. In particular,*

$$\mathcal{R}(a) \subset \operatorname{sp}_{\mathrm{ess}} T(a).$$

Proof. Suppose a does not vanish identically and $T(a)$ is normally solvable. Corollary 6.18 in conjunction with Lemma 6.14 shows that A is an injective or surjective semi-Fredholm operator. From Theorem 6.13 we infer that all perturbations of A by operators with small norm are also semi-Fredholm and injective or surjective.

Contrary to what we want, assume $0 \in \mathcal{R}(a)$. Then for every $\varepsilon > 0$, we can find a function $a_\varepsilon \in L^\infty(\Gamma) \setminus \{0\}$ such that a_ε vanishes on a set of positive measure and

$$\|a_\varepsilon P + Q - (aP + Q)\| \leq \|a_\varepsilon - a\|_\infty \|P\| < \varepsilon.$$

From the preceding paragraph we know that $A_\varepsilon := a_\varepsilon P + Q$ is semi-Fredholm and injective or surjective if only $\varepsilon > 0$ is sufficiently small.

Suppose first that A_ε is surjective. Let $g \in L^p(\Gamma, w)$ be a solution of the equation $(a_\varepsilon P + Q)g = 1$. Put $g_+ := Pg$ and $g_- := Qg$. We have $a_\varepsilon g_+ = 1 - g_-$, so $1 - g_-$ vanishes on a set of positive measure, and Theorem 6.1 implies that $1 - g_- = 0$ identically. But this is impossible, because $g_- \in \dot{L}_-^p(\Gamma, w)$ and $1 \notin \dot{L}_-^p(\Gamma, w)$.

Consequently, A_ε must be injective. Then A_ε^* has a dense image, and since A_ε^* is normally solvable together with A_ε, we conclude that A_ε^* is surjective. From (6.21) we know that $A_\varepsilon^* = H_\Gamma(Q a_\varepsilon I + P)H_\Gamma$. Hence, $Q a_\varepsilon I + P$ is surjective. Let $g \in L^q(\Gamma, w^{-1})$ be a solution of the equation $(Q a_\varepsilon I + P)g = 1$. Then $Pg = 1$ and $Q(a_\varepsilon g) = 0$. The function $f_+ := a_\varepsilon g$ therefore belongs to $L_+^q(\Gamma, w^{-1})$. Since a_ε vanishes on a set of positive measure, the Lusin-Privalov theorem tells us that $f_+ = 0$. We have $g = Pg + Qg = 1 + g_-$ with $g_- := Qg \in \dot{L}_-^q(\Gamma, w^{-1})$. Thus, $0 = a_\varepsilon g = a_\varepsilon(1 + g_-)$. Since $a_\varepsilon \neq 0$ on a set of positive measure, we may again have recourse to Theorem 6.1 to see that $1 + g_- = 0$ identically, which is impossible because $-1 \notin \dot{L}_-^q(\Gamma, w^{-1})$. This contradiction completes the proof. □

6.7   Hankel operators

Let Γ be a Carleson Jordan curve, $1 < p < \infty$, and $w \in A_p(\Gamma)$. The two operators

$$H(a) : \dot{L}_-^p(\Gamma, w) \to L_+^p(\Gamma, w), \ g \mapsto P(ag),$$
$$H(\tilde{a}) : L_+^p(\Gamma, w) \to \dot{L}_-^p(\Gamma, w), \ g \mapsto Q(ag)$$

are bounded for every $a \in L^\infty(\Gamma)$ and are called the *Hankel operators* generated by a. Notice that the Toeplitz operator $T(a)$ may be regarded as the operator

$$T(a) = PaP : \operatorname{Im} P \to \operatorname{Im} P,$$

while the Hankel operators $H(a)$ and $H(\tilde{a})$ are in this notation given by

$$H(a) = PaQ : \operatorname{Im} Q \to \operatorname{Im} P \ \text{ and } \ H(\tilde{a}) : QaP : \operatorname{Im} P \to \operatorname{Im} Q.$$

It is easily seen that if $\Gamma = \mathbf{T}$, $p = 2$, $w \equiv 1$, then the matrix representations of $H(a)$ and $H(\tilde{a})$ in the orthonormal bases

$$\left\{ \frac{1}{\sqrt{2\pi}} e^{in\theta} \right\}_{n=0}^\infty \ \text{ in } L_+^2(\mathbf{T}), \ \left\{ \frac{1}{\sqrt{2\pi}} e^{-in\theta} \right\}_{n=1}^\infty \ \text{ in } \dot{L}_-^2(\mathbf{T})$$

are given by

$$\begin{pmatrix} a_1 & a_2 & a_3 & \cdots \\ a_2 & a_3 & \cdots & \cdots \\ a_3 & \cdots & \cdots & \cdots \\ \cdots & \cdots & \cdots & \cdots \end{pmatrix} \ \text{ and } \ \begin{pmatrix} a_{-1} & a_{-2} & a_{-3} & \cdots \\ a_{-2} & a_{-3} & \cdots & \cdots \\ a_{-3} & \cdots & \cdots & \cdots \\ \cdots & \cdots & \cdots & \cdots \end{pmatrix},$$

respectively.

The role Hankel operators play in connection with Toeplitz operators is revealed by the formula

$$T(ab) = T(a)T(b) + H(a)H(\tilde{b}),\qquad(6.23)$$

which holds for arbitrary $a, b \in L^\infty(\Gamma)$ and is nothing but the obvious identity

$$PabP = PaPbP + PaQbP.$$

Put $H_\pm^\infty(\Gamma) := L^\infty(\Gamma) \cap E_\pm^1(\Gamma)$. It is well known that $H_\pm^\infty(\Gamma)$ coincides with $B_\pm(E^\infty(D_\pm))$ where $E^\infty(D_\pm)$ stands for the bounded analytic functions in $D_\pm$, but we will not make use of this interpretation. All we need know is that

$$H(a) = 0 \text{ for } a \in H_-^\infty(\Gamma), \quad H(\tilde{a}) = 0 \text{ for } a \in H_+^\infty(\Gamma).\qquad(6.24)$$

Indeed, if $a \in H_-^\infty(\Gamma)$ ($\subset L_-^q(\Gamma, w^{-1})$) and $f \in \dot{L}_-^p(\Gamma, w)$ then $af \in L^p(\Gamma, w) \cap \dot{E}_-^1(\Gamma)$ (Lemma 6.11) and hence $af \in \dot{L}_-^p(\Gamma, w)$ by (6.14), which implies that $H(a)f = P(af) = 0$. Analogously one can show that $H(\tilde{a}) = 0$ whenever $a \in H_+^\infty(\Gamma)$. Combining (6.23) and (6.24) we obtain that

$$T(a_- b a_+) = T(a_-)T(b)T(a_+) \text{ for all } a_\pm \in H_\pm^\infty(\Gamma), \ b \in L^\infty(\Gamma).\qquad(6.25)$$

Proposition 6.21. *If* $a \in C(\Gamma)$ *then* $H(a)$ *and* $H(\tilde{a})$ *are compact.*

Proof. By Mergelyan's theorem (see e.g. [70, p. 97]), the set $R(\Gamma)$ is uniformly dense in $C(\Gamma)$. Hence, there are $a_n \in R(\Gamma)$ such that $\|a - a_n\|_\infty \to 0$. Since

$$((a_n S - S a_n I)g)(t) = \frac{1}{\pi i} \int_\Gamma \frac{a_n(t) - a_n(\tau)}{\tau - t} g(\tau)\, d\tau \quad (t \in \Gamma),$$

the operators $a_n S - S a_n I$ are integral operators with continuous kernels and thus compact. Because

$$\left\| (aS - SaI) - (a_n S - S a_n I) \right\| \leq 2\|a - a_n\|_\infty \|S\| = o(1),$$

it follows that $aS - SaI = 2(aP - PaI)$ is compact. Thus, the operators

$$PaQ = -(aP - PaI)Q, \ QaP = Q(aP - PaI)$$

are also compact. $\qquad\square$

Corollary 6.22. *If* a *and* b *are in* $L^\infty(\Gamma)$ *and at least one of these two functions belongs to* $C(\Gamma)$, *then* $T(ab) - T(a)T(b)$ *is a compact operator.*

Proof. Immediate from (6.23) and Proposition 6.21. $\qquad\square$

6.8 Continuous symbols

We are now in a position to completely describe the spectra of Toeplitz operators with continuous symbols.

Let Γ be a Carleson Jordan curve, $1 < p < \infty$, and $w \in A_p(\Gamma)$. As above, we may without loss of generality assume that the origin belongs to the bounded component D_+ of $\mathbf{C} \setminus \Gamma$. For $n \in \mathbf{Z}$, define $\chi_n : \Gamma \to \mathbf{C}$ by $\chi_n(\tau) := \tau^n$.

Lemma 6.23. *The operator $T(\chi_1)$ is Fredholm on $L_+^p(\Gamma, w)$ and we have*

$$\alpha\big(T(\chi_1)\big) = 0, \quad \beta\big(T(\chi_1)\big) = 1.$$

Proof. If $g_+ \in L_+^p(\Gamma, w)$ and $0 = T(\chi_1)g_+ = P(\chi_1 g_+) = \chi_1 g_+$, then obviously $g_+ = 0$. Hence $\alpha(T(\chi_1)) = 0$. We now show that

$$\operatorname{Im} T(\chi_1) = \{g \in L_+^p(\Gamma, w) : (C_+ g)(0) = 0\} \tag{6.26}$$

where $C_+ : L_+^p(\Gamma, w) \to E^1(D_+)$ is the Cauchy operator given by (6.2). Since the map $g \mapsto (C_+ g)(0)$ is a bounded linear functional on $L_+^p(\Gamma, w)$, this will imply that $\operatorname{Im} T(\chi_1)$ is closed and that $\beta(T(\chi_1)) = 1$.

If $g = T(\chi_1)f_+ = \chi_1 f_+$ with $f_+ \in L_+^p(\Gamma, w)$, then $(C_+ g)(0) = \chi_1(0)f_+(0) = 0$. Conversely, suppose $g \in L_+^p(\Gamma, w)$ and $(C_+ g)(0) = 0$. We claim that $\chi_1^{-1} g \in L_+^p(\Gamma, w)$, which implies that $g \in \operatorname{Im} T(\chi_1)$ and thus proves (6.26).

Choose functions $g_n \in R(\Gamma)$ converging to g in $L^p(\Gamma, w)$. Then $h_n := Pg_n$ also converges to $g = Pg$ in $L_+^p(\Gamma, w)$. The functions $C_+ h_n - C_+ g$ are analytic in D_+ and hence, by Cauchy's formula,

$$
\begin{aligned}
|h_n(0)| &= |h_n(0) - (C_+ g)(0)| = \left| \frac{1}{2\pi i} \int_\Gamma \big(h_n(\tau) - g(\tau)\big) \tau^{-1} d\tau \right| \\
&\leq \frac{1}{2\pi} \frac{1}{\delta} \int_\Gamma \big| h_n(\tau) - g(\tau) \big| \, |d\tau|
\end{aligned}
$$

where $\delta := \min\{|\tau| : \tau \in \Gamma\}$. Consequently,

$$|h_n(0)| \leq \frac{1}{2\pi} \frac{1}{\delta} \|h_n - g\|_{p,w}^{1/p} \|w^{-1}\|_q^{1/q} = o(1)$$

and thus, $h_n - h_n(0) \to g$ in $L_+^p(\Gamma, w)$. Because, obviously, $\chi_1^{-1}(h_n - h_n(0)) \in L_+^p(\Gamma, w)$, we obtain that $\chi_1^{-1} g \in L_+^p(\Gamma, w)$. $\qquad \square$

Theorem 6.24. *Let* Γ *be a Carleson Jordan curve,* $1 < p < \infty$, $w \in A_p(\Gamma)$, *and suppose* $a \in C(\Gamma)$. *Then* $T(a)$ *is Fredholm on* $L_+^p(\Gamma, w)$ *if and only if* $a(\tau) \neq 0$ *for all* $\tau \in \Gamma$. *In that case*

$$\operatorname{Ind} T(a) = -\operatorname{wind} a$$

where $\operatorname{wind} a$ *denotes the winding number of the (naturally oriented) curve* $a(\Gamma)$ *about the origin. Thus,*

$$
\begin{aligned}
\operatorname{sp}_{\mathrm{ess}} T(a) &= a(\Gamma), \\
\operatorname{sp} T(a) &= a(\Gamma) \cup \left\{ \lambda \in \mathbf{C} \setminus a(\Gamma) : \operatorname{wind}(a - \lambda) \neq 0 \right\}.
\end{aligned}
$$

Proof. If a has no zeros on Γ then $a^{-1} \in C(\Gamma)$. By (6.23),

$$T(a^{-1})T(a) = I - H(a^{-1})H(\tilde{a}), \quad T(a)T(a^{-1}) = I - H(a)H(\tilde{a}^{-1}),$$

and since all occuring Hankel operators are compact due to Proposition 6.21, we see that $T(a^{-1})$ is a regularizer of $T(a)$. Hence, $T(a)$ is Fredholm. Conversely, if $T(a)$ is Fredholm then Theorem 6.20 implies that a cannot have zeros on Γ.

In view of Corollary 6.19, we are left with the index formula. Put $n := \operatorname{wind} a$. Since a is homotopic to χ_n within $GC(\Gamma)$, the invertible functions in $C(\Gamma)$, we deduce from what was already proved that $T(a)$ is homotopic to $T(\chi_n)$ within the set of Fredholm operators. The index is a homotopy invariant (Theorem 6.13 plus a compactness argument). Consequently, $\operatorname{Ind} T(a) = \operatorname{Ind} T(\chi_n)$. If $n \geq 0$, then $T(\chi_n) = (T(\chi_1))^n$ and using Theorem 6.12 and Lemma 6.23 we obtain that

$$\operatorname{Ind} T(a) = n \operatorname{Ind} T(\chi_1) = -n = -\operatorname{wind} a.$$

If $n < 0$, then $\chi_n \in H_-^\infty(\Gamma)$, whence $T(\chi_n)T(\chi_{|n|}) = I$ by (6.25). Again employing Theorem 6.12 and Lemma 6.23, we get

$$
\begin{aligned}
\operatorname{Ind} T(a) = \operatorname{Ind} T(\chi_n) &= \operatorname{Ind} I - \operatorname{Ind} T(\chi_{|n|}) \\
&= 0 - |n| \operatorname{Ind} T(\chi_1) = |n| = -n = -\operatorname{wind} a. \qquad \square
\end{aligned}
$$

6.9 Classical Toeplitz matrices

Let $L_+^p(\Gamma, w) = L_+^2(\mathbf{T})$ be the classical Hardy space on the complex unit circle. We know from Section 6.4 that in this case $T(a)$ may be given by the Toeplitz matrix (6.17). The projection P is now the orthogonal projection of $L^2(\mathbf{T})$ onto $L_+^2(\mathbf{T})$ and hence $\|P\| = 1$. This implies that

$$\|T(a)\| \leq \|a\|_\infty \quad \text{for every} \ a \in L^\infty(\mathbf{T}). \tag{6.27}$$

One can show that in (6.27) actually equality holds, but the next result is a consequence of solely (6.27).

A function $a \in L^\infty(\mathbf{T})$ is said to be *sectorial* if its essential range $\mathcal{R}(a)$ is contained in some open half-plane whose boundary passes through the origin.

Proposition 6.25 (Brown-Halmos). *If $a \in L^\infty(\mathbf{T})$ is sectorial then $T(a)$ is invertible on $L^2_+(\mathbf{T})$. Equivalently, for every $a \in L^\infty(\mathbf{T})$ we have the inclusion*

$$\operatorname{sp} T(a) \subset \operatorname{conv} \mathcal{R}(a), \tag{6.28}$$

where $\operatorname{conv} \mathcal{R}(a)$ *denotes the convex hull of* $\mathcal{R}(a)$.

Proof. If a is sectorial, then there are $\delta > 0$ and $c \in \mathbf{T}$ such that $\|1 - \delta ca\|_\infty < 1$. So (6.27) implies that

$$\|I - \delta c T(a)\|_\infty = \|T(1 - \delta ca)\|_\infty \le \|1 - \delta ca\|_\infty < 1,$$

which shows that $\delta c T(a)$ and thus the operator $T(a)$ itself are invertible. Finally, since $\lambda \notin \operatorname{conv} \mathcal{R}(a)$ if and only if $a - \lambda$ is sectorial, we get (6.28). $\qquad\square$

Let E be a measurable subset of $\mathbf{T}$ and denote by χ_E the characteristic function of E. Proposition 6.25 tells us that $\operatorname{sp} T(\chi_E) \subset [0, 1]$. The following proposition shows that actually $\operatorname{sp} T(\chi_E) = [0, 1]$.

Proposition 6.26 (Hartman-Wintner). *Let $a \in L^\infty(\mathbf{T})$ be a real-valued function. Then*

$$\operatorname{sp}_{\mathrm{ess}} T(a) = \operatorname{sp} T(a) = [m, M]$$

where $m := \operatorname*{ess\,inf}_{\tau \in \mathbf{T}} a(\tau)$ *and* $M := \operatorname*{ess\,sup}_{\tau \in \mathbf{T}} a(\tau)$.

Proof. By Proposition 6.25, it remains to show that $[m, M] \subset \operatorname{sp}_{\mathrm{ess}} T(a)$. From Theorem 6.20 we deduce that $\{m, M\} \subset \operatorname{sp}_{\mathrm{ess}} T(a)$. So let $m < \lambda < M$, put $b := a - \lambda$, and assume $T(b)$ is Fredholm with index $\varkappa$. Since $T(b)^* = T(\bar{b})$ where $\bar{b}(\tau) := \overline{b(\tau)}$ (which follows, for example, from the matrix representation (6.17)) and $\bar{b} = b$ in the case at hand, we have

$$\varkappa = \operatorname{Ind} T(b) = -\operatorname{Ind} T(b)^* = -\operatorname{Ind} T(b) = -\varkappa,$$

that is, $\varkappa = 0$. Now Corollary 6.19 implies that $T(b)$ is invertible. Let $g \in L^2_+(\mathbf{T})$ be the solution of the equation $T(b)g = 1$. Then by $bg = 1 + f$ with $f \in \dot{L}^2_-(\mathbf{T})$ and hence, for $n \ge 1$,

$$\int_0^{2\pi} b(e^{i\theta}) |g(e^{i\theta})|^2 e^{-in\theta} \, d\theta = \int_0^{2\pi} b(e^{i\theta}) g(e^{i\theta}) \overline{g(e^{i\theta})} e^{-in\theta} \, d\theta$$

$$= \int_0^{2\pi} \left(1 + f(e^{i\theta})\right) \overline{g(e^{i\theta})} e^{-in\theta} \, d\theta = 0$$

because the Fourier coefficients with positive indices of $(1 + f)\bar{g} \in L^2_-(\mathbf{T})$ vanish. Since $b|g|^2$ is real-valued, it follows that

$$\int_0^{2\pi} b(e^{i\theta})|g(e^{i\theta})|^2 e^{-in\theta} = 0 \quad \text{for all } n \in \mathbf{Z} \setminus \{0\}$$

and thus, $b|g|^2$ must be constant. But this is impossible, as $b|g|^2 = (a - \lambda)|g|^2$ takes on positive as well as negative values on sets of positive measure.   □

Given an (oriented) rectifiable Jordan curve Γ, we define $PC(\Gamma)$ as the set of all $a \in L^\infty(\Gamma)$ for which the one-sided limits

$$a(t \pm 0) := \lim_{\tau \to t \pm 0} a(\tau)$$

exist at each point $t \in \Gamma$; here $\tau \to t - 0$ means that τ approaches t following the orientation of Γ, while $\tau \to t + 0$ means that τ goes to t in the opposite direction. Functions in $PC(\Gamma)$ are called *piecewise continuous* functions. We remark that a function $a \in PC(\Gamma)$ may have at most countably many jumps and that for each $\delta > 0$ the set $\{t \in \Gamma : |a(t + 0) - a(t - 0)| > \delta\}$ is finite. We also note that $PC(\Gamma)$ is a C^*-subalgebra of $L^\infty(\Gamma)$.

If E is a subarc of $\mathbf{T}$, then $\chi_E \in PC(\mathbf{T})$. Proposition 6.26 shows that both the spectrum and the essential spectrum of $T(\chi_E)$ on $L^2_+(\mathbf{T})$ are the segment $[0, 1]$. The following proposition identifies these spectra for arbitrary functions in $PC(\mathbf{T})$ with only a single jump.

Proposition 6.27. *Let $a \in PC(\mathbf{T})$ have at most one jump, at $t \in \mathbf{T}$, say. Denote by $a^\#$ the closed, continuous, and naturally oriented curve which results from the (essential) range of a by filling in the line segment*

$$[a(t - 0), a(t + 0)].$$

Then $T(a)$ is Fredholm on $L^2_+(\mathbf{T})$ if and only if $0 \notin a^\#$, in which case $\operatorname{Ind} T(a)$ equals $-\operatorname{wind} a^\#$.

Proof. Suppose $0 \notin a^\#$. Let $b \in C(\mathbf{T} \setminus \{t\})$ be any function such that

$$b(t \pm 0) = a(t \pm 0), \ b(\mathbf{T}) \subset [a(t + 0), a(t - 0)].$$

Then b is sectorial and $a/b =: c$ is continuous and has no zeros on $\mathbf{T}$. We have $a = bc$, and so Corollary 6.22 implies that $T(a) = T(b)T(c) + K$ with some compact operator K. Since $T(b)$ is invertible by Proposition 6.25 and $T(c)$ is Fredholm by Theorem 6.24, we deduce from Theorems 6.12 and 6.13 that $T(a)$ is Fredholm.

We now prove the index formula. For $x \in [0, 1]$, change a to a function a_x such that $\mathcal{R}(a_x)$ is the union of $\mathcal{R}(a)$ and the segment $[a(t-0), a(t+0)(1-x) + a(t+0)x]$.

Clearly, this can be managed so that $a_x \in PC(\mathbf{T}) \cap C(\mathbf{T} \setminus \{t\})$, $a_0 = a$, the map $x \mapsto a_x$ is a continuous map of $[0,1]$ into $PC(\mathbf{T})$, and

$$a_x(t - 0) = a(t - 0)(1 - x) + a(t + 0)x, \quad a_x(t + 0) = a(t + 0).$$

Then a_1 is a continuous function with wind $a_1 = $ wind $a^{\#}$. From (6.27) we infer that the map $[0,1] \to \mathcal{B}(L_+^2(\mathbf{T}))$, $x \mapsto T(a_x)$ is continuous. By what was proved in the previous paragraph, the values of this map are Fredholm operators. From the homotopy invariance of the index and Theorem 6.24 we therefore get

$$\operatorname{Ind} T(a) = \operatorname{Ind} T(a_0) = \operatorname{Ind} T(a_1) = -\operatorname{wind} a_1 = -\operatorname{wind} a^{\#}.$$

Finally, assume $0 \in a^{\#}$ but $T(a)$ is Fredholm of index $\varkappa$. Slightly perturbing a we may obtain two functions b and c in $PC(\mathbf{T}) \cap C(\mathbf{T} \setminus \{t\})$ such that

$$0 \notin b^{\#}, \; 0 \notin c^{\#}, \; |\operatorname{wind} b^{\#} - \operatorname{wind} c^{\#}| = 1$$

and $\|b - a\|_{\infty}$, $\|c - a\|_{\infty}$ are as small as desired. It follows that $T(b)$ and $T(c)$ are Fredholm with different indices. However, this is impossible, since, by Theorem 6.13, the index $\varkappa$ of a Fredholm operator is stable under small perturbations. This contradiction completes the proof. $\qquad \square$

6.10 Separation of discontinuities

The following theorem reduces the case of finitely many discontinuities to the case of only a single discontinuity. On the one hand, this theorem is a baby version of Theorem 6.30, which will also be applicable to symbols with an infinite number of discontinuities. On the other hand, this theorem is the main tool for gaining information about the index (and thus about invertibility) of $T(a)$ from "local data"; notice that Theorem 6.30 will say nothing about the index.

Theorem 6.28. *Let Γ be a Carleson Jordan curve, $1 < p < \infty$, $w \in A_p(\Gamma)$, and $a \in L^{\infty}(\Gamma)$. Suppose $a = a_1 \ldots a_n$ where $a_1, \ldots, a_n$ are functions in $L^{\infty}(\Gamma)$ and suppose each point $t \in \Gamma$ possesses an open neighborhood $U_t \subset \Gamma$ such that at most one of the functions $a_1, \ldots, a_n$ has a discontinuity in U_t. Then $T(a) - T(a_1) \ldots T(a_n)$ is compact on $L^p(\Gamma, w)$. If $T(a_1), \ldots, T(a_n)$ are Fredholm, then so also is $T(a)$ and*

$$\operatorname{Ind} T(a) = \operatorname{Ind} T(a_1) + \ldots + \operatorname{Ind} T(a_n). \tag{6.29}$$

Proof. Clearly, it suffices to prove the theorem for $n = 2$. So assume $a, b \in L^{\infty}(\Gamma)$ and each point $t \in \Gamma$ has a neighborhood U_t such that $a|U_t$ or $b|U_t$ is continuous. Since Γ is compact, there is a finite collection $\{U_{t_i}\}$ of open neighborhoods covering Γ such that $a|U_{t_i}$ or $b|U_{t_i}$ is continuous. Let $\sum g_i = 1$ be a subordinate continuous partition of unity (see e.g. [181, 6.20]). Obviously, each g_i can be represented as $g_i = f_i^2$ where f_i is also continuous. By virtue of (6.23) we have

$$T(ab) - T(a)T(b) = \sum \left[T(af_i^2 b) - T(a)T(f_i^2)T(b) \right]$$

$$= \sum \left[T(af_i f_i b) - T(a)T(f_i)T(f_i)T(b) - T(a)H(f_i)H(\tilde{f}_i)T(b) \right]$$

$$= \sum \left[T(af_i)T(f_i b) + H(af_i)H\big((f_i b)^\sim\big) \right.$$
$$\left. - \big(T(af_i) - H(a)H(\tilde{f}_i)\big)\big(T(f_i b) - H(f_i)H(\tilde{b})\big) \right.$$
$$\left. - T(a)H(f_i)H(\tilde{f}_i)T(b) \right]$$

$$= \sum \left[H(af_i)H\big((f_i b)^\sim\big) - H(a)H(\tilde{f}_i)H(f_i)H(\tilde{b}) \right.$$
$$\left. + T(af_i)H(f_i)H(\tilde{b}) + H(a)H(\tilde{f}_i)T(f_i b) \right.$$
$$\left. - T(a)H(f_i)H(\tilde{f}_i)T(b) \right].$$

Since af_i or $f_i b$ is continuous, each term of the latter sum contains a Hankel operator with a continuous symbol and is therefore compact by Proposition 6.21. This shows that $T(ab) - T(a)T(b)$ is compact.

The remaining assertions follow from Theorems 6.12 and 6.13.   □

Corollary 6.29. *Let $a \in PC(\mathbf{T})$ have at most finitely many jumps and denote by $a^\#$ the continuous, closed, and naturally oriented curve which is obtained from the (essential) range of a by filling in the line segment $[a(t-0), a(t+0)]$ between the endpoints $a(t-0)$, $a(t+0)$ of each jump. The operator $T(a)$ is Fredholm on $L_+^2(\mathbf{T})$ if and only if $0 \notin a^\#$, in which case $\operatorname{Ind} T(a) = -\operatorname{wind} a^\#$.*

Proof. Let $0 \notin a^\#$ and let $t_1, \ldots, t_n$ be the points at which a has a jump. Choose functions $b_j \in PC(\mathbf{T}) \cap C(\mathbf{T} \setminus \{t_j\})$ such that

$$b_j(t_j \pm 0) = a(t_j \pm 0), \quad b_j(\mathbf{T}) \subset \big[a(t_j + 0), a(t_j - 0)\big].$$

Then $a = b_1 \ldots b_n c$ with some continuous function c having no zeros on $\mathbf{T}$. Combining Proposition 6.27 and Theorem 6.28 we arrive at the conclusion that $T(a)$ is Fredholm. The index formula can be verified by a homotopy argument as in the proof of Proposition 6.27, and then one can show as in the proof of Proposition 6.27 that $0 \notin a^\#$ whenever $T(a)$ is Fredholm.   □

6.11 Localization

Let Γ be a Carleson Jordan curve, $1 < p < \infty$, and $w \in A_p(\Gamma)$. Two functions $a, b \in L^\infty(\Gamma)$ are said to be *locally equivalent* at a point $t \in \Gamma$ if

$$\inf \left\{ \big\|(a-b)c\big\|_\infty : c \in C(\Gamma),\ c(t) = 1 \right\} = 0. \tag{6.30}$$

Obviously, if a and b are continuous, then a and b are locally equivalent at a point $t \in \Gamma$ if and only if $a(t) = b(t)$. Consequently, continuous functions are locally equivalent to constant functions. In case a and b are in $PC(\Gamma)$, they are locally equivalent at $t \in \Gamma$ if and only if $a(t \pm 0) = b(t \pm 0)$. In particular, a piecewise continuous function a is locally equivalent to the function

$$a(t - 0) + \big(a(t + 0) - a(t - 0)\big)\chi_t \tag{6.31}$$

where χ_t is the characteristic function of some semi-neighborhood of the point t and $\chi_t(t - 0) = 0$, $\chi_t(t + 0) = 1$.

Theorem 6.30 (Simonenko). *Let Γ be a Carleson Jordan curve, $1 < p < \infty$, and $w \in A_p(\Gamma)$. Let $a \in L^\infty(\Gamma)$ and suppose for each $t \in \Gamma$ we are given a function $a_t \in L^\infty(\Gamma)$ which is locally equivalent to a at t. If the operators $T(a_t)$ are Fredholm on $L^p_+(\Gamma, w)$ for all $t \in \Gamma$, then $T(a)$ is also Fredholm on $L^p_+(\Gamma, w)$.*

Proof. Put $\mathcal{B} := \mathcal{B}(L^p_+(\Gamma, w))$, $\mathcal{K} = \mathcal{K}(L^p_+(\Gamma, w))$, and for $A \in \mathcal{B}$ denote by A^π the coset $A + \mathcal{K}$ in the Calkin algebra $\mathcal{B}/\mathcal{K}$.

It is easily seen that (6.30) is equivalent to the requirement that for every $\varepsilon > 0$ there exists a function $c \in C(\Gamma)$ which is identically 1 on some open arc of Γ containing t, which is nonnegative on Γ, and for which $\|(a - b)c\|_\infty < \varepsilon$. Now fix $t \in \Gamma$. Since $T(a_t)$ is Fredholm, there is an $R_t \in \mathcal{B}$ such that $R_t^\pi T^\pi(a_t) = I^\pi$, and since a is locally equivalent to a_t at t, we can find a function $c_t \in C(\Gamma)$ such that $c|\gamma_t = 1$ for some open arc γ_t containing t, $c \ge 0$ on Γ, and

$$\big\|(T^\pi(a) - T^\pi(a_t))T^\pi(c_t)\big\| = \big\|T^\pi\big((a - a_t)c_t\big)\big\|$$
$$\le \|P\|\,\big\|(a - a_t)c\big\|_\infty < 1/\|R_t^\pi\| \tag{6.32}$$

(recall Corollary 6.22 for the first equality). Let $b_t \in C(\Gamma)$ be a nonnegative function such that $b_t(t) = 1$ and the support of b_t is a subset of γ_t. Then $c_t b_t = b_t$, hence $T^\pi(c_t)T^\pi(b_t) = T^\pi(b_t)$ by Corollary 6.22, and thus

$$
\begin{aligned}
R_t^\pi T^\pi(a)T^\pi(b_t) &= R_t^\pi\big(T^\pi(a) - T^\pi(a_t)\big)T^\pi(b_t) + R_t^\pi T^\pi(a_t)T^\pi(b_t) \\
&= \Big\{R_t^\pi\big(T^\pi(a) - T^\pi(a_t)\big)T^\pi(c_t) + I^\pi\Big\}T^\pi(b_t).
\end{aligned}
$$

The coset in the braces is invertible due to (6.32). Denoting its inverse by F_t^π, we get

$$F_t^\pi R_t^\pi T^\pi(a)T^\pi(b_t) = T^\pi(b_t) \tag{6.33}$$

We can find finitely many points $t_1, \ldots, t_n \in \Gamma$ such that $b := b_{t_1} + \ldots + b_{t_n} \in C(\Gamma)$ is positive and thus invertible. Put

$$R := \sum_{j=1}^{n} F_{t_j} R_{t_j} T(b_{t_j}).$$

We have

$$R^\pi T^\pi(a) = \sum_{j=1}^{n} F_{t_j}^\pi R_{t_j}^\pi T^\pi(b_{t_j})T^\pi(a),$$

and since $T^\pi(b_{t_j})T^\pi(a) = T^\pi(a)T^\pi(b_{t_j})$ by Corollary 6.22, we obtain from (6.33) that

$$R^\pi T^\pi(a) = \sum_{j=1}^{n} F_{t_j}^\pi R_{t_j}^\pi T^\pi(a)T^\pi(b_{t_j}) = \sum_{j=1}^{n} T^\pi(b_{t_j}) = T^\pi(b).$$

Because $T^\pi(b^{-1})T^\pi(b) = I^\pi$ by Corollary 6.22, it follows that $T(b^{-1})R$ is a left regularizer of $T(a)$. Analogously one can prove the existence of a right regularizer. Thus, $T(a)$ is Fredholm. $\qquad\square$

Corollary 6.31. *Let a be an arbitrary function in $PC(\mathbf{T})$. Fill in line segments between the endpoints of each jump and denote the resulting continuous, closed, and naturally oriented curve by $a^\#$. The operator $T(a)$ is Fredholm on $L^2_+(\mathbf{T})$ if and only if $0 \notin a^\#$, and in this case $\operatorname{Ind} T(a) = -\operatorname{wind} a^\#$.*

Proof. Suppose $0 \notin a^\#$. If $t \in \Gamma$ and $a_t \in PC(\mathbf{T})$ is any function such that

$$a_t(t \pm 0) = a(t \pm 0), \quad a_t(\mathbf{T}) \subset [a(t+0), a(t-0)],$$

then a is locally equivalent to a_t at t. Proposition 6.25 tells us that $T(a_t)$ is invertible, and Theorem 6.30 therefore shows that $T(a)$ is Fredholm. Since a may be uniformly approximated by functions $b \in PC(\mathbf{T})$ having only finitely many jumps and satisfying $0 \notin b^\#$, $\operatorname{wind} b^\# = \operatorname{wind} a^\#$, the index formula follows from Corollary 6.29. The "only if" part can be shown as in proof of Proposition 6.27 by a perturbation argument. $\qquad\square$

6.12 Wiener-Hopf factorization

In this section we prove a theorem which completely solves the problem whether a Toeplitz operator is Fredholm in analytical language.

Let Γ be a Carleson Jordan curve, $1 < p < \infty$, and $w \in A_p(\Gamma)$. If $T(a)$ is Fredholm then, by Theorem 6.20, the function a is automatically in $GL^\infty(\Gamma)$. So let us suppose from the beginning that $a \in GL^\infty(\Gamma)$. One says that a admits a *Wiener-Hopf factorization in $L^p(\Gamma, w)$* if a can be written in the form

$$a(\tau) = a_-(\tau)\tau^\varkappa a_+(\tau) \text{ for almost all } \tau \in \Gamma \qquad (6.34)$$

where $\varkappa$ is an integer and the functions $a_\pm$ enjoy the following properties:

$$a_- \in L^p_-(\Gamma, w), \ a_-^{-1} \in L^q_-(\Gamma, w^{-1}), \ a_+ \in L^q_+(\Gamma, w^{-1}), \ a_+^{-1} \in L^p_+(\Gamma, w), \qquad (6.35)$$

$$|a_+^{-1}|w \in A_p(\Gamma). \qquad (6.36)$$

Put $\chi_\varkappa(\tau) := \tau^\varkappa$. Since $a_+^{-1} = a^{-1}a_-\chi_\varkappa$ and $a^{-1}\chi_\varkappa \in GL^\infty(\Gamma)$, condition (6.36) is equivalent to the requirement that $|a_-|w \in A_p(\Gamma)$. Moreover, (6.36) is the same as saying that P be bounded on $L^p(\Gamma, |a_+^{-1}|w)$ and hence, taking into account (6.35), we may replace (6.36) by the condition that

$$\left\|a_+^{-1}P(a_+g)\right\|_{p,w} \le C_{p,w}\|g\|_{p,w} \quad \text{for all } g \in R(\Gamma). \tag{6.37}$$

Note that in (6.37) the function a_+g belongs to $L^q(\Gamma, w^{-1})$, so that $P(a_+g)$ is a well-defined function in $L^q(\Gamma, w^{-1})$. Also notice that $R(\Gamma)$ is dense in $L^p(\Gamma, w)$, which tells us that (6.37) is equivalent to the following: $a_+^{-1}P(a_+g)$ belongs to $L^p(\Gamma, w)$ for every $g \in R(\Gamma)$ and the map $R(\Gamma) \to L^p(\Gamma, w)$, $g \mapsto a_+^{-1}P(a_+g)$ extends to a bounded operator on $L^p(\Gamma, w)$.

Theorem 6.32 (Simonenko). *Let Γ be a Carleson curve, $1 < p < \infty$, $w \in A_p(\Gamma)$, and $a \in GL^\infty(\Gamma)$. Then $T(a)$ is Fredholm on $L_+^p(\Gamma, w)$ if and only if a admits a Wiener-Hopf factorization $a = a_-\chi_\varkappa a_+$ in $L^p(\Gamma, w)$. In that case the integer $\varkappa$ is uniquely determined and $\operatorname{Ind} T(a) = -\varkappa$.*

Proof. First suppose that a admits a Wiener-Hopf factorization in $L^p(\Gamma, w)$ with $\varkappa = 0$, i.e. let $a = a_-a_+$.

We show that $\operatorname{Ker} T(a) = \{0\}$. Indeed, if $T(a)g_+ = 0$ for $g_+ \in L_+^p(\Gamma, w)$, then $a_-a_+g_+ =: g_- \in \dot{L}^p(\Gamma, w)$. We have $a_+g_+ = a_-^{-1}g_-$. Since, by (6.35) and Lemma 6.11, $a_+g_+ \in E_+^1(\Gamma)$ and $a_-^{-1}g_- \in \dot{E}^1(\Gamma)$, we deduce from Theorem 6.4 that $a_+g_+ = 0$. Because $a_+ \ne 0$ a.e. by (6.36), it follows that $g_+ = 0$, that is, $\operatorname{Ker} T(a) = \{0\}$.

We now show that $T(a)$ is surjective. By (6.37), the map

$$R(\Gamma) \cap E_+^1(\Gamma) \to L_+^p(\Gamma, w), \quad g_+ \mapsto a_+^{-1}P(a_-^{-1}g_+) = a_+^{-1}P(a_+a^{-1}g_+) \tag{6.38}$$

extends to a bounded linear operator A on $L_+^p(\Gamma, w)$ (also recall (6.13)). For g_+ in $R(\Gamma) \cap E_+^1(\Gamma)$ we have

$$\begin{aligned}
T(a)Ag_+ &= P\big(a_-a_+a_+^{-1}P(a_-^{-1}g_+)\big)\\
&= P\big(a_-P(a_-^{-1}g_+)\big) = Pg_+ - P\big(a_-Q(a_-^{-1}g_+)\big) = Pg_+ = g_+,
\end{aligned}$$

and since both $T(a)$ and A are bounded, it results that $T(a)A = I$ on $L_+^p(\Gamma, w)$. This proves that $T(a)$ is surjective.

Thus, if $\varkappa = 0$ then $T(a)$ is invertible. If a has the Wiener-Hopf factorization $a = a_-\chi_\varkappa a_+$ then, by (6.25),

$$T(a) = T(a_-a_+)T(\chi_\varkappa) \ (\varkappa > 0) \quad \text{or} \quad T(a) = T(\chi_\varkappa)T(a_-a_+) \ (\varkappa < 0).$$

We have already proved that $T(a_-a_+)$ is invertible, and Theorem 6.24 implies that $T(\chi_\varkappa)$ is Fredholm of index $-\varkappa$. Consequently, $T(a)$ is Fredholm with index $-\varkappa$ due to Theorem 6.12. As a by-product we obtain that $\varkappa$ is uniquely determined.

To prove the "only if" portion, suppose $T(a)$ is Fredholm of index $-\varkappa$. Put $b := a\chi_{-\varkappa}$. Then, again by (6.25),

$$T(b) = T(\chi_{-\varkappa})T(a) \; (\varkappa > 0) \quad \text{or} \quad T(b) = T(a)T(\chi_{-\varkappa}) \; (\varkappa \leq 0).$$

Hence, by Theorems 6.24 and 6.12, $T(b)$ is Fredholm of index zero. Now Corollary 6.19 implies that $T(b)$ is invertible. From Lemma 6.14 we infer that $bP + Q$ is invertible on $L^p(\Gamma, w)$. Lemma 6.14 also gives the invertibility of $PbP + Q$ on $L^p(\Gamma, w)$. Since

$$PbI + Q = (I + PbQ)\,(PbP + Q)$$

and $(I + PbQ)^{-1} = I - PbQ$, it follows that $PbI + Q$ is invertible on $L^p(\Gamma, w)$. Using formula (6.21), we see that

$$b^{-1}P + Q = b^{-1}(P + bQ) = b^{-1}H_\Gamma(PbI + Q)^*H_\Gamma$$

is invertible on $L^q(\Gamma, w^{-1})$. Let $\varphi \in L^p(\Gamma, w)$ and $\psi \in L^q(\Gamma, w^{-1})$ be the solutions of the equations

$$(bP + Q)\varphi = 1, \; (b^{-1}P + Q)\psi = 1$$

and put $\varphi_+ := P\varphi$, $\psi_+ := P\psi$. Then $\varphi_+ \in L^p_+(\Gamma, w)$, $\psi_+ \in L^q_+(\Gamma, w^{-1})$, and

$$b\varphi_+ = 1 + h_-, \; b^{-1}\psi_+ = 1 + f_- \qquad (6.39)$$

with $h_- \in \dot{L}^p_-(\Gamma, w)$, $f_- \in \dot{L}^q_-(\Gamma, w^{-1})$. From (6.39) we get

$$\varphi_+\psi_+ = b\varphi_+b^{-1}\psi_+ = (1 + h_-)(1 + f_-) \qquad (6.40)$$

and hence, by Theorem 6.4, $\varphi_+\psi_+$ is some constant c. The analytic extension of the right-hand side of (6.40) is 1 at infinity, which implies that $c = 1$. Thus, $\varphi_+\psi_+ = 1 = (1 + h_-)(1 + f_-)$. Put $a_+ := \varphi_+^{-1}$ and $a_- := 1 - h_-$. Then $a = b\chi_\varkappa = a_-\chi_\varkappa a_+$ by (6.39) and we have

$$a_+ = \varphi_+^{-1} = \psi_+ \in L^q_+(\Gamma, w^{-1}), \; a_+^{-1} = \varphi_+ \in L^p_+(\Gamma, w),$$
$$a_- = 1 + h_- \in L^p_-(\Gamma, w), \; a_-^{-1} = (1 + h_-)^{-1} = 1 + f_- \in L^q_-(\Gamma, w^{-1}).$$

This shows that (6.34) and (6.35) are satisfied.

It remains to verify (6.36). Let $g \in R(\Gamma)$ and put $g_+ := Pg$, $g_- := Qg$. Since

$$T(b)\big(a_+^{-1}P(a_-^{-1}g_+)\big) = P\big(ba_+^{-1}P(a_-^{-1}g_+)\big)$$
$$= P\big(a_-P(a_-^{-1}g_+)\big) = Pg_+ - P\big(a_-Q(a_-^{-1}g_+)\big) = g_+$$

and $T(b)$ is invertible, we obtain that

$$\big\|a_+^{-1}P(a_-^{-1}g_+)\big\|_{p,w} \leq \|(T(b))^{-1}\| \, \|g_+\|_{p,w}.$$

Because $P(a_-^{-1}g_-) = 0$, it follows that

$$\left\| a_+^{-1} P(a_-^{-1}g) \right\|_{p,w} \le \left\| (T(b))^{-1} \right\| \, \|g_+\|_{p,w} \le \left\| (T(b))^{-1} \right\| \, \|P\| \, \|g\|_{p,w}$$

for all $g \in R(\Gamma)$. As $a_+^{-1} P a_+ I = a_+^{-1} P a_-^{-1} b I$, we finally get

$$\left\| a_+^{-1} P(a_+ g) \right\|_{p,w} = \left\| a_+^{-1} P a_-^{-1}(bg) \right\|_{p,w} \le \left\| (T(b))^{-1} \right\| \, \|P\| \, \|b\|_\infty \|g\|_{p,w}$$

whenever $bg \in R(\Gamma)$, which proves (6.37) and thus (6.36).   □

From the proof of the previous theorem we see that if $a \in GL^\infty(\Gamma)$ admits a Wiener-Hopf factorization $a = a_- a_+$ in $L^p(\Gamma, w)$, then the inverse operator of $T(a)$ is the extension to $L_+^p(\Gamma, w)$ of the operator given by (6.38).

Of course, given a function $a \in GL^\infty(\Gamma)$ it is by no means an easy task to decide whether it admits a Wiener-Hopf factorization in $L^p(\Gamma, w)$. For example, it is not at all obvious why a function $a \in PC(\mathbf{T}) \cap GL^\infty(\mathbf{T})$ admits such a factorization in $L^2(\mathbf{T})$ if and only if $0 \notin a^{\#}$. However, Theorem 6.32 becomes a powerful tool in combination with Theorem 6.30. The strategy is to choose the "local representatives" a_t as simple as possible, namely so that a Wiener-Hopf factorization of a_t may be constructed due to the special structure of a_t. We renounce giving an example here – the entire next chapter rests on this strategy and demonstrates the advantages of this approach as well as the obstacles one has to overcome when going this way.

6.13 Notes and comments

6.1–6.2. Detailed discussions of Smirnov classes and weighted Hardy spaces are in the books by Privalov [162], Duren [59], Gohberg and Krupnik [89], or Litvinchuk and Spitkovsky [137], for example. We learned Example 6.9 from [89]. For $\mu \in (0, 1)$, Lemma 6.10 is in [89, Chapter 2, Theorem 4.8]. The observation that Lemma 6.10 is also true for $\mu \in [1, \infty)$ and the proof given here are due to Grudsky [93]. We remark that for our purposes we really need Lemma 6.10 for $\mu \in [1, \infty)$.

6.3. Proofs of all results of this section can be found in Gohberg and Krupnik's book [89], in Mikhlin and Prössdorf's monograph [147], or in Gohberg, Goldberg, and Kaashoek's text [80].

6.4–6.9. Standard texts on one-dimensional Hardy space Toeplitz operators are the following books (in alphabetical order): Böttcher and Silbermann [23], Douglas [52], Gohberg and Krupnik [89], Mikhlin and Prössdorf [147], N.K. Nikolski [152]. Of course, the monographs by Clancey and Gohberg [33] and Litvinchuk and Spitkovsky [137] are essentially also books on (block) Toeplitz operators. These operators also play a dominant role in the two volumes by Gohberg, Goldberg, and Kaashoek [80].

The books [52] and [152] deal with Toeplitz operators on $L_+^2(\mathbf{T})$, in [23] the main focus is on operators on the spaces $L_+^p(\mathbf{T}, \varrho)$ with power weights ϱ, and [89],

[147] contain a complete theory for operators on the spaces $L_+^p(\Gamma, \varrho)$ with piecewise Lyapunov curves Γ and power weights ϱ. In Sections 6.4 to 6.9, we do nothing but checking whether some basic results of these books can be carried over to operators on $L_+^p(\Gamma, w)$ in case Γ is a Carleson curve and w is a Muckenhoupt weight.

Proposition 6.16 is in [89, Chapter 1, Theorem 7.1]. Theorem 6.17 and its two corollaries go back to Coburn [34] ($\Gamma = \mathbf{T}$, $p = 2$, $w = 1$) and Simonenko [196]; for Toeplitz operators with continuous symbols, such results were already established by Gohberg [78]. As for Theorem 6.20, we remark that Hartman and Wintner [99] proved that $\mathcal{R}(a) \subset \mathrm{sp}\,T(a)$ (in case $\Gamma = \mathbf{T}$ and $w = 1$), while Simonenko [196] showed that if $\lambda \in \mathcal{R}(a)$, then $T(a) - \lambda I$ cannot be semi-Fredholm.

Proposition 6.21 is essentially due to Mikhlin [146] (also see [89] and [147]). The compact Hankel operators on $L_+^2(\mathbf{T})$ were characterized by Hartman [98]. The result of Hartman is the starting point of the theory of Toeplitz operators with $C + H^\infty$ symbols (see [23], [52], [152], [161]).

Theorem 6.24 has a long history. That $T(a)$ is Fredholm of index $-\mathrm{wind}\,a$ whenever $a \in C(\Gamma)$ has no zeros is more or less explicit in works by F. Noether, S.G. Mikhlin, N.I. Muskhelishvili, F.D. Gakhov, V.V. Ivanov, M. Krein, A.P. Calderón, F. Spitzer, H. Widom, A. Devinatz, G. Fichera, and certainly others. In the form cited here (and under the assumption that Γ is a nice curve and w is at most a power weight), Theorem 6.24 was established by Gohberg [77], [78], Khvedelidze [122], [123], and Simonenko [191], [192]. The case of general weights $w \in A_p(\Gamma)$ was disposed of by Spitkovsky [200], and for general Carleson curves Γ and general weights $w \in A_p(\Gamma)$ the theorem was probably first explicitly stated in [15].

Proposition 6.25 was established by Simonenko [192], Widom [208], Brown and Halmos [29], and Devinatz [51]. Proposition 6.26 is Hartman and Wintner's [99].

Singular integral equations with piecewise Hölder continuous coefficients in classes of Hölder functions were already considered by Muskhelishvili [151] and Gakhov [72]. We remark in this connection that an equation with continuous coefficients on a bounded simple arc is equivalent to an equation with piecewise continuous coefficients on a Jordan curve. It was Khvedelidze [122], [123] who first studied equations with piecewise continuous coefficients on weighted Lebesgue spaces. Proposition 6.27 and its generalizations, Corollaries 6.29 and 6.31, are the culmination of the classical spectral theory of Toeplitz operators on $L_+^2(\mathbf{T})$. They were independently discovered by Simonenko [192], Widom [207], Devinatz [51], Shamir [189], and Gohberg [79].

6.10. Theorem 6.28 is due to Gohberg and Sementsul [92].

6.11. The foundation for the treatment of Toeplitz and singular integral operators with the help of local methods was laid by Simonenko [195], [196]. Theorem 6.30 is already in Simonenko's paper [192] and was independently also found by Douglas and Sarason [54]. Simonenko's method was elaborated by Gohberg and

Krupnik [89] to a powerful and easy-to-use local principle for studying invertibility in Banach algebras with nontrivial center (also see Douglas [52]).

6.12. The method of Wiener-Hopf factorization was introduced by N. Wiener and E. Hopf in 1931. What we call Wiener-Hopf factorization has its origin in the work by Gakhov [71], who solved the Riemann-Hilbert problem with Hölder continuous coefficients on smooth Jordan curves in this way (also see his book [72]). Note that for the homogeneous Riemann-Hilbert problem with a coefficient of vanishing winding number, this approach was already employed by Plemelj [159]. Consideration of more general classes of coefficients and of other spaces as well as the study of Wiener-Hopf integral equations necessitated essential modifications of the factorizations prevailing in the classical monographs by Gakhov [72] and Muskhelishvili [151]. For example, one had to work with factorizations whose factors are no longer functions in $L^\infty(\Gamma)$, which led to such conditions as (6.36).

Mark Krein [130] was the first to understand the operator theoretic essence and the Banach algebraic background of Wiener-Hopf factorization and to present the method in a crystal-clear manner. Theorem 6.32 was established by Simonenko [194], [196]. Papers by Widom [207], [208], [209] and Devinatz [51] contain results very close to Theorem 6.32. For an exhaustive discussion of Wiener-Hopf factorization we refer to the monographs by Gohberg and Krupnik [89], Clancey and Gohberg [33], and Litvinchuk and Spitkovsky [137].

Chapter 7
Piecewise continuous symbols

This chapter is the heart of the book. By first employing a localization theorem and subsequently constructing a Wiener-Hopf factorization for the symbols of the local representatives, we will completely identify the essential spectra of Toeplitz operators with piecewise continuous symbols. We know from the preceding chapter that the essential spectrum of a classical Toeplitz operator is the union of the essential range of the symbol and of line segments joining the endpoints of each jump. We will show that in the general case these line segments metamorphose into circular arcs, logarithmic double spirals, horns, spiralic horns, and eventually into what we call leaves. The shape of the leaves can be described in terms of the indicator functions.

7.1 Local representatives

Let Γ be a Carleson Jordan curve and fix $t \in \Gamma$. We first construct functions $g_{t,\gamma} \in PC(\Gamma)$ such that every function $b \in PC(\Gamma) \cap GL^\infty(\Gamma)$ is locally equivalent (in the sense of Section 6.11) to a nonzero multiple of $g_{t,\gamma}$ for some appropriate choice of the parameter γ.

The curve Γ divides the plane into a bounded component D_+ and an unbounded component D_-. Without loss of generality assume that the origin belongs to D_+. Let $\varphi : \mathbf{D} \to D_+$ and $\psi : \mathbf{C} \setminus \overline{\mathbf{D}} \to D_-$ be Riemann maps. Because Γ is locally connected, φ and ψ extend to homeomorphisms $\varphi : \overline{\mathbf{D}} \to \overline{D_+}$ and $\psi : \mathbf{C} \setminus \mathbf{D} \to \overline{D_-}$. We may suppose that $\varphi(1) = \psi(1) = t$, $\varphi(0) = 0$, $\psi(\infty) = \infty$. Put $\Lambda_0 := \varphi([0,1])$ and $\Lambda_\infty := \psi([1,\infty])$, where $[1,\infty] = [1,\infty) \cup \{\infty\}$. The curve $\Lambda_0 \cup \Lambda_\infty$ joins 0 to ∞ and meets Γ at exactly one point, namely t. Let $\arg z$ be any continuous branch of the argument in $\mathbf{C} \setminus (\Lambda_0 \cup \Lambda_\infty)$. For $\gamma \in \mathbf{C}$, define

$$z^\gamma := |z|^\gamma e^{i\gamma \arg z} \quad (z \in \mathbf{C} \setminus (\Lambda_0 \cup \Lambda_\infty)).$$

Thus, z^γ is an analytic function in $\mathbf{C} \setminus (\Lambda_0 \cup \Lambda_\infty)$. The restriction of z^γ to $\Gamma \setminus \{t\}$ will be denoted by $g_{t,\gamma}$. Clearly, $g_{t,\gamma}$ is continuous on $\Gamma \setminus \{t\}$, $g_{t,\gamma} \in PC(\Gamma)$, and

$$\frac{g_{t,\gamma}(t+0)}{g_{t,\gamma}(t-0)} = e^{i\gamma(\arg(t+0)-\arg(t-0))} = e^{-2\pi i \gamma}.$$

If b is a function in $PC(\Gamma)$ such that $b(t \pm 0) \neq 0$, we define $\gamma \in \mathbf{C}$ by

$$\operatorname{Re} \gamma := \frac{1}{2\pi} \arg \frac{b(t-0)}{b(t+0)}, \quad \operatorname{Im} \gamma := -\frac{1}{2\pi} \log \left| \frac{b(t-0)}{b(t+0)} \right|; \tag{7.1}$$

here we take any argument of $b(t-0)/b(t+0)$, which implies that any two choices of $\operatorname{Re} \gamma$ differ by an integer only. We have

$$e^{-2\pi i \gamma} = b(t+0)/b(t-0),$$

and hence, there is a constant $c \in \mathbf{C} \setminus \{0\}$ such that

$$b(t+0) = c g_{t,\gamma}(t+0), \quad b(t-0) = c g_{t,\gamma}(t-0), \tag{7.2}$$

which means that b is locally equivalent to $c g_{t,\gamma}$ at the point t.

For each $\varkappa \in \mathbf{Z}$, the function $g_{t,\gamma}$ admits the factorization

$$g_{t,\gamma}(\tau) = (1 - t/\tau)^{\varkappa - \gamma} \tau^\varkappa (\tau - t)^{\gamma - \varkappa} \quad (\tau \in \Gamma \setminus \{t\}) \tag{7.3}$$

with appropriate branches of $(1 - t/\tau)^{\varkappa - \gamma}$ and $(\tau - t)^{\gamma - \varkappa}$. Indeed, let $\arg z$ for $z \in \mathbf{C} \setminus (\Lambda_0 \cup \Lambda_\infty)$ be as above, take any continuous branch of $\arg(z - t)$ for $z \in \mathbf{C} \setminus \Lambda_\infty$, define

$$\arg(1 - t/z) = \arg((z-t)/z) := \arg(z-t) - \arg z \tag{7.4}$$

for $z \in \mathbf{C} \setminus (\Lambda_0 \cup \Lambda_\infty)$, and then put

$$(z - t)^\eta := |z - t|^\eta e^{i\eta \arg(z-t)}, \quad (1 - t/z)^\eta := |1 - t/z|^\eta e^{i\eta \arg(1-t/z)}$$

for $\eta \in \mathbf{C}$ and $z \in \mathbf{C} \setminus (\Lambda_0 \cup \Lambda_\infty)$. Since $\arg(1 - t/z)$ can be continuously continued across Λ_∞, the function $(1 - t/z)^\eta$ is well-defined for all $z \in \mathbf{C} \setminus \Lambda_0$. Obviously, $(z-t)^\eta$ and $(1-t/z)^\eta$ are analytic and nonzero in $\mathbf{C} \setminus \Lambda_\infty$ and $\mathbf{C} \setminus \Lambda_0$, respectively, and these functions are continuous on $D_+ \cup (\Gamma \setminus \{t\})$ and $D_- \cup (\Gamma \setminus \{t\})$, respectively. If $\tau \in \Gamma \setminus \{t\}$ then, by (7.4),

$$\begin{aligned}
&(1 - t/\tau)^{\varkappa - \gamma} \tau^\varkappa (\tau - t)^{\gamma - \varkappa} \\
&= |1 - t/\tau|^{\varkappa - \gamma} e^{i(\varkappa - \gamma) \arg(1 - t/\tau)} |\tau|^\varkappa e^{i\varkappa \arg \tau} |\tau - t|^{\gamma - \varkappa} e^{i(\gamma - \varkappa) \arg(\tau - t)} \\
&= |\tau|^\gamma e^{-i(\varkappa - \gamma) \arg \tau} e^{i\varkappa \arg \tau} = \tau^\gamma = g_{t,\gamma}(\tau),
\end{aligned}$$

which proves (7.3).

Lemma 7.1. *Let γ be a Carleson Jordan curve, $w \in A_p(\Gamma)$, and $t \in \Gamma$. Denote by α_t and β_t the indicator functions of Γ, p, w at $t \in \Gamma$. For $x, y \in \mathbf{R}$, put*

$$\varrho_-(\tau) := (1 - t/\tau)^{y+ix}, \quad \varrho_+(\tau) := (\tau - t)^{y+ix} \quad (\tau \in \Gamma \setminus \{t\}).$$

Then $|\varrho_-|w$ is a weight in $A_p(\Gamma)$ if and only if

$$-1/p - \alpha_t(x) < y < 1/q - \beta_t(x). \tag{7.5}$$

If (7.5) holds, then $\varrho_- \in L^p_-(\Gamma, w)$ and $\varrho_-^{-1} \in L^q_-(\Gamma, w^{-1})$. Analogously, $|\varrho_+|w$ belongs to $A_p(\Gamma)$ if and only if (7.5) is valid, in which case $\varrho_+ \in L^p_+(\Gamma, w)$ and $\varrho_+^{-1} \in L^q_+(\Gamma, w).$

Proof. Clearly, $|\varrho_+|w \in A_p(\Gamma)$ if and only if the point $y + ix$ belongs to the indicator set N_t, which, by Theorem 3.21, is equivalent to (7.5). If (7.5) is satisfied, then $|\varrho_+|w \in A_p(\Gamma)$ and hence, $|\varrho_+|w \in L^p(\Gamma)$ and $|\varrho_+|^{-1}w^{-1} \in L^q(\Gamma)$, that is, $\varrho_+ \in L^p(\Gamma, w)$ and $\varrho_+^{-1} \in L^q(\Gamma, w^{-1})$. To show that actually $\varrho_+ \in L^p_+(\Gamma, w)$ and $\varrho_+^{-1} \in L^q_+(\Gamma, w^{-1})$, we employ Lemma 6.10.

The function ϱ_+ is analytic in D_+ and continuous on $D_+ \cup (\Gamma \setminus \{t\})$. For $z \in D_+$,

$$|\varrho_+(z)| = |(z - t)^{y+ix}| = |z - t|^y e^{-x \arg(z-t)}. \tag{7.6}$$

Fix a small $\varepsilon > 0$ and put

$$D^+_\varepsilon(t) := D_+ \cap \{z \in \mathbf{C} : |z - t| < \varepsilon\}, \quad \Gamma^+_\varepsilon(t) := \partial D^+_\varepsilon(t) \setminus \Gamma.$$

Obviously, $\Gamma^+_\varepsilon(t)$ is an at most countable family of subarcs of the circle $\{z \in \mathbf{C} : |z - t| = \varepsilon\}$ and is therefore of the form

$$\Gamma^+_\varepsilon(t) = \bigcup_j \Gamma^+_{\varepsilon,j}(t), \quad \Gamma^+_{\varepsilon,j}(t) = \{t + \tau_j(\varepsilon)e^{i\varphi} : 0 < \varphi < \varphi_j\}$$

with certain $\tau_j(\varepsilon) \in \Gamma$ satisfying $|\tau_j(\varepsilon) - t| = \varepsilon$. If $z \in \Gamma^+_{\varepsilon,j}(t)$ then

$$\arg\left(\tau_j(\varepsilon) - t\right) < \arg(z - t) < \arg\left(\tau_j(\varepsilon) - t\right) + 2\pi,$$

and because $|z - t| = |\tau_j(\varepsilon) - t| = \varepsilon$, we obtain

$$\frac{\arg(\tau_j(\varepsilon) - t)}{-\log|\tau_j(\varepsilon) - t|} < \frac{\arg(z - t)}{-\log|z - t|} < \frac{\arg(\tau_j(\varepsilon) - t) + 2\pi}{-\log|\tau_j(\varepsilon) - t|}.$$

Thus, Theorem 1.10 implies that there is a constant $M \in (0, \infty)$ independent of j and ε such that

$$\left|\arg(z - t)/(-\log|z - t|)\right| \leq M \tag{7.7}$$

for all $z \in \Gamma_{\varepsilon,j}^+(t)$, all j, and all sufficiently small $\varepsilon > 0$. Consequently, (7.7) holds for all $z \in D_+$ sufficiently close to t. Combining (7.6) and (7.7) we get

$$|\varrho_+(z)| \leq |z-t|^y e^{M|x|(-\log|z-t|)} = |z-t|^{y-M|x|}$$

for all $z \in D_+$ in some neighborhood of t. Lemma 6.10 therefore shows that $\varrho_+ \in L_+^p(\Gamma, w)$. Analogously one can show that $\varrho_+^{-1} \in L_+^q(\Gamma, w)$.

The assertion concerning ϱ_- and ϱ_-^{-1} can be proved similarly. $\square$

The following proposition is, in a sense, the key result of this chapter. It provides a (preliminary) Fredholm criterion and an index formula for the Toeplitz operator generated by the local representative $g_{t,\gamma}$.

Proposition 7.2. *Let Γ be a Carleson Jordan curve, $1 < p < \infty$, and $w \in A_p(\Gamma)$. Denote by α_t and β_t the indicator functions of Γ, p, w at $t \in \Gamma$. The operator $T(g_{t,\gamma})$ is Fredholm on $L_+^p(\Gamma, w)$ if and only if*

$$\varkappa_t(\theta) := \frac{1}{p} - \operatorname{Re}\gamma + \theta\alpha_t(-\operatorname{Im}\gamma) + (1-\theta)\beta_t(-\operatorname{Im}\gamma) \notin \mathbf{Z} \text{ for all } \theta \in [0,1]. \quad (7.8)$$

If (7.8) holds then $[\varkappa_t(\theta)]$, the integral part of $\varkappa_t(\theta)$, is independent of $\theta \in [0,1]$ and

$$\operatorname{Ind} T(g_{t,\gamma}) = [\varkappa_t(\theta)], \quad (7.9)$$

where θ is any number in $[0,1]$.

Proof. Suppose (7.8) is satisfied. It is clear that then $[\varkappa_t(\theta)]$ does not depend on $\theta \in [0,1]$. Put $\varkappa := -[\varkappa_t(\theta)]$, $y := \varkappa - \operatorname{Re}\gamma$, $x := -\operatorname{Im}\gamma$. Then $\varkappa - \gamma = y + ix$ and from (7.3) we infer that

$$g_{t,\gamma}(\tau) = a_-(\tau)\tau^{\varkappa}a_+(\tau) \quad (\tau \in \Gamma \setminus \{t\}) \quad (7.10)$$

with

$$a_-(\tau) := (1 - t/\tau)^{y+ix}, \quad a_+(\tau) := (\tau - t)^{-y-ix}. \quad (7.11)$$

We claim that (7.10), (7.11) is a Wiener-Hopf factorization in $L^p(\Gamma, w)$, which, by Theorem 6.32, shows that $T(g_{t,\gamma})$ is Fredholm of index $-\varkappa$.

Condition (7.8) may be written in the form

$$\varkappa_t(\theta) := 1/p + y - \varkappa + \theta\alpha_t(x) + (1-\theta)\beta_t(x) \notin \mathbf{Z} \text{ for all } \theta \in [0,1],$$

and by the definition of $\varkappa$, this means that

$$-\varkappa < 1/p + y - \varkappa + \theta\alpha_t(x) + (1-\theta)\beta_t(x) < -\varkappa + 1 \text{ for all } \theta \in [0,1].$$

Letting $\theta = 1$ and $\theta = 0$, we get

$$-1/p - \alpha_t(x) < y \text{ and } y < 1/q - \beta_t(x),$$

respectively. Thus, Lemma 7.1 implies that conditions (6.35) and (6.36) are fulfilled which proves that (7.10), (7.11) is a Wiener-Hopf factorization in $L^p(\Gamma, w)$.

Conversely, suppose $T(g_{t,\gamma})$ is Fredholm of index $-\varkappa$ on $L_+^p(\Gamma, w)$. Then, again by Theorem 6.32, there is a factorization

$$g_{t,\gamma}(\tau) = b_-(\tau)\tau^\varkappa b_+(\tau)$$

such that $b_- \in L_-^p(\Gamma, w)$, $b_-^{-1} \in L_-^q(\Gamma, w^{-1})$, $b_+ \in L_+^q(\Gamma, w^{-1})$, $b_+^{-1} \in L_+^p(\Gamma, w)$, $|b_+^{-1}|w \in A_p(\Gamma)$. On the other hand, for every integer k we have

$$g_{t,\gamma}(\tau) = (1 - t/\tau)^{-k-\gamma}\tau^{-k}(\tau - t)^{\gamma+k}$$

by virtue of (7.3). On comparing these two factorizations we get

$$b_+^{-1}(\tau)(\tau - t)^{\gamma+k} = b_-(\tau)(1 - t/\tau)^{\gamma+k}\tau^{\varkappa+k}. \tag{7.12}$$

Since $|(\tau - t)^{\gamma+k}| = |\tau - t|^{k+\operatorname{Re}\gamma + r_t(\tau)\operatorname{Im}\gamma}$ with

$$r_t(\tau) := \arg(\tau - t)/\big(-\log|\tau - t|\big) = O(1)$$

(recall Theorem 1.10), it follows that

$$(\tau - t)^{\gamma+k} \in C(\Gamma) \subset L^q(\Gamma, w^{-1}), \quad (1 - t/\tau)^{\gamma+k} \in C(\Gamma) \subset L^q(\Gamma, w^{-1})$$

whenever k is a sufficiently large positive integer. The maximum modulus principle implies that for such k the functions $(z - t)^{\gamma+k}$ and $(1 - t/z)^{\gamma+k}$ are bounded for $z \in D_+$ and $z \in D_-$, respectively. Thus, by Lemma 6.10,

$$(\tau - t)^{\gamma+k} \in L_+^q(\Gamma, w^{-1}), \quad (1 - t/\tau)^{\gamma+k} \in L_-^q(\Gamma, w^{-1}).$$

Now Lemma 6.11 and equality (7.12) show that

$$h(\tau) := b_+^{-1}(\tau)(\tau - t)^{\gamma+k} \in E_+^1(\Gamma), \tag{7.13}$$
$$h(\tau)\tau^{-\varkappa-k} = b_-(\tau)(1 - t/\tau)^{\gamma+k} \in E_-^1(\Gamma) \tag{7.14}$$

if only $m := \varkappa + k$ is large enough.

For $j = 0, 1, \ldots, m$, put

$$h_j := \frac{1}{2\pi i} \int_\Gamma h(\tau)\tau^{-j-1} d\tau.$$

We claim that $g(\tau) := h(\tau) - \sum_{l=0}^m h_l \tau^l$ is a function in $E_+^1(\Gamma) \cap \dot{E}_-^1(\Gamma)$ and thus, by Theorem 6.4, vanishes identically. From (7.13), (7.14) and Theorems 6.2, 6.3 we infer that

$$\int_\Gamma g(\tau)\tau^k d\tau = 0 \tag{7.15}$$

for $k \in \{0, 1, 2, \ldots\}$ and $k \in \{-m-2, -m-3, \ldots\}$. If $k = -j-1$ with $0 \le j \le m$, then

$$\int_\Gamma g(\tau)\tau^k d\tau = \int_\Gamma h(\tau)\tau^{-j-1}d\tau - \sum_{l=0}^{m} h_l \int_\Gamma \tau^{l-j-1}d\tau = 2\pi i h_j - 2\pi i h_j = 0.$$

Consequently, (7.15) holds for all $k \in \mathbf{Z}$, which, again by Theorems 6.2 and 6.3, proves our claim.

Thus, $h(\tau)$ is a polynomial in τ of degree $s \le m = \varkappa + k$. Since $b_+(z)$ is finite in D_+, we see that $h(z) \ne 0$ for $z \in D_+$, and since $b_-(z) \ne 0$ for $z \in D_-$, we conclude that $h(z) \ne 0$ for $z \in D_-$. Finally, as

$$(\tau - t)^{\gamma + k}/h(\tau) = b_+(\tau) \in L^1(\Gamma),$$

the polynomial $h(\tau)$ cannot possess zeros on $\Gamma \setminus \{t\}$. In summary, $h(\tau) = (\tau - t)^s$ and

$$b_+(\tau) = (\tau - t)^{\gamma + k - s}. \tag{7.16}$$

Since, by assumption, $|b_+^{-1}|w \in A_p(\Gamma)$, we deduce from (7.16) and Lemma 7.1 with $y + ix = -\gamma - k + s$ that

$$-1/p - \alpha_t(-\operatorname{Im}\gamma) < -\operatorname{Re}\gamma - k + s < 1/q - \beta_t(-\operatorname{Im}\gamma),$$

i.e. $\quad k - s < 1/p - \operatorname{Re}\gamma + \alpha_t(-\operatorname{Im}\gamma) \le 1/p - \operatorname{Re}\gamma + \beta_t(-\operatorname{Im}\gamma) < k - s + 1,$

whence

$$k - s < 1/p - \operatorname{Re}\gamma + \theta\alpha_t(-\operatorname{Im}\gamma) + (1-\theta)\beta_t(-\operatorname{Im}\gamma) < k - s + 1 \quad \text{for all} \ \theta \in [0, 1].$$

This proves (7.8). $\qquad\square$

7.2 Fredholm criterion

Combining Proposition 7.2 with the local principle (Theorem 6.30), we can establish a Fredholm criterion for Toeplitz operators with piecewise continuous symbols in "analytical language".

Proposition 7.3. *Let Γ be a Carleson Jordan curve, $1 < p < \infty$, $w \in A_p(\Gamma)$, and $b \in PC(\Gamma)$. The Toeplitz operator $T(b)$ is Fredholm on $L_+^p(\Gamma, w)$ if and only if $b \in GL^\infty(\Gamma)$ and*

$$\varkappa_t(\theta) := \frac{1}{p} - \frac{1}{2\pi}\arg\frac{b(t-0)}{b(t+0)} +$$
$$+ \theta\,\alpha_t\left(\frac{1}{2\pi}\log\left|\frac{b(t-0)}{b(t+0)}\right|\right) + (1-\theta)\,\beta_t\left(\frac{1}{2\pi}\log\left|\frac{b(t-0)}{b(t+0)}\right|\right) \notin \mathbf{Z} \tag{7.17}$$

for all $t \in \Gamma$ and all $\theta \in [0, 1]$. If $T(b)$ is Fredholm and the set J_b of the points at which b has a jump is finite, then

$$\operatorname{Ind} T(b) = -\frac{1}{2\pi}\sum_\delta \{\arg b\}_\delta + \sum_{t \in J_b}\left([\varkappa_t(\theta)] + \frac{1}{2\pi}\arg\frac{b(t-0)}{b(t+0)}\right), \tag{7.18}$$

where δ ranges over the connected components of $\Gamma \setminus J_b$, $\{\arg b\}_\delta$ denotes the increment of $\arg b$ along δ, and $[\varkappa_t(\theta)]$ is the integral part of $\varkappa_t(\theta)$ (which does actually
not depend on θ).

Remark. It is clear that neither (7.17) nor (7.18) depend on the particular choice
of $\arg(b(t-0)/b(t+0))$.

Proof. Suppose $b \in GL^\infty(\Gamma)$. For $t \in \Gamma$, define $\gamma = \gamma_t \in \mathbf{C}$ by (7.1). We know
from (7.2) that b is locally equivalent to cg_{t,γ_t} at t, where $c = c_t \in \mathbf{C} \setminus \{0\}$ is some
constant.

If (7.17) holds for all $t \in \Gamma$ and all $\theta \in [0,1]$, we infer from (7.1) that

$$\varkappa_t(\theta) = 1/p - \operatorname{Re} \gamma_t + \theta \alpha_t(-\operatorname{Im} \gamma_t) + (1-\theta)\beta_t(-\operatorname{Im} \gamma_t) \notin \mathbf{Z}$$

for all $t \in \Gamma$ and all $\theta \in [0,1]$. So Proposition 7.2 implies that $T(g_{t,\gamma_t})$ is Fredholm
for all $t \in \Gamma$, and Theorem 6.30 then shows that $T(b)$ itself is Fredholm.

Conversely, if $T(b)$ is Fredholm then $b \in GL^\infty(\Gamma)$ due to Theorem 6.20. Fix
$t \in \Gamma$. Let $\gamma = \gamma_t \in \mathbf{C}$ be as above. If $\tau \in \Gamma \setminus \{t\}$, then g_{t,γ_t} is continuous
and nonzero at τ and hence, g_{t,γ_t} is locally equivalent to the nonzero constant
$g_{t,\gamma_t}(\tau)$ at τ. Clearly, $T(g_{t,\gamma_t}(\tau)) = g_{t,\gamma_t}(\tau)I$ is invertible and therefore Fredholm.
At the point t, the function g_{t,γ_t} is locally equivalent to $c_t^{-1}b$, and the operator
$T(c_t^{-1}b) = c_t^{-1}T(b)$ is Fredholm by assumption. Thus, Theorem 6.30 tells us that
$T(g_{t,\gamma_t})$ is Fredholm, and from Proposition 7.2 we so deduce that (7.17) must hold
for all $\theta \in [0,1]$.

We are left with the index formula. Since b is assumed to have only finitely
many jumps, we may write

$$b = a \prod_{t \in J_b} g_{t,\gamma_t}$$

where the functions g_{t,γ_t} are as above and a is an invertible function in $C(\Gamma)$. From
(7.17), Proposition 7.2, Theorems 6.24 and 6.28 we obtain

$$\operatorname{Ind} T(b) = \operatorname{Ind} T(a) + \sum_{t \in J_b} \operatorname{Ind} T(g_{t,\gamma_t}) = -\operatorname{wind} a + \sum_{t \in J_b} [\varkappa_t(\theta)].$$

Because

$$\frac{1}{2\pi} \sum_\delta \{\arg b\}_\delta = \frac{1}{2\pi} \sum_\delta \{\arg a\}_\delta + \frac{1}{2\pi} \sum_{t \in J_b} \sum_\delta \{\arg g_{t,\gamma_t}\}_\delta$$

$$= \operatorname{wind} a + \frac{1}{2\pi} \sum_{t \in J_b} \arg \frac{g_{t,\gamma_t}(t-0)}{g_{t,\gamma_t}(t+0)} = \operatorname{wind} a + \frac{1}{2\pi} \sum_{t \in J_b} \arg \frac{b(t-0)}{b(t+0)},$$

we arrive at the desired index formula. $\qquad \square$

7.3   Leaves and essential spectrum

The purpose of this section is to translate the Fredholm criterion contained in Proposition 7.3 into "geometrical language". Assume for a moment that $L^p(\Gamma, w)$ is $L^2(\mathbf{T})$. We then know from Corollaries 6.29 and 6.31 that the essential spectrum of $T(b)$ results from the essential range of b by filling in the line segment between the endpoints of the jumps. We will show that in the general case the essential spectrum of $T(b)$ is also the union of the essential range of b and certain sets joining the endpoints of the jumps. These sets will be called leaves.

Let Γ be a Carleson Jordan curve, $p \in (1, \infty)$, $w \in A_p(\Gamma)$, and $t \in \Gamma$. Denote by α_t and β_t the indicator functions of Γ, p, w at the point t. Put

$$Y(p, \alpha_t, \beta_t) := \left\{ \gamma = x + iy \in \mathbf{C} : \frac{1}{p} + \alpha_t(x) \leq y \leq \frac{1}{p} + \beta_t(x) \right\},$$

and given $z_1, z_2 \in \mathbf{C}$, let

$$\mathcal{L}(z_1, z_2; p, \alpha_t, \beta_t) := \left\{ M_{z_1, z_2}(e^{2\pi\gamma}) : \gamma \in Y(p, \alpha_t, \beta_t) \right\} \cup \{z_1, z_2\}$$

where M_{z_1, z_2} is the Möbius transform

$$M_{z_1, z_2}(\zeta) := (z_2\zeta - z_1)/(\zeta - 1).$$

We refer to $\mathcal{L}(z_1, z_2; p, \alpha_t, \beta_t)$ as the *leaf about (or between) z_1 and z_2 determined by p, α_t, β_t*.

From Theorem 3.31(a),(c) we know that $Y(p, \alpha_t, \beta_t)$ is a connected set containing points with arbitrary real parts. Hence, the set

$$\left\{ e^{2\pi\gamma} : \gamma \in Y(p, \alpha_t, \beta_t) \right\}$$

is connected and contains points arbitrarily close to the origin and to infinity. The Möbius transform M_{z_1, z_2} maps 0 and ∞ to z_1 and z_2, respectively. Thus, the leaf $\mathcal{L}(z_1, z_2; p, \alpha_t, \beta_t)$ is a connected set containing z_1 and z_2. More about leaves will be said in the next section.

For $a \in PC(\Gamma)$, denote by $\mathcal{R}(a)$ the *essential range* of a, i.e. let $\mathcal{R}(a)$ be the set $\bigcup_{t \in \Gamma}\{a(t-0), a(t+0)\}$. Let J_a stand for the set of all points at which a has a jump. Clearly, we may write

$$\mathcal{R}(a) = \bigcup_{t \in \Gamma \setminus J_a} \{a(t)\} \cup \bigcup_{t \in J_a} \{a(t-0), a(t+0)\}.$$

Theorem 7.4. *Let Γ be a Carleson Jordan curve, $p \in (1, \infty)$, $w \in A_p(\Gamma)$, and $a \in PC(\Gamma)$. Then the essential spectrum* $\mathrm{sp}_{\mathrm{ess}} T(a)$ *of the Toeplitz operator $T(a)$ on the Hardy space $L_+^p(\Gamma, w)$ is given by*

$$\mathrm{sp}_{\mathrm{ess}} T(a) = \mathcal{R}(a) \cup \bigcup_{t \in J_a} \mathcal{L}\big(a(t-0), a(t+0); p, \alpha_t, \beta_t\big), \tag{7.19}$$

where α_t and β_t are the indicator functions of Γ, p, w at t.

Proof. Theorem 6.20 shows that $\mathcal{R}(a) \subset \mathrm{sp}_{\mathrm{ess}} T(a)$. So let $\lambda \notin \mathcal{R}(a)$. For $t \in \Gamma$, put

$$\zeta_{t,\lambda} := \big(a(t-0) - \lambda\big)/\big(a(t+0) - \lambda\big).$$

Then $\zeta_{t,\lambda} \notin \{0, \infty\}$. Define X_t as the set

$$\bigcup_{\theta \in [0,1]} \Big\{\zeta \in \mathbf{C} \setminus \{0\} : \frac{1}{p} - \frac{1}{2\pi} \arg \zeta + \theta\alpha_t\Big(\frac{1}{2\pi} \log |\zeta|\Big) + (1-\theta)\beta_t\Big(\frac{1}{2\pi} \log |\zeta|\Big) \in \mathbf{Z}\Big\}.$$

From Proposition 7.3 we infer that $T(a) - \lambda I = T(a - \lambda)$ is not Fredholm if and only if there exists a $t \in \Gamma$ such that $\zeta_{t,\lambda} \in X_t$. Obviously, $\zeta_{t,\lambda} \in X_t$ if and only if $\lambda \in M_{a(t-0),a(t+0)}(X_t)$. Thus,

$$\mathrm{sp}_{\mathrm{ess}} T(a) = \mathcal{R}(a) \cup \bigcup_{t \in \Gamma} M_{a(t-0),a(t+0)}(X_t),$$

and it remains to show that $M_{z_1,z_2}(X_t) = \mathcal{L}(z_1, z_2; p, \alpha_t, \beta_t) \setminus \{z_1, z_2\}$ for $z_1, z_2 \in \mathbf{C}$. A number $\zeta \in \mathbf{C} \setminus \{0\}$ may be written in the form $\zeta = e^{2\pi x} e^{2\pi i y}$ with x and y in $\mathbf{R}$. We have $\zeta \in X_t$ if and only if

$$1/p - y + \theta\alpha_t(x) + (1-\theta)\beta_t(x) \in \mathbf{Z} \text{ for some } \theta \in [0, 1],$$

i.e. if and only if

$$1/p + \alpha_t(x) \leq y + \varkappa \leq 1/p + \beta_t(x) \text{ for some } \varkappa \in \mathbf{Z}.$$

Since $e^{2\pi i y} = e^{2\pi i(y+\varkappa)}$, it follows that

$$\begin{aligned} M_{z_1,z_2}(X_t) &= \Big\{M_{z_1,z_2}(e^{2\pi x} e^{2\pi i y}) : 1/p + \alpha_t(x) \leq y \leq 1/p + \beta_t(x)\Big\} \\ &= \mathcal{L}(z_1, z_2; p, \alpha_t, \beta_t) \setminus \{z_1, z_2\}, \end{aligned}$$

as desired. $\qquad\square$

7.4 Metamorphosis of leaves

We now describe the leaves $\mathcal{L}(z_1, z_2; p, \alpha_t, \beta_t)$ in some concrete situations. Obviously, if $z_1 = z_2 =: z$ then $\mathcal{L}(z_1, z_2; p, \alpha_t, \beta_t) = \{z\}$. So let $z_1 \neq z_2$. Since

$$M_{z_1,z_2}(\zeta) = z_1 + (z_2 - z_1)M_{0,1}(\zeta),$$

we have

$$\mathcal{L}(z_1, z_2; p, \alpha_t, \beta_t) = z_1 + (z_2 - z_1)\mathcal{L}(0, 1; p, \alpha_t, \beta_t),$$

i.e. $\mathcal{L}(z_1, z_2; p, \alpha_t, \beta_t)$ is the image of $\mathcal{L}(0, 1; p, \alpha_t, \beta_t)$ under the affine linear transformation which maps 0 and 1 to z_1 and z_2, respectively. In view of Theorem 7.4 we will nevertheless do everything with $z_1 = a(t-0)$ and $z_2 = a(t+0)$.

Throughout the following examples, we assume that $a(t-0) \neq a(t+0)$, we let δ_t^- and δ_t^+ be the spirality indices of Γ at t, and we let μ_t, ν_t stand for the indices of powerlikeness of $w \in A_p(\Gamma)$ at t.

Example 7.5: line segments. Suppose $\delta_t^- = \delta_t^+ = 0$, $\mu_t = \nu_t = 0$, and $p = 2$. This is, for example, the case if $L^p(\Gamma, w) = L^2(\mathbf{T})$. Theorem 3.29 implies that $\alpha_t(x) = \beta_t(x) = 0$ for all $x \in \mathbf{R}$, so

$$Y(p, \alpha_t, \beta_t) = \{\gamma = x + iy \in \mathbf{C} : y = 1/2\},$$

hence

$$\{e^{2\pi\gamma} : \gamma \in Y(p, \alpha_t, \beta_t)\} = \{-e^{2\pi x} : x \in \mathbf{R}\} = \{-r : r > 0\}$$

is the negative half-line, and thus the leaf $\mathcal{L}(a(t-0), a(t+0); p, \alpha_t, \beta_t)$ equals

$$\left\{\frac{a(t+0)r + a(t-0)}{r+1} : r > 0\right\} \cup \{a(t-0), a(t+0)\},$$

which is the line segment $[a(t-0), a(t+0)]$. In particular, we see that Theorem 7.4 indeed yields the Fredholm criterion of Corollary 6.31.   □

Example 7.6: circular arcs. Let $\delta_t^- = \delta_t^+ = 0$ and $\mu_t = \nu_t =: \lambda$. For instance, we are given this case if Γ has no helical points and w is a power weight. Now Theorem 3.29 shows that $\alpha_t(x) = \beta_t(x) = \lambda$ for all $x \in \mathbf{R}$, whence

$$Y(p, \alpha_t, \beta_t) = \{\gamma = x + iy \in \mathbf{C} : y = 1/p + \lambda\}.$$

Note that, by Theorem 3.31(b), $0 < 1/p + \lambda < 1$ whenever $w \in A_p(\Gamma)$. The set $\{e^{2\pi\gamma} : \gamma \in Y(p, \alpha_t, \beta_t)\}$ is the ray $\{e^{2\pi x}e^{2\pi i(1/p+\lambda)} : x \in \mathbf{R}\}$, and the Möbius transform $M_{a(t-0), a(t+0)}$ maps this ray to a certain circular arc between $a(t-0)$ and $a(t+0)$.

Given two points $z_1, z_2 \in \mathbf{C}$ and a number $\varphi \in (0, 1)$ we define

$$\mathcal{A}(z_1, z_2; \varphi) := \left\{z \in \mathbf{C} \setminus \{z_1, z_2\} : \arg\frac{z - z_1}{z - z_2} \in 2\pi\varphi + 2\pi\mathbf{Z}\right\} \cup \{z_1, z_2\}.$$

If $z_1 = z_2 =: z$, then $\mathcal{A}(z_1, z_2; \varphi)$ is simply $\{z\}$. If $z_1 \neq z_2$, then $\mathcal{A}(z_1, z_2; \varphi)$ is a circular arc between z_1 and z_2 whose shape is determined by φ: for $0 < \varphi < 1/2$ (resp. $1/2 < \varphi < 1$), it is the circular arc at the points of which the line segment $[z_1, z_2]$ is seen at the angle $2\pi\varphi$ (resp. $2\pi - 2\pi\varphi$) and which lies on the right (resp. left) of the straight line passing first z_1 and then z_2; we also have $\mathcal{A}(z_1, z_2; 1/2) = [z_1, z_2]$. A little thought reveals that

$$\mathcal{L}(a(t-0), a(t+0); p, \alpha_t, \beta_t) = \mathcal{A}(a(t-0), a(t+0); 1/p + \lambda).$$

Notice that the shape of the circular arc between $a(t-0)$ and $a(t+0)$ is entirely determined by $1/p + \lambda$. Clearly, if $1/p + \lambda = 1/2$ then the arc degenerates to a line segment. We also remark that if $\lambda = 0$ (e.g. if $L^p(\Gamma, w) = L^p(\mathbf{T})$), then the leaf is a line segment for $p = 2$ and a proper circular arc for all other $p \in (1, \infty)$.   □

Example 7.7: horns. Let $\delta_t^- = \delta_t^+ = 0$, but suppose μ_t and ν_t are only subject to the condition $-1/p < \mu_t \leq \nu_t < 1/q$ dictated by Theorem 3.31(b). This is the situation we encounter when considering "nice" curves with "arbitrary" weights. Theorem 3.29 tells us that $\alpha_t(x) = \mu_t$ and $\beta_t(x) = \nu_t$ for all $x \in \mathbf{R}$, implying that

$$Y(p, \alpha_t, \beta_t) = \{\gamma = x + iy \in \mathbf{C} : 1/p + \mu_t \leq y \leq 1/p + \nu_t\}$$

is a closed horizontal stripe of height $\nu_t - \mu_t < 1$. It therefore follows that the set $\{e^{2\pi\gamma} : \gamma \in Y(p, \alpha_t, \beta_t)\}$ is a sector with the vertex at the origin and that the leaf $\mathcal{L}(a(t-0), a(t+0); p, \alpha_t, \beta_t)$ is the closed set between two circular arcs. We refer to such sets as *horns*. Thinking of $Y(p, \alpha_t, \beta_t)$ as the union of the lines $\{x + iy \in \mathbf{C} : y = 1/p + \lambda\}$ $(\mu_t \leq \lambda \leq \nu_t)$, we obtain from Example 7.6 that

$$\mathcal{L}(a(t-0), a(t+0); p, \alpha_t, \beta_t) = \bigcup_{\lambda \in [\mu_t, \nu_t]} \mathcal{A}(a(t-0), a(t+0); 1/p + \lambda). \qquad \square$$

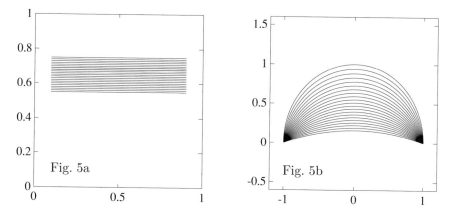

Fig. 5a   Fig. 5b

Fig. 5a shows part of the horizontal stripe $Y(p; \alpha_t, \beta_t) = \{x + iy \in \mathbf{C} : 0.55 \leq y \leq 0.75\}$; the horn $\mathcal{L}(-1, 1; p, \alpha_t, \beta_t)$ is plotted in Fig. 5b.

Example 7.8: logarithmic double spirals. Suppose $\delta_t^- = \delta_t^+ =: \delta$ and $\mu_t = \nu_t =: \lambda$. We have this constellation if, for example, Γ is comprised of two logarithmic spirals in a neighborhood of t and w is a power weight. Again having recourse to Theorem 3.29, we get $\alpha_t(x) = \beta_t(x) = \lambda + \delta x$ for all $x \in \mathbf{R}$. Thus,

$$Y(p, \alpha_t, \beta_t) = \{\gamma = x + iy \in \mathbf{C} : y = 1/p + \lambda + \delta x\}$$

is the straight line through $i(1/p + \lambda)$ with the slope δ. Consequently, the set $\{e^{2\pi\gamma} : \gamma \in Y(p, \alpha_t, \beta_t)\}$ equals

$$\{e^{2\pi i(1/p+\lambda)} e^{2\pi x(1+i\delta)} : x \in \mathbf{R}\} = \{e^{2\pi i(1/p+\lambda)} r^{1+i\delta} : r > 0\}.$$

This is a logarithmic spiral. The Möbius transform $M_{a(t-0),a(t+0)}$ maps this logarithmic spiral to a double spiral wriggling out of $a(t-0)$ and scrolling up at $a(t+0)$.

Given $z_1, z_2 \in \mathbf{C}$, $\delta \in \mathbf{R}$, and $\varphi \in (0,1)$, we denote by $\mathcal{S}(z_1, z_2; \delta, \varphi)$ the set

$$\left\{ z \in \mathbf{C} \setminus \{z_1, z_2\} : \arg \frac{z - z_1}{z - z_2} - \delta \log \left| \frac{z - z_1}{z - z_2} \right| \in 2\pi\varphi + 2\pi\mathbf{Z} \right\} \cup \{z_1, z_2\}$$

and call $\mathcal{S}(z_1, z_2; \delta, \varphi)$ a *logarithmic double spiral.* Notice that

$$\mathcal{S}(z_1, z_2; 0, \varphi) = \mathcal{A}(z_1, z_2; \varphi),$$

so that circular arcs (and line segments) are regarded as degenerate logarithmic double spirals. A little computation (or combination of Proposition 7.3 and Theorem 7.4) shows that

$$\mathcal{L}\big(a(t-0), a(t+0); p, \alpha_t, \beta_t\big) = \mathcal{S}\big(a(t-0), a(t+0); \delta; 1/p + \lambda\big). \qquad \square$$

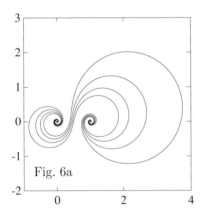

Fig. 6a

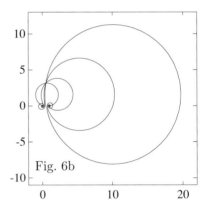

Fig. 6b

The logarithmic double spirals
$\mathcal{S}(0, 1, ; \delta; \lambda + \frac{1}{p})$ for $\delta = 4$
and $\lambda + \frac{1}{p} = \frac{5}{10}, \frac{6}{10}, \frac{7}{10}, \frac{8}{10}$.

The logarithmic double spirals
$\mathcal{S}(0, 1; \delta; \lambda + \frac{1}{p})$ for $\lambda + \frac{1}{p} = \frac{95}{100}$
and $\delta = \frac{1}{10}, 1, 3, 6$.

Example 7.9: spiralic horns. We now consider the case where $\delta_t^- = \delta_t^+ =: \delta$ and μ_t, ν_t are only required to satisfy the natural constraint $-1/p < \mu_t \le \nu_t < 1/q$. This case occurs, for example, if Γ is locally the union of two logarithmic spirals but $w \in A_p(\Gamma)$ may be "arbitrary". Theorem 3.29 says that $\alpha_t(x) = \mu_t + \delta x$ and $\beta_t(x) = \nu_t + \delta x$ for all $x \in \mathbf{R}$. Consequently,

$$Y(p, \alpha_t, \beta_t) = \left\{ \gamma = x + iy \in \mathbf{C} : \frac{1}{p} + \mu_t + \delta x \le y \le \frac{1}{p} + \nu_t + \delta x \right\}$$

is a closed stripe of height $\nu_t - \mu_t < 1$ and of slope δ. We may think of $Y(p, \alpha_t, \beta_t)$ as the union of straight lines of slope δ and may therefore argue as in the previous example to deduce that

$$\mathcal{L}\big(a(t-0), a(t+0); p, \alpha_t, \beta_t\big) = \bigcup_{\lambda \in [\mu_t, \nu_t]} \mathcal{S}\big(a(t-0), a(t+0); \delta, 1/p + \lambda\big).$$

Hence, the leaf is now the closed set between two logarithmic double spirals (with the same parameter δ). Such sets deserve to be called *spiralic horns*. □

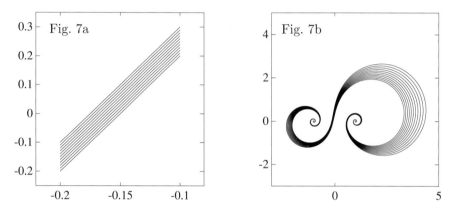

Fig. 7a shows a piece of the stripe $Y(p; \alpha_t, \beta_t) = \{x+iy \in \mathbf{C} : 0.6+4x \leq y \leq 0.7 + 4x\}$; the corresponding spiralic horn $\mathcal{L}(-1, 1; p, \alpha_t, \beta_t)$ is shown in Fig. 7b.

The preceding five examples describe all possible types of leaves in the case where $\delta_t^- = \delta_t^+$. We see that the metamorphosis of leaves for "nice" curves (i.e. curves with $\delta_t^- = \delta_t^+$) terminates with spiralic horns.

7.5 Logarithmic leaves

To understand the shape of the leaf $\mathcal{L}(a(t-0), a(t+0); p, \alpha_t, \beta_t)$ in the case where $\delta_t^- < \delta_t^+$, we first introduce the notion of a logarithmic leaf.

Let $1 < p < \infty$, let $z_1, z_2 \in \mathbf{C}$, and let $\delta_1, \delta_2, \mu_1, \mu_2, \nu_1, \nu_2$ be real numbers satisfying

$$\delta_1 \leq \delta_2, \quad 0 < \mu_1 \leq \nu_1 < 1, \quad 0 < \mu_2 \leq \nu_2 < 1. \tag{7.20}$$

Then

$$\min\{\mu_1 + \delta_1 x, \mu_2 + \delta_2 x\} \leq \max\{\nu_1 + \delta_1 x, \nu_2 + \delta_2 x\} \tag{7.21}$$

for all $x \in \mathbf{R}$. We denote by $Y^0(\delta_1, \delta_2; \mu_1, \mu_2, \nu_1, \nu_2)$ the set of all $\gamma = x + iy \in \mathbf{C}$ for which

$$\min\{\mu_1 + \delta_1 x, \mu_2 + \delta_2 x\} \leq y \leq \max\{\nu_1 + \delta_1 x, \nu_2 + \delta_2 x\}.$$

The *logarithmic leaf* $\mathcal{L}^0(z_1, z_2; \delta_1, \delta_2; \mu_1, \mu_2, \nu_1, \nu_2)$ is defined as the set

$$\{M_{z_1,z_2}(e^{2\pi\gamma}) : \gamma \in Y^0(\delta_1, \delta_2; \mu_1, \mu_2, \nu_1, \nu_2)\} \cup \{z_1, z_2\}.$$

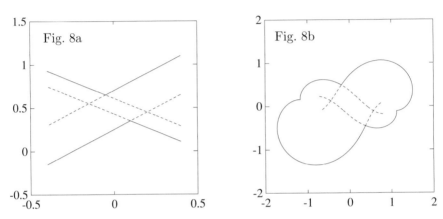

Fig. 8a shows the graphs of the functions in (7.21), Fig. 8b shows the boundary of the corresponding logarithmic leaf.

If $\delta_1 = \delta_2 =: \delta$, then $Y^0(\delta, \delta; \mu_1, \mu_2, \nu_1, \nu_2)$ degenerates to the stripe of all $\gamma = x + iy \in \mathbf{C}$ satisfying

$$\min\{\mu_1, \mu_2\} + \delta x \leq y \leq \max\{\nu_1, \nu_2\} + \delta x$$

and hence, by Examples 7.8 and 7.9,

$$\mathcal{L}^0(z_1, z_2; \delta, \delta; \mu_1, \mu_2, \nu_1, \nu_2) = \bigcup_{\lambda \in [\min\{\mu_1, \mu_2\}, \max\{\nu_1, \nu_2\}]} \mathcal{S}(z_1, z_2; \delta, \lambda)$$

is a spiralic horn for $\min\{\mu_1, \mu_2\} < \max\{\nu_1, \nu_2\}$ and a logarithmic double spiral for $\min\{\mu_1, \mu_2\} = \max\{\nu_1, \nu_2\}$; note that in the latter case we necessarily have $\mu_1 = \mu_2 = \nu_1 = \nu_2$ due to (7.20).

So let $\delta_1 < \delta_2$. Put

$$\alpha^0(x) := \min\{\mu_1 + \delta_1 x, \mu_2 + \delta_2 x\}, \quad \beta^0(x) := \max\{\nu_1 + \delta_1 x, \nu_2 + \delta_2 x\}.$$

From (7.20) we see that $\beta^0(0) - \alpha^0(0) < 1$, and the condition $\delta_1 < \delta_2$ implies that $\beta^0(x) - \alpha^0(x) \to +\infty$ as $x \to \pm\infty$. Thus, the equation $\beta^0(x) - \alpha^0(x) = 1$ has exactly two solutions: $x_-^0 < 0$ and $x_+^0 > 0$. Since

$$\{\gamma = x + iy \in Y^0(\delta_1, \delta_2; \mu_1, \mu_2, \nu_1, \nu_2) : x \le x_-^0\}$$
$$\supset \{\gamma = x + iy \in \mathbf{C} : x \le x_-^0, \alpha^0(x_-^0) \le y < \beta^0(x_-^0)\} =: \Pi_-^0$$

and

$$\{\gamma = x + iy \in Y^0(\delta_1, \delta_2; \mu_1, \mu_2, \nu_1, \nu_2) : x \ge x_+^0\}$$
$$\supset \{\gamma = x + iy \in \mathbf{C} : x \ge x_+^0, \alpha^0(x_+^0) \le y < \beta^0(x_+^0)\} =: \Pi_+^0,$$

and since the map $\gamma \mapsto M_{z_1,z_2}(e^{2\pi\gamma})$ has the period i, it follows that this map effects a bijection between

$$\{\gamma = x + iy \in Y^0(\delta_1, \delta_2; \mu_1, \mu_2, \nu_1, \nu_2) : x_0^- < x < x_0^+\} \cup \Pi_-^0 \cup \Pi_+^0 \qquad (7.22)$$

and the logarithmic leaf $\mathcal{L}^0(z_1, z_2; \delta_1, \delta_2; \mu_1, \mu_2, \nu_1, \nu_2)$ minus $\{z_1, z_2\}$. It is easily verified that Π_+^0 and Π_-^0 are mapped onto closed disks punctured at z_2 and z_1 and having the centers

$$\left(z_2 e^{2\pi x_\pm^0} - z_1 e^{-2\pi x_\pm^0}\right) / \left(2 \sinh(2\pi x_\pm^0)\right)$$

and the radii

$$|z_2 - z_1| / \left(\pm 2 \sinh(2\pi x_\pm^0)\right),$$

respectively. The first of the three sets in (7.22) is mapped onto something linking these two disks. A little thought shows that a logarithmic leaf is always bounded by pieces of at most four logarithmic double spirals.

Example 7.10: logarithmic leaves with a median separating point. Suppose $\mu_t = \nu_t =: \lambda$, but allow $\delta_t^- \le \delta_t^+$ to be arbitrary numbers. This case is encountered if Γ is an arbitrary Carleson Jordan curve and w is powerlike at t (and includes the case where $w = 1$ identically). From Theorem 3.30 we infer that $Y(p, \alpha_t, \beta_t)$ is the set of all $\gamma = x + iy \in \mathbf{C}$ such that

$$1/p + \lambda + \min\{\delta_t^- x, \delta_t^+ x\} \le y \le 1/p + \lambda + \max\{\delta_t^- x, \delta_t^+ x\}.$$

Thus, with the above notation,

$$Y(p, \alpha_t, \beta_t) = Y^0\left(\delta_t^-, \delta_t^+; \frac{1}{p} + \lambda, \frac{1}{p} + \lambda, \frac{1}{p} + \lambda, \frac{1}{p} + \lambda\right)$$
$$= \bigcup_{\delta \in [\delta_t^-, \delta_t^+]} \left\{x + iy \in \mathbf{C} : y = \frac{1}{p} + \lambda + \delta x\right\}.$$

Consequently $\mathcal{L}(a(t-0), a(t+0); p, \alpha_t, \beta_t)$ is the logarithmic leaf

$$
\mathcal{L}^0\left(a(t-0), a(t+0); \delta_t^-, \delta_t^+; \frac{1}{p}+\lambda, \frac{1}{p}+\lambda, \frac{1}{p}+\lambda, \frac{1}{p}+\lambda\right)
$$

$$
= \bigcup_{\delta \in [\delta_t^-, \delta_t^+]} \mathcal{S}\left(a(t-0), a(t+0); \delta, \frac{1}{p}+\lambda\right) \tag{7.23}
$$

(recall Example 7.8). Logarithmic leaves of this kind have two peculiarities. First, all logarithmic spirals participating in the union on the right of (7.23) contain the point

$$
s := M_{a(t-0), a(t+0)}(e^{2\pi i(1/p+\lambda)}).
$$

The set (7.23) is connected, while (7.23) minus the point s is disconnected. This is why we refer to s as a *separating point*. Secondly, since

$$
\left|s - a(t-0)\right| = \left|\frac{a(t+0)e^{2\pi i(1/p+\lambda)} - a(t-0)}{e^{2\pi i(1/p+\lambda)} - 1} - a(t-0)\right|
$$

$$
= \left|\frac{(a(t+0) - a(t-0))e^{2\pi i(1/p+\lambda)}}{e^{2\pi i(1/p+\lambda)} - 1}\right| = \frac{|a(t+0) - a(t-0)|}{2\sin(\pi/(1/p+\lambda))}
$$

and, analogously,

$$
\left|s - a(t+0)\right| = \frac{|a(t+0) - a(t-0)|}{2\sin(\pi/(1/p+\lambda))},
$$

the point s is at an equal distance to $a(t-0)$ and $a(t+0)$. We therefore call s a *median point* of the leaf. In summary, we have shown that if the weight $w \in A_p(\Gamma)$

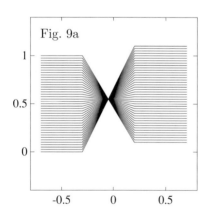

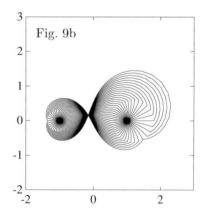

The set (7.22) with $\mu_1 = \mu_2 = \nu_1 = \nu_2 = 1/p + \lambda$ is indicated in Fig. 9a. The corresponding logarithmic leaf between -1 and 1 has a median separating point and is shown in Fig. 9b.

is powerlike at $t \in \Gamma$, then the leaf $\mathcal{L}(a(t-0), a(t+0)); p, \alpha_t, \beta_t)$ is a logarithmic leaf with a median separating point. As we can always find a Carleson Jordan curve Γ with prescribed spirality indices $\delta_t^- \leq \delta_t^+$ and a weight $w \in A_p(\Gamma)$ whose indices of powerlikeness are $\mu_t = \nu_t = \lambda$ with a given $\lambda \in (-1/p, 1/q)$, it follows that logarithmic leaves with a median separating point are the characteristic leaves for powerlike weights (on arbitrary curves).   □

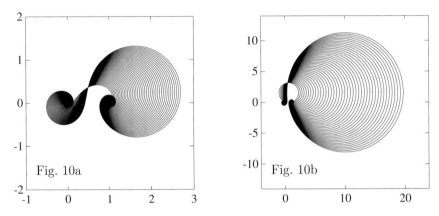

Two leaves with a separating median point: $\cup_{\delta \in [\delta_1, \delta_2]} \mathcal{S}(0, 1; \delta; \lambda + 1/p)$ for $[\delta_1, \delta_2] = [1/2, 4]$ (Fig. 10a) and $[\delta_1, \delta_2] = [1/10, 6]$ (Fig. 10b) in the case where $\lambda + 1/p = 7/10$.

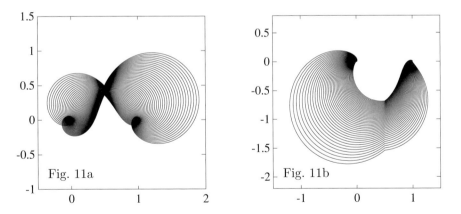

Two logarithmic leaves $\mathcal{L}(-1, 1; \delta_1, \delta_2; \mu_1, \mu_2, \nu_1, \nu_2)$ without separating points. Each of these leaves is bounded by pieces of four logarithmic double spirals.

Thus, until the present moment we can precisely describe the leaf $\mathcal{L}(a(t-0)$, $a(t+0), p, \alpha_t, \beta_t)$ if the curve is nice at t (Examples 7.5 to 7.9) or if the weight is nice at t (Example 7.10). To be more precise, we know that the leaf is a (possible degenerated) spiralic horn if the curve is spiralic at t, i.e. if $\delta_t^- = \delta_t^+$, and that the leaf is a logarithmic leaf with a median separating point in case the weight is powerlike at t. Using Theorem 3.37 one can show that every logarithmic leaf

$$\mathcal{L}^0\big(a(t-0), a(t+0); \delta_1, \delta_2; \mu_1, \mu_2, \nu_1, \nu_2\big)$$

is the leaf $\mathcal{L}(a(t-0), a(t+0); p, \alpha_t, \beta_t)$ for certain Γ, p, w, t. The really interesting thing, however, is the fact that there are leaves $\mathcal{L}(a(t-0), a(t+0); p, \alpha_t, \beta_t)$ beyond logarithmic leaves. In a sense, logarithmic leaves play the role for general leaves the "separating asymptotes" of Theorem 3.31(d) play for the indicator functions.

7.6 General leaves

We now describe the leaf $\mathcal{L}(a(t-0), a(t+0); p, \alpha_t, \beta_t)$ in the general case. In view of Section 7.4, it remains to consider the case where $\delta_t^- < \delta_t^+$. So suppose $\delta_t^- < \delta_t^+$.

By the discussion after Theorem 3.32, the convex function $\beta_t(x) - \alpha_t(x)$ is less than 1 for $x = 0$ and the equation $\beta_t(x) - \alpha_t(x) = 1$ has exactly two solutions, $x_t^- < 0$ and $x_t^+ > 0$. Let $\Pi_t := \Pi_t(\Gamma, p, w)$ denote the closed parallelogram spanned by the points (3.93) and (3.94), and let $\Pi_t^\pm := \Pi_t^\pm(\Gamma, p, w)$ stand for the horizontal half-stripes

$$\Pi_t^- := \left\{ x + iy \in \mathbf{C} : x < x_t^-, \ \frac{1}{p} + \alpha_t(x_t^-) \leq y \leq \frac{1}{p} + \beta_t(x_t^-) \right\},$$

$$\Pi_t^+ := \left\{ x + iy \in \mathbf{C} : x > x_t^+, \ \frac{1}{p} + \alpha_t(x_t^+) \leq y \leq \frac{1}{p} + \beta_t(x_t^+) \right\}.$$

The map $\gamma \mapsto M_{a(t-0),a(t+0)}(e^{2\pi\gamma})$ has the period i, and hence the leaf $\mathcal{L}(a(t-0)$, $a(t+0); p, \alpha_t, \beta_t)$ is the union of the two sets

$$\left\{ M_{a(t-0),a(t+0)}(e^{2\pi\gamma}) : \gamma \in \Pi_t^- \cup \Pi_t^+ \right\} \cup \left\{ a(t-0), a(t+0) \right\}, \tag{7.24}$$

$$\left\{ M_{a(t-0),a(t+0)}(e^{2\pi\gamma}) : \gamma = x + iy \in \Pi_t, \ \frac{1}{p} + \alpha_t(x) \leq y \leq \frac{1}{p} + \beta_t(x) \right\}. \tag{7.25}$$

Clearly, (7.24) is the union of two open disks. The set (7.25) is something linking these two disks and therefore the part of $Y(p, \alpha, \beta_t)$ contained in the parallelogram Π_t is the actually interesting part of $Y(p, \alpha_t, \beta_t)$. Notice in this connection that it is also precisely the indicator functions between x_t^- and x_t^+ which contain all information about the indicator set $N_t = N_t(\Gamma, p, w)$.

By Theorem 3.31(d), the asymptote to $y = 1/p + \beta_t(x)$ as $x \to +\infty$ has the equation $y = 1/p + \nu_t^+ + \delta_t^+ x$. The parallel to this asymptote through the point

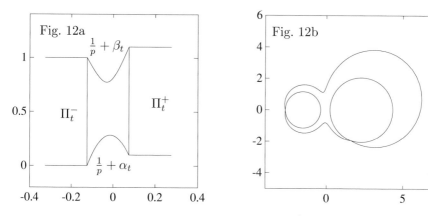

Fig. 12a shows the two half-stripes Π_t^- and Π_t^+ as well as the graphs of $y = 1/p + \alpha_t(x)$ and $y = 1/p + \beta_t(x)$ for $x \in [x_t^-, x_t^+]$. In Fig. 12b we see the two disks $\{M_{z,w}(e^{2\pi\gamma}) : \gamma \in \Pi_t^- \cup \Pi_t^+\}$ and the "linking set" given by (7.25).

$x_t^+ + i(1/p + \beta_t(x_t^+))$ has the equation

$$y = 1/p + \beta_t(x_t^+) + \delta_t^+(x - x_t^+).$$

Since β_t is convex, it follows that

$$\frac{1}{p} + \beta_t(x) \geq \frac{1}{p} + \beta_t(x_t^+) + \delta_t^+(x - x_t^+) \geq \frac{1}{p} + \nu_t^+ + \delta_t^+ x \qquad (7.26)$$

for $x \in [x_t^-, x_t^+]$. Analogously we get

$$\frac{1}{p} + \beta_t(x) \geq \frac{1}{p} + \beta_t(x_t^-) + \delta_t^-(x - x_t^-) \geq \frac{1}{p} + \nu_t^- + \delta_t^- x, \qquad (7.27)$$

$$\frac{1}{p} + \alpha_t(x) \leq \frac{1}{p} + \alpha_t(x_t^-) + \delta_t^+(x - x_t^-) \leq \frac{1}{p} + \mu_t^+ + \delta_t^+ x, \qquad (7.28)$$

$$\frac{1}{p} + \alpha_t(x) \leq \frac{1}{p} + \alpha_t(x_t^+) + \delta_t^-(x - x_t^+) \leq \frac{1}{p} + \mu_t^- + \delta_t^- x, \qquad (7.29)$$

for $x \in [x_t^-, x_t^+]$. Figure 13 shows the representation of the set

$$\left\{ x + iy \in \Pi_t : \frac{1}{p} + \alpha_t(x) \leq y \leq \frac{1}{p} + \beta_t(x) \right\} \cup \Pi_t^- \cup \Pi_t^+ \qquad (7.30)$$

as the union of subsets $A, B, \ldots, K$, where $A = \Pi_t^-$, $B = \Pi_t^+$, and the other subsets are obtained from the first set in (7.30) by dividing it into parts by the graphs of the functions in the middle of the estimates (7.26) to (7.29).

In Figure 14 we see the parts of the leaf $\mathcal{L}(a(t-0), a(t+0); p, \alpha_t, \beta_t)$ corresponding to the sets $A, B, \ldots, K$. Thus, a general leaf is a logarithmic leaf with a "halo". In Figure 14 the halo is $J \cup K$.

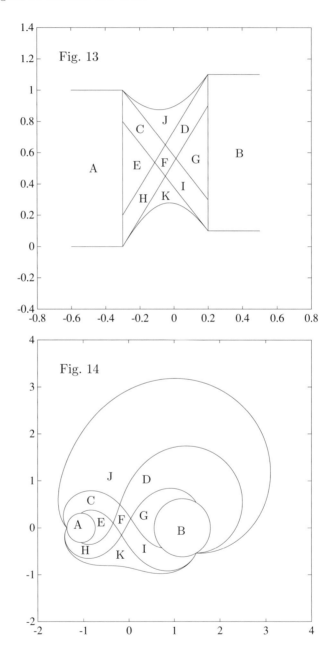

We know everything about the leaf $\mathcal{L}(a(t-0), a(t+0); p, \alpha_t, \beta_t)$ if only the indicator functions $\alpha_t(x)$ and $\beta_t(x)$ are available for $x \in [x_t^-, x_t^+]$. Theorems 3.33 and 3.37 completely describe all possible situations and thus all "halos" which may occur.

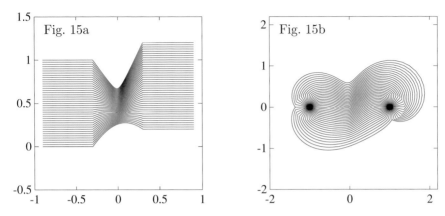

Figures 15a and 15b show us one more way to understand Figures 12, 13, 14. It is clearly seen that the leaf in Fig. 15b has no separating points and that the points -1 and 1 are inner points.

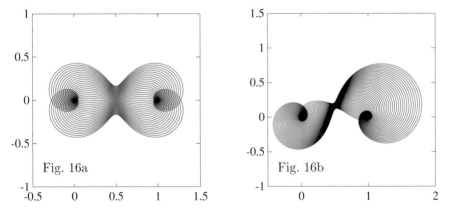

Leaves induced by indicator functions whose graphs are hyperbolas. We remark that the "middle peaks" of Fig. 11 metamorphose into lines looking (locally) like hyperbolas.

It is easily verified that the leaf $\mathcal{L}(a(t-0), a(t+0); p, \alpha_t, \beta_t)$ always contains the logarithmic leaf

$$\mathcal{L}^0\left(a(t-0), a(t+0); \delta_t^-, \delta_t^+; \frac{1}{p} + \mu_t^-, \frac{1}{p} + \mu_t^+, \frac{1}{p} + \nu_t^-, \frac{1}{p} + \nu_t^+\right)$$

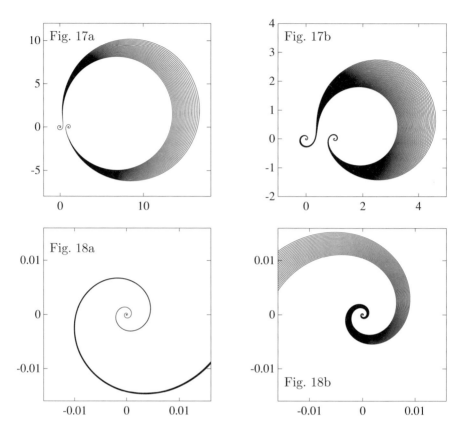

These figures nicely illustrate the beauty of leaves. Fig. 17a shows a
spiralic horn, in Fig. 17b we plotted a leaf emerging when choosing α_t
and β_t as hyperbolas, and thus a leaf containing a halo. Consequently,
the boundary of the set in Fig. 17a consists of two logarithmic double-
spirals, while (though this is hardly visible in the case at hand) no piece
of the boundary of the leaf in Fig. 17b is a piece of some logarithmic
double-spiral. When looking at Fig. 17 with a magnifying glass, which
is done in Fig. 18, we really see that the points 0 and 1 belong to the
boundary of the leaf in Fig. 17a/18a but are inner points of the leaf in
Fig. 17b/18b.

and that $\mathcal{L}(a(t-0), a(t+0); p, \alpha_t, \beta_t)$ is always a subset of the logarithmic leaf

$$\mathcal{L}^0\left(a(t-0), a(t+0); \delta_t^-, \delta_t^+; \frac{1}{p}+\mu_t, \frac{1}{p}+\mu_t, \frac{1}{p}+\nu_t, \frac{1}{p}+\nu_t\right)$$

where $\mu_t = \alpha_t(0)$ and $\nu_t = \beta_t(0)$. Both these estimates are crude, but they contain only the parameters $\delta_t^{\pm}, \mu_t^{\pm}, \nu_t^{\pm}, \mu_t, \nu_t$, which can be easier found than the complete indicator functions.

We conclude the discussion on the shape of leaves by characterizing some special leaves.

Proposition 7.11. *Let Γ be a Carleson Jordan curve, $p \in (1, \infty)$, $w \in A_p(\Gamma)$, and $t \in \Gamma$. Suppose $a(t-0) \neq a(t+0)$ and put*

$$\mathcal{L}_t := \mathcal{L}\big(a(t-0), a(t+0); p, \alpha_t, \beta_t\big).$$

(a) *The leaf $\mathcal{L}_t$ has no interior points if and only if $\delta_t^- = \delta_t^+$ and $\mu_t = \nu_t$, i.e. if and only if Γ is spiralic at t and w is powerlike at t.*

(b) *The points $a(t-0)$ and $a(t+0)$ belong to the boundary of $\mathcal{L}_t$ if and only if $\delta_t^- = \delta_t^+$, i.e. if and only if Γ is spiralic at t. In that case $\mathcal{L}_t$ is a (possibly degenerated) spiralic horn.*

(c) *The leaf $\mathcal{L}_t$ has a median separating point if and only if $\mu_t = \nu_t$, i.e. if and only if w is powerlike at t. In that case $\mathcal{L}_t$ is the logarithmic leaf*

$$\mathcal{L}_t = \mathcal{L}^0\Big(a(t-0), a(t+0); \delta_t^-, \delta_t^+; \frac{1}{p} + \lambda_t, \frac{1}{p} + \lambda_t, \frac{1}{p} + \lambda_t, \frac{1}{p} + \lambda_t\Big)$$

where $\lambda_t := \mu_t = \nu_t$.

Proof. (a),(b). If $\delta_t^- < \delta_t^+$, then $\mathcal{L}_t$ contains the two disks (7.24) and hence $a(t-0)$ and $a(t+0)$ are inner points of $\mathcal{L}_t$. So let $\delta_t^- = \delta_t^+$. If $\mu_t < \nu_t$, then Example 7.9 shows that $\mathcal{L}_t$ is a spiralic horn with inner points and that $a(t-0)$ and $a(t+0)$ lie on the boundary of $\mathcal{L}_t$. If $\mu_t = \nu_t$, then Example 7.8 implies that $\mathcal{L}_t$ is a logarithmic double spiral between $a(t-0)$ and $a(t+0)$.

(c). The "if" part is Example 7.10. To prove the "only if" portion, suppose $\mathcal{L}_t$ has a median separating point. If $\delta_t^- = \delta_t^+$, we infer from Examples 7.8 and 7.9 that then necessarily $\mu_t = \nu_t$. So let $\delta_t^- < \delta_t^+$. The existence of a separating point implies that the union of the graphs of the functions $y = 1/p + \alpha_t(x)$ and $y = 1/p + \beta_t(x)$ on $[x_t^-, x_t^+]$ is the two diagonals of the parallelogram Π_t. Let $x_t + i(1/p + y_t)$ be the point at which the diagonals intersect. The (unique) separation point is

$$s = M_{a(t-0),a(t+0)}\big(e^{2\pi x_t + 2\pi i(1/p + y_t)}\big),$$

and a straightforward computation shows that

$$\big|s - a(t-0)\big| = \frac{|a(t+0) - a(t-0)|e^{2\pi x_t}}{|e^{2\pi x_t}e^{2\pi i(1/p + y_t)} - 1|},$$

$$\big|s - a(t+0)\big| = \frac{|a(t+0) - a(t-0)|}{|e^{2\pi x_t}e^{2\pi i(1/p + y_t)} - 1|}.$$

Thus, s is at an equal distance to $a(t-0)$ and $a(t+0)$ if and only if $x_t = 0$. Consequently, $\mu_t = \alpha_t(0) = \beta_t(0) = \nu_t$. $\square$

7.7   Index and spectrum

The objective of this section is to establish an index formula for Fredholm Toeplitz operators with piecewise continuous symbols. In view of Corollary 6.19, this allows us to decide whether such operators are invertible.

Let $z_1, z_2 \in \mathbf{C}$ be two distinct points. Given $\varphi \in (0,1)$ and $\delta \in \mathbf{R}$, define the circular arc $\mathcal{A}(z_1, z_2; \varphi)$ and the logarithmic double spiral $\mathcal{S}(z_1, z_2; \delta, \varphi)$ as in Examples 7.6 and 7.8. In this section, we think of $\mathcal{A}(z_1, z_2; \varphi)$ as being oriented from z_2 to z_1, and we give $\mathcal{S}(z_1, z_2; \delta, \varphi)$ the orientation from z_1 to z_2. Then

$$\mathcal{K}_\delta := \mathcal{A}(z_1, z_2; \varphi) \cup \mathcal{S}(z_1, z_2; \delta, \varphi)$$

is a closed, continuous, oriented curve in the plane. For $z \notin \mathcal{K}_\delta$, we denote by wind $(\mathcal{K}_\delta, z)$ the winding number of $\mathcal{K}_\delta$ about z.

Lemma 7.12. *If $0 \notin \mathcal{K}_\delta$ then*

$$\text{wind}\,(\mathcal{K}_\delta, 0) = \left[\varphi - \frac{1}{2\pi} \arg \frac{z_1}{z_2}\right] - \left[\varphi - \frac{1}{2\pi}\left(\arg \frac{z_1}{z_2} - \delta \log \left|\frac{z_1}{z_2}\right|\right)\right], \qquad (7.31)$$

where $[\zeta]$ stands for the integral part of ζ.

Remark. The right-hand side of (7.31) is independent of the particular choice of $\arg(z_1/z_2)$.

Proof. We have

$$\mathcal{S}(z_1, z_2; \delta, \varphi) = M_{z_1, z_2}(S_\delta), \quad \mathcal{A}(z_1, z_2; \varphi) = M_{z_1, z_2}(S_0), \qquad (7.32)$$

where S_δ is the logarithmic spiral $\{e^{2\pi i \varphi} r^{1 + i\delta} : r > 0\}$ and S_0 is the ray $\{e^{2\pi i \varphi} r : r > 0\}$. For the sake of definiteness, suppose $\delta > 0$. The case $\delta < 0$ can be treated analogously.

Let $\mathcal{L}$ be the closure of the set $\{z \in \mathbf{C} \setminus \mathcal{K}_\delta : \text{wind}\,(\mathcal{K}_\delta, z) \neq 0\}$. From (7.32) we infer that the point $s := M_{z_1, z_2}(e^{2\pi i \varphi})$ separates $\mathcal{L}$. Denote by $\mathcal{L}_{z_1}$ and $\mathcal{L}_{z_2}$ the closures of the connected components of $\mathcal{L} \setminus \{s\}$ containing z_1 and z_2, respectively. Since $\delta > 0$, we have

$$z \in \mathcal{L}_{z_1} \setminus \mathcal{K}_\delta \Longrightarrow \text{wind}\,(\mathcal{K}_\delta, z) > 0, \qquad (7.33)$$
$$z \in \mathcal{L}_{z_2} \setminus \mathcal{K}_\delta \Longrightarrow \text{wind}\,(\mathcal{K}_\delta, z) < 0. \qquad (7.34)$$

Clearly, wind $(\mathcal{K}_\delta, z) = 0$ if $z \in \mathbf{C} \setminus \mathcal{L}$.

If η changes from 0 to δ, then the curves $\mathcal{K}_\eta$ change from $\mathcal{K}_0$ to $\mathcal{K}_\delta$. Denote by N the number of $\eta \in (0, \delta)$ such that $0 \in \mathcal{K}_\eta$. From (7.33) and (7.34) we obtain

$$\text{wind}\,(\mathcal{K}_\delta, 0) = N \ \text{ if } \ 0 \in \mathcal{L}_{z_1} \setminus \mathcal{K}_\delta, \qquad (7.35)$$
$$\text{wind}\,(\mathcal{K}_\delta, 0) = -N \ \text{ if } \ 0 \in \mathcal{L}_{z_2} \setminus \mathcal{K}_\delta. \qquad (7.36)$$

Assume $0 \in \mathcal{L}_{z_1} \setminus \mathcal{K}_\delta$. Since $M_{z_1,z_2}(\mathbf{D}) \supset \mathcal{L}_{z_1} \setminus \mathcal{K}_\delta$ and $M_{z_1,z_2}(z_1/z_2) = 0$, this implies that $|z_1/z_2| < 1$. We have $0 \in \mathcal{K}_\eta$ if and only if there is an $r > 0$ such that

$$M_{z_1,z_2}(e^{2\pi i \varphi} r^{1+i\eta}) = 0,$$

and because $M_{z_1,z_2}(z_1/z_2) = 0$, this happens if and only if $e^{2\pi i \varphi} r^{1+i\eta} = z_1/z_2$ for some $r > 0$, i.e. if and only if

$$\varphi - \frac{1}{2\pi}\left(\arg \frac{z_1}{z_2} - \eta \log \left|\frac{z_1}{z_2}\right|\right) \in \mathbf{Z}. \tag{7.37}$$

Let

$$\varphi - \frac{1}{2\pi}\left(\arg \frac{z_1}{z_2} - \delta \log \left|\frac{z_1}{z_2}\right|\right) =: l_\delta + \lambda_\delta \quad (l_\delta \in \mathbf{Z}, \lambda_\delta \in (0,1))$$

and

$$\varphi - \frac{1}{2\pi}\arg \frac{z_1}{z_2} =: l_0 + \lambda_0 \quad (l_0 \in \mathbf{Z}, \lambda_0 \in (0,1)).$$

Since $\log |z_1/z_2| < 0$, it follows that the number of $\eta \in (0,\delta)$ satisfying (7.37) is equal to the cardinality of the set $\{l_\delta + 1, \ldots, l_0\}$. This and (7.35) imply that $\mathrm{wind}\,(\mathcal{K}_\delta, 0) = l_0 - l_\delta$, which is (7.31).

If $0 \in \mathcal{L}_{z_2} \setminus \mathcal{K}_\delta$, we see analogously that $0 \in \mathcal{K}_\eta$ if and only if (7.37) is valid. However, now $\log |z_1/z_2| > 0$ and therefore the number of $\eta \in (0,\delta)$ satisfying (7.37) equals the cardinality of the set $\{l_0 + 1, \ldots, l_\delta\}$, i.e. equals $l_\delta - l_0$. Taking into account (7.36), we arrive again at the equality $\mathrm{wind}\,(\mathcal{K}_\delta, 0) = l_0 - l_\delta$. $\square$

Let $z_1, z_2 \in \mathbf{C}$ be distinct points and suppose the arc $\mathcal{A}(z_1, z_2; \varphi)$ does not pass through the origin. We denote by $\Delta^0(z_1, z_2; \varphi)$ the increment of $(\arg z)/(2\pi)$ as z traces out $\mathcal{A}(z_1, z_2; \varphi)$ from z_1 to z_2.

Lemma 7.13. *We have*

$$\Delta^0(z_1, z_2; \varphi) + \left[\varphi - \frac{1}{2\pi}\arg \frac{z_1}{z_2}\right] + \frac{1}{2\pi}\arg \frac{z_1}{z_2} = 0. \tag{7.38}$$

Proof. First of all, we remark that the left-hand side of (7.38) does not depend on the choice of $\arg(z_1/z_2)$. Cut the plane along the ray from the origin through z_2 and choose $\arg z \in [\arg z_2, \arg z_2 + 2\pi)$. Since $0 \notin \mathcal{A}(z_1, z_2; \varphi)$, we see that $\arg(z_1/z_2) \notin 2\pi\varphi + \mathbf{Z}$. If $\arg z_1 \in [\arg z_2, \arg z_2 + 2\pi\varphi)$ then (7.38) reads

$$-\frac{1}{2\pi}(\arg z_1 - \arg z_2) + 0 + \frac{1}{2\pi}(\arg z_1 - \arg z_2) = 0,$$

while if $\arg z_1 \in (\arg z_2 + 2\pi\varphi, \arg z_2 + 2\pi)$, the equality (7.38) is nothing but

$$1 - \frac{1}{2\pi}(\arg z_1 - \arg z_2) - 1 + \frac{1}{2\pi}(\arg z_1 - \arg z_2) = 0. \square$$

Theorem 7.14. *Let Γ be a Carleson Jordan curve, $p \in (1, \infty)$, and $w \in A_p(\Gamma)$. Suppose $a \in PC(\Gamma)$ and denote by J_a the points $t \in \Gamma$ at which a has a jump. If $\lambda \notin \mathrm{sp}_{\mathrm{ess}}\, T(a)$, then the index of $T(a) - \lambda I$ equals minus the winding number of the closed, continuous, and naturally oriented curve*

$$a^{\#} := a^{\#}_{\Gamma, p, w} := \mathcal{R}(a) \cup \bigcup_{t \in J_a} \mathcal{L}\left(a(t - 0), a(t + 0); p, \frac{\alpha_t + \beta_t}{2}, \frac{\alpha_t + \beta_t}{2}\right)$$

about the point λ.

Proof. Since both $\mathrm{Ind}\,(T(a) - \lambda I) = \mathrm{Ind}\, T(a - \lambda)$ and the integer $\mathrm{wind}\,(a^{\#}, \lambda) = \mathrm{wind}\,(a^{\#} - \lambda, 0)$ are stable under small perturbations of a in $L^{\infty}(\Gamma)$, we may assume that a has only finitely many jumps. In this case Proposition 7.3 with $b := a - \lambda$ and $J_b := J_a$ tells us that

$$\mathrm{Ind}\, T(b) = -\frac{1}{2\pi} \sum_{\delta} \{\arg b\}_{\delta} + \sum_{t \in J_b} \left([\varkappa_t(\theta_t)] + \frac{1}{2\pi} \arg \frac{b(t - 0)}{b(t + 0)} \right), \qquad (7.39)$$

where θ_t is an arbitrary number in $[0, 1]$. We have to show that the right-hand side of (7.39) equals $-\mathrm{wind}\,(b^{\#}, 0)$.

Put $J_b =: \{t_1, \ldots, t_n\}$, $b(t_j - 0) =: c_j$, $b(t_j + 0) =: d_j$. From Theorem 3.31(d) we know that, for each j, there are $\lambda_j \in (-1/p, 1/q)$ and $\delta_j \in \mathbf{R}$ such that

$$\alpha_{t_j}(x) \le \lambda_j + \delta_j x \le \beta_{t_j}(x) \quad \text{for all } x \in \mathbf{R}. \qquad (7.40)$$

Theorem 7.4 shows that $\mathcal{S}(c_j, d_j; \delta_j, 1/p + \lambda_j) \subset \mathcal{L}(c_j, d_j; p, \alpha_{t_j}, \beta_{t_j})$ does not contain the origin, and without loss of generality we may also assume that $0 \notin \mathcal{A}(c_j, d_j; 1/p + \lambda_j)$ for all j (otherwise we may slightly perturb b without violating (7.39)). By virtue of (7.40), we can find $\theta_j \in [0, 1]$ such that

$$\theta_j \alpha_{t_j}\left(\frac{1}{2\pi} \log \left|\frac{c_j}{d_j}\right|\right) + (1 - \theta_j)\beta_{t_j}\left(\frac{1}{2\pi} \log \left|\frac{c_j}{d_j}\right|\right) = \lambda_j + \frac{\delta_j}{2\pi} \log \left|\frac{c_j}{d_j}\right|.$$

This implies that the right-hand side of (7.39) may be replaced by

$$-\frac{1}{2\pi} \sum_{\delta} \{\arg b\}_{\delta} + \sum_{j=1}^{n} \left(\left[\varphi_j - \frac{1}{2\pi}\left(\arg \frac{c_j}{d_j} - \delta_j \log \left|\frac{c_j}{d_j}\right| \right) \right] + \frac{1}{2\pi} \arg \frac{c_j}{d_j} \right) \qquad (7.41)$$

where $\varphi_j := 1/p + \lambda_j$. Put $\mathcal{K}_{\delta_j} := \mathcal{A}(c_j, d_j; \varphi_j) \cup \mathcal{S}(c_j, d_j; \delta_j, \varphi_j)$. From Lemma 7.12 we infer that (7.41) equals

$$-\frac{1}{2\pi} \sum_{\delta} \{\arg b\}_{\delta} + \sum_{j=1}^{n} \left(\left[\varphi_j - \frac{1}{2\pi} \arg \frac{c_j}{d_j} \right] + \frac{1}{2\pi} \arg \frac{c_j}{d_j} - \mathrm{wind}\,(\mathcal{K}_{\delta_j}, 0) \right),$$

and Lemma 7.13 shows that this is

$$-\frac{1}{2\pi}\sum_{\delta}\{\arg b\}_{\delta} + \sum_{j=1}^{n}\left(-\Delta^{0}(c_{j},d_{j};\varphi_{j}) - \text{wind}\,(\mathcal{K}_{\delta_{j}},0)\right). \qquad (7.42)$$

Denoting by $\Delta(c_{j},d_{j};\delta_{j},\varphi_{j})$ the increment of $(\arg z)/(2\pi)$ as z traces out the double spiral $\mathcal{S}(c_{j},d_{j};\delta_{j},\varphi_{j})$, we have wind $(\mathcal{K}_{\delta_{j}},0) = \Delta(c_{j},d_{j};\delta_{j},\varphi_{j}) - \Delta^{0}(c_{j},d_{j};\varphi_{j})$. Thus, (7.42) is equal to

$$-\frac{1}{2\pi}\sum_{\delta}\{\arg b\} - \sum_{j=1}^{n}\Delta(c_{j},d_{j};\delta_{j},\varphi_{j}). \qquad (7.43)$$

Since $\mathcal{S}(c_{j},d_{j};\delta_{j},\varphi_{j})$ and $\mathcal{L}(c_{j},d_{j};p,(\alpha_{t_{j}}+\beta_{t_{j}})/2,(\alpha_{t_{j}}+\beta_{t_{j}})/2)$ are homotopic within the leaf $\mathcal{L}(c_{j},d_{j};p,\alpha_{t_{j}},\beta_{t_{j}})$ and thus give the same increment of the argument, we finally arrive at the conclusion that (7.43) equals

$$-\frac{1}{2\pi}\sum_{\delta}\{\arg b\}_{\delta} - \sum_{j=1}^{n}\Delta\left(c_{j},d_{j};p,\frac{\alpha_{t_{j}}+\beta_{t_{j}}}{2},\frac{\alpha_{t_{j}}+\beta_{t_{j}}}{2}\right),$$

which is just $-\text{wind}\,(b^{\#},0)$. $\qquad\square$

Of course, in Theorem 7.14 we may replace $(\alpha_{t}+\beta_{t})/2$ by any continuous function ψ_{t} such that $\alpha_{t}(x) \leq \psi_{t}(x) \leq \beta_{t}(x)$ for all $x \in \mathbf{R}$.

Corollary 7.15. *Let Γ be a Carleson Jordan curve, $p \in (1,\infty)$, $w \in A_{p}(\Gamma)$, and $a \in PC(\Gamma)$. Then the spectrum of $T(a)$ on $L_{+}^{p}(\Gamma,w)$ is*

$$\text{sp}\,T(a) = \text{sp}_{\text{ess}}\,T(a) \cup \{\lambda \in \mathbf{C} \setminus \text{sp}_{\text{ess}}\,T(a) : \text{wind}\,(a^{\#},\lambda) \neq 0\}$$

where $\text{sp}_{\text{ess}}\,T(a)$ and $a^{\#}$ are as in Theorems 7.4 and 7.14.

Proof. Combine Corollary 6.19 and Theorem 7.14. $\qquad\square$

7.8 Semi-Fredholmness

Let Γ be a Carleson Jordan curve, $p \in (1,\infty)$, and $w \in A_{p}(\Gamma)$. In accordance with Section 6.3, an operator $A \in \mathcal{B}(L_{+}^{p}(\Gamma,w))$ is said to be semi-Fredholm if it is normally solvable and

$$\alpha(A) := \dim \text{Ker}\,A < \infty \quad \text{or} \quad \beta(A) := \dim\left(L_{+}^{p}(\Gamma,w)/\text{Im}\,A\right) < \infty.$$

Let $\Phi_{+} := \Phi_{+}(L_{+}^{p}(\Gamma,w))$ and $\Phi_{-} := \Phi_{-}(L_{+}^{p}(\Gamma,w))$ denote the set of all normally solvable operators $A \in \mathcal{B}(L_{+}^{p}(\Gamma,w))$ for which $\beta(A) = \infty$ and $\alpha(A) = \infty$, respectively. Thus, $\text{Ind}\,A = \mp\infty$ for $A \in \Phi_{\pm}$.

It is well known that Φ_+ and Φ_- contain Toeplitz operators, i.e. that there exist Toeplitz operators which are semi-Fredholm but not Fredholm. For example, if B is bounded and analytic in D_+ (the interior of the curve Γ) and has infinitely many zeros in D_+, then $T(B) \in \Phi_+$ and $T(\overline{B}) \in \Phi_-$. The purpose of this section is to show that $\Phi_+ \cup \Phi_-$ does not contain Toeplitz operators with symbols in $PC(\Gamma)$.

In other words, we show that every semi-Fredholm Toeplitz operator with piecewise continuous symbol is automatically Fredholm. This phenomenon is well known for nice curves. The following proposition is a first generalization of this result, and its proof is based on the strategy which is usually pursued in the case of nice curves.

Proposition 7.16. *Let Γ be a spiralic Carleson Jordan curve, i.e. suppose $\delta_t^- = \delta_t^+$ for all $t \in \Gamma$. Let further $p \in (1, \infty)$ and $w \in A_p(\Gamma)$. If $a \in PC(\Gamma)$ and $T(a)$ semi-Fredholm on $L_+^p(\Gamma, w)$, then $T(a)$ is Fredholm.*

Proof. Suppose $T(a) \in \Phi_-$. Since the set Φ_- is stable under perturbations with small norm, we may assume that a has only finitely many jumps. If $a = 0$, then $T(a) = 0$ and hence $T(a)$ cannot be semi-Fredholm. So let $a \neq 0$. Theorem 6.20 implies that $a \in GPC(\Gamma)$, i.e. that a is invertible in the algebra $PC(\Gamma)$. From Theorem 7.4 and Example 7.8 we know that $T(a)$ is Fredholm on $L_+^{1+\varepsilon}(\Gamma)$ ($\varepsilon > 0$) if and only if

$$0 \notin \mathcal{R}(a) \cup \bigcup_{t \in J_a} \mathcal{L}\big(a(t-0), a(t+0); 1 + \varepsilon, \alpha_t, \beta_t\big) \tag{7.44}$$

where $\alpha_t(x) = \beta_t(x) = \delta_t x$ with $\delta_t := \delta_t^- = \delta_t^+$.

We claim that (7.44) is always satisfied if $\varepsilon > 0$ is sufficiently small. Indeed, (7.44) is violated if and only if there is a point $t \in J_a$ and a number $x \in \mathbf{R}$ such that

$$M_{a(t-0),a(t+0)}\big(e^{2\pi x} e^{2\pi i(1/(1+\varepsilon)+\delta_t x)}\big) = 0,$$

which happens if and only if

$$e^{2\pi x} = \left|\frac{a(t-0)}{a(t+0)}\right| \quad \text{and} \quad \frac{1}{1+\varepsilon} + \delta_t x \in \frac{1}{2\pi} \arg \frac{a(t-0)}{a(t+0)} + \mathbf{Z}. \tag{7.45}$$

Clearly, (7.45) cannot hold if $\varepsilon > 0$ is small enough. Thus, if $\varepsilon > 0$ is sufficiently small, then $T(a)$ must be Fredholm on $L_+^{1+\varepsilon}(\Gamma)$.

Put $r := p/(1+\varepsilon)$ and let $1/r + 1/s = 1$. Then

$$\int_\Gamma |f|^{1+\varepsilon} |d\tau| \leq \left(\int_\Gamma |f|^{(1+\varepsilon)r} w^{(1+\varepsilon)r} |d\tau|\right)^{1/r} \left(\int_\Gamma w^{-(1+\varepsilon)s} |d\tau|\right)^{1/s}$$

$$= \left(\int_\Gamma |f|^p w^p |d\tau|\right)^{(1+\varepsilon)/p} \left(\int_\Gamma w^{-(1+\varepsilon)s} |ds|\right)^{1/s}. \tag{7.46}$$

Since $w \in A_p(\Gamma)$, we deduce from Theorem 2.31 that $w^{1+\delta} \in A_{\tilde{p}}(\Gamma)$ for all sufficiently small $|\delta|$ and all $\tilde{p}$ in some neighborhood of p. This implies that $w^{-1-\delta} \in L^{\tilde{q}}(\Gamma)$ and thus $w^{-(1+\delta)\tilde{q}} \in L^1(\Gamma)$ for all $\tilde{q}$ in some neighborhood of q. Consequently, $w^{-(1+\varepsilon)s} \in L^1(\Gamma)$ whenever $\varepsilon > 0$ is sufficiently small, and from (7.46) we obtain that

$$L^p(\Gamma, w) \subset L^{1+\varepsilon}(\Gamma). \tag{7.47}$$

Since, by assumption, $T(a)$ has an infinite-dimensional kernel on $L^p_+(\Gamma, w)$, we infer from (7.47) that the kernel of $T(a)$ is also infinite-dimensional on $L^{1+\varepsilon}_+(\Gamma)$. This, however, contradicts the Fredholmness of $T(a)$ on $L^{1+\varepsilon}_+(\Gamma)$.

If $T(a) \in \Phi_+(L^p_+(\Gamma, w))$, then Theorem 6.20 shows that again $a \in GPC(\Gamma)$, and from Lemma 6.14 and (6.21) one easily gets that $T(a^{-1})$ must belong to $\Phi_-(L^q_+(\Gamma, w^{-1}))$. This can be led to a contradiction as above. $\qquad\square$

The previous proof is based on the fact that the set of all $p \in (0, \infty)$ for which $T(a)$ is Fredholm on $L^p_+(\Gamma, w)$ has cluster points at 1 and ∞. The following proposition shows why the preceding proof does not work in the general case.

Proposition 7.17. *Let Γ be a Carleson Jordan curve, $t \in \Gamma$, and suppose $\delta_t^- < \delta_t^+$. If a is any function in $GPC(\Gamma)$ such that*

$$\left|a(t-0)/a(t+0)\right| \geq e^{4\pi/(\delta_t^+ - \delta_t^-)} \tag{7.48}$$

then there are no $p \in (1, \infty)$ and no $w \in A_p(\Gamma)$ such that $T(a)$ is Fredholm on $L^p_+(\Gamma, w)$.

Proof. Let $p \in (1, \infty)$, $w \in A_p(\Gamma)$, and denote by α_t and β_t the indicator functions of Γ, p, w at t. We show that $0 \in \mathcal{L}(a(t-0), a(t+0); p, \alpha_t, \beta_t)$, which, by Theorem 7.4, gives the assertion.

Denote by Δ the sector $\Delta := \{\gamma = x + iy \in \mathbf{C} : 1 + \delta_t^- x \leq y \leq \delta_t^+ x\}$. If $x + iy \in \Delta$ then, by Theorem 3.31,

$$\frac{1}{p} + \alpha_t(x) \leq \frac{1}{p} + \mu_t^- + \delta_t^- x < 1 + \delta_t^- x \leq y$$

$$\leq \delta_t^+ x < \frac{1}{p} + \nu_t^+ + \delta_t^+ x \leq \frac{1}{p} + \beta_t(x),$$

whence $\Delta \subset Y(p, \alpha_t, \beta_t)$. Put $\Delta_0 := \{\gamma = x + iy \in \Delta : x \geq 2/(\delta_t^+ - \delta_t^-)\}$. Since $\Delta_0 \subset \Delta$, it follows that

$$M_{a(t-0),a(t+0)}(e^{2\pi\Delta_0}) \subset \mathcal{L}(a(t-0), a(t+0); p, \alpha_t, \beta_t).$$

Hence, we are left with proving that the set $M_{a(t-0),a(t+0)}(e^{2\pi\Delta_0})$ contains the origin. By (7.48), there is an $x_0 \geq 2/(\delta_t^+ - \delta_t^-)$ such that

$$e^{2\pi x_0} = \left|a(t-0)/a(t+0)\right|. \tag{7.49}$$

Because $\delta_t^+ x_0 - \delta_t^- x_0 - 1 \geq 1$, the segment $\{x + iy \in \Delta_0 : x = x_0\}$ is of length at least 1. Thus, there is a y_0 such that $x_0 + iy_0 \in \Delta_0$ and

$$e^{2\pi i y_0} = e^{i \arg(a(t-0)/a(t+0))}. \tag{7.50}$$

From (7.49) and (7.50) we obtain $M_{a(t-0),a(t+0)}(e^{2\pi(x_0+iy_0)}) = 0.$ $\qquad\square$

The next theorem disposes of the general case.

Theorem 7.18. *Let Γ be a Carleson Jordan curve, $p \in (1,\infty)$, and $w \in A_p(\Gamma)$. If $a \in PC(\Gamma)$ and $T(a)$ is semi-Fredholm on $L_+^p(\Gamma, w)$, then $T(a)$ is Fredholm on $L_+^p(\Gamma, w)$.*

Proof. Suppose $T(a)$ is semi-Fredholm but not Fredholm. For the sake of definiteness, assume $T(a) \in \Phi_-$; the case where $T(a) \in \Phi_+$ can be reduced to the Φ_- case as in the last paragraph of the proof of Proposition 7.16. Since Φ_- is an open subset of $\mathcal{B}(L_+^p(\Gamma, w))$, we may assume that a has only finitely many jumps. Hence, we may write $a = a_1 \ldots a_n$ where $a_j \in PC(\Gamma)$ is continuous on $\Gamma \setminus \{t_j\}$. Theorem 6.28 tells us that $T(a) - T(a_1) \ldots T(a_n)$ is compact and that the operators $T(a_1), \ldots, T(a_n)$ commute modulo compact operators. This implies that each operator $T(a_j)$ is semi-Fredholm and that at least one of them belongs to Φ_- (see, e.g., [89, Section 4.15]). Considering the latter operator instead of the original operator $T(a)$, we may assume that a itself has at most one jump, say at $t \in \Gamma$.

Since $a \in GPC(\Gamma)$ due to Theorem 6.20, we have $a = c g_{t,\gamma}$ where $c \in GC(\Gamma)$ and $g_{t,\gamma}$ is as in Section 7.1 with an appropriate $\gamma \in \mathbf{C}$. As $T(a) - T(c)T(g_{t,\gamma})$ is compact (Corollary 6.22) and $T(c)$ is Fredholm (Theorem 6.24), we see that $T(g_{t,\gamma}) \in \Phi_-$. In particular, there exist infinitely many linearly independent functions $\varphi_+ \in L_+^p(\Gamma, w)$ such that $g_{t,\gamma}\varphi_+ =: \varphi_- \in \dot{L}_-^p(\Gamma, w)$. Write $g_{t,\gamma}(\tau) = (1 - t/\tau)^{-\gamma}(\tau - t)^\gamma$ as in (7.3) (with $\varkappa = 0$). The equality $\varphi_- = g_{t,\gamma}\varphi_+$ can then be rewritten as

$$(\tau - t)^\gamma \varphi_+(\tau) = (1 - t/\tau)^\gamma \varphi_-(\tau) \quad (\tau \in \Gamma \setminus \{t\})$$

or

$$(\tau - t)^{\gamma+m} \varphi_+(\tau) = (1 - t/\tau)^\gamma (\tau - t)^m \varphi_-(\tau) \quad (\tau \in \Gamma \setminus \{t\}), \tag{7.51}$$

where m is any integer. As in the proof of Lemma 7.1 we see that

$$\left| \arg(z - t)/\left(-\log|z - t| \right) \right| \leq M < \infty$$

for all $z \in D_+$ sufficiently close to t. Thus, for these z we have

$$\left| (z - t)^{\gamma+m} \right| = |z - t|^{m + \operatorname{Re}\gamma} e^{-\operatorname{Im}\gamma \arg(z-t)} \leq |z - t|^{m + \operatorname{Re}\gamma - M|\operatorname{Im}\gamma|}.$$

Consequently, the function $(z - t)^{\gamma + m}$ is analytic and bounded in D_+ whenever $m > M|\operatorname{Im}\gamma| - \operatorname{Re}\gamma$. It can be shown similarly that the function $(1 - t/z)^\gamma (z - t)^m$ is analytic and bounded on bounded subsets of D_- if only m is large enough. At infinity, the latter function has a pole of order m. Thus, the function on the left of (7.51) belongs to $L_+^p(\Gamma, w)$ together with φ_+ and, up to a polynomial summand g of degree at most $m - 1$, the function on the right of (7.51) lies in $\dot{L}_-^p(\Gamma, w)$ together with φ_-. After subtracting $g(\tau)$ from both sides of (7.51) we therefore get a function in $L_+^p(\Gamma, w) \cap \dot{L}_-^p(\Gamma, w) = \{0\}$ (Theorem 6.4 and Corollary 6.8). Thus, both sides of (7.51) are a polynomial $g(\tau)$ of degree at most $m - 1$, and since $\varphi_+(\tau) = (\tau - t)^{-\gamma - m} g(\tau)$, we arrive at the conclusion that $\dim \operatorname{Ker} T(a) \leq m$. This contradiction shows that $T(a)$ must be Fredholm.   □

7.9   Notes and comments

7.1–7.2. Harold Widom [209] studied the Cauchy singular integral operators $S_{\mathbf{R}_+}$ on $L^p(\mathbf{R}_+)$ using Wiener-Hopf factorization and arrived at the conclusion that the spectrum of $S_{\mathbf{R}_+}$ is a circular arc between -1 and 1 whose shape is determined by the value of p. He so was the first to factorize the symbol of a local representative *and* to interpret the result in geometric language.

The idea of choosing $g_{t,\gamma}$ as in Section 7.1 and of factorizing it as in (7.3) is the foundation of Gohberg and Krupnik's papers [82], [83], [84] (also see [89]), in which they set up the edifice of Toeplitz operators with piecewise continuous symbols on $L_+^p(\Gamma, w)$ for piecewise Lyapunov curves Γ and power weights w. Thus, the strategy of Sections 7.1 and 7.2 is the one of Gohberg and Krupnik. The concrete plan of the proof of Proposition 7.2 is borrowed from Spitkovsky's article [200].

Wiener-Hopf factorization of $g_{t,\gamma}$ eventually leads to checking whether

$$|a_+^{-1}(\tau)|w(\tau) = |(\tau - t)^{\varkappa - \gamma}|w(\tau) = |\tau - t|^{\varkappa - \operatorname{Re}\gamma} e^{\operatorname{Im}\gamma \arg(\tau - t)} w(\tau) \qquad (7.52)$$

is a weight in $A_p(\Gamma)$. If Γ is piecewise Lyapunov (or more general, if Γ has no helical points), then $\arg(\tau - t) = O(1)$ and hence

$$|(\tau - t)^{\varkappa - \gamma}| = |\tau - t|^{\varkappa - \operatorname{Re}\gamma} e^{O(1)}.$$

One therefore has to decide whether $|\tau - t|^{\varkappa - \operatorname{Re}\gamma} w(\tau)$ belongs to $A_p(\Gamma)$. In the case where $w(\tau)$ is a power weight, say $w(\tau) = |\tau - t|^\lambda$ for simplicity, this happens if and only if $-1/p < \varkappa - \operatorname{Re}\gamma + \lambda < 1/q$, which is the basic inequality of the Gohberg/Krupnik/Widom theory. In the case of a general weight $w \in A_p(\Gamma)$, one arrives at characterizing the real numbers μ for which $|\tau - t|^\mu w(\tau)$ is a weight in $A_p(\Gamma)$, and this problem was solved by Spitkovsky [200]. We here consider general Carleson curves Γ and general weights $w \in A_p(\Gamma)$. In this situation (7.52) cannot be simplified and the problem is equivalent to finding just the indicator set

$$N_t(\Gamma, p, w) := \{\mu \in \mathbf{C} : |(\tau - t)^\mu|w(\tau) \in A_p(\Gamma)\}.$$

At this point it is clear that almost the entire Chapter 3 serves the formulation of Lemma 7.1 and Propositions 7.2 and 7.3, which were first established in [18].

7.3–7.6. In this generality, all results of Sections 7.3 to 7.6 are due to the authors [15], [16], [18]. In particular, the indicator functions and leaves were introduced only in [18]. Our survey [17] was written only after [18] (at the invitation of people interested in interpolation theory and submultiplicative functions) and is the first published work containing Theorem 7.4.

If Γ is a piecewise Lyapunov curve and w is a power weight, then the leaves are line segments or circular arcs (Examples 7.5 and 7.6) and in this case Theorem 7.4 is Gohberg and Krupnik's [84], [89]. For piecewise Lyapunov curves and arbitrary Muckenhoupt weights $w \in A_p(\Gamma)$, Theorem 7.4 was established by Spitkovsky [200]; in that case the leaves are horns (Example 7.7). We also remark that a result like Proposition 7.3 (but not like Theorem 7.4) is already in [2], [3] under the assumption that Γ is close to a logarithmic spiral and w is a power weight.

Spiralic horns (i.e. Theorem 7.4 in the situation of Example 7.9) first appeared in our paper [15]. The case of general Carleson curves without weights or with at most power weights was settled in [16], and in [18] we finally succeeded in disposing of the problem for general Carleson curves Γ and general Muckenhoupt weights $w \in A_p(\Gamma)$, that is, in formulating and proving Theorem 7.4 as it is cited here.

7.7. Theorem 7.14 and Corollary 7.15 are due to Gohberg and Krupnik [84], [89] for piecewise Lyapunov curves Γ and power weights w, to Spitkovsky [200] for piecewise Lyapunov curves Γ and arbitrary weights $w \in A_p(\Gamma)$, and were established in the form presented here in [18].

7.8. For piecewise Lyapunov curves Γ and power weights w, Proposition 7.16 and the idea of its proof are Gohberg and Krupnik's [89]. Proposition 7.17 as well as Theorem 7.18 and its proof first appeared in our paper [10] with Bishop and Spitkovsky.

Chapter 8
Banach algebras

Toeplitz operators are the elementary building stones of algebras of singular integral operators. This chapter is devoted to results which are needed in order to understand the architecture according to which the stones are assembled in the structure.

Singular operators are of local type, i.e. algebras of singular integral operators have a non-trivial center modulo compact operators. This enables us to make use of so-called local principles, which may be regarded as generalizations of Gelfand theory to non-commutative Banach algebras. Applying local principles to algebras of singular integral operators with piecewise continuous coefficients on Jordan curves leads to the study of Banach algebras generated by two idempotents (or "projections"), while the case of composed curves requires the investigation of Banach algebras generated by N idempotents which, fortunately, are subject to some additional conditions. In this chapter we establish an appropriate "N projections theorem" and we show how the $N = 2$ version of this theorem yields a symbol calculus for algebras associated with Jordan curves.

8.1 General theorems

Throughout what follows a *Banach algebra* is a complex Banach space $\mathcal{B}$ with an associative and distributive multiplication which satisfies $\|ab\| \leq \|a\| \, \|b\|$ for all $a, b \in \mathcal{B}$. If a Banach algebra has a unit element (which will frequently also be called the identity), then it is usually denoted by e or I. We always require that $\|e\| = \|I\| = 1$.

Let $\mathcal{B}$ be a Banach algebra with identity element e. An element $a \in \mathcal{B}$ is said to be *invertible* (in $\mathcal{B}$) if there is an element $b \in \mathcal{B}$ such that $ab = ba = e$. The *spectrum* of an element $a \in \mathcal{B}$ is defined by

$$\mathrm{sp}_{\mathcal{B}} a := \{\lambda \in \mathbf{C} : a - \lambda e \text{ is not invertible in } \mathcal{B}\}.$$

Sometimes, if the algebra $\mathcal{B}$ is clear from the context, we will abbreviate $\mathrm{sp}_{\mathcal{B}}a$ to $\mathrm{sp}\,a$.

Given two Banach algebras $\mathcal{A}$ ans $\mathcal{B}$, we say that a map $\varphi : \mathcal{A} \to \mathcal{B}$ is a *Banach algebra homomorphism* if φ is a bounded linear operator and $\varphi(ab) = \varphi(a)\varphi(b)$ for all $a, b \in \mathcal{A}$. Bijective Banach algebra homomorphisms are called *Banach algebra isomorphisms*.

A closed subalgebra J of a Banach algebra $\mathcal{B}$ is called a *closed two-sided ideal* of $\mathcal{B}$ if $ja \in J$ and $aj \in J$ whenever $j \in J$ and $a \in \mathcal{B}$. A closed two-sided ideal $J \subset \mathcal{B}$ is referred to as a *maximal ideal* if $J \neq \mathcal{B}$ and if there is no closed two-sided ideal $K \subset \mathcal{B}$ such that $J \subset K$, $J \neq K$, and $K \neq \mathcal{B}$. If J is a closed two-sided ideal of $\mathcal{B}$, then the *quotient algebra* $\mathcal{B}/J$ is a Banach algebra with the algebraic operations

$$\alpha(a + J) := \alpha a + J, \ (a + J) + (b + J) := (a + b) + J, \ (a + J)(b + J) := ab + J$$

and the norm $\|a + J\| := \mathrm{dist}\,(a, J)$.

Now let $\mathcal{B}$ be a commutative Banach algebra with identity e. The Banach algebra homomorphisms of $\mathcal{B}$ into $\mathbf{C}$ which send e to 1 are referred to as the *multiplicative linear functionals* of $\mathcal{B}$. Let $\mathcal{M}$ denote the set of all maximal ideals of $\mathcal{B}$ and let M stand for the set of all multiplicative linear functionals on $\mathcal{B}$. One can show that the map $M \to \mathcal{M}$, $\varphi \mapsto \mathrm{Ker}\,\varphi$ is bijective. Therefore no distinction is usually made between multiplicative linear functionals and maximal ideals. The formula $\hat{a}(m) = m(a)$ $(m \in M)$ assigns a function $\hat{a} : M \to \mathbf{C}$ to each $a \in \mathcal{B}$. This function is called the *Gelfand transform* of a. Let $\hat{\mathcal{B}}$ be the set $\{\hat{a} : a \in \mathcal{B}\}$. The *Gelfand topology* on M is the coarsest (weakest) topology on M that makes all the functions $\hat{a} \in \hat{\mathcal{B}}$ continuous. The set M equipped with the Gelfand topology is called the *maximal ideal space* of $\mathcal{B}$. One can show that M is a compact Hausdorff space. The map

$$\mathcal{G} : \mathcal{B} \to C(M), \ a \mapsto \hat{a}$$

is referred to as the *Gelfand map* of $\mathcal{B}$.

Theorem 8.1 (Gelfand). *Let $\mathcal{B}$ be a commutative Banach algebra with identity element and let M be the maximal ideal space of $\mathcal{B}$. An element $a \in \mathcal{B}$ is invertible if and only if $\hat{a}(m) \neq 0$ for all $m \in M$.*

In words: the Gelfand map is a Banach algebra homomorphism which preserves spectra. Theorem 8.1 is *proved* in every textbook on Banach algebras. $\qquad\square$

In Theorem 8.1, one associates with every element a of a commutative Banach algebra a collection of numbers, $\{\hat{a}(m)\}_{m \in M}$, in terms of which one can decide whether the given element is invertible or not. The idea behind so-called *local principles* is to associate with an element of a non-commutative Banach algebra a set of simpler objects which can answer for invertibility of the given element. One concrete realization of this strategy is the local principle of G.R. Allan and R. Douglas.

The *center* of a Banach algebra $\mathcal{B}$ is the set of all $z \in \mathcal{B}$ such that $za = az$ for all $a \in \mathcal{B}$.

Theorem 8.2 (Allan-Douglas). *Let $\mathcal{B}$ be a Banach algebra with identity e and let Z be a closed subalgebra of the center of $\mathcal{B}$ containing e. Denote the maximal ideal space of Z by Ω, and for each maximal ideal $\omega \in \Omega$, let J_ω be the smallest closed two-sided ideal of $\mathcal{B}$ which contains the set ω. An element $a \in \mathcal{B}$ is invertible in $\mathcal{B}$ if and only if the coset $a + J_\omega$ is invertible in $\mathcal{B}/J_\omega$ for every $\omega \in \Omega$.*

A *proof* of this theorem is in [23] or [96]. $\square$

We remark that if $J_\omega = \mathcal{B}$, then we consider $a + J_\omega$ as invertible by definition. The algebra $\mathcal{B}/J_\omega$ is referred to as the *local algebra of $\mathcal{B}$* at $\omega \in \Omega$, the coset $a + J_\omega$ is said to be the *local representative* of a at $\omega \in \Omega$, and the spectrum of $a + J_\omega$ in $\mathcal{B}/J_\omega$ is called the *local spectrum* of a at $\omega \in \Omega$.

If $\mathcal{B}$ itself is commutative, we can take $Z = \mathcal{B}$, and since then $\mathcal{B}/J_\omega = \mathcal{B}/\omega$ is isomorphic to $\mathbf{C}$ (Gelfand-Mazur theorem), Theorem 8.2 goes over into Theorem 8.1. Clearly, the larger the center of an algebra the finer one can "localize" in $\mathcal{B}$ using Theorem 8.2. In case the center of $\mathcal{B}$ is trivial, i.e. equals $\{\lambda e : \lambda \in \mathbf{C}\}$, Theorem 8.2 merely says that a is invertible if and only if a is invertible.

For a positive integer n, let $\mathbf{C}^{n \times n}$ denote the algebra of all complex $n \times n$ matrices. We identify $\mathbf{C}^{n \times n}$ with the Banach algebra $\mathcal{B}(\mathbf{C}^n)$ of all bounded linear operators on the (column-vectors) $\mathbf{C}^n$. Thus, the norm in $\mathbf{C}^{n \times n}$ is the operator norm in $\mathcal{B}(\mathbf{C}^n)$. Invertibility in $\mathbf{C}^{n \times n}$ may be studied via determinants. For a in $\mathbf{C}^{n \times n}$, the spectrum $\mathrm{sp}\, a$ is simply the set of eigenvalues of a.

The *standard polynomial F_{2n}* is defined by

$$F_{2n}(a_1, \ldots, a_{2n}) = \sum_{\sigma \in S_{2n}} (\mathrm{sign}\, \sigma) a_{\sigma(1)} \cdots a_{\sigma(2n)}$$

where S_{2n} is the group of all permutations of the set $\{1, 2, \ldots, 2n\}$. A Banach algebra $\mathcal{B}$ is said to be an *F_{2n}-algebra* if

$$F_{2n}(a_1, \ldots, a_{2n}) = 0 \quad \text{for all} \quad a_1, \ldots, a_{2n} \in \mathcal{B}.$$

For $n = 1$ we have
$$F_2(a_1, a_2) = a_1 a_2 - a_2 a_1$$

and thus, the F_2-algebras are just the commutative Banach algebras. The Amitsur-Levitzki theorem says that $\mathbf{C}^{n \times n}$ is an F_{2n}-algebra (see, e.g., [134, Theorem 20.1]). Finally, a Banach algebra $\mathcal{B}$ is referred to as an *F_{2n}^m-algebra* if

$$\left(F_{2n}(a_1, \ldots, a_{2n}) \right)^m = 0 \quad \text{for all} \quad a_1, \ldots, a_{2n} \in \mathcal{B}.$$

Clearly, an F_{2n}-algebra is an F_{2n}^m-algebra for every $m \geq 1$.

Theorem 8.3 (Krupnik-Finck-Roch). *Let $\mathcal{B}$ be an F_{2n}^m-algebra with identity element.*

(a) *For each maximal ideal J of $\mathcal{B}$, there exists a Banach algebra isomorphism*

$$\varphi_J : \mathcal{B}/J \to \mathbf{C}^{l \times l} \ \text{ with some } \ l = l(J) \le n.$$

(b) *An element $a \in \mathcal{B}$ is invertible if and only if $\varphi_J(a + J) \in \mathbf{C}^{l \times l}$ $(l = l(J))$ is invertible for every maximal ideal J of $\mathcal{B}$.*

A *proof* is in [134] for $m = 1$ and in [67] for general m.   □

Notice that if $m = n = 1$, then Theorem 8.3 coincides with Theorem 8.1.

8.2   Operators of local type

Let Γ be a composed Carleson curve, $p \in (1, \infty)$, and $w \in A_p(\Gamma)$. Also suppose that Γ is a compact subset of $\mathbf{C}$, i.e. that Γ is bounded and contains all its possible "endpoints". For brevity, we use the abbreviations

$$\mathcal{B}(\Gamma) := \mathcal{B}\big(L^p(\Gamma, w)\big), \quad \mathcal{K}(\Gamma) := \mathcal{K}\big(L^p(\Gamma, w)\big)$$

for the bounded and compact linear operators on $L^p(\Gamma, w)$. In these notation and in similar notations we will introduce in the following, we suppress the dependence on p and w; we rather think of Γ as an object which is a curve and carries a number p and a weight w.

Given a closed subalgebra E of $L^\infty(\Gamma)$, we let alg (S_Γ, E) stand for the smallest closed subalgebra of $\mathcal{B}(\Gamma)$ which contains the Cauchy singular integral operator S_Γ and all multiplication operators aI with $a \in E$. The operators in alg (S_Γ, E) are called *singular integral operators with coefficients in E*. Our aim is to establish Fredholm criteria for operators in alg $(S_\Gamma, PC(\Gamma))$.

Thus, we are interested in the invertibility of cosets $\pi(A) := A + \mathcal{K}(\Gamma)$ in the quotient algebra $\pi(\mathcal{B}(\Gamma)) := \mathcal{B}(\Gamma)/\mathcal{K}(\Gamma)$. Let

$$Z(\Gamma) := \big\{\pi(cI) : c \in C(\Gamma)\big\},$$
$$\Lambda(\Gamma) := \big\{A \in \mathcal{B}(\Gamma) : \pi(A)\pi(cI) = \pi(cI)\pi(A) \ \text{ for all } \ c \in C(\Gamma)\big\}.$$

In other words, $\Lambda(\Gamma)$ is the collection of all operators in $\mathcal{B}(\Gamma)$ which commute modulo compact operators with multiplications by continuous functions. Following [195], we refer to the operators in $\Lambda(\Gamma)$ as *operators of local type*.

Clearly, $\Lambda(\Gamma)$ is a closed subalgebra of $\mathcal{B}(\Gamma)$. The algebra $Z(\Gamma)$ is in general not a subalgebra of the center of $\mathcal{B}(\Gamma)/\mathcal{K}(\Gamma)$. However, $Z(\Gamma)$ is a closed subalgebra of the center of $\Lambda(\Gamma)/\mathcal{K}(\Gamma)$. This is the reason for narrowing down $\mathcal{B}(\Gamma)$ to $\Lambda(\Gamma)$. The following observation shows that $\Lambda(\Gamma)$ contains the operators we are interested in.

Proposition 8.4. *The operators in* $\mathrm{alg}\,(S_\Gamma, L^\infty(\Gamma))$ *are of local type.*

Proof. Multiplication operators are obviously of local type. When proving Proposition 6.21 we showed that $cS_\Gamma - S_\Gamma cI$ is compact for every $c \in C(\Gamma)$, i.e. that S_Γ is of local type. Since $\Lambda(\Gamma)$ is a closed algebra, we get the assertion. $\quad\square$

Let $A \in \Lambda(\Gamma)$ and suppose A is Fredholm. Then there is an operator $R \in \mathcal{B}(\Gamma)$ such that
$$\pi(R)\pi(A) = \pi(A)\pi(R) = \pi(I).$$
Since $\pi(A)$ commutes with the elements of $Z(\Gamma)$, so also does $\pi(R)$. Thus, an operator $A \in \Lambda(\Gamma)$ is Fredholm if and only if $\pi(A)$ is invertible in $\pi(\Lambda(\Gamma)) := \Lambda(\Gamma)/\mathcal{K}(\Gamma)$. Consequently, in order to decide whether operators of local type are Fredholm we may have recourse to Theorem 8.2 with $\mathcal{B} := \pi(\Lambda(\Gamma))$ and $Z := Z(\Gamma)$ as above.

The maximal ideal space of $Z(\Gamma)$ may be identified with Γ: the maximal ideal corresponding to $t \in \Gamma$ is the set
$$\{\pi(cI) : c \in C(\Gamma),\ c(t) = 0\}. \tag{8.1}$$
For $t \in \Gamma$, let $J_t(\Gamma)$ be the smallest closed two-sided ideal of $\pi(\Lambda(\Gamma))$ containing the set (8.1) and define
$$\mathcal{B}_t(\Gamma) := \pi\big(\Lambda(\Gamma)\big)/J_t(\Gamma), \quad \pi_t(A) := \pi(A) + J_t(\Gamma).$$

Theorem 8.2 now gives the following.

Proposition 8.5. *An operator A of local type is Fredholm on $L^p(\Gamma, w)$ if and only if $\pi_t(A)$ is invertible in $\mathcal{B}_t(\Gamma)$ for every $t \in \Gamma$. Equivalently,*
$$\mathrm{sp}_{\mathrm{ess}}\, A = \bigcup_{t \in \Gamma} \mathrm{sp}\,\pi_t(A). \tag{8.2}$$

$\quad\square$

We call $\mathcal{B}_t(\Gamma)$ the *local algebra* of Γ (and of p and w) at t, $\pi_t(A)$ the *local representative of A at t*, and $\mathrm{sp}\,\pi_t(A)$ the *local spectrum of A at t*. In (8.2), the essential spectrum of an operator is represented as the union of its local spectra. If $\varphi \in C(\Gamma)$ and $\varphi(t) = 1$, then $\pi(\varphi I) - \pi(I) = \pi\big((\varphi - 1)I\big) \in J_t(\Gamma)$ and hence $\pi_t(\varphi I) = \pi_t(I)$. It follows that
$$\pi_t(A) = \pi_t(\varphi I)\pi_t(A)\pi_t(\varphi I) = \pi_t(\varphi A \varphi I) \tag{8.3}$$
for every operator $A \in \Lambda(\Gamma)$. In particular, A and $\varphi A \varphi I$ have the same local spectrum at the point t. Since $\varphi A \varphi I$ does not depend on the properties of Γ and w outside the support $\mathrm{supp}\,\varphi$ of φ and because $\mathrm{supp}\,\varphi$ may be chosen as small as desired, it is clear that $\mathrm{sp}\,\pi_t(A)$ depends in fact only on the "local properties" of Γ and w at the point t.

The following result provides us with a simple description of the ideals $J_t(\Gamma)$.

Proposition 8.6. *Let Γ be a composed Carleson curve, $p \in (1, \infty)$, and $w \in A_p(\Gamma)$. Then for every $t \in \Gamma$,*

$$J_t(\Gamma) = \{\pi(cA) : c \in C(\Gamma),\ c(t) = 0,\ A \in \Lambda(\Gamma)\}. \tag{8.4}$$

Proof. Put $I_t := \{c \in C(\Gamma) : c(t) = 0\}$. By definition, the ideal $J_t(\Gamma)$ is the closure in $\Lambda(\Gamma)/\mathcal{K}(\Gamma)$ of all finite sums

$$\sum_{j=1}^{n} \pi(c_j A_j)\ \text{ with } c_j \in I_t,\ A_j \in \Lambda(\Gamma). \tag{8.5}$$

We claim that every sum (8.5) is actually of the form $\pi(cA)$ with $c \in I_t$ and $A \in \Lambda(\Gamma)$. Obviously, it suffices to prove this for $n = 2$.

Let $c_1, c_2 \in I_t$ and $A_1, A_2 \in \Lambda(\Gamma)$. Put $c := \sqrt{|c_1| + |c_2|}$. Then $c \in I_t$. Let $N_c := \{\tau \in \Gamma : c(\tau) = 0\}$ and for $j \in \{1, 2\}$, define

$$g_j(\tau) = \begin{cases} c_j(\tau)/c(\tau) & \text{if } \tau \notin N_c, \\ 0 & \text{if } \tau \in N_c. \end{cases}$$

If $\tau \notin N_c$ then

$$|g_j(\tau)| = \frac{|c_j(\tau)|}{c(\tau)} = \frac{|c_j(\tau)|}{|c_1(\tau)| + |c_2(\tau)|} c(\tau) \le c(\tau)$$

and hence, $g_j \in C(\Gamma)$. Since $c_1 = cg_1$ and $c_2 = cg_2$, it follows that $c_1 A_1 + c_2 A_2 = c(g_1 A_1 + g_2 A_2)$, which proves our claim.

The assertion will now follow once we have shown that the set on the right of (8.4) is closed. Let $\pi(B)$ be in the closure of the right-hand of (8.4). Given any converging series $\sum_{k=1}^{\infty} \varepsilon_k$ of positive numbers, we can find $A_k \in \Lambda(\Gamma)$ and $\varphi_k \in I_t$ such that $\|\pi(B - \varphi_k A_k)\| < \varepsilon_k/2$ for all k. Further, for each k there is an open neighborhood $U_k \subset \Gamma$ of t such that $\|\pi(A_k)\|\,|\varphi_k(\tau)| < \varepsilon_k/2$ for all $\tau \in U_k$. We may without loss of generality assume that $\overline{U}_{k+1} \subset U_k$ for all k. Finally, there are functions $\psi_k \in C(\Gamma)$ such that $0 \le \psi_k \le 1$, $\psi_k|\overline{U}_{k+1} = 1$, $\psi_k|\Gamma \setminus U_k = 0$ for all k. We therefore have $\|\pi(\varphi_k \psi_k I)\| \le \|\varphi_k \psi_k\|_\infty \le \sup_{\tau \in U_k} |\varphi_k(\tau)|$ and hence,

$$\|\pi(\psi_k B)\| \le \|\pi(B - \varphi_k A_k)\|\,\|\pi(\psi_k I)\| + \|\pi(A_k)\|\,\|\pi(\varphi_k \psi_k I)\| < \varepsilon_k.$$

Consequently, $\pi(E_n) := \pi(B) + \sum_{k=1}^{n} \pi(\psi_k B)$ is a Cauchy sequence. Let $\pi(E) \in \Lambda(\Gamma)/\mathcal{K}(\Gamma)$ denote the limit of $\pi(E_n)$ as $n \to \infty$. From the estimate

$$0 \le \left(1 + \sum_{k=1}^{n} \psi_k(\tau)\right)^{-1} - \left(1 + \sum_{k=1}^{n+m} \psi_k(\tau)\right)^{-1}$$

$$\le \begin{cases} \left(1 + \sum_{k=1}^{n} \psi_k(\tau)\right)^{-1} & \text{for } \tau \in U_{n+1} \\ 0 & \text{for } \tau \in \Gamma \setminus U_{n+1} \end{cases} \le (n+1)^{-1},$$

we see that $(1 + \sum_{k=1}^{n} \psi_k)^{-1}$ converges in $C(\Gamma)$ to some function $c \in C(\Gamma)$. Clearly, $c(t) = 0$ and thus $c \in I_t$. Since $\pi(B) = \pi(cE)$ by construction, the proof is complete. $\qquad\square$

8.3 Algebras generated by idempotents

The local algebra $\mathcal{B}_t(\Gamma) = \pi_t(\Lambda(\Gamma))$ is of very intricate structure. However, to study operators in $\mathrm{alg}\,(S_\Gamma, PC(\Gamma))$, we have mainly to deal with a certain subalgebra of $\mathcal{B}_t(\Gamma)$, namely, with the smallest closed subalgebra $\mathcal{A}_t(\Gamma)$ of $\mathcal{B}_t(\Gamma)$ containing the set $\{\pi_t(A) : A \in \mathrm{alg}\,(S_\Gamma, PC(\Gamma))\}$. It turns out that the algebra $\mathcal{A}_t(\Gamma)$ is much better than $\mathcal{B}_t(\Gamma)$: under certain circumstances, it is generated by a finite number of idempotents. Recall that an element p of a Banach algebra is called an *idempotent* (or, somewhat loosely, also a *projection*), if $p^2 = p$.

Suppose Γ is a Carleson Jordan curve. We know from Corollary 6.6 that then $S_\Gamma^2 = I$ and hence, $P_\Gamma := (I + S_\Gamma)/2$ is a projection. Clearly, $\mathrm{alg}\,(S_\Gamma, PC(\Gamma)) = \mathrm{alg}\,(P_\Gamma, PC(\Gamma))$. Since P_Γ is a projection, the element $r := \pi_t(P_\Gamma)$ is an idempotent element of the algebra $\mathcal{A}_t(\Gamma)$. Further, every function $a \in PC(\Gamma)$ may be written in the form

$$a(\tau) = a(t+0)\chi_t(\tau) + a(t-0)\big(1 - \chi_t(\tau)\big) + \varphi(\tau)$$

where χ_t is the characteristic function of any proper subarc of Γ starting at t and φ is continuous on Γ and zero at t. It follows that

$$
\begin{aligned}
\pi_t(aI) &= a(t+0)\pi_t(\chi_t I) + a(t-0)\big(\pi_t(I) - \pi_t(\chi_t I)\big) \\
&= a(t+0)s + a(t-0)(e-s)
\end{aligned}
$$

where $e := \pi_t(I)$ and $s := \pi_t(\chi_t I)$. Since $\chi_t^2 = \chi_t$, the element s is also an idempotent in $\mathcal{A}_t(\Gamma)$. In summary, we see that $\mathcal{A}_t(\Gamma)$ is generated by the identity e and the two idempotents r and s.

Things are more complicated for composed Carleson curves and we will devote Chapter 9 to this case. At the present moment, we leave the matter with the observation that algebras of singular integral operators with piecewise continuous coefficients motivate the study of Banach algebras generated by a finite number of idempotents.

The following "two projections theorem" is just the result we will employ for operators on Jordan curves.

Theorem 8.7. *Let $\mathcal{B}$ be a Banach algebra with identity e and let r and s be idempotents in $\mathcal{B}$. Let further $\mathcal{A}$ stand for the smallest closed subalgebra of $\mathcal{B}$ containing $e, r,$ and s. Put*

$$X := rsr + (e-r)(e-s)(e-r)$$

and suppose the points 0 and 1 are cluster points of the spectrum $\mathrm{sp}_\mathcal{B} X$. Define the map $\sigma_x : \{e, r, s\} \to \mathbf{C}^{2\times 2}$ for $x \in \mathbf{C} \setminus \{0, 1\}$ by

$$
\sigma_x(e) = \begin{pmatrix} 1 & 0 \\ 0 & 1 \end{pmatrix}, \quad
\sigma_x(r) = \begin{pmatrix} 1 & 0 \\ 0 & 0 \end{pmatrix}, \quad
\sigma_x(s) = \begin{pmatrix} x & x-1 \\ -x & 1-x \end{pmatrix} \tag{8.6}
$$

and for $x \in \{0, 1\}$ by

$$\sigma_x(e) = \begin{pmatrix} 1 & 0 \\ 0 & 1 \end{pmatrix}, \ \sigma_x(r) = \begin{pmatrix} 1 & 0 \\ 0 & 0 \end{pmatrix}, \ \sigma_x(s) = \begin{pmatrix} x & 0 \\ 0 & 1 - x \end{pmatrix}. \tag{8.7}$$

(a) *For each $x \in \mathrm{sp}_{\mathcal{B}} X$ the map σ_x extends to a Banach algebra homomorphism σ_x of $\mathcal{A}$ onto $\mathbf{C}^{2 \times 2}$.*

(b) *An element $a \in \mathcal{A}$ is invertible in $\mathcal{B}$ if and only if $\sigma_x(a)$ is invertible in $\mathbf{C}^{2 \times 2}$ for every $x \in \mathrm{sp}_{\mathcal{B}} X$.*

(c) *An element $a \in \mathcal{A}$ is invertible in $\mathcal{A}$ if and only if $\sigma_x(a)$ is invertible in $\mathbf{C}^{2 \times 2}$ for every $x \in \mathrm{sp}_{\mathcal{A}} X$.*

The proof is deferred to the next section. We here confine ourselves to the following remark. For $x \in \mathbf{C} \setminus \{0, 1\}$, let $\sqrt[4]{x/(1-x)}$ be any number with $\left(\sqrt[4]{x/(1-x)} \right)^4 = x/(1-x)$ and define $\sqrt[4]{(1-x)/x}$ as $\left(\sqrt[4]{x/(1-x)} \right)^{-1}$. Put

$$E_x = \mathrm{diag} \left(\sqrt[4]{x/(1-x)}, \ -\sqrt[4]{(1-x)/x} \right).$$

Then

$$E_x \begin{pmatrix} x & x - 1 \\ -x & 1 - x \end{pmatrix} E_x^{-1} = \begin{pmatrix} x & \sqrt{x(1-x)} \\ \sqrt{x(1-x)} & 1 - x \end{pmatrix}, \tag{8.8}$$

$$E_x \begin{pmatrix} 1 & 0 \\ 0 & 0 \end{pmatrix} E_x^{-1} = \begin{pmatrix} 1 & 0 \\ 0 & 0 \end{pmatrix}.$$

If $x \in \{0, 1\}$, then the matrix on the right of (8.8) coincides with the matrix for $\sigma_x(s)$ in (8.7). Thus, the conclusions (a),(b),(c) of Theorem 8.7 remain true with (8.6) and (8.7) replaced by

$$\sigma_x(e) = \begin{pmatrix} 1 & 0 \\ 0 & 1 \end{pmatrix}, \ \sigma_x(r) = \begin{pmatrix} 1 & 0 \\ 0 & 0 \end{pmatrix},$$

$$\sigma_x(s) = \begin{pmatrix} x & \sqrt{x(1-x)} \\ \sqrt{x(1-x)} & 1 - x \end{pmatrix}$$

for all $x \in \mathbf{C}$. We prefer the "root-free" version of the theorem because this version can be extended to the case of more than two idempotents without invoking nth roots.

This is also the right place to emphasize that there is no direct extension of Theorem 8.7 to algebras generated by the identity and three or more idempotents. Recall that a Banach algebra is said to be *separable* if it contains a countable dense subset. Evidently, every Banach algebra generated by finitely many elements is separable.

Theorem 8.8. *Every separable Banach algebra is isomorphic to a subalgebra of an algebra generated by the identity and three idempotents.*

A *proof* is in [172]. □

The preceding theorem tells us that reasonably generalizing Theorem 8.7 to algebras generated by a finite number of idempotents is impossible without imposing additional axioms on the idempotents. Such a system of axioms will be introduced in the next section. This system looks strange at the first glance, but its axioms are precisely satisfied for local algebras of singular integral operators.

8.4 An N projections theorem

The purpose of this section is to prove the following theorem.

Theorem 8.9. *Let $\mathcal{B}$ be a Banach algebra with identity I. Let $p_1, p_2, \ldots, p_{2N}$ and P be nonzero idempotents of $\mathcal{B}$ satisfying*

$$p_i p_j = \delta_{ij} p_j \ \text{for all} \ i, j, \quad p_1 + p_2 + \ldots + p_{2N} = I, \tag{8.9}$$

where δ_{ij} is the Kronecker delta, and

$$Q(p_{2i-1} + p_{2i})P = 0, \quad P(p_{2i} + p_{2i+1})Q = 0 \ \text{for all} \ i, \tag{8.10}$$

where $Q := I - P$ and $p_{2N+1} := p_1$. Let $\mathcal{A}$ stand for the smallest closed subalgebra of $\mathcal{B}$ containing P and $p_1, p_2, \ldots, p_{2N}$ (and thus I). Put

$$X := \sum_{i=1}^{N} (p_{2i-1} P p_{2i-1} + p_{2i} Q p_{2i}), \tag{8.11}$$

$$Y := \sum_{i=1}^{N} (p_{2i-1} P + p_{2i} Q) + \sum_{j=1}^{2N} (2j - 1) p_i. \tag{8.12}$$

(a) *For each $x \in \operatorname{sp}_{\mathcal{A}} X \setminus \{0, 1\}$ the map $\sigma_x : \{P, p_1, \ldots, p_{2N}\} \to \mathbf{C}^{2N \times 2N}$ given by*

$$\sigma_x(p_j) = \operatorname{diag}(0, \ldots, 0, 1, 0, \ldots, 0), \tag{8.13}$$

the 1 standing at the jth position, and

$$\sigma_x(P) = \operatorname{diag}(1, -1, 1, -1, \ldots, 1, -1) \times$$

$$\times \begin{pmatrix} x & x-1 & x-1 & x-1 & \cdots & x-1 & x-1 \\ x & x-1 & x-1 & x-1 & \cdots & x-1 & x-1 \\ x & x & x & x-1 & \cdots & x-1 & x-1 \\ x & x & x & x-1 & \cdots & x-1 & x-1 \\ \vdots & \vdots & \vdots & \vdots & \ddots & \vdots & \vdots \\ x & x & x & x & \cdots & x & x-1 \\ x & x & x & x & \cdots & x & x-1 \end{pmatrix} \tag{8.14}$$

extends to a Banach algebra homomorphism σ_x of $\mathcal{A}$ into $\mathbf{C}^{2N \times 2N}$.

(b) *For each $n = 4m - r$ with $m \in \{1, \ldots, N\}$ and $r \in \{0, 1, 2, 3\}$ define a map $G_n : \{P, p_1, \ldots, p_{2N}\} \to \mathbf{C}$ by*

$$G_{4m}(p_i) = \begin{cases} 1 & \text{if} \quad i = 2m \\ 0 & \text{if} \quad i \neq 2m \end{cases}, \qquad G_{4m}(P) = 0,$$

$$G_{4m-1}(p_i) = \begin{cases} 1 & \text{if} \quad i = 2m \\ 0 & \text{if} \quad i \neq 2m \end{cases}, \qquad G_{4m-1}(P) = 1,$$

(8.15)

$$G_{4m-2}(p_i) = \begin{cases} 1 & \text{if} \quad i = 2m - 1 \\ 0 & \text{if} \quad i \neq 2m - 1 \end{cases}, \qquad G_{4m-2}(P) = 0,$$

$$G_{4m-3}(p_i) = \begin{cases} 1 & \text{if} \quad i = 2m - 1 \\ 0 & \text{if} \quad i \neq 2m \end{cases}, \qquad G_{4m-3}(P) = 1.$$

If $n \in \mathrm{sp}_{\mathcal{A}} Y \cap \{1, 2, \ldots, 4N\}$, then G_n extends to a Banach algebra homomorphism of $\mathcal{A}$ into $\mathbf{C}$.

(c) *An element $A \in \mathcal{A}$ is invertible in $\mathcal{B}$ if and only if $\sigma_x(A)$ is invertible for all $x \in \mathrm{sp}_{\mathcal{B}} X \setminus \{0, 1\}$ and $G_n(A) \neq 0$ for all $n \in \mathrm{sp}_{\mathcal{B}} Y \cap \{1, 2, \ldots, 4N\}$.*

(d) *An element $A \in \mathcal{A}$ is invertible in $\mathcal{A}$ if and only if $\sigma_x(A)$ is invertible for all $x \in \mathrm{sp}_{\mathcal{A}} X \setminus \{0, 1\}$ and $G_n(A) \neq 0$ for all $n \in \mathrm{sp}_{\mathcal{A}} Y \cap \{1, 2, \ldots, 4N\}$.*

We remark that the condition (8.10) may also be written in the form

$$P(p_{2i-1} + p_{2i})P = (p_{2i-1} + p_{2i})P, \quad P(p_{2i} + p_{2i+1})P = P(p_{2i} + p_{2i+1}). \quad (8.16)$$

In a sense, (8.10) is the "Hankel formulation", while (8.16) is the "Toeplitz formulation" of the axiom. In what follows we use the convention that $p_k := p_r$ with $r \in \{1, \ldots, 2N\}$ if $k - r$ is divisible by $2N$. So (8.10) and (8.16) hold for all integers i.

Notice that the algebra $\mathcal{A}$ is actually generated by $2N + 1$ idempotents (or by $2N$ idempotents and the identity).

We now proceed to the proof of Theorem 8.9. Let $\mathcal{A}^0$ be the smallest (not necessarily closed) subalgebra of $\mathcal{B}$ which contains P and $p_1, \ldots, p_{2N}$.

Lemma 8.10. *The element X given by (8.11) belongs to the center of $\mathcal{A}^0$.*

Proof. It is clear that X commutes with each p_i. So it remains to show that $PX = XP$. Let us first prove that

$$X = \sum_{i=1}^{N} \Big((p_{2i} + p_{2i+1})Qp_{2i}Q + (p_{2i-1} + p_{2i})Pp_{2i-1}P \Big). \quad (8.17)$$

By (8.9), it suffices to show that

$$p_j X = \sum_{i=1}^{N} \left(p_j (p_{2i} + p_{2i+1}) Q p_{2i} Q + p_j (p_{2i-1} + p_{2i}) P p_{2i-1} P \right)$$

for $j \in \{1, \ldots, 2N\}$ or, equivalently, that

$$p_{2i} Q p_{2i} = p_{2i} Q p_{2i} Q + p_{2i} P p_{2i-1} P, \tag{8.18}$$

$$p_{2i-1} P p_{2i-1} = p_{2i-1} Q p_{2i-2} Q + p_{2i-1} P p_{2i-1} P \tag{8.19}$$

for all $i \in \{1, \ldots, N\}$. We have

$$p_{2i} Q p_{2i} Q + p_{2i} P p_{2i-1} P = p_{2i} Q - p_{2i} P p_{2i} Q + p_{2i} P p_{2i-1} P$$
$$= p_{2i} Q - p_{2i} P p_{2i} + p_{2i} P p_{2i} P + p_{2i} P p_{2i-1} P$$
$$= p_{2i} Q - p_{2i} P p_{2i} + p_{2i} P (p_{2i-1} + p_{2i}) P$$
$$= p_{2i} Q - p_{2i} P p_{2i} + p_{2i} P \quad \text{(by 8.16)}$$
$$= p_{2i} - p_{2i} P p_{2i} = p_{2i} (P + Q) p_{2i} - p_{2i} P p_{2i} = p_{2i} Q p_{2i},$$

which is (8.18). Analogously one can verify (8.19). This completes the proof of (8.17). Taking into account (8.16) and (8.17) we see that both PX and XP are equal to

$$\sum_{i=1}^{N} (p_{2i-1} + p_{2i}) P p_{2i-1} P. \qquad \square$$

Lemma 8.11. *Given $A \in \mathcal{A}^0$, there exist polynomials $R_{ij} \in \mathbf{C}[X]$ such that*

$$A = \sum_{i=1}^{2N} R_{ii}(X) p_i + \sum_{\substack{i,j=1 \\ i \neq j}}^{2N} R_{ij}(X) p_i P p_j. \tag{8.20}$$

Proof. Let $\mathcal{A}^1$ denote the set of all $A \in \mathcal{A}^0$ which can be written in the form (8.20). We first show that the generating elements of $\mathcal{A}^0$ belong to $\mathcal{A}^1$. This is clear for the idempotents p_i. Since

$$p_i P p_i = p_i P p_i p_i = \begin{cases} X p_i & \text{if } i \text{ is odd} \\ (I - X) p_i & \text{if } i \text{ is even,} \end{cases} \tag{8.21}$$

we get

$$P = \sum_{i,j=1}^{2N} p_i P p_j = \sum_{i=1}^{N} p_{2i} P p_{2i} + \sum_{i=1}^{N} p_{2i-1} P p_{2i-1} + \sum_{\substack{i,j=1 \\ i \neq j}}^{N} p_i P p_j$$

$$= \sum_{i=1}^{N} (I - X) p_{2i} + \sum_{i=1}^{N} X p_{2i-1} + \sum_{\substack{i,j=1 \\ i \neq j}}^{N} p_i P p_j,$$

which implies that $P \in \mathcal{A}^1$.

We now show that $\mathcal{A}^1$ is an algebra. Since the generating elements of $\mathcal{A}^0$ are in $\mathcal{A}^1$, this will prove that $\mathcal{A}^1 = \mathcal{A}^0$.

Obviously, $\mathcal{A}^1$ is a linear space. Furthermore, $p_i p_j$, $p_i p_j P p_k$, and $p_j P p_k p_i$ also belong to $\mathcal{A}^1$. Hence, we are left with verifying that $p_i P p_j p_k P p_l \in \mathcal{A}^1$ whenever $i \neq j$ and $k \neq l$. The latter element is 0 ($\in \mathcal{A}^1$) if $j \neq k$ and is $p_i P p_j P p_l$ for $j = k$. In case j is even, say $j = 2n$, we obtain from (8.16) that

$$
\begin{aligned}
p_i P p_{2n} P p_l &= p_i P(p_{2n-1} + p_{2n}) P p_l - p_i P p_{2n-1} P p_l \\
&= p_i(p_{2n-1} + p_{2n}) P p_l - p_i P p_{2n-1} P p_l,
\end{aligned}
\tag{8.22}
$$

while if j is odd, $j = 2n - 1$, we infer from (8.16) that

$$
\begin{aligned}
p_i P p_{2n-1} P p_l &= p_i P(p_{2n-2} + p_{2n-1}) P p_l - p_i P p_{2n-2} P p_l \\
&= p_i P(p_{2n-2} + p_{2n-1}) p_l - p_i P p_{2n-2} P p_l.
\end{aligned}
\tag{8.23}
$$

The first terms in (8.22) and (8.23) are clearly in $\mathcal{A}^1$ (recall (8.21) for $i = l$). Thus, (8.22) and (8.23) reduce the question whether $p_i P p_j P p_l$ is in $\mathcal{A}^1$ to the question whether $p_i P p_{j-1} P p_l$ belongs to $\mathcal{A}^1$. Repeating the argument we arrive at the problem whether $p_i P p_i P p_l$ is in $\mathcal{A}^1$. As

$$
p_i P p_i P p_l = p_i P p_i p_i P p_l = \begin{cases} X p_i P p_l & \text{if } i \text{ is odd} \\ (I - X) p_i P p_l & \text{if } i \text{ is even}, \end{cases}
$$

we see that $p_i P p_i P p_l$ is always in $\mathcal{A}^1$ (again recall (8.21) in case $i = l$). $\qquad\square$

Lemma 8.12. (a) *If $l > j > i$ or $j > i > l$ or $i > l > j$ then $p_i P p_j P p_l = (-1)^{j-1}(X - I) p_i P p_l$.*
(b) *If $l > i > j$ or $i > j > l$ or $j > l > i$ then $p_i P p_j P p_l = (-1)^{j-1} X p_i P p_l$.*
(c) *If $i \neq j$ then $p_i P p_j P p_i = (-1)^{j-i} X(X - I) p_i$.*

Proof. Let $j \neq i$ and $j \neq l$. We have

$$
p_i P p_j P p_l = p_i P(p_{j-1} + p_j) P p_l - p_i P p_{j-1} P p_l.
$$

If, in addition, $j - 1 \neq i$ and $j - 1 \neq l$, then $p_i P(p_{j-1} + p_j) P p_l = 0$ by virtue of (8.16), whence

$$
p_i P p_j P p_l = -p_i P p_{j-1} P p_l.
\tag{8.24}
$$

Now suppose the conditions of (a) are satisfied. Then there is a smallest positive integer k such that (everything modulo $2N$) $j \notin \{i, l\}, j-1 \notin \{i, l\}, \ldots, j - (k - 1) \notin \{i, l\}$, but $j - k = i$. Consequently, repeated application of (8.24) gives

$$
p_i P p_j P p_l = (-1)^{k-1} p_i P p_{j-(k-1)} P p_l
$$

and thus,

$$p_i P p_j P p_l = (-1)^{k-1}\big(p_i P(p_{j-k} + p_{j-(k-1)})P p_l - p_i P p_{j-k} P p_l\big)$$
$$= (-1)^{k-1}\big(p_i P(p_i + p_{i+1})P p_l - p_i P p_i P p_l\big).$$

Our assumptions imply that $l \neq i$ and $l \neq i+1$ (otherwise $j - (k-1)$ would be equal to l). Thus, taking into account (8.16) and (8.21) we get

$$p_i P p_j P p_l = \begin{cases} (-1)^{k-1}\big(p_i(p_i + p_{i+1})P p_l - p_i P p_i P p_l\big) & \text{if } i \text{ is odd} \\ (-1)^{k-1}\big(p_i P(p_i + p_{i+1})p_l - p_i P p_i P p_l\big) & \text{if } i \text{ is even} \end{cases}$$

$$= \begin{cases} (-1)^{k-1}(p_i P p_l - p_i P p_i P p_l) & \text{if } i \text{ is odd} \\ (-1)^{k-1}(p_i P p_i P p_l) & \text{if } i \text{ is even} \end{cases}$$

$$= \begin{cases} (-1)^{k-1}(I - X)p_i P p_l & \text{if } i \text{ is odd} \\ (-1)^{k-1}(-1)(I - X)p_i P p_l & \text{if } i \text{ is even.} \end{cases}$$

Replacing k by $j - i$ yields (a). The proofs of (b) and (c) are similar. $\square$

The element X belongs to the center of $\mathcal{A}^0$ (Lemma 8.10) and thus to the center of $\mathcal{A}$ itself. Denote by Z the smallest closed subalgebra of $\mathcal{A}$ which contains I and X. Since Z is singly generated, the maximal ideal space of Z is homeomorphic to $\mathrm{sp}_Z X$: the maximal ideal corresponding to $x \in \mathrm{sp}_Z X$ is the smallest closed two-sided ideal $I_x \subset Z$ containing $X - xI$. In accordance with Theorem 8.2, we associate with every $x \in \mathrm{sp}_Z X$ the smallest closed two-sided ideal J_x of $\mathcal{A}$ containing I_x. Obviously, $\mathrm{sp}_{\mathcal{A}} X \subset \mathrm{sp}_Z X$.

Lemma 8.13. *If $x \in \mathrm{sp}_{\mathcal{A}} X$ then $J_x \neq \mathcal{A}$, and if $x \in \mathrm{sp}_Z X \setminus \mathrm{sp}_{\mathcal{A}} X$ then $J_x = \mathcal{A}$.*

Proof. Let $x \in \mathrm{sp}_Z X$ and suppose $J_x \neq \mathcal{A}$. Then $X - xI + J_x = 0 + J_x$ is not invertible in $\mathcal{A}/J_x$ and hence, by Theorem 8.2, $X - xI$ is not invertible in $\mathcal{A}$. Thus, $x \in \mathrm{sp}_{\mathcal{A}} X$. Conversely, assume $x \in \mathrm{sp}_{\mathcal{A}} X$. Then, again by Theorem 8.2, there is a $y \in \mathrm{sp}_Z X$ such that $X - xI + J_y$ is not invertible in $\mathcal{A}/J_y$. If $y \neq x$, then $X - xI + J_y = (y-x)I + J_y$ is invertible in $\mathcal{A}/J_y$ and therefore $X - xI + J_x = 0 + J_x$ cannot be invertible in $\mathcal{A}/J_x$. This is impossible if $J_x = \mathcal{A}$, whence $J_x \neq \mathcal{A}$. $\square$

Theorem 8.2 and Lemma 8.13 imply that an element $A \in \mathcal{A}$ is invertible in $\mathcal{A}$ if and only if $A + J_x$ is invertible in $\mathcal{A}/J_x$ for all $x \in \mathrm{sp}_{\mathcal{A}} X$. Put $\mathcal{A}_x := \mathcal{A}/J_x$ and define $\pi_x : \mathcal{A} \to \mathcal{A}_x$ by $\pi_x(A) := A + J_x$.

Lemma 8.14. *If $x \in \mathrm{sp}_{\mathcal{A}} X \setminus \{0, 1\}$, then $\mathcal{A}_x$ is isomorphic to $\mathbf{C}^{2N \times 2N}$ and there is a Banach algebra isomorphism $\psi_x : \mathcal{A}_x \to \mathbf{C}^{2N \times 2N}$ such that*

$$\psi_x \circ \pi_x \big| \{P, p_1, \ldots, p_{2N}\} = \sigma_x \big| \{P, p_1, \ldots, p_{2N}\} \tag{8.25}$$

where σ_x is given by (8.13) and (8.14).

Proof. Consider the image $\pi_x(\mathcal{A}^0) \subset \mathcal{A}_x$ of the algebra $\mathcal{A}^0$. Every element $A \in \mathcal{A}^0$ can be written in the form (8.20). Since $\pi_x(X) = x\pi_x(I)$, it follows that $\pi_x(R(X)) = R(x)\pi_x(I)$ for every polynomial R. Consequently, every element of $\pi_x(\mathcal{A}^0)$ is a linear combination of the elements

$$\pi_x(p_i) \ (i = 1, \ldots, 2N) \quad \text{and} \quad \pi_x(p_i P p_j) \ (i, j = 1, \ldots, 2N; \ i \neq j). \tag{8.26}$$

Conversely, every linear combination of the elements (8.26) belongs to $\pi_x(\mathcal{A}^0)$. Thus, $\pi_x(\mathcal{A}^0)$ is a finite-dimensional space of dimension at most $(2N)^2$. In particular, $\pi_x(\mathcal{A}^0)$ is closed in $\mathcal{A}_x$. Because $\mathcal{A}^0$ is dense in $\mathcal{A}$, so is $\pi_x(\mathcal{A}^0)$ in $\mathcal{A}_x$. We therefore arrive at the conclusion that $\mathcal{A}_x = \pi_x(\mathcal{A}^0)$.

We claim that $\mathcal{A}_x$ has the dimendion $(2N)^2$ and that the elements (8.26) form a basis of $\mathcal{A}_x$. Given $i, j \in \{1, \ldots, 2N\}$, define $a_{ij} \in \mathcal{A}_x$ by

$$a_{ij} = \begin{cases} (-1)^{i-1}(x-1)^{-1}\pi_x(p_i P p_j) & \text{if} \quad i < j \\ (-1)^{i-1}x^{-1}\pi_x(p_i P p_j) & \text{if} \quad i > j \\ \pi_x(p_i) & \text{if} \quad i = j. \end{cases}$$

Using Lemma 8.12 one can straightforwardly verify that

$$a_{ij}a_{kl} = \delta_{jk}a_{il} \quad \text{for all} \ i, j, k, l. \tag{8.27}$$

Indeed, if, for example, $j = k$ and $j > i > l$ then

$$
\begin{aligned}
a_{ij}a_{jl} &= (-1)^{i-1}(x-1)^{-1}\pi_x(p_i P p_j)(-1)^{j-1}x^{-1}\pi_x(p_j P p_l) \\
&= (-1)^{i-1}(-1)^{j-1}x^{-1}(x-1)^{-1}\pi_x(p_i P p_j P p_l) \\
&= (-1)^{i-1}(-1)^{j-1}x^{-1}(x-1)^{-1}\pi_x((-1)^{j-1}(X-I)p_i P p_l) \\
&= (-1)^{i-1}x^{-1}\pi_x(p_i P p_l) = a_{il},
\end{aligned}
$$

and the other cases can be disposed of analogously.

Contrary to what we want, let us assume that the elements a_{ij} are linearly dependent. Then there are numbers c_{ij} such that

$$\sum_{i,j=1}^{2N} c_{ij}a_{ij} = 0 \tag{8.28}$$

and $c_{i_0 j_0} \neq 0$ for certain i_0 and j_0. Multiplying (8.28) from the left by $a_{k i_0}$ and from the right by $a_{j_0 k}$ and taking into account (8.27) we get

$$c_{i_0 j_0} a_{k i_0} a_{i_0 j_0} a_{j_0 k} = c_{i_0 j_0} a_{kk} = 0$$

and hence $a_{kk} = 0$ for all $k \in \{1, \ldots, 2N\}$. It follows that

$$\pi_x(I) = \pi_x\left(\sum_{k=1}^{2N} p_k\right) = \sum_{k=1}^{2N} a_{kk} = 0,$$

which is only possible if $J_x = \mathcal{A}$. This contradicts Lemma 8.13 and our assumption that $x \in \mathrm{sp}_{\mathcal{A}} X$.

In summary, we have shown that both the elements $\{a_{ij}\}_{i,j=1}^{2N}$ and the elements (8.26) constitute a basis of $\mathcal{A}_x$. Denote by E_{ij} the $2N \times 2N$ matrix whose i, j entry is 1 and the other entries of which are zero. From (8.27) we see that the map

$$\psi_x : \{a_{ij}\}_{i,j=1}^{2N} \to \mathbf{C}^{2N \times 2N}, \ a_{ij} \mapsto E_{ij}$$

extends to an algebra isomorphism ψ_x of $\mathcal{A}_x$ onto $\mathbf{C}^{2N \times 2N}$. By a theorem of Johnson (see, e.g. [103, p. 313]), surjective algebra homomorphisms onto semisimple algebras are always continuous. This implies that ψ_x is a Banach algebra isomorphism.

We are left with proving (8.25). Obviously,

$$\psi_x\big(\pi_x(p_i)\big) = \psi_x(a_{ii}) = E_{ii} = \sigma_x(p_i).$$

Further,

$$\pi_x(P) = \pi_x\left(\sum_{i,j=1}^{2N} p_i P p_j\right)$$

$$= \pi_x\left(\sum_{i<j} p_i P p_j\right) + \pi_x\left(\sum_{i>j} p_i P p_j\right) + \pi_x\left(\sum_i p_i P p_i\right)$$

$$= \sum_{i<j}(-1)^{i-1}(x-1)a_{ij} + \sum_{i>j}(-1)^{i-1}xa_{ij} + \pi_x\left(\sum_i p_i P p_i\right),$$

and (8.21) implies that

$$\pi_x\left(\sum_i p_i P p_i\right) = \sum_{i \text{ odd}} xa_{ii} + \sum_{i \text{ even}} (1-x)a_{ii}.$$

Thus,

$$\begin{aligned}
\psi_x\big(\pi_x(P)\big) =\ & \sum_{i<j}(-1)^{i-1}(x-1)E_{ij} + \sum_{i>j}(-1)^{i-1}xE_{ij} \\
& + \sum_{i \text{ odd}} xE_{ii} + \sum_{i \text{ even}} (1-x)E_{ii},
\end{aligned}$$

and this is easily seen to coincide with $\sigma_x(P)$. $\square$

The previous lemma proves part (a) of Theorem 8.9. We now turn to the local algebras $\mathcal{A}_x$ associated with the points x in $\mathrm{sp}_A X \cap \{0, 1\}$.

Lemma 8.15. If $x \in \mathrm{sp}_A X \cap \{0, 1\}$ then $\mathcal{A}_x$ is an F_2^{N+1}-algebra.

Proof. Instead of working with the polynomial $F_2^{N+1}(a,b) = (ab - ba)^{N+1}$ in two variables, which is nonlinear, let us consider the polynomial

$$F_2^{\langle N+1 \rangle}(a_1, b_1, \ldots, a_{N+1}, b_{N+1}) := \prod_{k=1}^{N+1} (a_k b_k - b_k a_k) \tag{8.29}$$

in $2(N+1)$ variables, which is linear in each variable. Notice that if $F_2^{\langle N+1 \rangle}$ annulates any $2(N+1)$ elements of $\mathcal{A}_x$, then $\mathcal{A}_x$ is an $F_2^{N+1}-$ algebra.

Since $\mathcal{A}_x$ is a linear space and $F_2^{\langle N+1 \rangle}$ is multilinear, it suffices to prove that (8.29) is zero for all choices of the elements a_k, b_k ($k = 1, \ldots, N+1$) among the possible basis elements (8.26) of the algebra $\mathcal{A}_x$. Lemma 8.12 shows that each commutator $a_k b_k - b_k a_k$ can be written as $c_k \pi_x(p_{i_k} P p_{j_k})$ with $i_k, j_k \in \{1, \ldots, 2N\}$, $i_k \neq j_k$ and $c_k \in \mathbf{C}$. Hence,

$$\prod_{k=1}^{N+1} (a_k b_k - b_k a_k) = c \pi_x \left(\prod_{k=1}^{N+1} p_{i_k} P p_{j_k} \right). \tag{8.30}$$

Clearly, at least two of the $2N+2$ elements $p_{i_1}, \ldots, p_{i_{N+1}}, p_{j_1}, \ldots, p_{j_{N+1}}$ must coincide. Since $i_k \neq j_k$, it follows that the product on the right of (8.30) is zero or contains a subproduct of the form $p_i P p_{l_1} P p_{l_2} \ldots P p_{l_r} P p_i$ with $r \geq 1$ and $i \notin \{l_1, \ldots, l_r\}$. In the latter case one can again use Lemma 8.12 to deduce that

$$\pi_x(p_i P p_{l_1} P p_{l_2} \ldots P p_{l_r} P p_i) = 0. \qquad \square$$

Combining Lemma 8.15 and Theorem 8.3, we obtain for each $x \in \operatorname{sp}_{\mathcal{A}} X \cap \{0, 1\}$ a collection of Banach algebra isomorphisms $\varphi_{x,J} : \mathcal{A}_x / J \to \mathbf{C}$ such that $\pi_x(A) \in \mathcal{A}_x$ is invertible if and only if

$$\varphi_{x,J}(\pi_x(A) + J) \neq 0$$

for all maximal ideals J of $\mathcal{A}_x$. The maps

$$G_{x,J} : \mathcal{A} \to \mathbf{C}, \quad A \mapsto \varphi_{x,J}(\pi_x(A) + J) \tag{8.31}$$

are nonzero Banach algebra homomorphisms.

Let $G : \mathcal{A} \to \mathbf{C}$ be any nonzero Banach algebra homomorphism. Then G maps idempotents to idempotents. Thus, if $p \in \mathcal{A}$ is an idempotent, then $G(p)$ is either 0 or 1. Since $G(I) = 1$ for every nonzero homomorphism G, we obtain that there is an $i_0 \in \{1, \ldots, 2N\}$ such that $G(p_{i_0}) = 1$ and by (8.9), $G(p_i) = 0$ for all $i \neq i_0$. Consequently, the restriction of G to $\{P, p_1, \ldots, p_{2N}\}$ coincides with one of the maps G_n ($n = 1, \ldots, 4N$) given by (8.15). A direct computation shows that

$$G_n(Y) = n. \tag{8.32}$$

As $\operatorname{sp} G_n(Y) \subset \operatorname{sp}_{\mathcal{A}} Y$, we see that $n \in \operatorname{sp}_{\mathcal{A}} Y \cap \{1, \ldots, 4N\}$.

Lemma 8.16. *If $n \in \mathrm{sp}_{\mathcal{A}} Y \cap \{1, \ldots, 4N\}$ then the map $G_n : \{P, p_1, \ldots, p_{2N}\} \to \mathbf{C}$ given by (8.15) extends to a Banach algebra homomorphism G_n of $\mathcal{A}$ onto $\mathbf{C}$.*

Proof. Let $n \in \mathrm{sp}_{\mathcal{A}} Y \cap \{1, \ldots, 4N\}$. We first show that if $x \in \mathrm{sp}_{\mathcal{A}} X \setminus \{0, 1\}$, then

$$n \notin \mathrm{sp}_{\mathcal{A}_x} \big(\pi_x(Y) \big). \tag{8.33}$$

By Lemma 8.14, this will follow as soon as we have shown that the $2N \times 2N$ matrix

$$\psi_x \big(\pi_x(Y) \big) - \mathrm{diag}\,(n, n, \ldots, n)$$

$$= \begin{pmatrix}
x & x-1 & x-1 & x-1 & \cdots & x-1 & x-1 \\
x & x & x-1 & x-1 & \cdots & x-1 & x-1 \\
x & x & x & x-1 & \cdots & x-1 & x-1 \\
x & x & x & x & \cdots & x-1 & x-1 \\
\vdots & \vdots & \vdots & \vdots & \ddots & \vdots & \vdots \\
x & x & x & x & \cdots & x & x-1 \\
x & x & x & x & \cdots & x & x
\end{pmatrix}$$

$$+ \,\mathrm{diag}\,(1 - n, 3 - n, \ldots, 4N - 1 - n)$$

is invertible. To see this, we evaluate the determinant of the slightly more general $M \times M$ matrix

$$\begin{pmatrix}
x + \lambda_1 & x-1 & x-1 & \cdots & x-1 & x-1 \\
x & x + \lambda_2 & x-1 & \cdots & x-1 & x-1 \\
x & x & x + \lambda_3 & \cdots & x-1 & x-1 \\
\vdots & \vdots & \vdots & \ddots & \vdots & \vdots \\
x & x & x & \cdots & x + \lambda_{M-1} & x-1 \\
x & x & x & \cdots & x & x + \lambda_M
\end{pmatrix} \tag{8.34}$$

where $\lambda_1, \ldots, \lambda_M, x \in \mathbf{C}$. Consider x as a variable and denote the determinant of (8.34) by $D(x)$. Subtracting in (8.34) the first row from all other rows and then the last column from the remaining columns, we get a matrix whose $1, N$ entry is $x - 1$ and the other entries of which are independent of x. Thus, $D(x)$ is a polynomial of first degree on x. Since $D(0) = \prod_{i=1}^{M} \lambda_i$ and $D(1) = \prod_{i=1}^{M} (1 + \lambda_i)$, it results that

$$D(x) = x \prod_{i=1}^{M} (1 + \lambda_i) + (1 - x) \prod_{i=1}^{M} \lambda_i. \tag{8.35}$$

Now put $M = 2N$ and $\lambda_i = 2i - 1 - n$. If n is odd, then one of the numbers λ_i equals zero but $\prod_{i=1}^{M} (1 + \lambda_i) \neq 0$. In case n is even, one of the numbers $1 + \lambda_i$ is zero while $\prod_{i=1}^{M} \lambda_i \neq 0$. Hence, in either case $D(x) \neq 0$ whenever $x \notin \{0, 1\}$. This proves our claim (8.33).

Since $n \in \mathrm{sp}_{\mathcal{A}} Y$, we deduce from Theorem 8.2 that

$$n \in \bigcup_{x \in \mathrm{sp}_{\mathcal{A}} X} \mathrm{sp}_{\mathcal{A}_x} \big(\pi_x(Y) \big),$$

whereas, by (8.33),

$$n \notin \bigcup_{x \in \mathrm{sp}_{\mathcal{A}} X \setminus \{0,1\}} \mathrm{sp}_{\mathcal{A}_x} \big(\pi_x(Y) \big).$$

Consequently,

$$n \in \bigcup_{x \in \mathrm{sp}_{\mathcal{A}} X \cap \{0,1\}} \mathrm{sp}_{\mathcal{A}_x} \big(\pi_x(Y) \big).$$

Let $n \in \mathrm{sp}_{\mathcal{A}_{x_0}} \big(\pi_{x_0}(Y) \big)$. From what was said after the proof of Lemma 8.15, there exists a maximal ideal J_0 of $\mathcal{A}_{x_0}$ such that

$$G_{x_0, J_0} \big(\pi_{x_0}(Y) \big) = n. \tag{8.36}$$

The map G_{x_0, J_0} is a Banach algebra homomorphism of $\mathcal{A}$ into $\mathbf{C}$. The restriction of G_{x_0, J_0} to $\{P, p_1, \ldots, p_{2N}\}$ is one of the maps (8.15). Comparing (8.32) and (8.36) we see that this restriction is just G_n.   □

Proof of Theorem 8.9. Part (a) is Lemma 8.14 and part (b) is Lemma 8.16. We also showed that if $A \in \mathcal{A}$, then

$$\mathrm{sp}_{\mathcal{A}} A = \bigcup_{x \in \mathrm{sp}_{\mathcal{A}} X \setminus \{0,1\}} \mathrm{sp}\, \sigma_x(A) \cup \bigcup_{x \in \mathrm{sp}_{\mathcal{A}} X \cap \{0,1\}} \mathrm{sp}_{\mathcal{A}_x} \pi_x(A). \tag{8.37}$$

For $x \in \mathrm{sp}_{\mathcal{A}} X \cap \{0,1\}$ we have $\mathrm{sp}_{\mathcal{A}_x} \pi_x(A) = \bigcup_J \{G_{x,J}(A)\}$, the union over all maximal ideals J of $\mathcal{A}_x$ (recall (8.31)), and for each J there is an $n \in \mathrm{sp}_{\mathcal{A}} Y \cap \{1, \ldots, 4N\}$ such that $G_{x,J}(A) = G_n(A)$. Thus,

$$\bigcup_{x \in \mathrm{sp}_{\mathcal{A}} X \cap \{0,1\}} \mathrm{sp}_{\mathcal{A}_x} \pi_x(A) \subset \bigcup_{n \in \mathrm{sp}_{\mathcal{A}} Y \cap \{1,\ldots,4N\}} \{G_n(A)\}. \tag{8.38}$$

If $n \in \mathrm{sp}_{\mathcal{A}} Y \cap \{1, \ldots, 4N\}$, then G_n extends to a Banach algebra homomorphism of $\mathcal{A}$ into $\mathbf{C}$ due to Lemma 8.16. Therefore $\{G_n(A)\} \subset \mathrm{sp}_{\mathcal{A}} A$ and hence

$$\bigcup_{n \in \mathrm{sp}_{\mathcal{A}} Y \cap \{1,\ldots,4N\}} \{G_n(A)\} \subset \mathrm{sp}_{\mathcal{A}} A. \tag{8.39}$$

Combining (8.37), (8.38), (8.39) we get

$$\mathrm{sp}_{\mathcal{A}} A = \bigcup_{x \in \mathrm{sp}_{\mathcal{A}} X \setminus \{0,1\}} \mathrm{sp}\, \sigma_x(A) \cup \bigcup_{n \in \mathrm{sp}_{\mathcal{A}} Y \cap \{1,\ldots,4N\}} \{G_n(A)\},$$

which completes the proof of part (d).

We now prove part (c). Denote by $\Lambda(\mathcal{B})$ the set of all elements $B \in \mathcal{B}$ such that $BX = XB$. Obviously, $\Lambda(\mathcal{B})$ is a Banach algebra. We have $\mathcal{A} \subset \Lambda(\mathcal{B})$ and

$$\operatorname{sp}_{\mathcal{B}} A = \operatorname{sp}_{\Lambda(\mathcal{B})} A \quad \text{for every} \quad A \in \mathcal{A}. \tag{8.40}$$

For $x \in \operatorname{sp}_{\mathcal{A}} X$, define $J_x =: J_x^{\mathcal{A}} \subset \mathcal{A}$ as above (before Lemma 8.13). In addition, if $x \in \operatorname{sp}_{\mathcal{B}} X$ let $J_x^{\mathcal{B}} \subset \Lambda(\mathcal{B})$ be the smallest closed two-sided ideal of $\Lambda(\mathcal{B})$ containing $X - xI$. Put

$$\mathcal{A}_x := \mathcal{A}/J_x^{\mathcal{A}}, \quad \mathcal{B}_x := \Lambda(\mathcal{B})/J_x^{\mathcal{B}},$$

and denote by $\pi_x^{\mathcal{A}} : \mathcal{A} \to \mathcal{A}_x$, $\pi_x^{\mathcal{B}} : \Lambda(\mathcal{B}) \to \mathcal{B}_x$ the canonical homomorphisms. Finally, let

$$\tilde{\mathcal{A}}_x := \pi_x^{\mathcal{B}}(\mathcal{A}).$$

Theorem 8.2, Lemma 8.13 (with $\mathcal{B}$ in place of $\mathcal{A}$), and (8.40) show that

$$\operatorname{sp}_{\mathcal{B}} A = \bigcup_{x \in \operatorname{sp}_{\mathcal{B}} X} \operatorname{sp}_{\mathcal{B}_x} \pi_x^{\mathcal{B}}(A) \tag{8.41}$$

for every $A \in \mathcal{A}$. The coset $\pi_x^{\mathcal{B}}(A)$ belongs to $\tilde{\mathcal{A}}_x$. Proceeding exactly as in the proof of Lemma 8.14, we see that if $x \in \operatorname{sp}_{\mathcal{B}} X \setminus \{0, 1\}$, then $\tilde{\mathcal{A}}_x$ is isomorphic to $\mathbf{C}^{2N \times 2N}$ and there is a Banach algebra isomorphism $\tilde{\psi}_x : \tilde{\mathcal{A}}_x \to \mathbf{C}^{2N \times 2N}$ such that

$$\tilde{\psi}_x \circ \pi_x^{\mathcal{B}} | \{P, p_1, \ldots, p_{2N}\} = \sigma_x | \{P_x, p_1, \ldots, p_{2N}\}$$

where σ_x is given by (8.13) and (8.14). Thus, if $x \in \operatorname{sp}_{\mathcal{B}} X \setminus \{0, 1\}$ then $\operatorname{sp}_{\tilde{\mathcal{A}}_x} \pi_x^{\mathcal{B}}(A) = \operatorname{sp} \sigma_x(A)$, and since $\operatorname{sp} \sigma_x(A)$ is a finite set, it follows that

$$\operatorname{sp}_{\mathcal{B}_x} \pi_x^{\mathcal{B}}(A) = \operatorname{sp} \sigma_x(A) \quad \text{for} \quad x \in \operatorname{sp}_{\mathcal{B}} X \setminus \{0, 1\}. \tag{8.42}$$

The proof of Lemma 8.15 shows that $\tilde{\mathcal{A}}_x$ is an F_2^{N+1}-algebra for $x \in \operatorname{sp}_{\mathcal{B}} X \cap \{0, 1\}$, and therefore we obtain as above that there is a subset $\mu_x \subset \{1, \ldots, 4N\}$ such that

$$\operatorname{sp}_{\tilde{\mathcal{A}}_x} \pi_x^{\mathcal{B}}(A) = \bigcup_{n \in \mu_x} \{G_n(A)\} \quad \text{for} \quad x \in \operatorname{sp}_{\mathcal{B}} X \cap \{0, 1\},$$

whence

$$\operatorname{sp}_{\mathcal{B}_x} \pi_x^{\mathcal{B}}(A) = \bigcup_{n \in \mu_x} \{G_n(A)\} \quad \text{for} \quad x \in \operatorname{sp}_{\mathcal{B}} X \cap \{0, 1\}. \tag{8.43}$$

Combining (8.41), (8.42), (8.43) we get

$$\operatorname{sp}_{\mathcal{B}} A = \bigcup_{x \in \operatorname{sp}_{\mathcal{B}} X \setminus \{0,1\}} \operatorname{sp} \sigma_x(A) \cup \bigcup_{n \in \mu} \{G_n(A)\} \tag{8.44}$$

with some set $\mu \subset \{1, \ldots, 4N\}$.

We are left with showing that $\mu = \operatorname{sp}_{\mathcal{B}} Y \cap \{1, \ldots, 4N\}$. For every $\lambda \in \mathbf{C}$, the matrix $\sigma_x(Y - \lambda I)$ is the matrix (8.34) with $M = 2N$ and $\lambda_i = 2i - 1 - \lambda$. From (8.35) we infer that every eigenvalue λ of $\sigma_x(Y)$ solves the equation

$$x \prod_{i=1}^{2N} (2i - \lambda) + (1 - x) \prod_{i=1}^{2N} (2i - 1 - \lambda) = 0.$$

Consequently, $\operatorname{sp}\sigma_x(Y) \cap \{1,\ldots,4N\} = \emptyset$ whenever $x \notin \{0,1\}$. Therefore (8.44) and (8.32) give

$$\operatorname{sp}_{\mathcal{B}}Y \cap \{1,\ldots,4N\} = \left(\bigcup_{n\in\mu}\{G_n(Y)\}\right)\cap\{1,\ldots,4N\} = \bigcup_{n\in\mu}\{n\} = \mu. \qquad \square$$

In the concrete applications we will encounter, $\operatorname{sp}_{\mathcal{A}}X$ and $\operatorname{sp}_{\mathcal{B}}X$ coincide and are connected sets containing the points 0 and 1. In this case Theorem 8.9 can be somewhat simplified.

Lemma 8.17. *If the points 0 and 1 are cluster points of* $\operatorname{sp}_{\mathcal{A}}X$ *(and thus belong to* $\operatorname{sp}_{\mathcal{A}}X$*), then* $\{1,\ldots,4N\} \subset \operatorname{sp}_{\mathcal{A}}Y$*.*

Proof. From Theorem 8.9 we know that

$$\bigcup_{x\in\operatorname{sp}_{\mathcal{A}}X\setminus\{0,1\}} \operatorname{sp}\sigma_x(Y) \subset \operatorname{sp}_{\mathcal{A}}Y. \tag{8.45}$$

For $x \in \{0,1\}$, define $\sigma_x(p_j)$ and $\sigma_x(P)$ by (8.13) and (8.14) and put

$$\sigma_x(Y) := \sum_{i=1}^{N}\left(\sigma_x(p_{2i-1})\sigma_x(P) + \sigma_x(p_{2i})\sigma_x(Q)\right) + \sum_{j=1}^{2N}(2j-1)\sigma_x(p_j).$$

The map $\operatorname{sp}_{\mathcal{A}}X \to \mathbf{C}^{2N\times 2N}$, $x \mapsto \sigma_x(Y)$ is continuous and the eigenvalues of a matrix depend continuously on the matrix (see, e.g., [106, Appendix D]). Since $\operatorname{sp}_{\mathcal{A}}Y$ is closed and 0 and 1 are cluster points of $\operatorname{sp}_{\mathcal{A}}X \setminus \{0,1\}$, we therefore obtain from (8.45) that

$$\operatorname{sp}\sigma_0(Y) \cup \operatorname{sp}\sigma_1(Y) \subset \operatorname{sp}_{\mathcal{A}}Y. \tag{8.46}$$

The matrices $\sigma_0(Y)$ and $\sigma_1(Y)$ are triangular. The diagonal of $\sigma_0(Y)$ equals $\operatorname{diag}(G_1(Y), G_3(Y),\ldots,G_{4N-1}(Y))$ and the diagonal of $\sigma_1(Y)$ is $\operatorname{diag}(G_2(Y), G_4(Y),\ldots,G_{4N}(Y))$, where

$$G_n(Y) := \sum_{i=1}^{N}\left(G_n(p_{2i-1})G_n(P) + G_n(p_{2i})G_n(Q)\right) + \sum_{j=1}^{2N}(2j-1)G_n(p_j) = n.$$

Hence,

$$\operatorname{sp}\sigma_0(Y) \cup \operatorname{sp}\sigma_1(Y) = \bigcup_{n=1}^{4N}\{G_n(Y)\} = \{1,\ldots,4N\}.$$

This and (8.46) give the assertion. $\qquad \square$

Corollary 8.18. *Let the situation be as in Theorem 8.9. In addition, assume that the points 0 and 1 are cluster points of* $\operatorname{sp}_{\mathcal{B}}X$*. Define the map* σ_x *of* $\{P, p_1,\ldots,p_{2N}\}$ *to* $\mathbf{C}^{2N\times 2N}$ *by (8.13) and (8.14) for* $x \in \mathbf{C}\setminus\{0,1\}$*, and let*

$$\sigma_0(p_j) = \sigma_1(p_j) = \operatorname{diag}(0,\ldots,0,1,0,\ldots,0), \tag{8.47}$$

the 1 at the jth position,

$$\sigma_0(P) = \operatorname{diag}(0, 1, 0, 1, \ldots, 0, 1), \quad \sigma_1(P) = \operatorname{diag}(1, 0, 1, 0, \ldots, 1, 0). \quad (8.48)$$

(a) *For each $x \in \operatorname{sp}_{\mathcal{A}} X$ the map σ_x extends to a Banach algebra homomorphisms of $\mathcal{A}$ into $\mathbf{C}^{2N \times 2N}$.*

(b) *An element $A \in \mathcal{A}$ is invertible in $\mathcal{B}$ if and only if $\sigma_x(A)$ is invertible for every $x \in \operatorname{sp}_{\mathcal{B}} X$.*

(c) *An element $A \in \mathcal{A}$ is invertible in $\mathcal{A}$ if and only if $\sigma_x(A)$ is invertible for every $x \in \operatorname{sp}_{\mathcal{A}} X$.*

Proof. First of all we remark that $\operatorname{sp}_{\mathcal{B}} X \subset \operatorname{sp}_{\mathcal{A}} X$, so that 0 and 1 are also cluster points of $\operatorname{sp}_{\mathcal{A}} X$.

(a) Theorem 8.9(a) shows that σ_x extends for $x \in \operatorname{sp}_{\mathcal{A}} X \setminus \{0, 1\}$. Combining Lemmas 8.16 and 8.17 we conclude that the maps G_n extend to Banach algebra homomorphisms of $\mathcal{A}$ onto $\mathbf{C}$ for all $n \in \{1, \ldots, 4N\}$. Thus, the two maps

$$A \mapsto \operatorname{diag}(G_1(A), G_3(A), \ldots, G_{4N-1}(A)), \quad (8.49)$$
$$A \mapsto \operatorname{diag}(G_2(A), G_4(A), \ldots, G_{4N}(A)) \quad (8.50)$$

are well-defined Banach algebra homomorphisms of $\mathcal{A}$ into $\mathbf{C}^{2N \times 2N}$. The restrictions of these maps to $\{P, p_1, \ldots, p_{2N}\}$ coincide with σ_0 and σ_1, respectively. Consequently, σ_0 and σ_1 extend to Banach algebra homomorphisms of $\mathcal{A}$ into $\mathbf{C}^{2N \times 2N}$.

(b) From Theorem 8.9(c) and Lemma 8.17 we infer that

$$\operatorname{sp}_{\mathcal{B}} A = \bigcup_{x \in \operatorname{sp}_{\mathcal{B}} X \setminus \{0,1\}} \operatorname{sp} \sigma_x(A) \cup \bigcup_{n=1}^{4N} \{G_n(A)\}.$$

By part (a), $\sigma_0(A)$ and $\sigma_1(A)$ are equal to the diagonal matrices on the right of (8.49) and (8.50), respectively. Thus,

$$\bigcup_{n=1}^{4N} \{G_n(A)\} = \operatorname{sp} \sigma_0(A) \cup \operatorname{sp} \sigma_1(A).$$

(c) This can be proved as part (b). □

Proof of Theorem 8.7. Given e, r, s as in Theorem 8.7, put $N = 2$, $I := e$, $p_1 := r$, $p_2 := e - r$, $P := s$ and apply Corollary 8.18. Clearly, (8.9) and (8.10) are satisfied. The element X of Theorem 8.7 coincides with the X given by (8.11), the matrices (8.6) are the matrices (8.13) and (8.14), and the matrices (8.7) are the matrices (8.47) and (8.48). □

8.5   Algebras associated with Jordan curves

In this section we establish a symbol calculus for deciding whether an operator in $\text{alg}\,(S_\Gamma, PC(\Gamma))$ is Fredholm in case Γ is a Carleson Jordan curve. Of course, we are also given a $p \in (1, \infty)$ and a weight $w \in A_p(\Gamma)$, and all operators are considered on $L^p(\Gamma, w)$.

For $t \in \Gamma$, define the local algebra $\mathcal{B}_t := \mathcal{B}_t(\Gamma)$ as in Section 8.2 and the subalgebra $\mathcal{A}_t := \mathcal{A}_t(\Gamma)$ as in Section 8.3. Also as in the latter section, let

$$e = \pi_t(I), \quad r = \pi_t(P_\Gamma), \quad s = \pi_t(\chi_t I),$$

where χ_t is the characteristic function of any subarc of Γ such that $\chi_t(t-0) = 0$ and $\chi_t(t+0) = 1$. In order to make use of Theorem 8.7, we have to determine the spectrum of

$$X = rsr + (e-r)(e-s)(e-r) = \pi_t\Big(P_\Gamma \chi_t P_\Gamma + Q_\Gamma(1 - \chi_t)Q_\Gamma\Big).$$

Theorem 8.19. *We have*

$$\text{sp}_{\mathcal{B}_t} X = \text{sp}_{\mathcal{A}_t} X = \mathcal{L}(0, 1; p, \alpha_t, \beta_t)$$

where α_t and β_t are the indicator functions of Γ, p, w at t.

Proof. Denote by Φ the set of Fredholm operators on $L^p(\Gamma, w)$ and by $G\mathcal{B}_t$ the group of invertible elements in $\mathcal{B}_t = \mathcal{B}_t(\Gamma)$. Since

$$X - \lambda e = r(s - \lambda e)r + (e-r)(e-s-\lambda e)(e-r)$$

and

$$rar + (e-r)b(e-r) = (rar + e - r)\big((e-r)b(e-r) + r\big)$$
$$= \big((e-r)b(e-r) + r\big)(rar + e - r),$$

we get $\text{sp}_{\mathcal{B}_t} X = S_1 \cup S_2$ where

$$S_1 := \big\{\lambda \in \mathbf{C} : r(s - \lambda e)r + e - r \notin G\mathcal{B}_t\big\},$$
$$S_2 := \big\{\lambda \in \mathbf{C} : (e-r)(e-s-\lambda e)(e-r) + r \notin G\mathcal{B}_t\big\}.$$

Let $\varphi_t : \Gamma \to \mathbf{C}$ be any function which is continuous on $\Gamma \setminus \{t\}$ and for which $\varphi_t(t-0) = 0$ and $\varphi_t(t+0) = 1$. Clearly, $\pi_t(\varphi_t I) = \pi_t(\chi_t I)$ and $\pi_\tau(\varphi_t I) = \varphi_t(\tau)\pi_\tau(I)$ for $\tau \in \Gamma \setminus \{t\}$. Thus, from Proposition 8.5 we deduce that

$$\begin{aligned}
S_0 \; &:= \; \big\{\lambda \in \mathbf{C} : P_\Gamma(\varphi_t - \lambda)P_\Gamma + Q_\Gamma \notin \Phi\big\} \\
&= \; \bigcup_{\tau \in \Gamma} \Big\{\lambda \in \mathbf{C} : \pi_\tau\big(P_\tau(\varphi_t - \lambda)P_\Gamma + Q_\Gamma\big) \notin G\mathcal{B}_\tau\Big\} \\
&= \; \Big\{\lambda \in \mathbf{C} : \pi_t\big(P_\Gamma(\chi_t - \lambda)P_\Gamma + Q_\Gamma\big) \notin G\mathcal{B}_t\Big\} \qquad (8.51) \\
&\quad \cup \bigcup_{\tau \in \Gamma \setminus \{t\}} \Big\{\lambda \in \mathbf{C} : (\varphi_t(\tau) - \lambda)\pi_\tau(P_\Gamma) + \pi_\tau(Q_\Gamma) \notin G\mathcal{B}_\tau\Big\}. \qquad (8.52)
\end{aligned}$$

If $\varphi_t(\tau) - \lambda \neq 0$ then $(\varphi_t(\tau) - \lambda)^{-1} P_\Gamma + Q_\Gamma$ is the inverse of $(\varphi_t(\tau) - \lambda) P_\Gamma + Q_\Gamma$ and hence $(\varphi_t(\tau) - \lambda) \pi_\tau(P_\Gamma) + \pi_\tau(Q_\Gamma)$ is invertible. If $\varphi_t(\tau) - \lambda = 0$, then

$$(\varphi_t(\tau) - \lambda) \pi_\tau(P_\Gamma) + \pi_\tau(Q_\Gamma) = \pi_\tau(Q_\Gamma)$$

and Theorem 8.7 gives

$$\mathrm{sp}_{\mathcal{B}_\tau} \pi_\tau(Q_\Gamma) = \mathrm{sp}_{\mathcal{B}_\tau}(e - r) = \mathrm{sp} \begin{pmatrix} 0 & 0 \\ 0 & 1 \end{pmatrix} = \{0, 1\},$$

which implies that $\pi_\tau(Q_\Gamma)$ cannot be invertible. Thus, the set (8.52) is

$$\bigcup_{\tau \in \Gamma \setminus \{t\}} \{\varphi_t(\tau)\} = \varphi_t(\Gamma \setminus \{t\}).$$

Because (8.51) is nothing but S_1, we arrive at the equality

$$S_0 = S_1 \cup \varphi_t(\Gamma \setminus \{t\}). \tag{8.53}$$

On the other hand, Lemma 6.14 shows that

$$S_0 = \Big\{ \lambda \in \mathbf{C} : T(\varphi_t - \lambda) \text{ is not Fredholm} \Big\} = \mathrm{sp}_{\mathrm{ess}} T(\varphi_t),$$

and Theorem 7.4 so implies that

$$\begin{aligned} S_0 &= \{0, 1\} \cup \varphi_t(\Gamma \setminus \{t\}) \cup \mathcal{L}(0, 1; p, \alpha_t, \beta_t) \\ &= \mathcal{L}(0, 1; p, \alpha_t, \beta_t) \cup \varphi_t(\Gamma \setminus \{t\}). \end{aligned} \tag{8.54}$$

Comparing (8.53) and (8.54) we obtain

$$S_1 \cup \varphi_t(\Gamma \setminus \{t\}) = \mathcal{L}(0, 1; p, \alpha_t, \beta_t) \cup \varphi_t(\Gamma \setminus \{t\}).$$

As φ_t is an arbitrary continuous function on $\Gamma \setminus \{t\}$ which is only subject to the restrictions $\varphi_t(t - 0) = 0$ and $\varphi_t(t + 0) = 1$, it follows that $S_1 = \mathcal{L}(0, 1; p, \alpha_t, \beta_t)$.

To identify S_2, put $1 - \varphi_t =: \psi_t$ and $1 - \chi_t := \varrho_t$. As above, we arrive at the equality

$$S_2 \cup \psi_t(\Gamma \setminus \{t\}) = \Big\{ \lambda \in \mathbf{C} : Q_\Gamma(\varrho_t - \lambda) Q_\Gamma + P_\Gamma \notin \Phi \Big\} \cup \psi_t(\Gamma \setminus \{t\}). \tag{8.55}$$

Proposition 6.16, Lemma 6.14, and Theorem 6.20 show that $Q_\Gamma(\varrho_t - \lambda) Q_\Gamma + P_\Gamma$ cannot be Fredholm if $\lambda \in \mathcal{R}(\varrho_t) = \{0, 1\}$. If $\lambda \notin \{0, 1\}$, repeated application of Lemma 6.14 gives

$$\begin{aligned} Q_\Gamma(\varrho_t - \lambda) Q_\Gamma + P_\Gamma \in \Phi &\iff (\varrho_t - \lambda) Q_\Gamma + P_\Gamma \in \Phi \\ &\iff Q_\Gamma + (\varrho_t - \lambda)^{-1} P_\Gamma \in \Phi \iff P_\Gamma(\varrho_t - \lambda)^{-1} P_\Gamma + Q_\Gamma \in \Phi \\ &\iff T\big((\varrho_t - \lambda)^{-1}\big) \text{ is Fredholm on } L_+^p(\Gamma, w). \end{aligned}$$

Thus,

$$\left\{\lambda \in \mathbf{C} : Q_\Gamma(\varrho_t - \lambda)Q_\Gamma + P_\Gamma \notin \Phi\right\}$$
$$= \{0,1\} \cup \left\{\lambda \notin \{0,1\} : T\big((\varrho_t - \lambda)^{-1}\big) \text{ is not Fredholm}\right\}, \qquad (8.56)$$

and because $(\varrho_t - \lambda)^{-1}(t - 0) = 1/(1 - \lambda)$, $(\varrho_t - \lambda)^{-1}(t + 0) = -1/\lambda$, we deduce from Theorem 7.4 that (8.56) is

$$\{0,1\} \cup \left\{\lambda \notin \{0,1\} : 0 \in \mathcal{L}\Big(\frac{1}{1-\lambda}, \frac{-1}{\lambda}; p, \alpha_t, \beta_t\Big)\right\}. \qquad (8.57)$$

Since

$$0 \in \mathcal{L}\Big(\frac{1}{1-\lambda}, -\frac{1}{\lambda}; p, \alpha_t, \beta_t\Big) \setminus \{0,1\}$$
$$\Longleftrightarrow M_{1/(1-\lambda),-1/\lambda}(e^{2\pi\gamma}) = 0 \text{ for some } \gamma \in Y(p, \alpha_t, \beta_t)$$
$$\Longleftrightarrow e^{2\pi\gamma} = \lambda/(\lambda - 1) \text{ for some } \gamma \in Y(p, \alpha_t, \beta_t)$$
$$\Longleftrightarrow \lambda = M_{0,1}(e^{2\pi\gamma}) \text{ for some } \gamma \in Y(p, \alpha_t, \beta_t)$$
$$\Longleftrightarrow \lambda \in \mathcal{L}(0, 1; p, \alpha_t, \beta_t),$$

we see that (8.57) is equal to $\{0,1\} \cup \big(\mathcal{L}(0,1;p,\alpha_t,\beta_t) \setminus \{0,1\}\big) = \mathcal{L}(0,1;p,\alpha_t,\beta_t)$. This and (8.55) give $S_2 \cup \psi_t\big(\Gamma \setminus \{t\}\big) = \mathcal{L}(0,1;p,\alpha_t,\beta_t) \cup \psi_t\big(\Gamma \setminus \{t\}\big)$, whence $S_2 = \mathcal{L}(0,1;p,\alpha_t,\beta_t)$. In summary, we have proved that $\mathrm{sp}_{\mathcal{B}_t} X = \mathcal{L}(0,1;p,\alpha_t,\beta_t)$. A leaf does not separate the plane, and therefore $\mathrm{sp}_{\mathcal{A}_t} X = \mathrm{sp}_{\mathcal{B}_t} X$. □

Theorem 8.20. *Let Γ be a Carleson Jordan curve, $p \in (1,\infty)$, and $w \in A_p(\Gamma)$. For $t \in \Gamma$, denote by α_t and β_t the indicator functions of Γ, p, w at t. Define the "leaf bundle" $\mathcal{M}$ by*

$$\mathcal{M} := \mathcal{M}_{\Gamma,p,w} := \bigcup_{t\in\Gamma} \big(\{t\} \times \mathcal{L}(0,1;p,\alpha_t,\beta_t)\big). \qquad (8.58)$$

For $(t,x) \in \Gamma \times \mathbf{C}$ and $a \in PC(\Gamma)$ put

$$\mathrm{Sym}_{t,x}(aI) = \begin{pmatrix} a(t+0) & 0 \\ 0 & a(t-0) \end{pmatrix}, \qquad (8.59)$$

for $(t,x) \in \Gamma \times \big(\mathbf{C} \setminus \{0,1\}\big)$ let

$$\mathrm{Sym}_{t,x}(P_\Gamma) = \begin{pmatrix} x & x(1-x) \\ 1 & 1-x \end{pmatrix}, \qquad (8.60)$$

and for $(t,x) \in \Gamma \times \{0,1\}$ set

$$\mathrm{Sym}_{t,x}(P_\Gamma) = \begin{pmatrix} x & 0 \\ 0 & 1-x \end{pmatrix}. \qquad (8.61)$$

(a) *For each $(t, x) \in \mathcal{M}$ the map* $\mathrm{Sym}_{t,x} : \{aI : a \in PC(\Gamma)\} \cup \{P_\Gamma\} \to \mathbf{C}^{2\times2}$ *given by (8.59), (8.60), (8.61) extends to a Banach algebra homomorphism*

$$\mathrm{Sym}_{t,x} : \mathrm{alg}\left(S_\Gamma, PC(\Gamma)\right) \to \mathbf{C}^{2\times2}.$$

(b) *An operator $A \in \mathrm{alg}\left(S_\Gamma, PC(\Gamma)\right)$ is Fredholm on $L^p(\Gamma, w)$ if and only if*

$$\det\left(\mathrm{Sym}_{t,x}(A)\right) \neq 0 \quad \text{for all} \quad (t, x) \in \mathcal{M}.$$

(c) *If an operator $A \in \mathrm{alg}\left(S_\Gamma, PC(\Gamma)\right)$ is Fredholm, then it has a regularizer in the algebra $\mathrm{alg}\left(S_\Gamma, PC(\Gamma)\right)$.*

Proof. Fix $t \in \Gamma$ and apply Theorem 8.7 to $\mathcal{A} := \mathcal{A}_t(\Gamma)$, $\mathcal{B} := \mathcal{B}_t(\Gamma)$, $r := \pi_t(P_\Gamma)$, $s := \pi_t(\chi_t I)$. From Theorems 8.7(a) and 8.19 we know that for each $x \in \mathcal{L}(0, 1; p, \alpha_t, \beta_t)$ the map $\sigma_x : \{e, r, s\} \to \mathbf{C}^{2\times2}$ defined by (8.6) and (8.7) extends to a Banach algebra homomorphism σ_x of $\mathcal{A}_t(\Gamma)$ into $\mathbf{C}^{2\times2}$. Therefore

$$\sigma_x \circ \pi_t : \mathrm{alg}\left(S_\Gamma, PC(\Gamma)\right) \to \mathbf{C}^{2\times2}, \quad A \mapsto \sigma_x\left(\pi_t(A)\right)$$

is a well-defined Banach algebra homomorphism for each point $(t, x) \in \mathcal{M}$. Theorem 8.7(b) in conjunction with Theorem 8.19 implies that

$$\mathrm{sp}_{\mathcal{B}_t(\Gamma)} \pi_t(A) = \bigcup_{x \in \mathcal{L}(0,1;p,\alpha_t,\beta_t)} \mathrm{sp}\,\sigma_x\left(\pi_t(A)\right),$$

and Proposition 8.5 then gives $\mathrm{sp}_{\mathrm{ess}}\, A = \bigcup_{(t,x)\in\mathcal{M}} \mathrm{sp}\,\sigma_x(\pi_t(A))$, i.e. A is Fredholm if and only if $\det(\sigma_x(\pi_t(A))) \neq 0$ for all $(t, x) \in \mathcal{M}$. Combining Theorems 8.2, 8.7(c), 8.19 we also see that $\pi(\mathrm{alg}\left(S_\Gamma, PC(\Gamma)\right))$ is inverse closed in the Calkin algebra.

If $a \in PC(\Gamma)$, then $\pi_t(aI) = a(t+0)s + a(t-0)(e - s)$. Hence, for $x \notin \{0, 1\}$ we get from (8.6) that

$$\sigma_x \pi_t(aI) = a(t+0) \begin{pmatrix} x & x-1 \\ -x & 1-x \end{pmatrix} + a(t-0) \begin{pmatrix} 1-x & 1-x \\ x & x \end{pmatrix}$$

$$= \begin{pmatrix} a(t+0)x + a(t-0)(1-x) & (a(t-0) - a(t+0))(1-x) \\ (a(t-0) - a(t+0))x & a(t+0)(1-x) + a(t-0)x \end{pmatrix}, \quad (8.62)$$

while for $x \in \{0, 1\}$ we infer from (8.7) that

$$\sigma_0 \pi_t(aI) = \mathrm{diag}\left(a(t-0), a(t+0)\right), \quad \sigma_1 \pi_t(aI) = \mathrm{diag}\left(a(t+0), a(t-0)\right). \quad (8.63)$$

Also by (8.6) and (8.7), we have

$$\sigma_x \pi_t(P_\Gamma) = \begin{pmatrix} 1 & 0 \\ 0 & 0 \end{pmatrix} \quad (8.64)$$

for all x. For $x \notin \{0, 1\}$, let

$$D_x := \begin{pmatrix} 1 & 1-x \\ -1 & x \end{pmatrix}, \quad D_x^{-1} = \begin{pmatrix} x & x-1 \\ 1 & 1 \end{pmatrix}.$$

From (8.62) and (8.64) we obtain

$$D_x^{-1}\big(\sigma_x \pi_t(aI)\big)D_x = \begin{pmatrix} a(t+0) & 0 \\ 0 & a(t-0) \end{pmatrix} = \mathrm{Sym}_{t,x}(aI),$$

$$D_x^{-1}\big(\sigma_x \pi_t(P_\Gamma)\big)D_x = \begin{pmatrix} x & x(1-x) \\ 1 & 1-x \end{pmatrix} = \mathrm{Sym}_{t,x}(P_\Gamma).$$

for $x \notin \{0, 1\}$. Finally, let

$$D_0 := \begin{pmatrix} 0 & 1 \\ 1 & 0 \end{pmatrix} \quad \text{and} \quad D_1 := \begin{pmatrix} 1 & 0 \\ 0 & 1 \end{pmatrix}.$$

Using (8.63) and (8.64) we see that

$$D_x^{-1}\big(\sigma_x \pi_t(aI)\big)D_x = \mathrm{Sym}_{t,x}(aI), \quad D_x^{-1}\big(\sigma_x \pi_t(P_\Gamma)\big)D_x = \mathrm{Sym}_{t,x}(P_\Gamma)$$

is also true for $x \in \{0, 1\}$. Since $A \mapsto D_x^{-1}(\sigma_x \pi_t(A))D_x$ is a Banach algebra homomorphism together with $\sigma_x \circ \pi_t$ and since $\mathrm{sp}\, D_x^{-1}\big(\sigma_x \pi_t(A)\big)D_x = \mathrm{sp}\,\sigma_x \pi_t(A)$, we get all assertions. $\qquad\square$

The previous proof shows that Theorem 8.20 remains literally true with (8.59), (8.60), (8.61) replaced by

$\mathrm{Sym}_{t,x}(aI)$
$$= \begin{pmatrix} a(t+0)x + a(t-0)(1-x) & (a(t-0) - a(t+0)(1-x) \\ (a(t-0) - a(t+0))x & a(t+0)(1-x) + a(t-0)x \end{pmatrix} \quad (8.65)$$

for $(t, x) \in \Gamma \times (\mathbf{C} \setminus \{0, 1\})$,

$\mathrm{Sym}_{t,x}(aI) = \mathrm{diag}\,\big(a(t+0)x + a(t-0)(1-x), a(t+0)(1-x) + a(t-0)x\big)$ (8.66)

for $(t, x) \in \Gamma \times \{0, 1\}$,

$$\mathrm{Sym}_{t,x}(P_\Gamma) = \begin{pmatrix} 1 & 0 \\ 0 & 0 \end{pmatrix} \quad (8.67)$$

for all $(t, x) \in \Gamma \times \mathbf{C}$, respectively.

Example 8.21. If Γ is a Jordan curve then $S_\Gamma^2 = I$ by Corollary 6.6 and hence $\text{sp}\, S_\Gamma = \text{sp}_{\text{ess}}\, S_\Gamma = \{-1, 1\}$. Notice that Theorem 8.20 yields

$$\det \text{Sym}_{t,x}(S_\Gamma - \lambda I) = \det \text{Sym}_{t,x}(2P_\Gamma - (1+\lambda)I)$$

$$= \begin{cases} \det \begin{pmatrix} 2x - 1 - \lambda & 2x(1-x) \\ 2 & 2(1-x) - 1 - \lambda \end{pmatrix} = \lambda^2 - 1 & \text{if } x \notin \{0, 1\}, \\[2mm] \det \begin{pmatrix} 2x - 1 - \lambda & 0 \\ 0 & 2(1-x) - 1 - \lambda \end{pmatrix} = \lambda^2 - 1 & \text{if } x \in \{0, 1\}, \end{cases}$$

and so also gives $\text{sp}_{\text{ess}}\, S_\Gamma = \{-1, 1\}$. Thus, the leaves do not yet appear in the spectrum of the Cauchy singular integral operator. It is only piecewise continuous coefficients which make visible the leaves. $\square$

Example 8.22. Let Γ be a Carleson Jordan curve, $p \in (1, \infty)$, and $w \in A_p(\Gamma)$. Let $\eta \subset \Gamma$ be a simple arc and denote the endpoints of η by t_1 and t_2. Suppose η and Γ are oriented from t_1 to t_2. We want to determine the spectrum of $A :=$ $\chi_\eta S_\Gamma \chi_\eta + \chi_{\Gamma \setminus \eta}$ on $L^p(\Gamma, w)$.

We compute $\text{Sym}_{t,x}(A)$ using Theorem 8.20. We have

$$\text{Sym}_{t,x}(S_\Gamma) = \begin{pmatrix} 2x - 1 & 2x(1-x) \\ 2 & 1 - 2x \end{pmatrix} =: C_x \quad \text{if } x \notin \{0, 1\},$$

$$\text{Sym}_{t,x}(S_\Gamma) = \begin{pmatrix} 2x - 1 & 0 \\ 0 & 1 - 2x \end{pmatrix} =: D_x \quad \text{if } x \in \{0, 1\},$$

and hence if $x \notin \{0, 1\}$, then $\text{Sym}_{t,x}(A)$ equals

$$\begin{pmatrix} 1 & 0 \\ 0 & 1 \end{pmatrix} C_x \begin{pmatrix} 1 & 0 \\ 0 & 1 \end{pmatrix} + \begin{pmatrix} 0 & 0 \\ 0 & 0 \end{pmatrix}, \quad \begin{pmatrix} 0 & 0 \\ 0 & 0 \end{pmatrix} C_x \begin{pmatrix} 0 & 0 \\ 0 & 0 \end{pmatrix} + \begin{pmatrix} 1 & 0 \\ 0 & 1 \end{pmatrix}, \quad (8.68)$$

$$\begin{pmatrix} 1 & 0 \\ 0 & 0 \end{pmatrix} C_x \begin{pmatrix} 1 & 0 \\ 0 & 0 \end{pmatrix} + \begin{pmatrix} 0 & 0 \\ 0 & 1 \end{pmatrix}, \quad \begin{pmatrix} 0 & 0 \\ 0 & 1 \end{pmatrix} C_x \begin{pmatrix} 0 & 0 \\ 0 & 1 \end{pmatrix} + \begin{pmatrix} 1 & 0 \\ 0 & 0 \end{pmatrix}, \quad (8.69)$$

for $t \in \eta \setminus \{t_1, t_2\}$, $t \in (\Gamma \setminus \eta) \setminus \{t_1, t_2\}$, $t = t_1$, $t = t_2$, respectively. The eigenvalues of these matrices are

$$\{-1, 1\}, \ \{1\}, \ \{2x - 1, 1\}, \ \{1 - 2x, 1\} \quad (8.70)$$

respectively. For $x \in \{0, 1\}$, we have to replace C_x by D_x in (8.68) and (8.69), and the eigenvalues of the resulting matrices are again given by (8.70). Thus, $\text{sp}_{\text{ess}}(A)$ is

$$\{-1, 1\} \cup \left(2\mathcal{L}(0, 1; p, \alpha_{t_1}, \beta_{t_1}) - 1 \right) \cup \left(1 - 2\mathcal{L}(0, 1; p, \alpha_{t_2}, \beta_{t_2}) \right)$$
$$= \mathcal{L}(-1, 1; p, \alpha_{t_1}, \beta_{t_1}) \cup \mathcal{L}(1, -1; p, \alpha_{t_2}, \beta_{t_2}). \quad \square$$

We conclude this chapter by describing the maximal ideal spaces of two commutative Banach subalgebras of $\text{alg}\,(S_\Gamma, PC(\Gamma))/\mathcal{K}(\Gamma)$: the algebra of (cosets of) singular integral operators with continuous coefficients and the "Toeplitz algebra".

Lemma 8.23. *If* Γ *is a Carleson Jordan curve,* $p \in (1,\infty)$*, and* $w \in A_p(\Gamma)$*, then the set* $\mathcal{K}(L^p(\Gamma,w))$ *of all compact operators is contained in* $\mathrm{alg}\,(S_\Gamma, C(\Gamma))$*, the smallest closed subalgebra of* $\mathcal{B}(L^p(\Gamma,w))$ *containing* S_Γ *and* $\{aI : a \in C(\Gamma)\}$*.*

Proof. Without loss of generality assume $|\Gamma| = 2\pi$. Let $\eta : \mathbf{T} \to \Gamma$ be a homeomorphism such that $|\eta'(z)| = 1$ for almost all $z \in \mathbf{T}$. Clearly, each portion $\mathbf{T}(z,\varepsilon)$ ($z \in \mathbf{T}, \varepsilon > 0$) is contained in an arc $\gamma_{z,\varepsilon} \subset \mathbf{T}$ of length at most $2\pi\varepsilon$. We have

$$\frac{1}{\varepsilon} \int\limits_{\mathbf{T}(z,\varepsilon)} ((w \circ \eta)(\zeta))^p |d\zeta| = \frac{1}{\varepsilon} \int\limits_{\eta(\mathbf{T}(z,\varepsilon))} w(\tau)^p |d\tau|, \tag{8.71}$$

and $\eta(\mathbf{T}(z,\varepsilon))$ is a subset of $\eta(\gamma_{z,\varepsilon})$ the length of which is $|\gamma_{z,\varepsilon}| \le 2\pi\varepsilon$. Therefore $\eta(\mathbf{T}(z,\varepsilon)) \subset \Gamma(\eta(z), 2\pi\varepsilon)$ and hence (8.71) is not greater than

$$\frac{1}{\varepsilon} \int\limits_{\Gamma(\eta(z),2\pi\varepsilon)} w(\tau)^p |d\tau| = 2\pi\frac{1}{2\pi\varepsilon} \int\limits_{\Gamma(\eta(z),2\pi\varepsilon)} w(\tau)^p |d\tau|.$$

Analogously,

$$\frac{1}{\varepsilon} \int\limits_{\mathbf{T}(z,\varepsilon)} ((w \circ \eta)(\zeta))^{-q} |d\zeta| \le 2\pi\frac{1}{2\pi\varepsilon} \int\limits_{\Gamma(\eta(z),2\pi\varepsilon)} w(\tau)^{-q} |d\tau|.$$

Thus, $\varrho := w \circ \eta \in A_p(\mathbf{T})$.

Theorem 8(c) of [107] says that every function in $L^p(\mathbf{T}, \varrho)$ can be approximated in the norm of $L^p(\mathbf{T}, \varrho)$ by the partial sums of its Fourier series if (and only if) $\varrho \in A_p(\mathbf{T})$. Because the map $f \mapsto f \circ \eta$ is an isometric isomorphism of $L^p(\Gamma, w)$ onto $L^p(\mathbf{T}, \varrho)$, it follows that there is a sequence of finite-rank operators on $L^p(\Gamma, w)$ converging strongly to the identity operator. Consequently, every compact operator on $L^p(\Gamma, w)$ can be approximated in the operator norm by finite-rank operators. A finite-rank operator on $L^p(\Gamma, w)$ is of the form

$$(Kf)(t) = \sum_{j=1}^m a_j(t) \int\limits_\Gamma b_j(\tau) f(\tau)\, d\tau \quad (t \in \Gamma) \tag{8.72}$$

with $a_j \in L^p(\Gamma, w)$ and $b_j \in L^q(\Gamma, w^{-1})$. Since, by Lemma 4.4, $C(\Gamma)$ is dense in $L^p(\Gamma, w)$ and $L^q(\Gamma, w^{-1})$, every operator of the form (8.72) can be approximated in the operator norm by operators of the same form with a_j and b_j in $C(\Gamma)$. As the operator (8.72) is obviously equal to

$$K = \sum_{j=1}^m a_j (S_\Gamma \chi I - \chi S_\Gamma) b_j I$$

with $\chi(\tau) := \tau$ ($\tau \in \Gamma$), we see that $K \in \mathrm{alg}\,(S_\Gamma, C(\Gamma))$ whenever $a_j, b_j \in C(\Gamma)$. $\qquad\square$

Corollary 8.24. *Let Γ be a Carleson Jordan curve, $p \in (1, \infty)$, and $w \in A_p(\Gamma)$. Then*

$$\text{alg}\,(S_\Gamma, C(\Gamma)) = \{aP_\Gamma + bQ_\Gamma + K : a, b \in C(\Gamma),\ K \in \mathcal{K}(\Gamma)\} \qquad (8.73)$$

where $\mathcal{K}(\Gamma) := \mathcal{K}(L^p(\Gamma, w))$. The algebra

$$\pi(\text{alg}\,(S_\Gamma, C(\Gamma))) := \text{alg}\,(S_\Gamma, C(\Gamma))/\mathcal{K}(\Gamma)$$

is commutative, its maximal ideal space is homeomorphic to $\Gamma \times \{0, 1\}$, and the Gelfand map

$$\mathcal{G} : \pi(\text{alg}\,(S_\Gamma, C(\Gamma))) \to C(\Gamma \times \{0, 1\})$$

is for $(t, x) \in \Gamma \times \{0, 1\}$ given by

$$(\mathcal{G}\pi(aP_\Gamma + bQ_\Gamma))(t, x) = a(t)(1 - x) + b(t)x.$$

An operator $A \in \text{alg}\,(S_\Gamma, C(\Gamma))$ is Fredholm if and only if $(\mathcal{G}\pi(A))(t, x) \neq 0$ for all $(t, x) \in \Gamma \times \{0, 1\}$.

Proof. Denote the set on the right of (8.73) by $\mathcal{D}$. Lemma 8.23 implies that $\mathcal{D} \subset \text{alg}\,(S_\Gamma, C(\Gamma))$. From Proposition 6.21 we infer that $\mathcal{D}$ is an algebra, and hence the inclusion $\text{alg}\,(S_\Gamma, C(\Gamma)) \subset \mathcal{D}$ will follow as soon as we have shown that $\mathcal{D}$ is closed. This in turn is an immediate consequence of the estimate

$$\|aP_\Gamma + bQ_\Gamma\|_{\text{ess}} := \|\pi(aP_\Gamma + bQ_\Gamma)\| \geq \max\{\|a\|_\infty, \|b\|_\infty\}. \qquad (8.74)$$

The proof of (8.74) is simple: Theorem 8.20 with (8.66) and (8.67) gives

$$\text{Sym}_{t,0}(aP_\Gamma + bQ_\Gamma) = \text{diag}\,(a(t), b(t)), \qquad (8.75)$$

hence $a(\Gamma) \cup b(\Gamma) \subset \text{sp}_{\text{ess}}\,(aP_\Gamma + bQ_\Gamma)$, and since the spectral radius does not exceed the norm, we arrive at (8.74). This completes the proof of (8.73).

From (8.73) and Proposition 6.21 we deduce that $\pi(\text{alg}\,(S_\Gamma, C(\Gamma)))$ is commutative. Let Ω be its maximal ideal space. Because $\text{sp}_{\text{ess}}\,K = \{0\}$ for every compact operator K, Theorem 8.20 and Lemma 8.23 imply that the diagonal matrix $\text{Sym}_{t,0}(K)$ is the zero matrix whenever K is compact. Thus, by (8.74) and (8.75), the map

$$\pi(aP_\Gamma + bQ_\Gamma) \mapsto a(t)(1 - x) + b(t)x \qquad (8.76)$$

is a well-defined multiplicative linear functional for every $(t, x) \in \Gamma \times \{0, 1\}$. Consequently, $\Omega \supset \Gamma \times \{0, 1\}$.

Theorems 8.2, 8.7(c), and 8.19 tell us that an operator $A \in \text{alg}\,(S_\Gamma, C(\Gamma))$ is Fredholm if and only if $\pi(A)$ is invertible in $\pi(\text{alg}\,(S_\Gamma, C(\Gamma)))$. Equivalently, the spectrum of $\pi(A)$ in $\pi(\text{alg}\,(S_\Gamma, C(\Gamma)))$ coincides with the essential spectrum of A. If $A = aP_\Gamma + bQ_\Gamma$, the latter spectrum is equal to $a(\Gamma) \cup b(\Gamma)$. This easily implies that $\Gamma \times \{0, 1\} \supset \Omega$.

Hence, $\Omega = \Gamma \times \{0, 1\}$ and the map (8.76) gives the Gelfand map $\mathcal{G}$. Theorem 8.1 completes the proof. $\square$

Proposition 8.25. *Let* Γ *be a Carleson Jordan curve,* $p \in (1, \infty)$*, and* $w \in A_p(\Gamma)$*. If* $a, b \in PC(\Gamma)$ *then* $T(a)T(b) - T(b)T(a) \in \mathcal{K}(L_+^p(\Gamma, w))$*.*

Proof. A little thought in conjunction with Theorem 6.28 reveals that it suffices to prove the assertion for the case where a and b have only one jump, at α and β, respectively. If $\alpha \neq \beta$, we may write

$$T(a)T(b) - T(b)T(a) = \big[T(ba) - T(b)T(a)\big] - \big[T(ab) - T(a)T(b)\big]$$

and use Theorem 6.28 to conclude that the commutator is compact. So let $\alpha = \beta$. Then there exists a constant $\lambda \in \mathbf{C}$ and a function $c \in C(\Gamma)$ such that $a = \lambda b + c$. Hence

$$T(a)T(b) - T(b)T(a) = T(\lambda b + c)T(b) - T(b)T(\lambda b + c)$$
$$= T(c)T(b) - T(b)T(c) = \big[T(bc) - T(b)T(c)\big] - \big[T(cb) - T(c)T(b)\big],$$

and the compactness of the commutator follows again from Theorem 6.28. $\square$

Corollary 8.26. *Let* Γ *be a Carleson Jordan curve,* $p \in (1, \infty)$*, and* $w \in A_p(\Gamma)$*. Denote by* $\mathrm{alg}\, T(PC(\Gamma))$ *the smallest closed subalgebra of* $\mathcal{B}(L_+^p(\Gamma, w))$ *containing the set* $\{T(a) : a \in PC(\Gamma)\}$*. We have*

$$\mathcal{K}(L_+^p(\Gamma, w)) \subset \mathrm{alg}\, T(PC(\Gamma)), \tag{8.77}$$

the algebra

$$\pi\big(\mathrm{alg}\, T(PC(\Gamma))\big) := \mathrm{alg}\, T(PC(\Gamma)) / \mathcal{K}(L_+^p(\Gamma, w))$$

is commutative, its maximal ideal space is homeomorphic to the leaf bundle $\mathcal{M}$ *given by (8.58) (with an exotic topology), and the Gelfand transform*

$$\mathcal{G} : \pi\big(\mathrm{alg}\, T(PC(\Gamma))\big) \to C(\mathcal{M})$$

acts by the rule

$$(\mathcal{G}\pi(T(a)))(t, x) = a(t - 0)(1 - x) + a(t + 0)x$$

for $(t, x) \in \mathcal{M}$*. An operator* A *in the algebra* $\mathrm{alg}\, T(PC(\Gamma))$ *is Fredholm if and only if* $(\mathcal{G}\pi(A))(t, x) \neq 0$ *for all* $(t, x) \in \mathcal{M}$*.*

Proof. Let $\mathcal{F}$ be the smallest closed subalgebra of $\mathrm{alg}\,(S_\Gamma, P\dot{C}(\Gamma))$ containing the set

$$\{P_\Gamma a P_\Gamma + Q_\Gamma : a \in PC(\Gamma)\},$$

and denote by $\pi(\mathcal{F})$ the image of $\mathcal{F}$ in the Calkin algebra. Lemma 8.23 implies that $\mathcal{F}$ contains all operators of the form $P_\Gamma K P_\Gamma + Q_\Gamma$ with $K \in \mathcal{K}(L^p(\Gamma, w))$. This easily gives (8.77) and shows that the map

$$\pi\big(\mathrm{alg}\, T(PC(\Gamma))\big) \to \pi(\mathcal{F}), \ \pi\big(T(a)\big) \mapsto \pi(P_\Gamma a P_\Gamma + Q_\Gamma)$$

is a well-defined Banach algebra isomorphism. Proposition 8.25 tells us that $\pi(\mathcal{F})$ is commutative. Theorem 8.20 with the matrices (8.65), (8.66), (8.67) gives

$$\mathrm{Sym}_{t,x}(P_\Gamma a P_\Gamma + Q_\Gamma) = \mathrm{diag}\left(a(t+0)x + a(t-0)(1-x), 1\right)$$

for all $(t, x) \in \mathcal{M}$. The proof can now be finished by the arguments of the proof of Corollary 8.24.

$\square$

8.6 Notes and comments

8.1. Proofs of Theorem 8.1 are in every text on Banach algebras and also in [52], [80], and [181], for example.

Theorems like Theorem 8.2 are known as central decompositions of algebras and are well known especially to workers in C^*-algebras. In the form cited here, the theorem was established by Allan [5]. In the case where both $\mathcal{B}$ and Z are C^*-algebras, the theorem was independently found by Douglas [52, Theorem 7.47], who was also the first to understand the relevance of this theorem to the spectral theory of Toeplitz operators. Theorem 8.2 is closely related to the local principle of Gohberg and Krupnik [89, Chapter 5, Theorems 1.1 and 1.2], which in turn has its roots in Simonenko's local method [195]. For instance, we can show that at least in the case where Z is a C^*-algebra, Theorem 8.2 is a special case of the local principle of Gohberg and Krupnik. Note that in many applications $\mathcal{B}$ is no C^*-algebra, although Z is isometrically isomorphic to $C(\Gamma)$ and is therefore a C^*-algebra. We remark that localization with Theorem 8.2 usually delivers better local algebras and local representatives than Gohberg-Krupnik localization. The point is that the latter principle requires construction of a covering system of localizing classes, which is often very easy but accompanied with some arbitrariness, while Allan-Douglas localization does everything automatically (and thus, in a sense, canonically) once we have only fed in the central subalgebra Z.

For F_{2n}-algebras we refer to the monographs by Rowen [178] and Krupnik [134]. Theorem 8.3 was established by Krupnik [133] for $m = 1$ and by Finck and Roch [67] for general m.

8.2. The algebra $\Lambda(\Gamma)$ and the name "operators of local type" were introduced by Simonenko [195] (also see his book [199] with Chin Ngok Min), local algebras and the notion of the "local spectrum" go back to Douglas [52], [53]. Proposition 8.5 is Simonenko's result in the language of Douglas. Proposition 8.6 is taken from the paper [187] by Semenyuta and Khevelev.

8.3–8.4. C^*-algebras and Banach algebras generated by idempotents have been a big business for a long time, and a detailed discussion of this topic would go beyond the scope of this book. For example, abstract C^*-algebras generated by two idempotents were studied by Halmos [97], Pedersen [156], Power [161], Raeburn and Sinclair [170], Vasilevski and Spitkovsky [205], [201].

The idea of proceeding as in the first two paragraphs of Section 8.3, i.e. the idea of first localizing and then employing a two projections theorem to the local algebra, is due to Douglas [53]. He considered operators on $L^2(\mathbf{T})$, in which case the local algebras $\mathcal{A}_t(\mathbf{T})$ were C^*-algebras, and the two projections theorems one had at that time worked just for C^*-algebras. The elegance of Douglas' approach has attracted many researchers into Toeplitz and singular integral operators. As the metamorphosis of line segments into circular arcs was the most exciting phenomenon of the Banach space theory of Toeplitz and singular integral operators and as localization worked equally well in the C^*-algebra and Banach algebra cases, the great challenge was to extend the existing two projections theorems to Banach algebras. In the late eighties, Roch and Silbermann [173] were indeed able to prove a two projections theorem for Banach algebras and thus to extend Douglas' method to operators on $L^p(\Gamma, w)$ provided Γ is a piecewise Lyapunov Jordan curve and w is a power weight. At the beginning of the nineties, Spitkovsky [200] discovered the appearance of horns in the local spectra of Toeplitz operators in the case of general weights $w \in A_p(\Gamma)$. In contrast to circular arcs, horns are heavy sets, that is, sets which may arise from other sets by filling in holes. This circumstance caused serious problems with the inverse closedness of Banach algebras generated by two idempotents in larger algebras. These obstacles were overcome by Finck, Roch, Silbermann [68] and Gohberg, Krupnik [90] and resulted in Theorem 8.7, which is the theorem of [68], complemented by results of [90].

It has been well known for a long time that there are no useful N projections theorems for $N \geq 3$ unless the projections are subject to additional axioms. See, for example, the paper by Vasilev [204] and the book by Samoilenko [182]. Theorem 8.8 is due to Roch et al. [172] and nicely illustrates this fact.

With Theorems 7.4 and 8.7, everything was clear for singular integral operators on $L^p(\Gamma, w)$ if only Γ is a Jordan curve. Consideration of composed curves required an appropriate N projections theorem, and workers in singular integral operators were all aware of this need. Consequently, about two years ago, the search for such a theorem was begun by a couple of groups. After a year or so, it was many people who had an N projections theorem, and moreover, they all had (almost) the same theorem, namely Theorem 8.9. We remark that this should not come as too much a surprise, since the conditions (8.10) are the deciding axioms of Theorem 8.9 and such conditions already appeared in Gohberg and Krupnik's papers [86], [88], which were known to all participants of the run. In the end and at the request of the other authors, Steffen Roch, a member of one of the groups, unified these approaches and styles, closed the gaps, and brought the things to the form in which they are presented in Section 8.4. Thus, Section 8.4 is essentially a part of [172].

Corollary 8.18 is Proposition 9(b) of [172]. In [172], the matrices $\sigma_x(P)$ are defined by (8.14) for all $x \in \mathrm{sp}_{\mathcal{A}} X$. However, in [172] we did actually not prove that, with this definition, σ_0 and σ_1 extend to Banach algebra homomorphisms.

This defect is remedied by Corollary 8.18, in which $\sigma_x(P)$ is given by (8.14) for $x \in \mathrm{sp}_A X \setminus \{0, 1\}$ and by (8.48) for $x \in \{0, 1\}$.

8.5. In the case of piecewise Lyapunov curves, the results of this section are due to Gohberg, Krupnik [87], [89] (power weights) and Finck, Roch, Silbermann [68] (Muckenhoupt weights). For Carleson curves, they were established by the authors in [15], [16], [18].

Some (hand-drawn) pictures of leaf bundles can be found in the paper [13], which is an extended version of a plenary lecture given by one of the authors at the DMV Jahrestagung in Duisburg, 1995.

Chapter 9
Composed curves

The purpose of this chapter is to establish Fredholm criteria for operators in the algebra $\text{alg}\,(S_\Gamma, PC(\Gamma))$ in case Γ is an arbitrary composed Carleson curve. Localization techniques give us the essential spectrum of an operator in $\text{alg}\,(S_\Gamma, PC(\Gamma))$ as the union of its local spectra.

The simplest situation beyond Jordan curves is the one in which Γ is a so-called flower, i.e. the union of a finite number N of Jordan curves with exactly one common point, the center of the flower. In this case the local algebras may be studied by the N projections theorem of the previous chapter at the center of the flower and by the 2 projections theorem at the remaining points.

A general composed curve looks like a so-called star in a neighborhood of each of its points, that is, like the union of finitely many simple arcs with exactly one common point. We will show that every Carleson star can be extended to a Carleson flower and that every Muckenhoupt weight on a Carleson star can be extended to a Muckenhoupt weight on the corresponding Carleson flower. Once this is done, local algebras associated with stars may be embedded into local algebras at the centers of flowers.

Throughout this chapter, all curves are assumed to be compact subsets of the plane. In particular, by a simple arc we always mean a curve homeomorphic to the line segment $[0, 1]$.

9.1 Extending Carleson stars

We call a composed curve γ a *Carleson star* if either γ is a simple Carleson arc or γ is comprised of a finite number $N \geq 2$ of simple Carleson arcs $\eta_1, \ldots, \eta_N$ having exactly one point t in common. In the latter case the point t is required to be an endpoint of each of the arcs η_j and is referred to as the *center* of the star. When considering a simple arc as a star, we take one of its two endpoints as the center.

A composed curve F is said to be a *Carleson flower* if it is either a Carleson Jordan curve or a finite union of $N \geq 2$ Carleson Jordan curves with exactly one common point. Thus, a Carleson flower F may be written as

$$F = \Gamma_1 \cup \ldots \cup \Gamma_N$$

where $\Gamma_1, \ldots, \Gamma_N$ are Carleson Jordan curves and $\Gamma_i \cap \Gamma_j = \{t\}$ for $i \neq j$. If $N \geq 2$, we refer to t as the *center* of the flower. Each point of a Jordan curve may be its center if the curve is considered as a flower.

We always think of Jordan curves as being oriented counter-clockwise. Accordingly, a flower F always consists of a finite number of counter-clockwise oriented Jordan curves. A star γ is comprised of a finite number of oriented simple arcs. In the following, the notation $\gamma \subset F$ does not only mean that γ is a subset of F but also that the simple arcs of γ have the orientation inherited from F.

The behavior of the Cauchy singular integral operator on Carleson flowers is much better than on Carleson stars. The point is that if F is a Carleson flower, then

$$S_F^2 = I$$

and hence $P_F := (I + S_F)/2$ is a projection (see Theorem 9.15 below). This is not true for the Cauchy singular integral operator on a Carleson star. However, if γ is a Carleson star and $F \supset \gamma$ is a Carleson flower, then the essential spectrum of S_γ equals the essential spectrum of

$$\chi_\gamma S_F \chi_\gamma I + \chi_{F \setminus \gamma} I \,.$$

Notice that the latter operator belongs to $\mathrm{alg}\,(S_F, PC(F))$, i.e., studying the sole operator S_γ naturally leads to the investigation of singular integral operators on F with *piecewise continuous* coefficients.

Thus, we arrive at the problem of extending Carleson stars γ to Carleson flowers F. For "nice" stars γ, it is more or less obvious that they can be extended to "nice" flowers F. However, a clean treatment of the problem in the general case requires several deep results from geometric function theory.

The *hyperbolic metric* in the complex unit disk $\mathbf{D}$ and the upper complex half-plane $\mathbf{C}_+ := \{z \in \mathbf{C} : \mathrm{Im}\, z > 0\}$ is given by

$$|dz|_H := \frac{2}{1 - |z|^2} |dz| \text{ and } |dz|_H := \frac{1}{\mathrm{Im}\, z} |dz|,$$

respectively. The maximal hyperbolic geodesics of $\mathbf{D}$ are circular arcs intersecting the boundary $\mathbf{T} = \partial \mathbf{D}$ at a right angle (diameters of $\mathbf{D}$ are regarded as degenerate circular arcs), and the maximal hyperbolic geodesics of $\mathbf{C}_+$ are half-circles centered at the boundary $\mathbf{R} = \partial \mathbf{C}_+$ (including half-lines perpendicular to $\mathbf{R}$).

The following theorem is just the result we need to solve our extension problem.

Theorem 9.1 (Zinsmeister). *Suppose Ω is a simply connected domain in the extended complex plane $\mathbf{C} \cup \{\infty\}$ and the boundary $\partial\Omega$ is a Carleson curve. Let $\varphi : \mathbf{D} \to \Omega$ be a Riemann map of the unit disk $\mathbf{D}$ onto Ω. If $\sigma \subset \mathbf{D}$ is a Carleson curve, then $\varphi(\sigma) \subset \Omega$ is also a Carleson curve.*

This result is a special case of Proposition 4 of [211]. The *proof* is based on the Hayman-Wu theorem (for which there are now a variety of proofs [102], [65], [75], [154]). Note that in [211] the term "Carleson domain" does not mean a domain bounded by a Carleson curve (although such a domain is an example of one), but something more general.   □

Theorem 9.2. *If γ is a Carleson star, then there exists a Carleson flower F such that $\gamma \subset F$.*

Proof. Suppose γ is made up of the simple arcs $\eta_1, \ldots, \eta_N$ with the common endpoint t. Denote by y_j the other endpoint of η_j $(j = 1, \ldots, N)$. Put $\Omega := \mathbf{C} \setminus \gamma$ and let $\varphi : \mathbf{D} \to \Omega$ be a Riemann map. Because $\partial\Omega = \gamma$ is rectifiable, it is certainly locally connected, and hence, φ extends continuously to $\mathbf{D} \cup \mathbf{T}$. Each point $y_j \in \{y_1, \ldots, y_N\}$ has exactly one preimage $z_j \in \mathbf{T}$, and the center t has exactly N preimages $t_1, \ldots, t_N$ on $\mathbf{T}$ which separate the points $z_1, , \ldots, z_N$. We assume everything is labeled so that as we traverse $\mathbf{T}$ in the clockwise (sic!) direction, we encounter the points in the order $t_1, z_1, t_2, z_2, \ldots, t_N, z_N$. Put $t_{N+1} := t_1$.

For $j \in \{1, \ldots, N\}$, let σ_j be the hyperbolic geodesic of $\mathbf{D}$ connecting z_j to t_{j+1} if η_j is directed away from t, and let σ_j stand for the hyperbolic geodesic of $\mathbf{D}$ connecting t_j to z_j in case η_j is an incoming arc. Then $\sigma := \sigma_1 \cup \ldots \cup \sigma_N$ is clearly a union of Carleson curves. We may assume that no point of σ is mapped to ∞ (otherwise replace φ by $\varphi \circ \psi$ with an appropriate Möbius transform $\psi : \mathbf{D} \to \mathbf{D}$). Thus, by Theorem 9.1, $\varphi(\sigma) \cup \gamma$ is a Carleson flower extending the star γ.   □

We should note that in the preceding proof the use of hyperbolic geodesics was completely irrelevant and was only made for concreteness and with a preview to the next section. Any choice of N simple Carleson arcs connecting the desired points without crossing each other would have worked just as well. Moreover, every possible completion of γ to a Carleson flower arises in this way. In fact, for any simply connected domain $\Omega \subset \mathbf{C} \cup \{\infty\}$ whose boundary is a Carleson curve, any Riemann map φ of $\mathbf{D}$ onto Ω, and any Carleson curve Γ, the preimage $\varphi^{-1}(\Gamma \cap \Omega)$ is always a Carleson curve in the disk. If $\Gamma \subset \Omega$, this was established in [66], the general case is in [11].

An independent proof of Theorem 9.2, i.e. a proof which does not have recourse to Theorem 9.1, will be given in the next section.

9.2   Extending Muckenhoupt weights

Extending a Carleson star γ to a Carleson flower F is not sufficient for our purposes. Our problem is rather as follows: we are given a Carleson star γ and a weight $w \in A_p(\gamma)$, and we have to construct a Carleson flower F and a weight $W \in A_p(F)$ such that $\gamma \subset F$ and $W|\gamma = w$. In this section we show that this can always be managed. All we will need in the forthcoming sections is Theorem 9.13. The entire section is devoted to the proof of this theorem and readers who are convinced in the truth of Theorem 9.13 may safely skip the rest of this section.

We first state some auxiliary results. In what follows, we denote by $|\Gamma|$ and $\mathrm{dist}\,(z,\Gamma)$ the Euclidean length of Γ and the Euclidean distance between z and Γ, while $|\Gamma|_H$ stands for the hyperbolic length of Γ.

Theorem 9.3 (Koebe's distortion theorem). *Let $\Omega \subset \mathbf{C}\cup\{\infty\}$ be a simply connected domain, let $\partial\Omega$ be the boundary of Ω, and let $\varphi : \mathbf{D} \to \Omega$ be a Riemann map. Then for all $z \in \mathbf{D}$,*

$$\frac{1}{4}\left(1 - |z|^2\right)|\varphi'(z)| \le \mathrm{dist}\,\left(\varphi(z), \partial\Omega\right) \le \left(1 - |z|^2\right)|\varphi'(z)|. \tag{9.1}$$

Moreover, if $z, z_0 \in \mathbf{D}$ and the hyperbolic distance between z and z_0 in $\mathbf{D}$ is $h(z, z_0)$, then

$$\frac{e^{2h(z,z_0)} - 1}{4e^{2h(z,z_0)}} \le \frac{|\varphi(z) - \varphi(z_0)|}{\mathrm{dist}\,\left(\varphi(z_0), \partial\Omega\right)} \le e^{2h(z,z_0)} - 1. \tag{9.2}$$

For a *proof* of (9.1) see, e.g., [160, p. 22] and for a proof of (9.2) see [75, Lemma 3.1]. □

In the following we frequently write $A_\tau \simeq B_\tau$ to indicate that the ratio A_τ/B_τ is bounded and bounded away from zero as τ ranges over some set. For example, (9.1) implies that $|\varphi'(z)|(1 - |z|^2) \simeq \mathrm{dist}\,(\varphi(z), \partial\Omega)$.

Suppose now that the boundary $\partial\Omega$ of a simply connected domain $\Omega \subset \mathbf{C} \cup \{\infty\}$ is rectifiable and equip $\partial\Omega$ with Lebesgue length measure. Given a measurable subset $E \subset \partial\Omega$ and a point $z \in \Omega$, we denote by $\omega(z, E, \Omega)$ the *harmonic measure* of E at z:

$$\omega(z, E, \Omega) := \left|\varphi_z^{-1}(E)\right|/(2\pi)$$

where $\varphi_z : \mathbf{D} \to \Omega$ is a Riemann map taking 0 to z. Equivalently, $\omega(z, E, \Omega)$ may be defined as the value at z of the harmonic extension of the characteristic function of E into Ω.

Theorem 9.4 (Lavrentiev's estimate). *Let $\Omega \subset \mathbf{C}$ be a simply connected domain with rectifiable boundary $\partial\Omega$, let $z \in \Omega$, and let E be a measurable subset of $\partial\Omega$. If $r = \mathrm{dist}\,(z, \partial\Omega)$, then*

$$\omega(z, E, \Omega) \le C\frac{\log(|\partial\Omega|/r)}{|\log(|E|/r)| + 1}$$

with some universal constant $C \in (0, \infty)$.

A *proof* is in [162, pp. 125–127] and also in [12]. □

A function f analytic in the unit disk $\mathcal{D}$ is called a *Bloch function* if

$$\sup_{z \in \mathbf{D}} \left(1 - |z|^2\right)|f'(z)| < \infty$$

and is said to be in the *class S* if f is univalent (= injective), $f(0) = 0$, and $f'(0) = 1$.

Theorem 9.5. *A function $f : \mathbf{D} \to \mathbf{C}$ is a Bloch function if and only if there is a constant $c \in \mathbf{C}$ and a function $h \in S$ such that $f(z) = c \log h'(z) + f(0)$ for all $z \in \mathbf{D}$.*

Theorem 9.6. *There exists a function $K : (0, \infty) \to (0, \infty)$ with the following property: if $h \in S$ and $\varepsilon > 0$, then there is a measurable subset $E \subset \mathbf{T}$ such that $|E| < \varepsilon$ and*

$$\int_0^1 |h'(r\tau)| \, dr \leq K(\varepsilon) \text{ for all } \tau \in \mathbf{T} \setminus E.$$

Theorems 9.5 and 9.6 are Theorems 9.4 and 10.8 of [160], respectively. □

Finally, we will also need the so-called A_∞ condition. Two nonnegative Borel measures ν and μ on $\mathbf{C}$ are said to be *comparable* if there are numbers $a, b \in (0, 1)$ such that if Δ is a disk and $E \subset \Delta$ is Borel, then

$$\nu(E) < a\nu(\Delta) \implies \mu(E) < b\mu(\Delta).$$

Theorem 9.7. (a) *Comparability of measures is an equivalence relation.*

(b) *If ν and μ are comparable, then there exist constants $d \in (0, \infty)$ and $\delta \in (0, \infty)$ such that*

$$d\left(\frac{\nu(E)}{\nu(\Delta)}\right)^\delta \leq \frac{\mu(E)}{\mu(\Delta)} \leq 1 \tag{9.3}$$

for all disks Δ and all Borel subsets $E \subset \Delta$.

(c) *If γ is a composed Carleson curve, $p \in (1, \infty)$, and $w \in A_p(\gamma)$, then the measures ν and μ given by*

$$\nu(E) := |\gamma \cap E|, \quad \mu(E) := \int_{\gamma \cap E} w(\tau)^p |d\tau| \tag{9.4}$$

are comparable.

Parts (a) and (b) follow from [36, Lemma 5.1] and part (c) can be shown by slightly modifying the argument of the proof of [203, Theorem 18 on p. 9] (see the notes and comments to Section 2.7). □

Inequality (9.3) for the measures (9.4) is called the A_∞ *condition* for the weight w (or the triple Γ, p, w). By virtue of part (a), we may in (9.3) change the roles of ν and μ. This gives the inequality

$$\frac{\mu(E)}{\mu(\Delta)} \leq c\left(\frac{\nu(E)}{\nu(\Delta)}\right)^\gamma \tag{9.5}$$

with constants $c \in (0, \infty)$ and $\gamma \in (0, \infty)$. Often it is (9.5) which is referred to as the A_∞ condition for w.

Taking Theorems 9.3 to 9.7 for granted, we now proceed to the problem of extending Muckenhoupt weights on Carleson stars to Muckenhoupt weights on Carleson flowers. The desired result is Theorem 9.13, the construction of the flower and the concrete extension of the weight are described after the proof of Lemma 9.11.

We say that $\{\eta_n\}_{n \in \mathbf{Z}}$ is a partition of the simple arc η if each η_n itself is a simple arc, each pair η_i, η_k ($i \neq k$) have at most an endpoint in common, and $\eta = \bigcup_{n \in \mathbf{Z}} \eta_n$.

Lemma 9.8. *Let $\sigma \subset \mathbf{D}$ be a maximal hyperbolic geodesic and let $\tau \in \mathbf{T}$ be anyone of the two arcs of $\mathbf{T}$ between the endpoints of σ. There exist partitions $\{I_n\}_{n \in \mathbf{Z}}$ of σ and $\{J_n\}_{n \in \mathbf{Z}}$ of τ such that if we denote the endpoints of I_n by z_n and z_{n+1}, then*

$$|I_n|_H = 1 \text{ for all } n \in \mathbf{Z}, \tag{9.6}$$

$$|J_n| \simeq \text{dist}(z_n, J_n) \simeq 1 - |z_n|, \tag{9.7}$$

$$\omega(z_n, J_n, \mathbf{D}) > 4/(10\pi). \tag{9.8}$$

Proof. Without loss of generality assume that -1 is an endpoint of σ. Consider the Riemann maps

$$\varphi : \mathbf{D} \to \mathbf{C}_+, \varphi(z) := i\frac{1-z}{1+z} \text{ and } \psi : \mathbf{C}_+ \to \mathbf{D}, \ \psi(w) := \frac{i-w}{i+w}.$$

Clearly, $\varphi(\sigma)$ is a half-line of the form $\{w \in \mathbf{C}_+ : \text{Re}\, w = a\}$. For $n \in \mathbf{Z}$, put

$$w_n := a + ie^n, \ z_n := \psi(w_n),$$

and let $I_n \subset \sigma$ be the arc with the endpoints z_n and z_{n+1}. Further, let

$$v_n := a + \varepsilon e^n, \ t_n := \psi(v_n)$$

where $\varepsilon := 1$ if ψ maps $\{x \in \mathbf{R} : x > a\}$ onto τ and $\varepsilon := -1$ if ψ maps $\{x \in \mathbf{R} : x < a\}$ onto τ. Denote by $J_n \in \mathbf{T}$ the arc whose endpoints are t_n and t_{n+1}.

It can be easily verified that the line segment $[w_n, w_{n+1}]$ has hyperbolic length 1 in $\mathbf{C}_+$. Since hyperbolic length is a conformal invariant, we obtain (9.6). A straightforward computation gives

$$|J_n| \simeq |t_{n+1} - t_n| \simeq e^{-|n|}, \ 1 - |z_n| \simeq e^{-|n|}, \ \text{dist}(z_n, J_n) \simeq |z_n - t_n| \simeq e^{-|n|},$$

which proves (9.7). It is well known that $\pi\omega(z_n, J_n, \mathbf{D})$ coincides with the angle at which the line segment $[v_n, v_{n+1}]$ is seen at w_n. Thus,

$$w(z_n, J_n, \mathbf{D}) = \frac{1}{\pi}\left(\arctan e - \frac{\pi}{4}\right) > \frac{4}{10\pi}. \qquad \square$$

Lemma 9.9. *Let $\Omega \subset \mathbf{C} \cup \{\infty\}$ be a simply connected domain with rectifiable boundary and let $\varphi : \mathbf{D} \to \Omega$ be a Riemann map. Let $\sigma \subset \mathbf{D}$ be a hyperbolic geodesic, suppose $\varphi(\sigma)$ is bounded, and let $\{z_n\}, \{J_n\}$ be as in Lemma 9.8. Then there exist a constant $M \in (0, \infty)$ and measurable sets $F_n \subset J_n$ such that*

$$|F_n| \geq (3/4)\,|J_n|, \tag{9.9}$$
$$w(z_n, F_n, \mathbf{D}) > 3/(10\pi), \tag{9.10}$$
$$\left|\varphi(t) - \varphi(z_n)\right| \leq M|\varphi'(z_n)|\left(1 - |z_n|^2\right) \text{ for all } t \in F_n. \tag{9.11}$$

Proof. Let K be the function of Theorem 9.6. Define $\varrho_n : \mathbf{D} \to \mathbf{D}$ by $\varrho_n(z) := (z + z_n)/(1 + \overline{z}_n z)$. Then ϱ_n is a Möbius transform of $\mathbf{D}$ onto itself mapping 0 to z_n. Further, let $\sigma_n : \mathbf{C} \to \mathbf{C}$ be the linear map given by

$$\sigma_n(z) := (z - \varphi(z_n))/\left(\varphi'(z_n)\left(1 - |z_n|^2\right)\right).$$

The composition $h_n := \sigma_n \circ \varphi \circ \varrho_n$ is univalent on $\mathbf{D}$, $h_n(0) = 0$, and $h_n'(0) = 1$. Thus, $h_n \in S$ and so Theorem 9.6 implies that there are $E_n \subset \mathbf{T}$ such that $|E_n| < \varepsilon$ and

$$|h_n(\tau)| = \left|h_n(\tau) - h_n(0)\right| \leq \int_0^1 \left|h_n'(r\tau)\right| dr \leq K(\varepsilon)$$

for all $\tau \in \mathbf{T} \setminus E_n$. Consequently,

$$\begin{aligned}\left|\varphi(\varrho_n(\tau)) - \varphi(z_n)\right| &= |h_n(\tau)|\,|\varphi'(z_n)|\left(1 - |z_n|^2\right) \\ &\leq K(\varepsilon)|\varphi'(z_n)|\left(1 - |z_n|^2\right)\end{aligned} \tag{9.12}$$

whenever $\tau \in \mathbf{T} \setminus E_n$. Put $F_n := J_n \setminus \varrho_n(E_n)$. From (9.12) we infer that

$$\left|\varphi(t) - \varphi(z_n)\right| \leq K(\varepsilon)|\varphi'(z_n)|\left(1 - |z_n|^2\right) \text{ for } t \in F_n. \tag{9.13}$$

We have

$$w(z_n, \varrho_n(E_n), \mathbf{D}) = \frac{1}{2\pi}\left|\varrho_n^{-1}(\varrho_n(E_n))\right| = \frac{1}{2\pi}|E_n| < \frac{\varepsilon}{2\pi},$$

and since $w(z_n, J_n, \mathbf{D}) > 4/(10\pi)$ by (9.8), it follows that $w(z_n, F_n, \mathbf{D}) > 3/(10\pi)$ whenever $\varepsilon \leq 2/10$. If $t \in J_n$, then $|t - z_n| \leq C(1 - |z_n|^2)$ with some constant $C \in (0, \infty)$ independent of n by virtue of (9.7). Thus, for $t \in J_n$,

$$\left|(\varrho_n^{-1})'(t)\right| = \frac{1 - |z_n|^2}{|1 - \overline{z}_n t|^2} = \frac{1 - |z_n|^2}{|t - z_n|^2} \geq \frac{1}{C^2}\frac{1}{1 - |z_n|^2},$$

whence

$$\varepsilon > |E_n| = \int\limits_{\varrho_n(E_n)} \left|(\varrho_n^{-1})'(t)\right| dt$$

$$\geq \int\limits_{\varrho_n(E_n)\cap J_n} \left|(\varrho_n^{-1})'(t)\right| dt \geq \frac{1}{C^2}\frac{1}{1-|z_n|^2}\left|\varrho_n(E_n)\cap J_n\right|.$$

Combining the latter inequality and (9.7) we get

$$\left|\varrho_n(E_n)\cap J_n\right| \leq C^2\varepsilon\left(1-|z_n|^2\right) \leq D\varepsilon|J_n|$$

with some constant $D \in (0,\infty)$ independent of ε and n. Thus,

$$|F_n| \geq (1-D\varepsilon)|J_n| \geq (3/4)|J_n|$$

if only $\varepsilon \leq 1/(4D)$. This completes the proof of (9.9) and (9.10). The estimate
(9.13) gives (9.11) with $M = K(\varepsilon_0)$ where $\varepsilon_0 < \min\{2/10, 1/(4D)\}$.   □

Lemma 9.10.   *There exists a function $H : (1,\infty) \to [0,1)$ such that $H(s) \to 0$ as
$s \to \infty$ and such that if $\Omega' \subset \mathbf{C}$ is any simply connected domain with rectifiable
boundary, $x \in \Omega'$, and $r = \mathrm{dist}\,(x,\partial\Omega')$, then*

$$\omega(x, \partial\Omega'\setminus D(x, sr), \Omega') \leq H(s) \ \ \text{for all} \ \ s > 1, \tag{9.14}$$

where $D(x,\varrho)$ stands for the disk $\{z \in \mathbf{C} : |z-x| < \varrho\}$.

Proof. We start with the function K given by Theorem 9.6. Clearly, we may assume
that K is monotonically decreasing and that $K(\varepsilon) \to +\infty$ as $\varepsilon \to 0$ (otherwise
replace K by $K^*(\varepsilon) := (1/\varepsilon) + \sup\{K(\eta) : \eta \geq \varepsilon\}$). Denote by $\psi : \mathbf{D} \to \Omega'$ a
Riemann map such that $\psi(0) = x$ and $\psi'(0) > 0$. By Theorem 9.3, $(1/4)\psi'(0) \leq
r \leq \psi'(0)$. Put $f(z) := (\psi(z)-x)/\varphi'(0)$. Since f is univalent, $f(0) = 0$, $f'(0) = 1$,
it follows from the definition of K that there is a measurable subset $E \subset \mathbf{T}$ such
that $|E| < \varepsilon$ and

$$|f(t) - f(0)| \leq \int\limits_0^1 |f'(rt)|\,dr \leq K(\varepsilon)$$

whenever $t \in \mathbf{T} \setminus E$. Thus,

$$|\psi(t) - x| \leq K(\varepsilon)\psi'(0) \leq 4K(\varepsilon)r \tag{9.15}$$

for all $t \in \mathbf{T} \setminus E$.

   If $s > 4K(\varepsilon)$ and $|\psi(t)-x| \geq sr$, then (9.15) implies that $t \in E$. Consequently,

$$\omega\big(x, \partial\Omega'\setminus D(x, sr), \Omega'\big) = \frac{1}{2\pi}\left|\psi^{-1}\big(\partial\Omega'\setminus D(x, sr)\big)\right| \leq \frac{1}{2\pi}|E| < \frac{\varepsilon}{2\pi}. \tag{9.16}$$

For $s > 1$, define $H(s) := \inf\{\varepsilon/(2\pi) : 4K(\varepsilon) < s\}$. With this definition, we may rewrite (9.16) in the form (9.14).

It remains to prove that $H(s) \to 0$ as $s \to \infty$. The function H is obviously monotonically decreasing, because if $s_1 < s_2$, then

$$2\pi H(s_1) = \inf\{\varepsilon : 4K(\varepsilon) < s_1\} \geq \inf\{\varepsilon : 4K(\varepsilon) < s_2\} = 2\pi H(s_2).$$

If H would not go to zero, there were $\delta > 0$ and $s_0 \in (1, \infty)$ such that $\delta \leq \inf\{\varepsilon : 4K(\varepsilon) < s\}$ for all $s > s_0$. However, this is impossible since

$$\bigcup_{s > s_0}\{\varepsilon : 4K(\varepsilon) < s\} = (0, \infty). \qquad \square$$

Now let γ be a Carleson star. Put $\Omega := \mathbf{C} \setminus \gamma$ and let $\varphi : \mathbf{D} \to \Omega$ be a Riemann map. Let $\sigma \subset \mathbf{D}$ be a maximal hyperbolic geodesic and assume $\varphi(\sigma)$ does not contain the point at infinity. Let z_n, I_n, J_n, F_n be as in Lemmas 9.8 and 9.9. Define

$$A_n := \varphi(F_n), \quad \Gamma_n := \varphi(I_n), \quad x_n := \varphi(z_n), \quad r_n := \mathrm{dist}\,(x_n, \gamma).$$

Lemma 9.11. *There are constants $M_1, M_2 \in (0, \infty)$ such that*

$$A_n \subset \gamma \cap D(x_n, M_1 r_n) \text{ for all } n \in \mathbf{Z}, \tag{9.17}$$
$$|\Gamma_n| \leq M_2|A_n| \text{ for all } n \in \mathbf{Z}, \tag{9.18}$$
$$|A_n| \simeq r_n. \tag{9.19}$$

Proof. From (9.11) and (9.1) we infer that if $t \in F_n$, then

$$\big|\varphi(t) - \varphi(z_n)\big| \leq M|\varphi'(z_n)|(1 - |z_n|^2) \leq 4M\mathrm{dist}\,(x_n, \gamma) = 4Mr_n,$$

whence $A_n \subset D(x_n, 4Mr_n)$. Thus, we get (9.17) with $M_1 = 4M$.

Since γ is a Carleson curve, we have $|\gamma \cap D(x, r)| \leq C_\gamma r$ for all $x \in \gamma$. Taking into account the proof of (1.3), we obtain that $|\gamma \cap D(x, r)| \leq 2C_\gamma r$ for all $x \in \mathbf{C}$. Using (9.17) we therefore see that

$$|A_n| \leq \big|\gamma \cap D(x_n, M_1 r_n)\big| \leq 2M_1 C_\gamma r_n. \tag{9.20}$$

We now show that $|A_n| \geq cr_n$ with some constant $c \in (0, \infty)$ independent of n. Let H be the function of Lemma 9.10 and choose $N_0 > 0$ so that $N_0 M > 1$ and $H(N_0 M_1) = H(4N_0 M) < 2/(10\pi)$.

Suppose first that $r_n < (\mathrm{diam}\,\gamma)/(2N_0 M_1)$. Then the boundary of the disk $D(x_n, N_0 M_1 r_n)$ necessarily hits γ. Hence, the connected component Ω'_n of the set

$\Omega \cap D(x_n, N_0 M_1 r_n)$ containing x_n is a bounded simply connected domain. We claim that

$$\omega(x_n, A_n, \Omega'_n) \geq \omega(x_n, A_n, \Omega) - \omega(x_n, \Sigma_n, \Omega'_n) \tag{9.21}$$

where $\Sigma_n := \partial \Omega'_n \setminus \gamma$. By the maximum principle for harmonic functions, (9.21) will follow as soon as we have shown that

$$\omega^*(\zeta, A_n, \Omega'_n) \geq \omega(\zeta, A_n, \Omega) - \omega^*(\zeta, \Sigma_n, \Omega'_n) \tag{9.22}$$

for almost all $\zeta \in \partial \Omega'_n$, where $\omega^*(\zeta, E, G)$ stands for the boundary value of $\omega(z, E, G)$ at $\zeta \in \partial G$. If $\zeta \in A_n$, then (9.22) reads $1 \geq 1 - \omega^*(\zeta, \Sigma_n, \Omega'_n)$, which is certainly true. For $\zeta \in \partial \Omega'_n \cap (\gamma \setminus A_n)$, inequality (9.22) is $0 \geq 0 - \omega^*(\zeta, \Sigma_n, \Omega'_n)$, which is also true. Finally, if $\zeta \in \partial \Omega'_n \setminus \gamma$, then (9.22) requires that $0 \geq \omega^*(\zeta, A_n, \Omega) - 1$, which is again satisfied. This proves our claim (9.21).

Since the points of Σ_n have the distance $N_0 M_1 r_n$ to x_n, we deduce from Lemma 9.10 that

$$\omega(x_n, \Sigma_n, \Omega'_n) \leq H(N_0 M_1) < 2/(10\pi). \tag{9.23}$$

From (9.10) and the conformal invariance of harmonic measure we obtain that $\omega(x_n, A_n, \Omega) > 3/(10\pi)$. This together with (9.21) and (9.23) shows that

$$\omega(x_n, A_n, \Omega'_n) > 1/(10\pi). \tag{9.24}$$

Since γ is a Carleson curve, we have

$$|\partial \Omega'_n| \leq |\Sigma_n| + |\gamma \cap D(x_n, N_0 M_1 r_n)| \leq (2\pi + 2C_\gamma) N_0 M_1 r_n. \tag{9.25}$$

Combining (9.24) and (9.25) with Theorem 9.4 we obtain

$$\frac{1}{10\pi} < C \frac{\log((2\pi + 2C_\gamma) N_0 M_1)}{|\log(|A_n|/r_n)| + 1},$$

whence

$$|A_n|/r_n > e\left((2\pi + 2C_\gamma) N_0 M_1\right)^{-10\pi C} =: c_1 > 0. \tag{9.26}$$

Clearly, there are only finitely many n for which $r_n \geq (\operatorname{diam} \gamma)/(2N_0 M_1)$. For these n we have $|A_n|/r_n \geq c_2 > 0$ with some constant c_2 independent of n. This along with (9.26) gives the desired estimate $|A_n| \geq c r_n$ with $c := \min\{c_1, c_2\}$. Taking into account (9.20), we so arrive at (9.19).

We finally prove (9.18). If $x \in \Gamma_n$, then $x = \varphi(z)$ with $z \in I_n$. By (9.6), the hyperbolic distance between z and z_n is at most 1. Therefore (9.2) yields

$$|\varphi(z) - \varphi(z_n)| \leq (e^2 - 1)\operatorname{dist}(\varphi(z_n), \gamma) = (e^2 - 1)r_n,$$

and because

$$\operatorname{dist}(\varphi(z), \gamma) \leq |\varphi(z) - \varphi(z_n)| + \operatorname{dist}(\varphi(z_n), \gamma) = |\varphi(z) - \varphi(z_n)| + r_n,$$

we get the estimate

$$\text{dist}\,(\varphi(z),\gamma) \leq e^2 r_n \ \text{ for all } \ z \in I_n. \tag{9.27}$$

By virtue of (9.1),

$$|\Gamma_n| = \int_{\Gamma_n} |dw| = \int_{I_n} |\varphi'(z)|\,|dz| \simeq \int_{I_n} \frac{\text{dist}\,(\varphi(z),\gamma)}{1-|z|^2}\,|dz|,$$

and hence, due to (9.27) and (9.6),

$$|\Gamma_n| \leq c_0 e^2 r_n \int_{I_n} \frac{2\,|dz|}{1-|z|^2} = c_0 e^2 r_n \tag{9.28}$$

with some $c_0 \in (0,\infty)$ independent of n. Combining (9.19) and (9.28) we get (9.18). $\qquad\square$

We now extend the Carleson star γ to a flower F as in the proof of Theorem 9.2; without invoking Theorem 9.1 we do not yet know whether F is also Carleson. Since γ is comprised of N simple Carleson arcs, $\gamma = \eta_1 \cup \ldots \cup \eta_N$, we have N hyperbolic geodesics $\sigma_1,\ldots,\sigma_N \subset \mathbf{D}$ and N subarcs $\tau_1,\ldots,\tau_N \subset \mathbf{T}$. For each $j \in \{1,\ldots,N\}$, we construct the sets $I_n^{(j)}, J_n^{(j)}, \Gamma_n^{(j)}, A_n^{(j)}$ as above. Put

$$\{\Gamma_n\}_{n\in\mathbf{Z}} := \bigcup_{m\in\mathbf{Z}}\bigcup_{j=1}^{N}\{\Gamma_m^{(j)}\}, \quad \{A_n\}_{n\in\mathbf{Z}} := \bigcup_{m\in\mathbf{Z}}\bigcup_{j=1}^{N}\{A_m^{(j)}\}$$

and suppose the enumeration is so that Γ_n is associated with A_n: if $\Gamma_n = \Gamma_m^{(j)}$ then $A_n = A_m^{(j)}$. Further, for $\Gamma_n = \Gamma_m^{(j)}$ let $x_n := \varphi(z_m^{(j)})$ and $r_n := \text{dist}\,(x_n,\gamma)$. Thus,

$$\gamma \supset \bigcup_{n\in\mathbf{Z}} A_n, \quad F \setminus \gamma = \bigcup_{n\in\mathbf{Z}} \Gamma_n.$$

For each $n \in \mathbf{Z}$, set

$$w_n := \left(\frac{1}{|A_n|} \int_{A_n} w(t)^p |dt|\right)^{1/p}$$

and then define a weight $\tilde{w}$ on $F \setminus \gamma$ by $\tilde{w}(x) := w_n$ for $x \in \Gamma_n$. Finally, let $W := w\chi_\gamma + \tilde{w}\chi_{F\setminus\gamma}$. Obviously, W is an extension of w from γ to F. We have to show that $W \in A_p(F)$, which automatically also implies that F is Carleson.

Lemma 9.12. *Let $x \in F \setminus \gamma$ and let $r \le (1/10)\, \mathrm{dist}\,(x, \gamma)$. Then at most $2N$ of the arcs Γ_n hit $D(x, r)$ and there are constants $0 < B_1 \le B_2 < \infty$ independent of x and r such that*

$$B_1 \le w_n/w_k \le B_2 \quad \text{whenever} \ \ \Gamma_n \cap D(x,r) \ne \emptyset \ \ \text{and} \ \ \Gamma_k \cap D(x,r) \ne \emptyset. \quad (9.29)$$

Proof. We have $x = \varphi(z_0)$ where z_0 lies on some hyperbolic geodesic σ belonging to $\{\sigma_1, \ldots, \sigma_N\}$. If z is any point in $\mathbf{D}$ such that $\varphi(z) \in D(x, r)$, then $\big|\varphi(z) - \varphi(z_0)\big| < r \le (1/10)\, \mathrm{dist}\,(x, \gamma)$, while (9.2) implies that

$$\frac{|\varphi(z) - \varphi(z_0)|}{\mathrm{dist}\,(\varphi(z_0), \gamma)} \ge \frac{e^{2h(z, z_0)} - 1}{4 e^{2h(z, z_0)}}.$$

Therefore $(e^{2h(z, z_0)} - 1)/(4 e^{2h(z, z_0)}) \le 1/10$, whence

$$h(z, z_0) < \frac{1}{2} \log \frac{10}{6} < \frac{1}{2} \log e = \frac{1}{2}.$$

Since the arcs $I_n \subset \sigma_1 \cup \ldots \cup \sigma_n$ all have hyperbolic length 1, it follows that at most $2N$ arcs $\Gamma_n = \varphi(I_n)$ can hit $D(x, r)$.

Now suppose $\Gamma_n \cap D(x, r) \ne \emptyset$. We claim that

$$r_n \simeq \mathrm{dist}\,(x, \gamma) \qquad (9.30)$$

From (9.1) we know that

$$r_n \simeq |\varphi'(z_n)|\big(1 - |z_n|\big), \quad \mathrm{dist}\,(x, \gamma) \simeq |\varphi'(z_0)|\big(1 - |z_0|\big). \qquad (9.31)$$

Let $\tilde{\sigma}_n$ be the maximal hyperbolic geodesic of $\mathbf{D}$ through the points z_n and z_0. Divide $\tilde{\sigma}_n$ into arcs $\tilde{I}_n$ as in the proof of Lemma 9.8. The arc I_n has the endpoints z_n and z_{n+1}, and at least one point of this arc lies in the hyperbolic disk $\{z : h(z, z_0) < 1/2\}$. As $h(z_n, z_{n+1}) = 1$, it results that $h(z_0, z_n) < 3/2$. This shows that the points z_n and z_0 belong to arcs $\tilde{I}_m$ and $\tilde{I}_k$ with $|m - k| \le 2$. Consequently, $1 - |z_n| \simeq e^{-|m|} \simeq e^{-|k|} \simeq 1 - |z_0|$ (also see the proof of Lemma 9.8). Due to (9.31), the relation (9.30) will therefore follow as soon as we have shown that $|\varphi'(z_n)| \simeq |\varphi'(z_0)|$.

Theorem 9.5 with $h(z) := (\varphi(z) - \varphi(0))/\varphi'(0)$ implies that $\log \varphi'(z)$ is a Bloch function. Hence

$$\left| \frac{d}{dz} \log \varphi'(z) \right| \le \frac{2C}{1 - |z|^2} \quad \text{for all} \ \ z \in \mathbf{D}$$

with some constant $C \in (0, \infty)$. We conclude that

$$\big|\log |\varphi'(z_n)| - \log |\varphi'(z_0)|\big| \le \big|\log \varphi'(z_n) - \log \varphi'(z_0)\big| = \left| \int_{z_0}^{z_n} \frac{d}{dz} \log \varphi'(z)\, dz \right|$$

$$\le C \int_{z_0}^{z_n} \frac{2|dz|}{1 - |z|^2} = C \int_{z_0}^{z_n} |dz|_H = C h(z, z_0) < (3/2)C =: C_0,$$

whence $e^{-C_0} \le |\varphi'(z_n)|/\varphi'(z_0)| \le e^{C_0}$, which completes the proof of (9.30).

Let A_n be the subset of γ associated with Γ_n. By (9.19) and (9.30), there exist constants $0 < M_3 \leq M_4 < \infty$ such that

$$M_3 r_x \leq |A_n| \leq M_4 r_x, \quad M_3 r_x \leq r_n \leq M_4 r_x \tag{9.32}$$

where $r_x := \operatorname{dist}(x, \gamma)$. By (9.17),

$$A_n \subset \gamma \cap D(x_n, M_1 r_n) \subset \gamma \cap D(x, |x_n - x| + M_1 r_n),$$

and we have $|x_n - x| < r + |\Gamma_n|$ and $r \leq r_x/10$. From (9.18) and (9.32) we infer that $|\Gamma_n| \leq M_2|A_n| \leq M_2 M_4 r_x$. Thus, $A_n \subset \gamma \cap D(x, M_5 r_x)$ with $M_5 := 1/10 + M_2 M_4 + M_1 M_4$. Define the measures ν and μ by (9.4). Since

$$\nu(A_n) = |A_n| \geq M_3 r_x, \quad \nu(D(x, M_5 r_x)) = |\gamma \cap D(x, M_5 r_x)| \leq 2C_\gamma M_5 r_x,$$

we obtain from (9.3) that

$$d\left(\frac{M_3}{2C_\gamma M_5}\right)^\delta \leq d\left(\frac{\nu(A_n)}{\nu(D(x, M_5 r_x))}\right)^\delta \leq \frac{\mu(A_n)}{\mu(D(x, M_5 r_x))} \leq 1.$$

Hence, $\mu(A_n) \simeq \mu(D(x, M_5 r_x))$ for all n such that $\Gamma_n \cap D(x, r) \neq \emptyset$, while $|A_n| \simeq r_x$ for all these n by virtue of (9.32). This implies that

$$\mu(A_n) \simeq \mu(A_k), \ |A_n| \simeq |A_k| \text{ if } \Gamma_n \cap D(x, r) \neq \emptyset \text{ and } \Gamma_k \cap D(x, r) \neq \emptyset,$$

whence

$$w_n := \left(\frac{\mu(A_n)}{|A_n|}\right)^{1/p} \simeq \left(\frac{\mu(A_k)}{|A_k|}\right)^{1/p} =: w_k. \qquad \square$$

Here now is the main result of this section.

Theorem 9.13. *Let γ be a Carleson star, $p \in (1, \infty)$, and $w \in A_p(\gamma)$. There exist a Carleson flower F and a weight $W \in A_p(F)$ such that $\gamma \subset F$ and $W|\gamma = w$.*

Proof. We have already constructed F and W. To show that $W \in A_p(F)$, suppose first that $x \in F \setminus \gamma$ and $r \leq (1/10) \operatorname{dist}(x, \gamma)$. Then the point $z_0 := \varphi^{-1}(x)$ lies on some hyperbolic geodesic $\sigma \in \{\sigma_1, \ldots, \sigma_N\}$. In the same way we proved (9.30), one can show that if $\varphi(z) \in D(x, r)$, then

$$|\varphi'(z)|(1 - |z|^2) \simeq \operatorname{dist}(x, \gamma) =: r_x. \tag{9.33}$$

Moreover, for such z we obtain from (9.2) that

$$\frac{r}{r_x} > \frac{|\varphi(z) - \varphi(z_0)|}{\operatorname{dist}(\varphi(z_0), \gamma)} \geq \frac{e^{2h(z, z_0)} - 1}{4e^{2h(z, z_0)}},$$

whence, by the inequality $\log(1 + a) < a$,

$$h(z, z_0) < \frac{1}{2} \log \frac{r_x}{r_x - 4r} < \frac{1}{2} \frac{4r}{r_x - 4r} < \frac{2r}{(1 - 4/10)r_x} = \frac{10}{3} \frac{r}{r_x}.$$

Thus, if $\varphi(z) \in D(x,r)$ then z belongs to the hyperbolic disk

$$D_H := \{z : h(z,z_0) < 10r/(3r_x)\}.$$

This observation together with (9.33) shows that

$$|\varphi(\sigma) \cap D(x,r)| \leq \int_{\sigma \cap D_H} |\varphi'(z)|\,|dz| = \int_{\sigma \cap D_H} |\varphi'(z)|\big(1 - |z|^2\big)\frac{|dz|}{1 - |z|^2}$$

$$\leq M_6 r_x \int_{\sigma \cap D_H} |dz|_H \leq M_6 r_x \frac{20}{3}\frac{r}{r_x} = \frac{20}{3}M_6 r$$

with some $M_6 \in (0,\infty)$ independent of z and x. As we have N geodesics, we get $|F \cap D(x,r)| \leq (20/3)M_6 N r =: M_7 r$. The weight W is piecewise constant on the set $F \cap D(x,r)$. By Lemma 9.12, the ratio of the different values of the weight on $F \cap D(x,r)$ lies between B_1 and B_2. Thus,

$$\left(\frac{1}{r}\int_{F \cap D(x,r)} W(t)^p\,|dt|\right)^{1/p} = \left(\frac{1}{r}\sum_n \int_{\Gamma_n \cap D(x,r) \neq \emptyset} w_n^p\,|dt|\right)^{1/p}$$

$$\leq \left(\frac{1}{r}B_2^p w_{n_0}^p|F \cap D(x,r)|\right)^{1/p} \leq M_7^{1/p}B_2 w_{n_0} \qquad (9.34)$$

and similarly,

$$\left(\frac{1}{r}\int_{F \cap D(x,r)} W(t)^{-q}\,|dt|\right)^{1/q} \leq M_7^{1/q}B_1^{-1}w_{n_0}^{-1}. \qquad (9.35)$$

Clearly, the product of (9.34) and (9.35) is uniformly bounded.

Now assume that either $x \in \gamma$ or $x \in F \setminus \gamma$ and $r > (1/10)r_x$ where $r_x := \mathrm{dist}\,(x,\gamma)$. Obviously, it is sufficient to consider only the case where $(F \setminus \gamma) \cap D(x,r) \neq \emptyset$. Let $\{\Gamma_n\}$ be the arcs which hit $D(x,r)$ and let $\{A_n\}$ be the corresponding subsets of γ. Pick anyone of the arcs Γ_n. There is a $z \in I_n$ such that $\varphi(z) \in D(x,r)$. For $z \in I_n$ we have, by (9.2) and the definition of z_n, x_n, r_n,

$$\frac{e^{2h(z,z_n)} - 1}{4e^2} \leq \frac{|\varphi(z) - x_n|}{r_n} \leq e^{2h(z,z_n)} - 1$$

and

$$\frac{e^{2h(z,z_n)} - 1}{4e^2} \leq \frac{|\varphi(z) - x_n|}{\mathrm{dist}\,(\varphi(z),\gamma)} \leq e^{2h(z,z_n)} - 1,$$

whence

$$r_n/(4e^2) \leq \mathrm{dist}\,(\varphi(z),\gamma) \leq 4e^2 r_n. \qquad (9.36)$$

Since $A_n \subset D(x_n, M_1 r_n)$ due to (9.17) and $|\varphi(z) - x_n| \le |\Gamma_n| \le M_8 r_n$ by virtue of (9.18) and (9.19), it follows that $A_n \subset D(\varphi(z), (M_1 + M_8) r_n) =: D(\varphi(z), M_9 r_n)$. Because $|\varphi(z) - x| < r$, we so see that $A_n \subset D(x, r + M_9 r_n)$. From (9.36) and the assumption that $r > (1/10) r_x$ we infer that

$$r_n \le 4e^2 \mathrm{dist}\left(\varphi(z), \gamma\right) \le 4e^2(r + r_x) \le 44e^2 r.$$

Consequently,

$$A_n \subset D(x, M_{10} r) \tag{9.37}$$

with $M_{10} := 1 + 44e^2 M_9$. Put

$$(F \setminus \gamma)(x, r) := (F \setminus \gamma) \cap D(x, r), \ \gamma(x, r) := \gamma \cap D(x, r).$$

We will prove that

$$\int\limits_{(F \setminus \gamma)(x, r)} \tilde{w}(t)^p |dt| \le M_2 \int\limits_{\gamma(x, M_{10} r)} w(t)^p |dt|, \tag{9.38}$$

$$\int\limits_{(F \setminus \gamma)(x, r)} \tilde{w}(t)^{-q} |dt| \le M_2 \int\limits_{\gamma(x, M_{10} r)} w(t)^{-q} |dt|, \tag{9.39}$$

where M_2 is the constant from (9.18). These two estimates imply that

$$\left(\frac{1}{r} \int\limits_{F \cap D(x,r)} W(t)^p |dt| \right)^{1/p} \left(\frac{1}{r} \int\limits_{F \cap D(x,r)} W(t)^{-q} |dt| \right)^{1/q}$$

$$= \frac{1}{r} \left(\int\limits_{(F \setminus \gamma)(x,r)} \tilde{w}(t)^p |dt| + \int\limits_{\gamma(x,r)} w(t)^p |dt| \right)^{1/p} \left(\int\limits_{(F \setminus \gamma)(x,r)} \tilde{w}(t)^{-q} |dt| + \int\limits_{\gamma(x,r)} w(t)^{-q} |dt| \right)^{1/q}$$

$$\le (1 + M_2) \left(\frac{1}{r} \int\limits_{\gamma(x, M_{10} r)} w(t)^p |dt| \right)^{1/p} \left(\frac{1}{r} \int\limits_{\gamma(x, M_{10} r)} w(t)^{-q} |dt| \right)^{1/q},$$

and this is uniformly bounded since $w \in A_p(\gamma)$.

From (9.18) and (9.37) we obtain

$$\int\limits_{(F \setminus \gamma)(x, r)} \tilde{w}(t)^p |dt| \le \sum_n w_n^p |\Gamma_n| = \sum_n \left(\frac{1}{|A_n|} \int\limits_{A_n} w(t)^p |dt| \right) |\Gamma_n|$$

$$\le M_2 \sum_n \int\limits_{A_n} w(t)^p |dt| \le M_2 \int\limits_{\gamma(x, M_{10} r)} w(t)^p |dt|,$$

which is (9.38). By Jensen's inequality, $(\int f\, dm)^{-\alpha} \le \int f^{-\alpha} dm$ for every $\alpha > 0$. Thus, again using (9.18) and (9.37) we get

$$\int\limits_{(F\backslash\gamma)(x,r)} \tilde{w}(t)^{-q}|dt| \le \sum_n w_n^{-q}|\Gamma_n| = \sum_n \left(\frac{1}{|A_n|}\int\limits_{A_n} w(t)^p|dt|\right)^{-q/p}|\Gamma_n|$$

$$\le \sum_n \left(\frac{1}{|A_n|}\int\limits_{A_n} w(t)^{-q}|dt|\right)|\Gamma_n| \le M_2 \sum_n \int\limits_{A_n} w(t)^{-q}|dt| \le M_2 \int\limits_{\gamma(x,M_{10}r)} w(t)^{-q}|dt|,$$

and this is (9.39).   □

9.3   Operators on flowers

The property of flowers which distinguishes them from other composed curves is uncovered by Theorem 9.15. To prove this theorem, we need one more approximation result.

Lemma 9.14. *Let* Γ *be a composed rectifiable curve,* $p \in (1,\infty)$, *and* $w \in L^p(\Gamma)$. *Then the set* $R(\Gamma)$ *of all rational functions without poles on* Γ *is dense in* $L^p(\Gamma, w)$.

Proof. Suppose first that $\Gamma =: \eta$ is a simple arc. We know from Lemma 4.4 that $C(\eta)$ is dense in $L^p(\eta, w)$. Since $w^p \in L^1(\eta)$, this easily implies that $C_0(\eta)$, the set of functions in $C(\eta)$ vanishing at the endpoints of η, is also dense in $L^p(\eta, w)$.

If Γ is a composed curve, we may write $\Gamma = \eta_1 \cup \ldots \cup \eta_m$ where $\eta_1, \ldots, \eta_m$ are simple arcs each pair of which have at most endpoints in common. Let $C_0(\eta_j)$ be the set of all functions in $C(\Gamma)$ which vanish identically on $\Gamma \setminus \eta_j$ and are zero at the endpoints of η_j. Put

$$E_0(\Gamma) := \{f_1 + \ldots + f_m : f_j \in C_0(\eta_j)\}.$$

From what was already shown we infer that $E_0(\Gamma)$ is dense in $L^p(\Gamma, w)$. Since $E_0(\Gamma) \subset C(\Gamma)$ and since, by Mergelyan's theorem ([70, p.119]), $R(\Gamma)$ is dense in $C(\Gamma)$, it results that $R(\Gamma)$ is dense in $L^p(\Gamma, w)$.   □

Theorem 9.15. *If* F *is a Carleson flower,* $p \in (1, \infty)$, *and* $W \in A_p(F)$, *then* $S_F^2 f = f$ *and hence* $P_F^2 f = P_F f$ *for all* $f \in L^p(F, W)$.

Proof. Let F be comprised of the Carleson Jordan curves $\Gamma_1, \ldots, \Gamma_N$. Denote by D_j the bounded connected component of $\mathbf{C} \setminus \Gamma_j$ and let D_∞ stand for the unbounded connected component of $\mathbf{C} \setminus F$. For $j \in \{1, \ldots, N\}$, let $R_0^-(D_j)$ be the set of all functions in $R(F)$ which are analytic outside D_j and which vanish at infinity. Further, let $R^-(D_\infty)$ denote the functions in $R(F)$ which are analytic in $D_1 \cup \ldots \cup D_N$.

Every function $r \in R(F)$ is of the form $r = r_1 + \ldots + r_N + r_\infty$ with $r_j \in R_0^-(D_j)$ for $j = 1, \ldots, N$ and $r_\infty \in R^-(D_\infty)$. Denote by χ_j^0 the characteristic function of Γ_j. By decomposing r_j into partial fractions and using Lemma 6.5 one can show that if $j \in \{1, \ldots, N\}$, then

$$\chi_k^0 S_F \chi_i^0 r_j = \begin{cases} -\chi_k^0 r_j & \text{if} \quad k = i = j, \\ \chi_k^0 r_j & \text{if} \quad k = i \neq j, \\ -2\chi_k^0 r_j & \text{if} \quad k \neq i = j, \\ 0 & \text{if} \quad k \neq i \neq j, \end{cases} \tag{9.40}$$

while

$$\chi_k^0 S_F \chi_i^0 r_\infty = \begin{cases} \chi_k^0 r_\infty & \text{if} \quad k = i, \\ 0 & \text{if} \quad k \neq i; \end{cases} \tag{9.41}$$

we remark that in these equalities the notation $\chi_k^0 f = \chi_k^0 g$ means that $f(t) = g(t)$ for all $t \in \Gamma_k$ at which Γ_k has a tangent, implying that $f(t) = g(t)$ for almost all $t \in \Gamma_k$. For $j \in \{1, \ldots, N\}$ we deduce from (9.40) that

$$\begin{aligned} S_F r_j &= \sum_{k,i} \chi_k^0 S_F \chi_i^0 r_j = \chi_j^0 S_F \chi_j^0 r_j + \sum_{k=i\neq j} \chi_k^0 S_F \chi_i^0 r_j + \sum_{k\neq i=j} \chi_k^0 S_F \chi_i^0 r_j \\ &= -\chi_j^0 r_j + \sum_{k\neq j} \chi_k^0 r_j - 2 \sum_{k\neq j} \chi_k^0 r_j = -r_j. \end{aligned} \tag{9.42}$$

Thus, $S_F^2 r_j = r_j$ for $j \in \{1, \ldots, N\}$. Similarly, using (9.41) one obtains that

$$S_F r_\infty = r_\infty, \tag{9.43}$$

whence $S_F^2 r_\infty = r_\infty$. Consequently, $S_F^2 r = r$ for all $r \in R(F)$. Lemma 9.14 finally implies that $S_F^2 f = f$ for all $f \in L^p(F, W)$. $\qquad\square$

Let $F = \Gamma_1 \cup \ldots \cup \Gamma_N$ be a Carleson flower with Carleson Jordan curves $\Gamma_1, \ldots, \Gamma_N$. Denote the center of F by t. For sufficiently small $\varepsilon > 0$, the connected component $F_0(t, \varepsilon)$ of the portion $F(t, \varepsilon)$ which contains t can be written as

$$F_0(t, \varepsilon) = \bigcup_{i=1}^{N} (\eta_{2i-1}^* \cup \eta_{2i}^*)$$

where $\eta_{2i-1}^* \subset \Gamma_i$ and $\eta_{2i}^* \subset \Gamma_i$ are outgoing and incoming arcs, respectively. Let χ_j^* be the characteristic function of η_j^* $(j \in \{1, \ldots, 2N\})$.

Given $p \in (1, \infty)$ and $W \in A_p(F)$, construct the local algebra $\mathcal{B}_t(F)$ as in Section 8.2, put

$$p_j := \pi_t(\chi_j^* I), \quad P := \pi_t(P_F), \quad Q := \pi_t(I - P_F), \tag{9.44}$$

and let $\mathcal{A}_t(F)$ be the smallest closed subalgebra of $\mathcal{B}_t(F)$ containing the elements $P, p_1, \ldots, p_{2N}$.

Theorem 9.16. *The elements* (9.44) *satisfy the hypothesis* (8.9) *and* (8.10) *of Theorem 8.9.*

We remark that for piecewise Lyapunov flowers (carrying a power weight) such a result was first established by Gohberg and Krupnik [86], [88] (also see [89]) and that just this result motivates the axiom (8.10).

Proof. Condition (8.9) is evidently satisfied. Let us prove the first set of equalities in (8.10). We have $\chi_{2i-1}^* + \chi_{2i}^* = \chi_i^0 + c_i$ where χ_i^0 is the characteristic function of Γ_i and $c_i \in PC(F)$ vanishes on $F_0(t,\varepsilon)$. Hence, $\pi_t(c_i I) = 0$ and

$$Q(p_{2i-1} + p_{2i})P = \pi_t\big(Q_F(\chi_i^0 + c_i)P_F\big) = \pi_t(Q_F\chi_i^0 P_F), \qquad (9.45)$$

where $Q_F := I - P_F$. We claim that

$$T := Q_F \chi_i^0 P_F = 0. \qquad (9.46)$$

Let $r_j \in R_0^-(D_j)$ be as in the proof of Theorem 9.15. Then $P_F r_j = (1/2)(r_j + S_F r_j) = 0$ due to (9.42) and thus, $T r_j = 0$ for $j = 1, \ldots, N$. On the other hand, if $r_\infty \in R^-(D_\infty)$ then (9.43) and (9.41) imply that

$$T r_\infty = Q_F \chi_i^0 P_F r_\infty = Q_F \chi_i^0 r_\infty = \chi_i^0 r_\infty - P_F \chi_i^0 r_\infty = \chi_i^0 r_\infty - \chi_i^0 r_\infty = 0.$$

Taking into account Lemma 9.14 we therefore get (9.46) and may conclude that (9.45) vanishes, as desired.

We now prove that $P(p_{2i} + p_{2i+1})Q = 0$ for every $i \in \{1, \ldots, N\}$. Denote by $\tilde{\eta}_j^*$ ($j \in \{1, \ldots, 2N\}$) the arc which results from η_j^* by changing the orientation and consider the Carleson star

$$\tilde{F}_0(t,\varepsilon) := \bigcup_{i=1}^{N} (\tilde{\eta}_{2i}^* \cup \tilde{\eta}_{2i+1}^*).$$

By Theorem 9.13, we may extend $\tilde{F}_0(t,\varepsilon)$ to a Carleson flower $\tilde{F}$ and $w|\tilde{F}_0(t,\varepsilon)$ to a weight $W \in A_p(\tilde{F})$ so that $\tilde{F} = \tilde{\Gamma}_1 \cup \ldots \cup \tilde{\Gamma}_N$ with Carleson Jordan curves $\tilde{\Gamma}_1, \ldots, \tilde{\Gamma}_N$ and so that $\tilde{\eta}_{2i}^*$ and $\tilde{\eta}_{2i+1}^*$ are the outgoing and incoming arcs of $\tilde{\Gamma}_i$, respectively. From what was already proved we know that

$$\pi_t(Q_{\tilde{F}})\pi_t(\chi_{2i}^* I + \chi_{2i+1}^* I)\pi_t(P_{\tilde{F}}) = 0. \qquad (9.47)$$

Due to the change of orientation we have

$$\pi_t(P_{\tilde{F}}) = \pi_t\big((I + S_{\tilde{F}})/2\big) = \pi_t\big((I - S_F)/2\big) = \pi_t(Q_F)$$

and thus $\pi_t(Q_{\tilde{F}}) = \pi_t(P_F)$. Consequently, by (9.47),

$$0 = \pi_t(P_F)\pi_t(\chi_{2i}^* I + \chi_{2i+1}^* I)\pi_t(Q_F) = P(p_{2i} + p_{2i+1})Q. \qquad \square$$

In order to apply Theorem 8.9 to the algebra $\mathcal{B} := \mathcal{B}_t(F)$ we need know the local spectrum

$$\mathcal{X}_t(F) := \mathrm{sp}_{\mathcal{B}_t(F)}(X) \qquad (9.48)$$

where X is the element (8.11).

Theorem 9.17. *If* Γ *is a Carleson Jordan curve,* $p \in (1, \infty)$, *and* $w \in A_p(\Gamma)$, *then for every* $t \in \Gamma$,

$$\mathcal{X}_t(\Gamma) = \mathcal{L}(0, 1; p, \alpha_t, \beta_t)$$

where α_t *and* β_t *are the indicator functions of* Γ, p, w *at* t.

Proof. If $\varepsilon > 0$ is sufficiently small, then $\Gamma_0(t, \varepsilon) = \eta_1^* \cup \eta_2^*$ with an outgoing arc η_1^* and an incoming arc η_2^*. The element (8.11) is

$$X = p_1 P p_1 + p_2 Q p_2 = \pi_t(\chi_1^* P_\Gamma \chi_1^* I + \chi_2^* Q_\Gamma \chi_2^* I). \tag{9.49}$$

Put $e := \pi_t(I)$, $r := \pi_t(P_\Gamma)$, $s := \pi_t(\chi_1^* I)$ and apply Theorem 8.7 to the algebra $\mathcal{B} := \mathcal{B}_t(\Gamma)$. The element X appearing in Theorem 8.7 is

$$rsr + (e - r)(e - s)(e - r)$$

and, at the first glance, this is *not* the X given by (9.49): the element (9.49) is

$$srs + (e - s)(e - r)(e - s)$$

. However, actually we have

$$rsr + (e - r)(e - s)(e - r) = e - (r - s)^2 = srs + (e - s)(e - r)(e - s),$$

which reveals that both elements are the same. Theorem 8.19 shows that the spectrum of $rsr + (e - r)(e - s)(e - r)$ equals $\mathcal{L}(0, 1; p, \alpha_t, \beta_t)$. Consequently, $\mathcal{X}_t(\Gamma)$ also equals $\mathcal{L}(0, 1; p, \alpha_t, \beta_t)$.   □

In what follows we have frequently to change curves and thus local algebras. This requires some caution, and we will therefore denote the canonical homomorphism $\pi_t : \Lambda(\Gamma) \to \mathcal{B}_t(\Gamma)$ by π_t^Γ.

Lemma 9.18. *Let* $\Gamma_1, \ldots, \Gamma_N$ *be Carleson Jordan curves and let* $F = \Gamma_1 \cup \ldots \cup \Gamma_N$ *be a Carleson flower. Let* t *be the center of* F *and denote the characteristic function of* Γ_j *by* χ_j^0. *Suppose* $p \in (1, \infty)$ *and* $W \in A_p(F)$. *Let further* A_j *be operators in* $\mathcal{B}(\Gamma_j) := \mathcal{B}(L^p(\Gamma_j, W|\Gamma_j))$ *and denote the extension* $A_j \oplus 0$ *of* A_j *by zero to an operator on* $\mathcal{B}(F) := \mathcal{B}(L^p(F, W))$ *also by* A_j. *Define* $A \in \mathcal{B}(F)$ *by* $A = \sum_{j=1}^N \chi_j^0 A_j \chi_j^0 I$. *Then the following hold:*

(a) $A \in \mathcal{K}(F) \Longleftrightarrow A_j \in \mathcal{K}(\Gamma_j)$ *for all* j;

(b) $A \in \Lambda(F) \Longleftrightarrow A_j \in \Lambda(\Gamma_j)$ *for all* j;

(c) $\operatorname{sp} \pi_t^F(A) = \bigcup_{j=1}^N \operatorname{sp} \pi_t^{\Gamma_j}(A_j)$.

Proof. Part (a) is obvious, part (b) can be verified straightforwardly, and part (c) can be easily proved using Proposition 8.6.   □

Here is a preliminary result about the spectrum (9.48) at the center of a flower.

Proposition 9.19. *Let $\Gamma_1, \ldots, \Gamma_N$ be Carleson Jordan curves and let $F = \Gamma_1 \cup \ldots \cup \Gamma_N$ be a Carleson flower with the center t. Suppose $p \in (1, \infty)$ and $W \in A_p(F)$. Then*

$$\mathcal{X}_t(F) = \bigcup_{i=1}^{N} \mathcal{L}(0, 1; p, \alpha_t^i, \beta_t^i) \tag{9.50}$$

where α_t^i and β_t^i are the indicator functions of the triple $\Gamma_i, p, W|\Gamma_i$ at t.

Proof. By (9.48), (8.11), (9.44), we have

$$
\begin{aligned}
\mathcal{X}_t(F) &= \operatorname{sp} \pi_t^F \left(\sum_{i=1}^{N} (\chi_{2i-1}^* P_F \chi_{2i-1}^* I + \chi_{2i}^* Q_F \chi_{2i}^* I) \right) \\
&= \operatorname{sp} \pi_t^F \left(\sum_{i=1}^{N} \chi_i^0 (\chi_{2i-1}^* P_{\Gamma_i} \chi_{2i-1}^* + \chi_{2i}^* Q_{\Gamma_i} \chi_{2i}^*) \chi_i^0 I \right).
\end{aligned}
$$

Lemma 9.18(c) therefore gives

$$\mathcal{X}_t(F) = \bigcup_{i=1}^{N} \operatorname{sp} \pi_t^{\Gamma_i} (\chi_{2i-1}^* P_{\Gamma_i} \chi_{2i-1}^* I + \chi_{2i}^* Q_{\Gamma_i} \chi_{2i}^* I) = \bigcup_{i=1}^{N} \mathcal{X}_t(\Gamma_i),$$

and Theorem 9.17 completes the proof. $\square$

Corollary 9.26 will show that the leaves in the union (9.50) are all one and the same set.

Corollary 9.20. *Under the hypotheses of Proposition 9.19, the set $\mathcal{X}_t(F)$ is a connected set containing the points 0 and 1.*

Proof. We know from Section 7.3 that each of the leaves in (9.50) is a connected set which contains 0 and 1. $\square$

At this point we might apply Corollary 8.18 to establish the analogue of Theorem 8.20 for operators on flowers. However, we will immediately turn to operators on general composed curves.

9.4   Local algebras

Let Γ be a composed Carleson curve, $p \in (1, \infty)$, $w \in A_p(\Gamma)$, and $t \in \Gamma$. If $\varepsilon > 0$ is sufficiently small then the connected component $\gamma := \Gamma_0(t, \varepsilon)$ of the portion $\Gamma(t, \varepsilon)$ is a star comprised of $N(t)$ simple Carleson arcs $\eta_1, \ldots, \eta_{N(t)}$:

$$\gamma := \Gamma_0(t, \varepsilon) = \eta_1 \cup \ldots \cup \eta_{N(t)}. \tag{9.51}$$

Put $\varepsilon_i(t) = 1$ if η_i is an outgoing arc, put $\varepsilon_i(t) = -1$ if η_i is an incoming arc, and let

$$E(t) := \operatorname{diag}\big(\varepsilon_1(t), \ldots, \varepsilon_{N(t)}(t)\big). \tag{9.52}$$

For a function $a \in PC(\Gamma)$ and for $i \in \{1, \ldots, N\}$ define

$$a_i(t) := \lim_{\tau \to t, \tau \in \eta_i} a(\tau). \tag{9.53}$$

By Theorem 9.13, we can find a Carleson flower F consisting of $N(t)$ Carleson Jordan curves $\Gamma_1, \ldots, \Gamma_{N(t)}$ and a weight $W \in A_p(F)$ such that $\gamma \subset F$ and $W|\gamma = w|\gamma$. We may assume that $\eta_i \subset \Gamma_i$. If $\varepsilon > 0$ is small enough, then $(\Gamma_i)_0(t, \varepsilon) = \eta^*_{2i-1} \cup \eta^*_{2i}$ where η^*_{2i-1} stands for the outgoing arc and η^*_{2i} denotes the incoming arc. Thus, if η_i is outgoing then $\eta_i = \eta^*_{2i-1}$ and η^*_{2i} is not contained in γ, while if η_i is incoming then $\eta_i = \eta^*_{2i}$ and η^*_{2i-1} is not a subset of γ. Clearly,

$$F_0(t, \varepsilon) = \bigcup_{i=1}^{N(t)} (\Gamma_i)_0(t, \varepsilon) = \bigcup_{i=1}^{N(t)} (\eta^*_{2i-1} \cup \eta^*_{2i}).$$

Put $\delta_j(t) = 1$ if $\eta^*_j \subset \gamma$ and $\delta_j(t) = 0$ if η^*_j is not contained in γ. Equivalently, if η_i is outgoing then $\delta_{2i-1}(t) = 1$ and $\delta_{2i}(t) = 0$, and if η_i is incoming then $\delta_{2i-1}(t) = 0$ and $\delta_{2i}(t) = 1$. Let

$$D_t := \operatorname{diag}\big(\delta_1(t), \delta_2(t), \ldots, \delta_{2N(t)}(t)\big). \tag{9.54}$$

At the point t we may consider six algebras: the three local algebras

$$\mathcal{B}_t(\Gamma) := \mathcal{B}_t(\Gamma, p, w), \quad \mathcal{B}_t(\gamma) := \mathcal{B}_t(\gamma, p, w|\gamma), \quad \mathcal{B}_t(F) := \mathcal{B}_t(F, p, W)$$

and their subalgebras $\mathcal{A}_t(\Gamma)$, $\mathcal{A}_t(\gamma)$, $\mathcal{A}_t(F)$.

The aim of this section is to prove the following theorem. We remark that this theorem is a preliminary result in the sense that all we know about $\mathcal{X}_t(F)$ is (9.50). In the next section we will introduce the leaf $\mathcal{L}_t(\Gamma)$ of Γ (and p and w) at the point t and will show that $\mathcal{X}_t(F) = \mathcal{L}_t(\Gamma)$.

Theorem 9.21. *For $x \in \mathbf{C}$ define the map*

$$\operatorname{Sym}_{t,x} : \{S_\Gamma\} \cup \{aI : a \in PC(\Gamma)\} \to \mathbf{C}^{N(t) \times N(t)} \tag{9.55}$$

by the following formulas:

$$\operatorname{Sym}_{t,x}(S_\Gamma) := E(t) \begin{pmatrix} 2x-1 & 2x-2 & 2x-2 & \cdots & 2x-2 \\ 2x & 2x-1 & 2x-2 & \cdots & 2x-2 \\ 2x & 2x & 2x-1 & \cdots & 2x-2 \\ \vdots & \vdots & \vdots & \ddots & \vdots \\ 2x & 2x & 2x & \cdots & 2x-1 \end{pmatrix} \; \text{if } x \notin \{0,1\}, \tag{9.56}$$

$$\operatorname{Sym}_{t,x}(S_\Gamma) := (2x-1)E(t) \; \text{if } x \in \{0,1\}, \tag{9.57}$$

$$\operatorname{Sym}_{t,x}(aI) := \operatorname{diag}\big(a_1(t), \ldots, a_{N(t)}(t)\big) \; \text{for all } x. \tag{9.58}$$

Let $\mathcal{X}_t(F)$ be the set (9.50).

(a) *For each $x \in \mathcal{X}_t(F)$ the map* $\mathrm{Sym}_{t,x}$ *extends to a Banach algebra homomorphism*

$$\mathrm{Sym}_{t,x} : \mathrm{alg}\left(S_\Gamma, PC(\Gamma)\right) \to \mathbf{C}^{N(t) \times N(t)}.$$

(b) *If $A \in \mathrm{alg}\left(S_\Gamma, PC(\Gamma)\right)$ then*

$$\mathrm{sp}_{\mathcal{B}_t(\Gamma)} \pi_t^\Gamma(A) = \mathrm{sp}_{\mathcal{A}_t(\Gamma)} \pi_t^\Gamma(A) = \bigcup_{x \in \mathcal{X}_t(F)} \mathrm{sp}\,\mathrm{Sym}_{t,x}(A),$$

i.e. $\mathcal{A}_t(\Gamma)$ is inverse closed in $\mathcal{B}_t(\Gamma)$ and the element $\pi_t^\Gamma(A)$ is invertible in the local algebra $\mathcal{B}_t(\Gamma)$ if and only if the $N(t) \times N(t)$ matrices $\mathrm{Sym}_{t,x}(A)$ are invertible for all $x \in \mathcal{X}_t(F)$.

We begin by connecting the local algebras $\mathcal{B}_t(\Gamma)$ and $\mathcal{B}_t(F)$ via the local algebra $\mathcal{B}_t(\gamma)$.

Lemma 9.22. *The map* $\Phi_t^\pi : \mathcal{B}_t(\Gamma) \to \mathcal{B}_t(\gamma),\ \pi_t^\Gamma(A) \mapsto \pi_t^\gamma(\chi_\gamma A \chi_\gamma I)$ *is a well-defined Banach algebra isomorphism.*

Proof. It is easily seen that the map $\Phi : \Lambda(\Gamma) \to \mathcal{B}_t(\gamma),\ A \mapsto \pi_t^\gamma(\chi_\gamma A \chi_\gamma I)$ is a Banach algebra homomorphism and that $\mathcal{K}(\Gamma) \subset \mathrm{Ker}\,\Phi$. Therefore

$$\Phi^\pi : \Lambda(\Gamma)/\mathcal{K}(\Gamma) \to \mathcal{B}_t(\gamma),\ \pi^\Gamma(A) \mapsto \pi_t^\gamma(\chi_\gamma A \chi_\gamma I)$$

is a well-defined Banach algebra homomorphism. From Proposition 8.6 one readily gets the inclusion $J_t(\Gamma) \subset \mathrm{Ker}\,\Phi^\pi$, which implies that Φ_t^π is also a well-defined Banach algebra homomorphism. The bijectivity of Φ_t^π can be verified straightforwardly with the help of Proposition 8.6.   □

Lemma 9.23. *The map* $\Psi_t^\pi : \mathcal{B}_t(\gamma) \to \mathcal{B}_t(F),\ \pi_t^\gamma(A) \mapsto \pi_t^F(\chi_\gamma A \chi_\gamma I)$ *is a well-defined injective Banach algebra homomorphism enjoying the property that $\pi_t^\gamma(A)$ is invertible in $\mathcal{B}_t(\gamma)$ if and only if $\pi_t^F(\chi_\gamma A \chi_\gamma I) + \pi_t^F(\chi_{F\setminus\gamma} I)$ is invertible in $\mathcal{B}_t(F)$.*

Proof. As in the proof of Lemma 9.22 one can show by successive factorization that Ψ_t^π is a well-defined Banach algebra homomorphism. Notice, however, that Ψ_t^π maps the identity $\pi_t^\gamma(I)$ of $\mathcal{B}_t(\gamma)$ to $\pi_t^F(\chi_\gamma I)$, which is not the identity of $\mathcal{B}_t(F)$. The injectivity of Ψ_t^π follows easily from Proposition 8.6.

If $A \in \Lambda(\gamma)$ and $\pi_t^\gamma(A)$ is invertible, then there is an operator $B \in \Lambda(\gamma)$ such that $\pi_t^\gamma(BA) = \pi_t^\gamma(AB) = \pi_t^\gamma(I)$. Again using Proposition 8.6 one can show directly that $\pi_t^\gamma(\chi_\gamma B \chi_\gamma I + \chi_{F\setminus\gamma} I)$ is the inverse of $\pi_t^\gamma(\chi_\gamma A \chi_\gamma I + \chi_{F\setminus\gamma} I)$. Conversely, if $A \in \Lambda(\gamma)$ and $\pi_t^F(\chi_\gamma A \chi_\gamma I + \chi_{F\setminus\gamma} I)$ is invertible from the left, we can find an operator $B \in \Lambda(F)$ such that $\pi_t^F(B) \pi_t^F(\chi_\gamma A \chi_\gamma I + \chi_{F\setminus\gamma} I) = \pi_t^F(I)$. Using the injectivity of Ψ_t^π we obtain $\pi_t^\gamma(\chi_\gamma B \chi_\gamma I) \pi_t^\gamma(A) = \pi_t^\gamma(I)$, which shows that $\pi_t^\gamma(\chi_\gamma B \chi_\gamma I)$ is a left inverse of $\pi_t^\gamma(A)$. Right invertibility can be considered analogously.   □

Lemma 9.24. *Let D_t be the matrix (9.54). For $x \in \mathbf{C} \setminus \{0, 1\}$, denote by M_x the $2N(t) \times 2N(t)$ matrix*

$$
M_x := \operatorname{diag}(1, -1, 1, -1, \ldots, 1, -1) \times
$$

$$
\times
\begin{pmatrix}
x & x-1 & x-1 & x-1 & \cdots & x-1 & x-1 \\
x & x-1 & x-1 & x-1 & \cdots & x-1 & x-1 \\
x & x & x & x-1 & \cdots & x-1 & x-1 \\
x & x & x & x-1 & \cdots & x-1 & x-1 \\
\vdots & \vdots & \vdots & \vdots & \ddots & \vdots & \vdots \\
x & x & x & x & \cdots & x & x-1 \\
x & x & x & x & \cdots & x & x-1
\end{pmatrix},
\tag{9.59}
$$

and for $x \in \{0, 1\}$, let M_x be the $2N(t) \times 2N(t)$ matrix

$$
M_x := \operatorname{diag}(x, 1-x, x, 1-x, \ldots, x, 1-x).
\tag{9.60}
$$

(a) *For each $x \in \mathcal{X}_t(F)$ the map*

$$
H_{t,x} : \{P_\Gamma\} \cup \{aI : a \in PC(\Gamma)\} \to \mathbf{C}^{2N(t) \times 2N(t)}
$$

given by

$$
H_{t,x}(aI) := D_t \operatorname{diag}\big(a_1(t), a_1(t), \ldots, a_{N(t)}(t), a_{N(t)}(t)\big) D_t,
\tag{9.61}
$$

$$
H_{t,x}(P_\Gamma) := D_t M_x D_t
\tag{9.62}
$$

extends to a Banach algebra homomorphism of $\operatorname{alg}(S_\Gamma, PC(\Gamma))$ to $\mathbf{C}^{2N(t) \times 2N(t)}$.

(b) *For $A \in \operatorname{alg}(S_\Gamma, PC(\Gamma))$ and $x \in \mathcal{X}_t(F)$, put*

$$
\operatorname{Sym}_{t,x}^0(A) := H_{t,x}(A) + I - D_t.
\tag{9.63}
$$

Then

$$
\operatorname{sp}_{\mathcal{B}_t(\Gamma)} \pi_t^\Gamma(A) = \operatorname{sp}_{\mathcal{A}_t(\Gamma)} \pi_t^\Gamma(A) = \bigcup_{x \in \mathcal{X}_t(F)} \operatorname{sp} \operatorname{Sym}_{t,x}^0(A)
\tag{9.64}
$$

for every $A \in \operatorname{alg}(S_\Gamma, PC(\Gamma))$.

Proof. Combining Lemmas 9.22 and 9.23 we see that the map

$$
H_t := \Psi_t^\pi \circ \Phi_t^\pi \circ \pi_t^\Gamma : \Lambda(\Gamma) \to \mathcal{B}_t(F), \quad A \mapsto \pi_t^F(\chi_\gamma A \chi_\gamma I)
$$

is a Banach algebra homomorphism with the property that $\pi_t^\Gamma(A)$ is invertible in $\mathcal{B}_t(\Gamma)$ if and only if $H_t(A) + \pi_t^F(\chi_{F \setminus \gamma} I)$ is invertible in $\mathcal{B}_t(F)$. It is easily seen that Lemmas 9.22 and 9.23 remain literally true with $\mathcal{B}$ replaced by $\mathcal{A}$. Thus, $\pi_t^\Gamma(A)$ is invertible in $\mathcal{A}_t(\Gamma)$ if and only if $H_t(A) + \pi_t^F(\chi_{F \setminus \gamma} I)$ is invertible in $\mathcal{A}_t(F)$.

Clearly, H_t maps alg $(S_\Gamma, PC(\Gamma))$ into $\mathcal{A}_t(F)$. The algebra $\mathcal{A}_t(F)$ is generated by $P := \pi_t^F(P_F)$ and $p_i := \pi_t^F(\chi_i^* I)$ ($i = 1, 2, \ldots, 2N(t)$). Theorem 9.16 and Corollary 9.20 show that the hypotheses of Corollary 8.18 are satisfied for $\mathcal{B} := \mathcal{B}_t(F)$ and $\mathcal{A} := \mathcal{A}_t(F)$. We have $\mathrm{sp}_{\mathcal{B}}(X) = \mathcal{X}_t(F)$ by definition (9.48). Invoking Theorem 8.19 and repeating the proof of Theorem 9.17 or taking into account that the leaf $\mathcal{X}_t(\Gamma_i)$ does not separate the plane, we arrive at the conclusion that

$$\mathrm{sp}_{\mathcal{A}_t(\Gamma_i)} \pi_t^{\Gamma_i}(\chi_{2i-1}^* P_{\Gamma_i} \chi_{2i-1}^* I + \chi_{2i}^* Q_{\Gamma_i} \chi_{2i}^* I) = \mathcal{X}_t(\Gamma_i).$$

Since Lemma 9.18(c) is also valid for the algebras $\mathcal{A}$, the proof of Proposition 9.19 finally gives $\mathrm{sp}_{\mathcal{A}}(X) = \mathcal{X}_t(F)$.

Corollary 8.18 implies that for each $x \in \mathcal{X}_t(F)$ the map

$$\sigma_{t,x} : \{P, p_1, \ldots, p_{2N(t)}\} \to \mathbf{C}^{2N(t) \times 2N(t)}$$

which sends p_j to the matrices (8.13) and P to the matrix M_x extends to a Banach algebra homomorphism $\sigma_{t,x} : \mathcal{A}_t(F) \to \mathbf{C}^{2N(t) \times 2N(t)}$. Moreover, since $\mathrm{sp}_{\mathcal{B}}(X) = \mathrm{sp}_{\mathcal{A}}(X)$, the algebra $\mathcal{A}_t(F)$ is inverse closed in $\mathcal{B}_t(F)$, and an element $\pi_t^F(A)$ is invertible in $\mathcal{B}_t(F)$ if and only if $\sigma_{t,x}(\pi_t^F(A))$ is an invertible matrix for every $x \in \mathcal{X}_t(F)$.

Define $H_{t,x}$ as the map

$$\sigma_{t,x} \circ H_t : \mathrm{alg}\,(S_\Gamma, PC(\Gamma)) \to \mathbf{C}^{2N(t) \times 2N(t)}, \quad A \mapsto \sigma_{t,x}\big(\pi_t^F(\chi_\gamma A \chi_\gamma I)\big) \qquad (9.65)$$

and let $\mathrm{Sym}^0_{t,x}$ denote the map given by

$$\mathrm{alg}\,(S_\Gamma, PC(\Gamma)) \to \mathbf{C}^{2N(t) \times 2N(t)}, \quad A \mapsto H_{t,x}(A) + \sigma_{t,x}\big(\pi_t^F(\chi_{F \setminus \gamma} I)\big). \qquad (9.66)$$

We know that $H_{t,x}$ is a Banach algebra homomorphism and we know that $\pi_t^\Gamma(A)$ is invertible if and only if $\mathrm{Sym}^0_{t,x}(A)$ is invertible for all $x \in \mathcal{X}_t(F)$. All assertions of Lemma 9.24 will therefore follow once we have shown that the map $H_{t,x}$ defined by (9.65) sends aI and P_Γ to the matrices (9.61) and (9.62), respectively, and that the map (9.66) is nothing but the map given by (9.63).

Obviously, $\sigma_{t,x}(\pi_t^F(\chi_\gamma I)) = D_t$ and $\sigma_{t,x}(\pi_t^F(\chi_{F \setminus \gamma} I)) = I - D_t$. Moreover, the element $\sigma_{t,x}(\pi_t^F(\chi_\gamma a \chi_\gamma I))$ equals

$$\sum_{i=1}^{N(t)} \sigma_{t,x}\Big(\pi_t^F\big(\delta_{2i-1}(t)a_i(t)\chi_{2i-1}^* I + \delta_{2i}(t)a_i(t)\chi_{2i}^* I\big)\Big)$$

$$= D_t \,\mathrm{diag}\,\big(a_1(t), a_1(t), \ldots, a_{N(t)}(t), a_{N(t)}(t)\big),$$

which is (9.61), and $\sigma_{t,x}(\pi_t^F(\chi_\gamma P_\Gamma \chi_\gamma I))$ is equal to

$$\sigma_{t,x}\big(\pi_t^F(\chi_\gamma I)\big)\sigma_{t,x}\big(\pi_t^F(P_F)\big)\sigma_{t,x}\big(\pi_t^F(\chi_\gamma I)\big) = D_t M_x D_t,$$

which coincides with (9.62). $\qquad \qquad \square$

Proof of Theorem 9.21. From (9.61) and (9.62) we infer that

$$H_{t,x}(A) = D_t H_{t,x}(A) D_t \qquad (9.67)$$

for every $A \in \mathrm{alg}\,(S_\Gamma, PC(\Gamma))$. By appropriately changing rows and columns of the matrix $H_{t,x}(A)$ one can produce a matrix of the form $\begin{pmatrix} * & 0 \\ 0 & 0 \end{pmatrix}$ where $*$ is an $N(t) \times N(t)$ matrix. To be precise, let Π_t be the permutation matrix for which

$$\Pi_t D_t \Pi_t^{-1} = \begin{pmatrix} I & 0 \\ 0 & 0 \end{pmatrix}.$$

From (9.67) we then get

$$\Pi_t H_{t,x}(A) \Pi_t^{-1} = \begin{pmatrix} I & 0 \\ 0 & 0 \end{pmatrix} \Pi_t H_{t,x}(A) \Pi_t^{-1} \begin{pmatrix} I & 0 \\ 0 & 0 \end{pmatrix}.$$

Denote the nonzero $N(t) \times N(t)$ block of the latter matrix by $\mathrm{Sym}_{t,x}(A)$. Thus,

$$\Pi_t H_{t,x}(A) \Pi_t^{-1} = \begin{pmatrix} \mathrm{Sym}_{t,x}(A) & 0 \\ 0 & 0 \end{pmatrix}.$$

Since $H_{t,x}$ is a Banach algebra homomorphism, the map

$$\mathrm{Sym}_{t,x} : \mathrm{alg}\,(S_\Gamma, PC(\Gamma)) \to \mathbf{C}^{N(t) \times N(t)}$$

must also be a Banach algebra homomorphism. Because

$$\Pi_t \, \mathrm{Sym}_{t,x}^0(A) \Pi_t^{-1} = \Pi_t \big(H_{t,x}(A) + I - D_t \big) \Pi_t^{-1}$$
$$= \begin{pmatrix} \mathrm{Sym}_{t,x}(A) & 0 \\ 0 & 0 \end{pmatrix} + \begin{pmatrix} 0 & 0 \\ 0 & I \end{pmatrix} = \begin{pmatrix} \mathrm{Sym}_{t,x}(A) & 0 \\ 0 & I \end{pmatrix},$$

we see that (9.64) is true with $\mathrm{Sym}_{t,x}^0$ replaced by $\mathrm{Sym}_{t,x}$.

What we are left with is to show that $\mathrm{Sym}_{t,x}(S_\Gamma)$ and $\mathrm{Sym}_{t,x}(aI)$ are just the matrices (9.56), (9.57) and (9.58), respectively. This is easy for aI:

$$\Pi_t H_{t,x}(aI) \Pi_t^{-1} = \mathrm{diag}\,(a_1(t), \ldots, a_{N(t)}, 0, \ldots, 0)$$

and thus $\mathrm{Sym}_{t,x}(aI) = \mathrm{diag}\,(a_1(t), \ldots, a_{N(t)})$. A little computation shows that if $x \notin \{0,1\}$, then

$$\Pi_t H_{t,x}(P_\Gamma) \Pi_t^{-1} = \Pi_t D_t \Pi_t^{-1} \Pi_t M_x \Pi_t^{-1} \Pi_t D_t \Pi_t^{-1}$$

has the north-west $N(t) \times N(t)$ block

$$E(t) \begin{pmatrix} x + \frac{\varepsilon_1(t)-1}{2} & x - 1 & \cdots & x - 1 \\ x & x + \frac{\varepsilon_2(t)-1}{2} & \cdots & x - 1 \\ \vdots & \vdots & \ddots & \vdots \\ x & x & \cdots & x + \frac{\varepsilon_{N(t)}(t)-1}{2} \end{pmatrix},$$

which implies that $\mathrm{Sym}_{t,x}(S_\Gamma) = 2\,\mathrm{Sym}_{t,x}(P_\Gamma) - I$ is the matrix (9.56). The diagonal of (9.56) is (9.57). $\qquad \square$

9.5   Symbol calculus

We are now in a position to establish a symbol calculus for deciding whether an operator in $\mathrm{alg}\,(S_\Gamma, PC(\Gamma))$ is Fredholm in case Γ is an arbitrary composed Carleson curve carrying a weight $w \in A_p(\Gamma)$. Recall that in Chapter 8 we did this for Carleson Jordan curves Γ.

Suppose η is a simple Carleson arc and t is an endpoint of η. The arc η has an orientation. We denote by η_t the arc η with the orientation in which t is the starting point. Thus, if t is the starting point in the original orientation of η then $P_{\eta_t} = P_\eta$, while if t is the terminating point in the original orientation of η then

$$P_{\eta_t} = (I + S_{\eta_t})/2 = (I - S_\eta)/2 =: Q_\eta.$$

Note that P_{η_t} depends only on the point t and the arc η as a set and is independent of the original orientation of η.

Theorem 9.25. *Let η be a simple Carleson arc, $p \in (1, \infty)$, and $w \in A_p(\eta)$.*

(a) *There exists a Carleson Jordan curve Γ and a weight $W \in A_p(\Gamma)$ such that $\eta \subset \Gamma$ and $W|\eta = w$.*

(b) *Let Γ be any Carleson Jordan curve and $W \in A_p(\Gamma)$ be any weight such that $\eta \subset \Gamma$ and $W|\eta = w$. If t is any endpoint of η, then*

$$\mathrm{sp}\,\pi_t^\eta(P_{\eta_t}) = \mathcal{L}(0, 1; p, \alpha_t^\Gamma, \beta_t^\Gamma) \tag{9.68}$$

where $\alpha_t^\Gamma, \beta_t^\Gamma$ are the indicator functions of Γ, p, W at t and sp refers to the spectrum in $\mathcal{B}_t(\eta)$ or in the (inverse closed) subalgebra $\mathcal{A}_t(\eta)$.

Proof. Part (a) is nothing but a special case of Theorem 9.13. So let the situation be as in part (b). To compute $\mathrm{sp}\,\pi_t^\eta(P_{\eta_t})$ we apply Theorem 9.21. If t is the starting point of η, then $\eta_t = \eta$ and hence

$$\mathrm{sp}\,\pi_t^\eta(P_{\eta_t}) = \mathrm{sp}\,\pi_t^\eta\left(\frac{I + S_\eta}{2}\right) = \bigcup_{x \in \mathcal{X}_t(\Gamma)} \mathrm{sp}\,\left(\frac{1 + 1(2x - 1)}{2}\right) = \bigcup_{x \in \mathcal{X}_t(\Gamma)} \mathrm{sp}\,(x) = \mathcal{X}_t(\Gamma).$$

In case t is the terminating point of η, we have

$$\mathrm{sp}\,\pi_t^\eta(P_{\eta_t}) = \mathrm{sp}\,\pi_t^\eta\left(\frac{I - S_\eta}{2}\right)$$

$$= \bigcup_{x \in \mathcal{X}_t(\Gamma)} \mathrm{sp}\,\left(\frac{1 - (-1)(2x-1)}{2}\right) = \bigcup_{x \in \mathcal{X}_t(\Gamma)} \mathrm{sp}\,(x) = \mathcal{X}_t(\Gamma).$$

Thus, in both cases (9.68) follows from Theorem 9.17. $\square$

Theorem 9.25(b) shows that the leaf $\mathcal{L}(0,1;p,\alpha_t^\Gamma,\beta_t^\Gamma)$ does actually not depend on the orientation of η and on the concrete extensions of η and w to Γ and W. We denote this leaf by $\mathcal{L}_t(\eta) := \mathcal{L}_t(\eta,p,w)$ and call it the *leaf of the simple Carleson arc η (and of p and w) at the endpoint t.* Equivalently, we might define

$$\mathcal{L}_t(\eta) = \mathrm{sp}\,\pi_t^\eta(P_{\eta_t}), \qquad (9.69)$$

but it is only Theorem 9.25 which reveals that $\mathcal{L}_t(\eta)$ is in fact a leaf.

Corollary 9.26. *Let η_1 and η_2 be simple Carleson arcs which have a common endpoint t but no other common points. Let further $p \in (1,\infty)$, $w_1 \in A_p(\eta_1)$, $w_2 \in A_p(\eta_2)$. If $\eta_1 \cup \eta_2$ is a Carleson curve and the weight w given by $w|\eta_1 = w_1$ and $w|\eta_2 = w_2$ belongs to $A_p(\eta_1 \cup \eta_2)$, then*

$$\mathcal{L}_t(\eta_1,p,w_1) = \mathcal{L}_t(\eta_2,p,w_2).$$

Proof. Denote by η_1^- the arc η_1 with the orientation in which t is the terminating point and by η_2^+ the arc η_2 with the orientation in which t is the starting point. By virtue of Theorem 9.13, there exists a Carleson Jordan curve Γ and a weight $W \in A_p(\Gamma)$ such that $\eta_1^- \cup \eta_2^+ \subset \Gamma$ and $W|\eta_1 \cup \eta_2 = w$. By the definition of $\mathcal{L}_t(\eta)$,

$$\mathcal{L}_t(\eta_1,p,w_1) = \mathcal{L}_t(\eta_1^-,p,w_1) = \mathcal{L}(0,1;p,\alpha_t^\Gamma,\beta_t^\Gamma)$$
$$= \mathcal{L}_t(\eta_2^+,p,w_2) = \mathcal{L}_t(\eta_2,p,w_2). \qquad \square$$

Now let $\gamma = \eta_1 \cup \ldots \cup \eta_N$ be a Carleson star centered at t and let $w \in A_p(\gamma)$. Then $w|\eta_i \cup \eta_j \in A_p(\eta_i \cup \eta_j)$ for each pair i,j. We therefore infer from Corollary 9.26 that

$$\mathcal{L}_t(\eta_1) = \ldots = \mathcal{L}_t(\eta_N). \qquad (9.70)$$

We denote the set (9.70) by $\mathcal{L}_t(\gamma) := \mathcal{L}_t(\gamma,p,w)$ and call it the *leaf of Carleson star γ (and of p and w) at its center t.* Notice again that $\mathcal{L}_t(\gamma)$ is independent of the orientations of the arcs γ is comprised of.

Finally, let Γ be a general composed Carleson curve, $p \in (1,\infty)$, and $w \in A_p(\Gamma)$. For each $t \in \Gamma$, the connected component $\Gamma_0(t,\varepsilon)$ of the portion $\Gamma(t,\varepsilon)$ which contains t is a Carleson star if only $\varepsilon > 0$ is sufficiently small. We may therefore define the leaf $\mathcal{L}_t(\Gamma_0(t,\varepsilon))$. From (9.69) and (8.3) we see that $\mathcal{L}_t(\Gamma_0(t,\varepsilon))$ does not depend on ε. We denote $\mathcal{L}_t(\Gamma_0(t,\varepsilon))$ by

$$\mathcal{L}_t(\Gamma) := \mathcal{L}_t(\Gamma,p,w)$$

and refer to it as the *leaf of the composed Carleson curve Γ (and of p and w) at the point t.* Obviously, $\mathcal{L}_t(\Gamma)$ does not depend on the orientations of the arcs comprising Γ.

Here now is the main result of this chapter. Recall that the multiplicity $N(t)$ of a point $t \in \Gamma$ is given by (9.51).

Theorem 9.27. *Suppose Γ is a composed Carleson curve, $p \in (1, \infty)$, and $w \in A_p(\Gamma)$. For $(t, x) \in \Gamma \times \mathbf{C}$, define the map*

$$\mathrm{Sym}_{t,x} : \{S_\Gamma\} \cup \{aI : a \in PC(\Gamma)\} \to \mathbf{C}^{N(t) \times N(t)}$$

by (9.56), (9.57), (9.58).

(a) *For each (t, x) in the leaf bundle*

$$\mathcal{M} := \mathcal{M}(\Gamma, p, w) := \bigcup_{t \in \Gamma} (\{t\} \times \mathcal{L}_t(\Gamma))$$

the map $\mathrm{Sym}_{t,x}$ extends to a Banach algebra homomorphism of the algebra $\mathrm{alg}\,(S_\Gamma, PC(\Gamma))$ into $\mathbf{C}^{N(t) \times N(t)}$.

(b) *If $A \in \mathrm{alg}\,(S_\Gamma, PC(\Gamma))$ then*

$$\mathrm{sp}_{\mathcal{B}_t(\Gamma)} \pi_t(A) = \mathrm{sp}_{\mathcal{A}_t(\Gamma)} \pi_t(A) = \bigcup_{x \in \mathcal{L}_t(\Gamma)} \mathrm{sp}\,\mathrm{Sym}_{t,x}(A).$$

(c) *An operator $A \in \mathrm{alg}\,(S_\Gamma, PC(\Gamma))$ is Fredholm on $L^p(\Gamma, w)$ if and only if $\mathrm{Sym}_{t,x}(A)$ is invertible for all $(t, x) \in \mathcal{M}$. Equivalently*

$$\mathrm{sp}_{\mathrm{ess}}\, A = \bigcup_{(t,x) \in \mathcal{M}} \mathrm{sp}\,\mathrm{Sym}_{t,x}(A).$$

(d) *If an operator $A \in \mathrm{alg}\,(S_\Gamma, PC(\Gamma))$ is Fredholm, then it has a regularizer in $\mathrm{alg}\,(S_\Gamma, PC(\Gamma))$.*

Proof. Theorem 9.21 tells us that parts (a) and (b) of the present theorem are true with $\mathcal{X}_t(F)$ $(F = F_t)$ in place of $\mathcal{L}_t(\Gamma)$. From Proposition 9.19 we know that

$$\mathcal{X}_t(F) = \mathcal{L}(0, 1; p, \alpha_t^1, \beta_t^1) \cup \ldots \cup \mathcal{L}(0, 1; p, \alpha_t^{N(t)}, \beta_t^{N(t)}).$$

By the definition of the leaf at the endpoint of a simple arc, this may be written in the form

$$\mathcal{X}_t(F) = \mathcal{L}_t(\eta_1) \cup \ldots \cup \mathcal{L}_t(\eta_{N(t)}),$$

and Corollary 9.26 shows that $\mathcal{L}_t(\eta_1) = \ldots = \mathcal{L}_t(\eta_{N(t)}) =: \mathcal{L}_t(\Gamma)$. Thus, $\mathcal{X}_t(F) = \mathcal{L}_t(\Gamma)$.

Part (c) is immediate from part (b) and Proposition 8.5. Finally, applying Theorem 8.2 to the algebra $\pi(\mathrm{alg}\,(S_\Gamma, PC(\Gamma))$, we obtain part (d) from part (b). $\qquad \square$

In the case where Γ is a Carleson Jordan curve we can use Theorem 8.20 or Theorem 9.27. They assign the same diagonal matrices $\operatorname{diag}(a(t+0), a(t-0))$ to aI and $\operatorname{diag}(x, 1-x)$ to S_Γ for $x \in \{0, 1\}$. However, for $x \notin \{0, 1\}$, the symbol matrices of S_Γ are

$$S_1(x) := \begin{pmatrix} x & x(1-x) \\ 1 & 1-x \end{pmatrix} \quad \text{and} \quad S_2(x) := \begin{pmatrix} x & x-1 \\ -x & 1-x \end{pmatrix},$$

respectively. Given $x \notin \{0, 1\}$, denote by y any number such that $y^2 = -x$ and put $C := \operatorname{diag}(y, y^{-1})$. Since $CDC^{-1} = D$ for every diagonal matrix D and $CS_2(x)C^{-1} = S_1(x)$, we see that both theorems are equivalent.

9.6 Essential spectrum of the Cauchy singular integral

We conclude this chapter with computing the essential spectrum of the Cauchy singular integral operator S_Γ. If Γ is a Carleson Jordan curve, then $\operatorname{sp}_{\mathrm{ess}} S_\Gamma = \{-1, 1\}$ by Example 8.21. Now let η be a simple Carleson arc with the endpoints t_1 and t_2 and suppose η is oriented from t_1 to t_2. By Theorem 9.13, we may think of η as a subarc of some Carleson Jordan curve, and Example 8.22 then gives

$$\operatorname{sp}_{\mathrm{ess}} S_\eta = \mathcal{L}(-1, 1; p, \alpha_{t_1}, \beta_{t_1}) \cup \mathcal{L}(1, -1; p, \alpha_{t_2}, \beta_{t_2}).$$

The same result is, of course, also delivered by Theorem 9.27, which yields

$$\begin{aligned}
\operatorname{sp} \pi_{t_1}(S_\eta) &= 2\mathcal{L}_{t_1}(\eta) - 1 = \mathcal{L}(-1, 1; p, \alpha_{t_1}, \beta_{t_1}), \\
\operatorname{sp} \pi_{t_2}(S_\eta) &= 1 - 2\mathcal{L}_{t_2}(\eta) = \mathcal{L}(1, -1; p, \alpha_{t_2}, \beta_{t_2}), \\
\operatorname{sp} \pi_t(S_\eta) &= \{-1, 1\} \text{ if } t \in \eta \setminus \{t_1, t_2\}
\end{aligned}$$

and tells us that $\operatorname{sp}_{\mathrm{ess}} S_\eta$ is the union of these sets.

We now turn to the case of a general composed Carleson curve. Let $t \in \Gamma$ and suppose in the representation (9.51) we have $N_0(t)$ outgoing and $N_i(t)$ incoming arcs. The difference

$$\varepsilon(t) := N_0(t) - N_i(t) = \varepsilon_1(t) + \ldots + \varepsilon_{N(t)}(t)$$

is called the *valency* of t. The following theorem shows that the local spectrum $\operatorname{sp} \pi_t(S_\Gamma)$ is a union of $|\varepsilon(t)|$ leaves arising from the leaf $\mathcal{L}_t(\Gamma)$ in a special manner.

For a continuous function $f : \mathbf{R} \to \mathbf{R}$, define $T_n f : \mathbf{R} \to \mathbf{R}$ by $(T_n f)(x) := f(nx)$.

Theorem 9.28. *Let Γ be a composed Carleson curve, $p \in (1, \infty)$, and $w \in A_p(\Gamma)$. For $t \in \Gamma$, let α_t, β_t be any continuous functions such that $\mathcal{L}_t(\Gamma) = \mathcal{L}(0, 1; p, \alpha_t, \beta_t)$. Given integers $n \geq 1$ and $k \geq 0$, put*

$$\mathcal{L}_{t,k,n} := \mathcal{L}\left(0, 1; p, \frac{1}{n}\left(\frac{1}{p} - \frac{n}{p} + k + T_n \alpha_t\right), \frac{1}{n}\left(\frac{1}{p} - \frac{n}{p} + k + T_n \beta_t\right)\right). \quad (9.71)$$

Then the local spectrum $\operatorname{sp}\pi_t(S_\Gamma)$ *equals*

$$\{-1,1\} \quad \text{if} \quad \varepsilon(t)=0, \tag{9.72}$$

$$\bigcup_{k=0}^{\varepsilon(t)-1} (2\mathcal{L}_{t,k,\varepsilon(t)}-1) \quad \text{if} \quad \varepsilon(t)>0, \tag{9.73}$$

$$\bigcup_{k=0}^{|\varepsilon(t)|-1} \left(1-2\mathcal{L}_{t,k,|\varepsilon(t)|}\right) \quad \text{if} \quad \varepsilon(t)<0. \tag{9.74}$$

We have $\operatorname{sp}_{\mathrm{ess}}(S_\Gamma) = \bigcup_{t\in\Gamma}\operatorname{sp}\pi_t(S_\Gamma)$.

Proof. Put $N := N(t)$, $\varepsilon := \varepsilon(t)$, $\varepsilon_j := \varepsilon_j(t)$, and denote the matrix (9.56) by A_x. From Theorem 9.27 we deduce that

$$\operatorname{sp}\pi_t(S_\Gamma) = \bigcup_{x\in\mathcal{L}(0,1;p,\alpha_t,\beta_t)} \operatorname{sp}A_x \tag{9.75}$$

(note that if $x \in \{0,1\}$, then (9.56) is triangular with the diagonal (9.57)). Let $\Delta_N(\varepsilon_1,\ldots,\varepsilon_N;\lambda) := \det(A_x - \lambda I)$. If $\varepsilon_j + \varepsilon_{j+1} = 0$ for some j, we first add the jth row of $A_x - \lambda I$ to the $j+1$st row and then subtract the $j+1$st column from the jth column to obtain

$$\Delta_N(\varepsilon_1,\ldots,\varepsilon_N;\lambda) = (\lambda^2-1)\Delta_{N-2}(\varepsilon_1,\ldots,\varepsilon_{j-1},\varepsilon_{j+2},\ldots,\varepsilon_N;\lambda).$$

Repeating this procedure we see that $\Delta_N(\varepsilon_1,\ldots,\varepsilon_N;\lambda)$ equals

$$\begin{cases} (\lambda^2-1)^{(N-\varepsilon)/2}\Delta_\varepsilon(1,\ldots,1;\lambda) & \text{if} \quad \varepsilon>0, \\ (\lambda^2-1)^{(N-|\varepsilon|)/2}\Delta_{|\varepsilon|}(-1,\ldots,-1;\lambda) & \text{if} \quad \varepsilon<0. \end{cases} \tag{9.76}$$

If $\varepsilon = 0$, then $\Delta_N(\varepsilon_1,\ldots,\varepsilon_N;\lambda) = (\lambda^2-1)^{N/2}$, which gives (9.72). So assume $\varepsilon > 0$. Using the identity

$$\begin{vmatrix} c & a & a & \cdots & a \\ b & c & a & \cdots & a \\ b & b & c & \cdots & a \\ \vdots & \vdots & \vdots & \ddots & \vdots \\ b & b & b & \cdots & c \end{vmatrix}_{\varepsilon\times\varepsilon} = \frac{a}{a-b}(c-b)^\varepsilon + \frac{b}{b-a}(c-a)^\varepsilon$$

we arrive at the equality

$$\Delta_\varepsilon(1,\ldots,1;\lambda) = x(1-\lambda)^\varepsilon - (x-1)(-1-\lambda)^\varepsilon. \tag{9.77}$$

If $x = 0$ then (9.77) vanishes for $\lambda = -1$, and if $x = 1$ then (9.77) is zero for $\lambda = 1$. Hence,

$$\operatorname{sp}A_0 \cup \operatorname{sp}A_1 = \{-1,1\}. \tag{9.78}$$

Let $x \notin \{0,1\}$. Then (9.76) and (9.77) give

$$\{-1,1\} \cup \operatorname{sp} A_x = \{-1,1\} \cup \left\{\lambda \in \mathbf{C} : \left(\frac{\lambda+1}{\lambda-1}\right)^{\varepsilon} = \frac{x}{x-1}\right\}. \tag{9.79}$$

We have

$$\left\{\frac{x}{x-1} : x \in \mathcal{L}(0,1;p,\alpha_t,\beta_t) \setminus \{0,1\}\right\} = \left\{e^{2\pi z} : z \in Y(p,\alpha_t,\beta_t)\right\}$$

and therefore, by (9.75), (9.78), (9.79),

$$\operatorname{sp} \pi_t(S_\Gamma) = \{-1,1\} \cup \left\{\operatorname{sp} A_x : x \in \mathcal{L}(0,1;p,\alpha_t,\beta_t) \setminus \{0,1\}\right\}$$

$$= \{-1,1\} \cup \bigcup_{k=0}^{\varepsilon-1} \left\{\lambda \in \mathbf{C} : \frac{\lambda+1}{\lambda-1} = e^{2\pi(z+ik)/\varepsilon} \text{ for some } z \in Y(p,\alpha_t,\beta_t)\right\}.$$

Since $1/p + \alpha_t(x) \le y \le 1/p + \beta_t(x)$ if and only if

$$\frac{1}{p} + \left(\frac{1}{p\varepsilon} - \frac{1}{p} + \frac{k}{\varepsilon} + \frac{\alpha_t(\varepsilon x/\varepsilon)}{\varepsilon}\right) \le \frac{y+k}{\varepsilon} \le \frac{1}{p} + \left(\frac{1}{p\varepsilon} - \frac{1}{p} + \frac{k}{\varepsilon} + \frac{\beta_t(\varepsilon x/\varepsilon)}{\varepsilon}\right),$$

we see that $z \in Y(p,\alpha_t,\beta_t)$ if and only if

$$\frac{z+ik}{\varepsilon} \in Y\left(p, \frac{1}{\varepsilon}\left(\frac{1}{p} - \frac{\varepsilon}{p} + k + T_\varepsilon \alpha_t\right), \frac{1}{\varepsilon}\left(\frac{1}{p} - \frac{\varepsilon}{p} + k + T_\varepsilon \beta_t\right)\right) =: Y_k.$$

Because

$$\left\{\lambda \in \mathbf{C} : \frac{\lambda+1}{\lambda-1} = e^{2\pi\zeta} \text{ for some } \zeta \in Y_k\right\}$$

$$= \left\{\lambda \in \mathbf{C} : \lambda = 2\frac{e^{2\pi\zeta}}{e^{2\pi\zeta}-1} - 1 \text{ for some } \zeta \in Y_k\right\},$$

we finally get (9.73). The case $\varepsilon < 0$ may be tackled analogously or may be reduced to the case $\varepsilon > 0$ by changing the orientation of Γ and taking into account that $S_{-\Gamma} = -S_\Gamma$. $\qquad\square$

We remark that there is an alternative description of the sets (9.73) and (9.74). Namely, with

$$M_{0,1}(\mathcal{L}_t(\Gamma)) := \left\{\frac{x}{x-1} : x \in \mathcal{L}_t(\Gamma) \setminus \{0,1\}\right\}$$

we infer from (9.79) that

$$\operatorname{sp} \pi_t(S_\Gamma) = \{-1,1\} \cup \left\{\lambda \in \mathbf{C} : \left(\frac{\lambda+1}{\lambda-1}\right)^{\varepsilon(t)} \in M_{0,1}(\mathcal{L}_t(\Gamma))\right\}$$

if $\varepsilon(t) > 0$ and

$$\mathrm{sp}\,\pi_t(S_\Gamma) = \{-1, 1\} \cup \left\{\lambda \in \mathbf{C} : \left(\frac{\lambda - 1}{\lambda + 1}\right)^{|\varepsilon(t)|} \in M_{0,1}(\mathcal{L}_t(\Gamma))\right\}$$

if $\varepsilon(t) < 0$. Thus, up to Möbius transforms, the leaves comprising the local spectrum of S_Γ at t are the $|\varepsilon(t)|$th "roots" of the leaf $\mathcal{L}_t(\Gamma)$, where $\varepsilon(t)$ is the valency of the point $t \in \Gamma$.

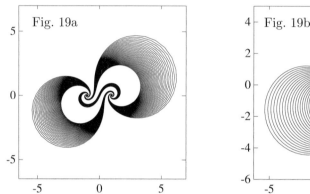
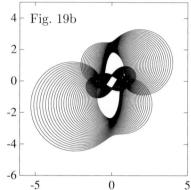

Figures 19a and 19b show examples of local spectra $\mathrm{sp}\,\pi_t(S_\Gamma)$ of the operator S_Γ at the center of a Carleson star Γ with the valency 3 and 4, respectively. In Figure 19a we see the union of three spiralic horns, in Figure 19b we have a union of four leaves whose indicator functions look like hyperbolas. The points -1 and 1 belong to the boundary of the set in Figure 19a and are interior points of the set plotted in Figure 19b.

9.7   Notes and comments

Figures 20 to 22 summarize things in pictures. Notice that passage from Carleson Jordan curves to simple Carleson arcs requires the tools of *this* chapter.

9.1–9.2. The material of these two sections is taken from the paper by Bishop, Spitkovsky, and the authors [10].

9.3–9.4. We here again follow [10]. Theorems 9.15 and 9.16 are essentially due to Gohberg and Krupnik [86], [88], Theorems 9.17 and 9.21 were established in [10]. A result like Theorem 9.21 is already in [172]; there the hypotheses included that $w = 1$ and that Γ may locally be extended to a Carleson flower.

9.5. Theorem 9.25 is from [10], its Corollary 9.26 can implicitly already be found in [91] under the assumption that η_1 and η_2 are Lyapunov arcs.

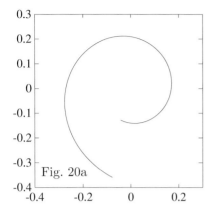

Fig. 20a

Figure 20a shows a simple Lyapunov arc Γ, Figures 20b and 20c show what the Gohberg, Krupnik, Widom and Spitkovsky theories tell us about the essential spectrum of S_Γ in the presence of a power weight and of a non--powerlike Muckenhoupt weight, respectively. Note that a weight alone, even if it is very complicated, cannot produce anything beyond circular arcs and horns.

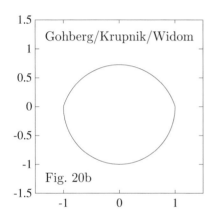

Gohberg/Krupnik/Widom

Fig. 20b

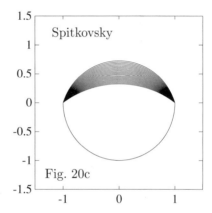

Spitkovsky

Fig. 20c

In the form stated here, Theorem 9.27 first appeared in [10]. However, we wish to emphasize that this theorem has grown out of a long development and is therefore only the end of a chain of results by many mathematicians.

It was once more Gohberg and Krupnik [86], [87], [88] who first realized the possibility of a symbol calculus and who were also the first to understand the structure of the symbol matrices (without having a two projections theorem at their disposal). Of course, they considered composed Lyapunov curves and power weights. In this case their symbol calculus does the same as Theorem 9.27, but it is more complicated.

For the sake of simplicity, let us for a moment suppose that Γ is a Lyapunov Jordan curve. Gohberg and Krupnik define their symbol on the cylinder $\Gamma \times [0, 1]$ and put all information about the leaf $\mathcal{L}_t(\Gamma)$ into the function $x \mapsto \mathrm{Sym}_{t,x}(A)$. This works well if the leaves themselves are curves (e.g. circular arcs), but it is clear that difficulties arise as soon as the leaves become heavy sets. Instead of

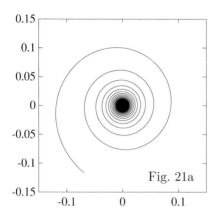

Fig. 21a

A simple Carleson arc Γ scolling up like a spiral at one of its endpoints is shown in Figure 21a (to make visible the whirl, we took a logarithmic scale for the radius in polar coordinates). The essential spectrum of S_Γ is in Figure 21b (power weights) and in Figure 21c (non-powerlike weights). General Carleson curves lead to the emergence of logarithmic spirals, but curves alone cannot destroy logarithmic spirality.

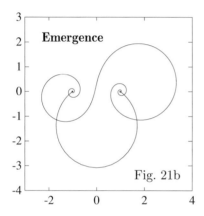

Emergence

Fig. 21b

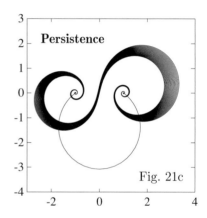

Persistence

Fig. 21c

this, *we* define the symbol on the leaf bundle $\mathcal{M}$ and hence, for each A, the map $x \mapsto \mathrm{Sym}_{t,x}(A)$ is one and the same function. Redefining the Gohberg-Krupnik symbol to a symbol on $\mathcal{M}$ gives

$$\mathrm{Sym}_{t,x}^{GK}(aI) = \begin{pmatrix} a(t+0) & 0 \\ 0 & a(t-0) \end{pmatrix}, \quad \mathrm{Sym}_{t,x}^{GK}(P_\Gamma) = \begin{pmatrix} x & \sqrt{x(1-x)} \\ \sqrt{x(1-x)} & 1-x \end{pmatrix},$$

whereas Theorem 9.27 yields

$$\mathrm{Sym}_{t,x}(aI) = \begin{pmatrix} a(t+0) & 0 \\ 0 & a(t-0) \end{pmatrix}, \quad \mathrm{Sym}_{t,x}(P_\Gamma) = \begin{pmatrix} x & x-1 \\ -x & 1-x \end{pmatrix}$$

for $x \notin \{0,1\}$. We know the enigma's resolution (see the remark after Theorem 8.7): letting $E_x := \mathrm{diag}\left(\sqrt[4]{x/(1-x)},\ -\sqrt[4]{(1-x)/x} \right)$ we have $\mathrm{Sym}_{t,x}^{GK}(A) = E_x \mathrm{Sym}_{t,x}(A) E_x^{-1}$. Thus, the differences between the Gohberg-Krupnik symbol and

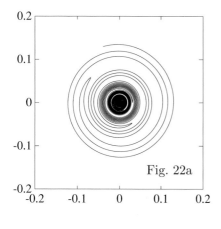

Fig. 22a

Figure 22a shows a simple Carleson arc Γ scrolling up like an oscillating spiral (again with a logarithmic scale for the radius). The essential spectrum of S_Γ is plotted in Figure 22b (power weight) and Figure 22c (non-powerlike weight). While in Figure 22b the leaf is bounded by two logarithmic spirals, no piece of the boundary of the leaf in Figure 22c is a piece of such a spiral. Thus, curves and weights together are able to destroy logarithmic spirality.

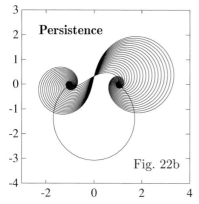

Persistence

Fig. 22b

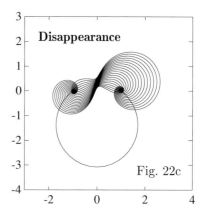

Disappearance

Fig. 22c

the symbol delivered by Theorem 9.27 are minor ones if Γ is a Jordan curve. However, the differences become serious when passing to composed curves.

In the case where Γ is a composed curve, the Gohberg-Krupnik theory was considerably extended and improved by many mathematicians, including Costabel [42], [43], Duduchava [57], Nyaga [153], Rabinovich [163], [164], Plamenevski and Senichkin [157], [158]. Note that in the work of Rabinovich, Plamenevski, and Senichkin "slow oscillations" are allowed in the curve, the weights, and the coefficients; the spirality indices of the curve are nevertheless zero.

In 1990, Roch and Silbermann [174] presented a round and very elegant theory of singular integral operators with piecewise continuous coefficients (and Carleman shifts) on composed Lyapunov curves Γ with power weights. Their approach was based on establishing an isometry between the local algebra $\mathcal{A}_t(\Gamma)$ and the local algebra $\mathcal{A}_t(\Gamma')$ where Γ' (the "straightening" of Γ) is a finite union of rays with the common point t. The algebra $\mathcal{A}_t(\Gamma')$ was in turn studied by so-called

Mellin techniques (see Section 10.6). In this connection, we also refer to Roch and Silbermann's papers [175] and [176].

The case of composed Lyapunov curves with arbitrary Muckenhoupt weights was disposed of by Gohberg, Krupnik, and Spitkovsky [91] by combining the original techniques of [86], [87], [88] and Spitkovsky's result [200]. The "general contour" in the title of [91] is misleading.

The difference between the method of [91] and the approach of [53], [173] motivated the search for an N projections theorem. More about this is said in the notes and comments to Sections 8.3 and 8.4.

9.6. We here encounter a curiosity: although plenty of people have dealt with symbol calculi for operators in $\mathrm{alg}(S_\Gamma, PC(\Gamma))$, we know of no reference in which such a symbol calculus was applied to the problem of finding the essential spectrum of S_Γ itself (even in the case of a composed Lyapunov curve with a power weight). A result like Theorem 9.28 appeared in the paper [19] by Rabinovich and the authors for the first time. A similar result was independently also found by S. Roch (private communication) and is in [172]. Theorem 9.28 as it is cited here was obtained only in Bishop, Spitkovsky, and the authors' article [10].

Chapter 10
Further results

This chapter contains extensions, generalizations, and analogues of certain results of the previous chapters. For lack of space, the results of the present chapter are not equipped with full proofs, but each section is a short introductory survey on a more or less autonomous topic with precise references to the literature.

We here confine ourselves to accentuating three sections. First, in Section 10.4 we completely describe the spectrum (and not only the essential spectrum) of the Cauchy singular integral S_Γ on $L^p(\Gamma, w)$ in case Γ is an arbitrary composed Carleson curve and w is an arbitrary weight in $A_p(\Gamma)$. Secondly, although the approach of Section 10.6 meets with certain barriers, it is undoubtedly the most promising strategy for tackling higher dimensions. Finally, Section 10.7 shows that there are problems all data of which are Lyapunov and which nevertheless lead to studying singular integral operators on spiralic Carleson curves.

In Sections 10.1 to 10.5 we consistently assume that a composed curve is a compact set, i.e. that a composed curve is bounded and contains the endpoints of the arcs it is comprised of.

10.1 Matrix case

Let Γ be a composed Carleson curve, $p \in (1, \infty)$, and $w \in A_p(\Gamma)$. We denote by $L_n^p(\Gamma, w)$ the direct sum of n copies of $L^p(\Gamma, w)$ and provide $L_n^p(\Gamma, w)$ with the norm

$$\|f\| = \|(f_1, \ldots, f_n)\| := \left(\|f_1\|^p + \ldots + \|f_n\|^p \right)^{1/p}$$

or any norm equivalent to this one. The operator S_Γ is defined on $L_n^p(\Gamma, w)$ elementwise, that is,

$$S_\Gamma(f_1, \ldots, f_n) := (S_\Gamma f_1, \ldots, S_\Gamma f_n).$$

We let $PC_n(\Gamma) := PC_{n \times n}(\Gamma)$ stand for the algebra of all $n \times n$ matrix functions with entries in $PC(\Gamma)$. Writing the elements of $L_n^p(\Gamma, w)$ as columns, we can define the multiplication operator aI for $a \in PC_n(\Gamma)$ as multiplication by the matrix function a.

Let $\mathrm{alg}\,(S_\Gamma, PC_n(\Gamma))$ stand for the smallest closed subalgebra of $L_n^p(\Gamma, w)$ containing S_Γ and the set $\{aI : a \in PC_n(\Gamma)\}$. The methods and results of Chapter 8 and 9 extend to the matrix setting with obvious adjustments. The only nontrivial point is the N projections theorem. A matrix version of Theorem 8.9 is in [172]. We confine ourselves to formulating the matrix analogue of Corollary 8.18.

Theorem 10.1. *Let $\mathcal{B}$ be a Banach algebra with identity e. Suppose $\mathcal{C}$ is a closed subalgebra of $\mathcal{B}$ which contains e and is isomorphic to $\mathbf{C}^{n \times n}$. Let further P and $p_1, \ldots, p_{2N}$ be nonzero elements of $\mathcal{B}$ such that*

$$p_j c = c p_j \quad \text{and} \quad cP = Pc \quad \text{for all} \ \ j = 1, \ldots, 2N \ \ \text{and all} \ \ c \in \mathcal{C},$$
$$p_i p_j = \delta_{ij} p_i \quad \text{for all} \ \ i, j \ \ \text{and} \ \ p_1 + \ldots + p_{2N} = e,$$
$$Q(p_{2i-1} + p_{2i})P = P(p_{2i} + p_{2i+1})Q = 0 \quad \text{for all} \ \ i = 1, \ldots, N,$$

where $Q := e - P$ and $p_{2N+1} := p_1$. Denote by $\mathcal{A}$ the smallest closed subalgebra of $\mathcal{B}$ containing $\mathcal{C} \cup \{P, p_1, \ldots, p_{2N}\}$. Put

$$X := \sum_{i=1}^{N} (p_{2i-1} P p_{2i-1} + p_{2i} Q p_{2i})$$

and suppose that 0 and 1 are cluster points of $\mathrm{sp}_{\mathcal{B}} X$. Let $\varphi : \mathcal{C} \to \mathbf{C}^{n \times n}$ be any isomorphism, let 0 and I stand for the $n \times n$ zero and identity matrices, respectively, and put

$$E := \mathrm{diag}\,(I, -I, I, -I, \ldots, I, -I).$$

For $x \in \mathbf{C}$, define the map $\sigma_x : \mathcal{C} \cup \{P, p_1, \ldots, p_{2N}\} \to \mathbf{C}^{2Nn \times 2Nn}$ as follows. Let

$$\sigma_x(c) = \mathrm{diag}\,(\varphi(c), \ldots, \varphi(c)) \quad (c \in \mathcal{C}),$$
$$\sigma_x(p_j) = \mathrm{diag}\,(0, \ldots, 0, I, 0, \ldots, 0), \ \text{the} \ I \ \text{at the} \ j\text{th position},$$

for all $x \in \mathbf{C}$. For $x \in \mathbf{C} \setminus \{0, 1\}$, let $\sigma_x(P)$ be the matrix

$$E \begin{pmatrix}
xI & (x-1)I & (x-1)I & (x-1)I & \cdots & (x-1)I & (x-1)I \\
xI & (x-1)I & (x-1)I & (x-1)I & \cdots & (x-1)I & (x-1)I \\
xI & xI & xI & (x-1)I & \cdots & (x-1)I & (x-1)I \\
xI & xI & xI & (x-1)I & \cdots & (x-1)I & (x-1)I \\
\vdots & \vdots & \vdots & \vdots & \ddots & \vdots & \vdots \\
xI & xI & xI & xI & \cdots & xI & (x-1)I \\
xI & xI & xI & xI & \cdots & xI & (x-1)I
\end{pmatrix},$$

and for $x \in \{0, 1\}$, put

$$\sigma_x(P) = \mathrm{diag}\,(xI, (1-x)I, \ldots, xI, (1-x)I).$$

Then the following hold.

(a) *For each $x \in \mathrm{sp}_{\mathcal{A}}X$ the map σ_x extends to a Banach algebra homomorphism of $\mathcal{A}$ into $\mathbf{C}^{2Nn \times 2Nn}$.*

(b) *An element $A \in \mathcal{A}$ is invertible in $\mathcal{B}$ if and only if $\sigma_x(A)$ is invertible for every $x \in \mathrm{sp}_{\mathcal{B}}X \, (\subset \mathrm{sp}_{\mathcal{A}}X)$.*

(c) *An element $A \in \mathcal{A}$ is invertible in $\mathcal{A}$ if and only if $\sigma_x(A)$ is invertible for every $x \in \mathrm{sp}_{\mathcal{A}}X$.*

A result of this type was first proved by Finck, Roch, Silbermann [68] for $N = 1$. In the form cited here, Theorem 10.1 was explicitly stated in [172].

Having Theorem 10.1 at our disposal, we can derive the following matrix analogue of Theorem 9.27. The notation is as in the beginning of Section 9.4.

Theorem 10.2. *Suppose Γ is a composed Carleson curve, $p \in (1, \infty)$, and $w \in A_p(\Gamma)$. For $t \in \Gamma$, let*

$$E(t) := \mathrm{diag}\,(\varepsilon_1(t)I, \ldots, \varepsilon_{N(t)}(t)I),$$

and for $(t, x) \in \Gamma \times \mathbf{C}$, define the map

$$\mathrm{Sym}_{t,x} : \{S_\Gamma\} \cup \{aI : a \in PC_n(\Gamma)\} \to \mathbf{C}^{N(t)n \times N(t)n}$$

as follows. Let

$$\mathrm{Sym}_{t,x}(S_\Gamma) = E(t) \begin{pmatrix} (2x-1)I & (2x-2)I & (2x-2)I & \cdots & (2x-2)I \\ 2xI & (2x-1)I & (2x-2)I & \cdots & (2x-2)I \\ 2xI & 2xI & (2x-1)I & \cdots & (2x-2)I \\ \vdots & \vdots & \vdots & \ddots & \vdots \\ 2xI & 2xI & 2xI & \cdots & (2x-1)I \end{pmatrix}$$

for $x \in \mathbf{C} \setminus \{0, 1\}$, put

$$\mathrm{Sym}_{t,x}(S_\Gamma) = (2x-1)E(t)$$

for $x \in \{0, 1\}$, and set

$$\mathrm{Sym}_{t,x}(aI) = \mathrm{diag}\,(a_1(t), \ldots, a_{N(t)}(t))$$

for all $x \in \mathbf{C}$. We then have the following.

(a) *For each (t, x) in the leaf bundle*

$$\mathcal{M} := \mathcal{M}(\Gamma, p, w) := \bigcup_{t \in \Gamma} (\{t\} \times \mathcal{L}_t(\Gamma))$$

the map $\mathrm{Sym}_{t,x}$ extends to a Banach algebra homomorphism of the algebra $\mathrm{alg}\,(S_\Gamma, PC_n(\Gamma))$ into $\mathbf{C}^{N(t)n \times N(t)n}$.

(b) *An operator $A \in \mathrm{alg}\,(S_\Gamma, PC_n(\Gamma))$ is Fredholm on $L_n^p(\Gamma, w)$ if and only if $\mathrm{Sym}_{t,x}(A)$ is invertible for every $(t, x) \in \mathcal{M}$.*

(c) *If an operator $A \in \mathrm{alg}\,(S_\Gamma, PC_n(\Gamma))$ is Fredholm on $L_n^p(\Gamma, w)$, then it has a regularizer in $\mathrm{alg}\,(S_\Gamma, PC_n(\Gamma))$.*

Here is a generalization of Theorem 7.18.

Theorem 10.3. *If* Γ *is a composed Carleson curve,* $p \in (1, \infty)$*, and* $w \in A_p(\Gamma)$*, then all semi-Fredholm operators in* $\mathrm{alg}\,(S_\Gamma, PC_n(\Gamma))$ *are Fredholm.*

The proof follows from combining Theorem 7.18 with the construction of the linear dilation of operators in $\mathrm{alg}\,(S_\Gamma, PC_n(\Gamma))$ as in [86], [91].

We finally formulate a result for block Toeplitz operators. Given $a \in PC_n(\Gamma)$, the operator $T(a)$ is defined on $H_n^p(\Gamma, w) := P_\Gamma L_n^p(\Gamma, w)$ by $f \mapsto P_\Gamma(af)$.

Theorem 10.4. *Let* Γ *be a Carleson Jordan curve,* $p \in (1, \infty), w \in A_p(\Gamma)$*, and* $a \in PC_n(\Gamma)$*. The operator* $T(a)$ *is Fredholm on* $H_n^p(\Gamma, w)$ *if and only if*

$$\det\big(xa(t - 0) + (1 - x)a(t + 0)\big) \neq 0$$

for all (t, x) *in the leaf bundle*

$$\mathcal{M} := \mathcal{M}(\Gamma, p, w) := \bigcup_{t \in \Gamma} \big(\{t\} \times \mathcal{L}(0, 1; p, \alpha_t, \beta_t)\big),$$

where α_t, β_t *are the indicator functions at* t*. Equivalently,*

$$\mathrm{sp}_{\mathrm{ess}}\, T(a) = \bigcup_{(t,x) \in \mathcal{M}} \mathrm{sp}\,\big(xa(t - 0) + (1 - x)a(t + 0)\big).$$

Since Lemma 6.14 remains true in the matrix case, Theorem 10.4 follows from applying Theorem 10.2 to the operator $A = aP_\Gamma + Q_\Gamma$. Another (more elementary) proof can be based on Theorem 7.4 and the following fact [85] (also see [33, p. 171] or [137, p. 199]): if $a \in PC_n(\Gamma)$ is an invertible matrix function and a has jumps on only a finite set Λ, then $a = fbg$ where f and g are invertible matrix functions in $C_n(\Gamma) := C_{n \times n}(\Gamma)$ and b is an invertible upper-triangular matrix function in $PC_n(\Gamma)$ which has its jumps only on the set Λ.

For composed Lyapunov curves, Theorems 10. 3 and 10.4 as well as a theorem doing the same as Theorem 10.2 are due to Gohberg and Krupnik [86] (power weights) and Gohberg, Krupnik, and Spitkovsky [91] (Muckenhoupt weights). In the form cited here, Theorems 10.2 to 10.4 were obtained in [10].

10.2   Index formulas

Once Theorems 7.14 and 10.2 are established, one can derive index formulas for Fredholm operators in $\mathrm{alg}\,(S_\Gamma, PC_n(\Gamma))$ using the method developed by Gohberg and Krupnik for operators on composed Lyapunov curves. For such curves, index formulas are in [86], [88], [89] (power weights) and in [91] (general weights). In this section we state an index formula for Fredholm operators in $\mathrm{alg}\,(S_\Gamma, PC_n(\Gamma))$ in the case of general curves (and weights). This formula was first given in [10] and differs from those in [86], [88], [89], [91]: it is based on Theorems 9.27/10.2 and thus on Theorems 8.9/10.1, and the latter theorems were not known at the time the afore-mentioned papers were written.

Let Γ be a composed Carleson curve, $p \in (1, \infty)$, and $w \in A_p(\Gamma)$. We may write $\Gamma = \eta_1 \cup \ldots \cup \eta_m$ where $\eta_1, \ldots, \eta_m$ are simple oriented arcs each pair of which have at most endpoints in common. A point $t \in \Gamma$ is called a *node* if t is an endpoint of al least one of the arcs $\eta_1, \ldots, \eta_m$. The set of all nodes is denoted by T, and we also put $E := \{\eta_1, \ldots, \eta_m\}$. Note that the partition of Γ into simple oriented arcs is not unique. Hence, neither are the sets E and T. For example, if Γ is a Jordan curve, we may represent it as the union of m simple oriented arcs of every number $m \geq 2$. In what follows we always assume that we are given a fixed representation $\Gamma = \eta_1 \cup \ldots \cup \eta_m$. In that case E and T are uniquely determined. Of course, in practice one strives to keep m minimal.

Denote by $\Phi(S_\Gamma, PC_n(\Gamma))$ the set of all operators in $\mathrm{alg}\,(S_\Gamma, PC_n(\Gamma))$ which are Fredholm on $L_n^p(\Gamma, w)$. We will construct functions φ_η $(\eta \in E)$ and φ_t $(t \in T)$ of $\Phi(S_\Gamma, PC_n(\Gamma))$ into $\mathbf{R}$ such that

$$\mathrm{Ind}\, A = \sum_{\eta \in E} \varphi_\eta(A) + \sum_{t \in T} \varphi_t(A) \tag{10.1}$$

for every $A \in \Phi(S_\Gamma, PC_n(\Gamma))$. It should be emphasized that in fact φ_η and φ_t (and thus also $\mathrm{Ind}\, A$) depend on Γ, p, w: $\varphi_\eta = \varphi_{\eta, \Gamma, p, w}$, $\varphi_t = \varphi_{t, \Gamma, p, w}$. We also remark that $\varphi_\eta(A)$ and $\varphi_t(A)$ need not be integers. The sum on the right of (10.1) will however always be an integer.

Let first $\eta \in E = \{\eta_1, \ldots, \eta_m\}$, denote the starting point of η by t_1 and its terminating point by t_2, and put $\eta^0 := \eta \setminus \{t_1, t_2\}$.

For $A \in \mathrm{alg}\,(S_\Gamma, PC_n(\Gamma))$ and $(t, x) \in \mathcal{M}(\Gamma, p, w)$, define $\mathrm{Sym}_{t,x}(A)$ as in Theorem 10.2. If

$$(t, x) \in \mathcal{M}(\eta^0, p, w) := \bigcup_{t \in \eta^0} (\{t\} \times \mathcal{L}_t(\Gamma)),$$

then $\mathrm{Sym}_{t,x}(A)$ is a $2n \times 2n$ matrix and hence

$$\mathrm{Sym}_{t,x}(A) = \begin{pmatrix} A_{11}(t, x) & A_{12}(t, x) \\ A_{21}(t, x) & A_{22}(t, x) \end{pmatrix}$$

with $n \times n$ matrices $A_{ij}(t, x)$. In case $A \in \{S_\Gamma\} \cup \{aI : a \in PC_n(\Gamma)\}$, it can be straightforwardly verified that

$$A_{22}(t - 0, 0) = A_{22}(t, 0), \quad A_{11}(t + 0, 1) = A_{11}(t, 1), \tag{10.2}$$
$$A_{22}(t - 0, 1) = A_{22}(t, 1), \quad A_{11}(t + 0, 0) = A_{11}(t, 0). \tag{10.3}$$

Since $\mathrm{Sym}_{t,0}(A)$ and $\mathrm{Sym}_{t,1}(A)$ are block-diagonal, the four maps

$$A \mapsto A_{22}(t, 0), \quad A \mapsto A_{11}(t, 1),$$
$$A \mapsto A_{22}(t, 1), \quad A \mapsto A_{11}(t, 0)$$

are Banach algebra homomorphisms of $\text{alg}\,(S_\Gamma, PC_n(\Gamma))$ into $\mathbf{C}^{n \times n}$ for each $t \in \eta^0$. Consequently, the equalities (10.2), (10.3) hold for every operator A in $\text{alg}\,(S_\Gamma, PC_n(\Gamma))$.

Now suppose that $A \in \Phi(S_\Gamma, PC_n(\Gamma))$. Then $\text{Sym}_{t,x}(A)$ is invertible for all $(t, x) \in \mathcal{M}(\eta^0, p, w)$, and again taking into account that $\text{Sym}_{t,0}(A)$ and $\text{Sym}_{t,1}(A)$ are block-diagonal, we see that $A_{11}(t, 0)$ and $A_{22}(t, 1)$ are invertible. For (t, x) in $\mathcal{M}(\eta^0, p, w)$, put

$$A(t, x) := \frac{\det(\text{Sym}_{t,x}(A))}{\det(A_{11}(t, 0) A_{22}(t, 1))}. \tag{10.4}$$

Obviously,

$$A(t, 0) = \frac{\det A_{22}(t, 0)}{\det A_{22}(t, 1)}, \quad A(t, 1) = \frac{\det A_{11}(t, 1)}{\det A_{11}(t, 0)}. \tag{10.5}$$

For $t \in \eta^0$, define $A_0(t) := A(t, 0)$. One can show that the limits

$$A_0(t_1 + 0) := \lim_{t \to t_1} A_0(t) \quad \text{and} \quad A_0(t_2 - 0) := \lim_{t \to t_2} A_0(t) \tag{10.6}$$

exist and are finite and nonzero. This in conjunction with (10.5) and (10.2), (10.3) implies that A_0 is an invertible function in $PC(\eta)$ and that

$$A_0(t - 0) = A(t, 0), \quad A_0(t + 0) = A(t, 1) \quad \text{for all } t \in \eta^0. \tag{10.7}$$

Denote by $J[A_0]$ the points of η^0 at which A_0 has a jump. Theorem 10.2 shows that if $t \in J[A_0]$, then $A(t, x) \neq 0$ for all $x \in \mathcal{L}_t(\Gamma)$. Let $\mathcal{L}_t^\#$ be any continuous curve joining 0 to 1 and staying within $\mathcal{L}_t(\Gamma)$. The leaf $\mathcal{L}_t(\Gamma)$ is of the form $\mathcal{L}(0, 1; p, \alpha_t, \beta_t)$ with continuous functions $\alpha_t, \beta_t : \mathbf{R} \to \mathbf{R}$ such that $\alpha_t(\xi) \leq \beta_t(\xi)$ for all $\xi \in \mathbf{R}$. Thus, we could for example take $\mathcal{L}_t^\# := \mathcal{L}(0, 1; p, \psi_t, \psi_t)$ where $\psi_t : \mathbf{R} \to \mathbf{R}$ is any continuous function satisfying $\alpha_t(\xi) \leq \psi_t(\xi) \leq \beta_t(\xi)$ for all $\xi \in \mathbf{R}$. Clearly, $\psi_t = \alpha_t$, $\psi_t = \beta_t$, or $\psi_t = (\alpha_t + \beta_t)/2$ are canonical choices. From Theorem 3.31 we infer that there are even $a_t, b_t \in \mathbf{R}$ such that $\alpha_t(\xi) \leq a_t + b_t\xi \leq \beta_t(\xi)$ for all $\xi \in \mathbf{R}$. Letting $\psi_t(\xi) = a_t + b_t\xi$ and $\mathcal{L}_t^\# := \mathcal{L}(0, 1; p, \psi_t, \psi_t)$, we obtain a (possibly degenerated) logarithmic double spiral $\mathcal{L}_t^\# \subset \mathcal{L}_t(\Gamma)$.

For each point $t \in J[A_0]$, join the endpoints (10.7) of the jump of A_0 at t by the continuous curve $\{A(t, x) : x \in \mathcal{L}_t^\#\}$ and orient this curve from $A(t, 0)$ to $A(t, 1)$. This construction gives us a continuous and oriented curve

$$A_\eta^\# := A_0\left(\eta^0 \setminus J[A_0]\right) \cup \bigcup_{t \in J[A_0]} \{A(t, x) : x \in \mathcal{L}_t^\#\}$$

joining the points (10.6). The origin does not belong to $A_\eta^\#$. Let $\{\arg z\}_{z \in A_\eta^\#}$ denote the increment of any continuous branch of the argument as $A_\eta^\#$ is traced out from $A_0(t_1 + 0)$ to $A_0(t_2 - 0)$. We finally put

$$\varphi_\eta(A) := -\frac{1}{2\pi} \{\arg z\}_{z \in A_\eta^\#} \tag{10.8}$$

and have thus defined the functions φ_η in (10.1).

Now let $t \in T$. If $A \in \Phi(S_\Gamma, PC_n(\Gamma))$ then the $N(t)n \times N(t)n$ matrix $\mathrm{Sym}_{t,x}(A)$ given by Theorem 10.2 is invertible for all $x \in \mathcal{L}_t(\Gamma)$. Choose a curve $\mathcal{L}_t^\# \subset \mathcal{L}_t(\Gamma)$ joining 0 to 1 as above. Then

$$A_t^\# := \big\{ \det \mathrm{Sym}_{t,x}(A) : x \in \mathcal{L}_t^\# \big\}$$

is a continuous curve joining $\det \mathrm{Sym}_{t,0}(A)$ to $\det \mathrm{Sym}_{t,1}(A)$ which does not pass through the origin. Denote by $\{\arg z\}_{z \in A_t^\#}$ the increment of any continuous argument as $A_t^\#$ is traced out from $\det \mathrm{Sym}_{t,0}(A)$ to $\det \mathrm{Sym}_{t,1}(A)$ and define

$$\varphi_t(A) := -\frac{1}{2\pi} \{\arg z\}_{z \in A_t^\#}. \tag{10.9}$$

We so also know the functions φ_t in (10.1).

Theorem 10.5. *Let Γ be a composed Carleson curve, $p \in (1, \infty)$, and $w \in A_p(\Gamma)$. If $A \in \mathrm{alg}\,(S_\Gamma, PC_n(\Gamma))$ is Fredholm on $L_n^p(\Gamma, w)$, then the index of A is given by*

$$\mathrm{Ind}\, A = -\frac{1}{2\pi} \sum_{\eta \in E} \{\arg z\}_{z \in A_\eta^\#} - \frac{1}{2\pi} \sum_{t \in T} \{\arg z\}_{z \in A_t^\#}. \tag{10.10}$$

The main steps of the proof are in [10].

In the case where Γ is a Jordan curve, the right-hand side of (10.10) may be written in another form. Given $A \in \Phi(S_\Gamma, PC_n(\Gamma))$, define $A(t, x)$ for $(t, x) \in \mathcal{M}(\Gamma, p, w)$ by (10.4) and put $A_0(t) := A(t, 0)$ for $t \in \Gamma$. Let $\mathcal{L}_t^\# \subset \mathcal{L}_t(\Gamma)$ be as above. Denote by $A_\Gamma^\#$ the closed, continuous, and naturally oriented curve resulting from the essential range of A_0 by filling in the curve $\{A(t, x) : x \in \mathcal{L}_t^\#\}$ between $A(t, 0) = A_0(t - 0)$ and $A(t, 1) = A_0(t + 0)$ for each $t \in J[A_0]$ (recall (10.7)). The curve $A_\Gamma^\#$ does not pass through the origin, and hence its winding number $\mathrm{wind}\,(A_\Gamma^\#, 0)$ about the origin is well-defined. Again notice that $A_\Gamma^\#$ and thus $\mathrm{wind}\,(A_\Gamma^\#, 0)$ actually depend also on p and w.

Theorem 10.6. *Let Γ be a Carleson Jordan curve, $p \in (1, \infty)$, $w \in A_p(\Gamma)$, and suppose $A \in \mathrm{alg}\,(S_\Gamma, PC_n(\Gamma))$ is Fredholm on $L_n^p(\Gamma, w)$. Then*

$$\mathrm{Ind}\, A = -\mathrm{wind}\,(A_\Gamma^\#, 0). \tag{10.11}$$

Let us at least verify that the right-hand sides of (10.10) and (10.11) coincide. For simplicity, choose two points t_1 and t_2 on Γ, let η_1 be the positively oriented arc from t_1 to t_2 and denote by η_2 the positively oriented arc from t_2 to t_1. Then $E = \{\eta_1, \eta_2\}$ and $T = \{t_1, t_2\}$. We have to show that

$$\varphi_{\eta_1}(A) + \varphi_{\eta_2}(A) + \varphi_{t_1}(A) + \varphi_{t_2}(A) = -\mathrm{wind}\,(A_\Gamma^\#, 0). \tag{10.12}$$

For $t \in \Gamma$, we put $c_t := \det(A_{11}(t,0)A_{22}(t,1))$. The curve $A_{\eta_1}^{\#}$ joins $A_0(t_1 + 0) = A(t_1, 1)$ to $A_0(t_2 - 0) = A(t_2, 0)$ and coincides with $A_{\Gamma}^{\#}$ between these two points. The curve $A_{t_2}^{\#}$ joins $\det \mathrm{Sym}_{t_2,0}(A) = c_{t_2} A(t_2, 0)$ to $\det \mathrm{Sym}_{t_2,1}(A) = c_{t_2} A(t_2, 1)$ along

$$\left\{ \det \mathrm{Sym}_{t_2,x}(A) : x \in \mathcal{L}_{t_2}^{\#} \right\} = c_{t_2} \left\{ A(t_2, x) : x \in \mathcal{L}_{t_2}^{\#} \right\},$$

while $A_{\Gamma}^{\#}$ joins $A_0(t_2 - 0) = A(t_2, 0)$ to $A_0(t_2 + 0) = A(t_2, 1)$ along $\{A(t_2, x) : x \in \mathcal{L}_{t_2}^{\#}\}$. Clearly, the increment of $\arg z$ as z traces out $\{c_{t_2} A(t_2, x) : x \in \mathcal{L}_{t_2}^{\#}\}$ is equal to the increment of $\arg z$ as z traverses $\{A(t_2, x) : x \in \mathcal{L}_{t_2}^{\#}\}$. The arc η_2 and the point t_1 can be considered analogously. In summary, it follows that the left-hand side of (10.12) is

$$-\frac{1}{2\pi} \{\arg z\}_{z \in A_{\Gamma}^{\#}} = -\mathrm{wind}\,(A_{\Gamma}^{\#}, 0),$$

which completes the proof.

Applying Theorem 10.6 and Lemma 6.14 to the operator $aP_{\Gamma} + Q_{\Gamma}$, we arrive at the following result.

Theorem 10.7. *Let Γ be a Carleson Jordan curve, $p \in (1, \infty)$, $w \in A_p(\Gamma)$, and $a \in PC_n(\Gamma)$. If the Toeplitz operator $T(a)$ is Fredholm on $H_n^p(\Gamma, w) := P_{\Gamma} L_n^p(\Gamma, w)$, then its index is minus the winding number about the origin of the closed, continuous, and naturally oriented curve*

$$(\det a)(\Gamma \setminus \Lambda_a) \cup \bigcup_{t \in \Lambda_a} \left\{ \det \big(xa(t - 0) + (1 - x)a(t + 0) \big) : x \in \mathcal{L}_t^{\#} \right\},$$

where

$$\Lambda_a := \{ t \in \Gamma : a(t - 0) \neq a(t + 0) \}$$

and $(\det a)(\Gamma \setminus \Lambda_a)$ stands for the essential range of $\det a$.

We have deduced Theorem 10.6 from Theorem 10.5 and Theorem 10.7 from Theorem 10.6. We should note that in fact the proof given in [10] proceeds in the reverse direction: first Theorem 10.7 is established, with its help then Theorem 10.6 is proved, and finally Theorem 10.5 is derived from Theorem 10.6.

We remark that the proof of Theorem 10.5 also makes essential use of a result by Alexei Karlovich [112], which states that if the operators $A^{(k)} \in \mathrm{alg}(S_{\Gamma}, PC_n(\Gamma))$ converge uniformly to $A \in \mathrm{alg}\,(S_{\Gamma}, PC_n(\Gamma))$, then the functions $\det(\mathrm{Sym}\,(A^{(k)}))$, $\det A_{11}^{(k)}$, $\det A_{22}^{(k)}$ converge uniformly on the set $\mathcal{M} = \mathcal{M}(\Gamma, p, w)$ to the functions $\det(\mathrm{Sym}\,(A))$, $\det A_{11}$, $\det A_{22}$, respectively, and that all these functions are bounded (although the function $\mathrm{Sym}_{t,x} A$ may be an unbounded function of $(t, x) \in \mathcal{M}$).

10.3 Kernel and cokernel dimensions

The following theorem and its corollary generalize Theorem 6.17 and Corollaries 6.18 and 6.19.

Theorem 10.8. *Let Γ be a composed Carleson curve, $p \in (1, \infty)$, and $w \in A_p(\Gamma)$. For a function $c \in L^\infty(\Gamma)$, define $N_c := \{t \in \Gamma : c(t) = 0\}$. Let a and b be functions in $L^\infty(\Gamma)$ and consider the operators*

$$A := aP_\Gamma + bQ_\Gamma \in \mathcal{B}(L^p(\Gamma, w)), \quad A^* := H_\Gamma(Q_\Gamma a + P_\Gamma b)H_\Gamma \in \mathcal{B}(L^q(\Gamma, w^{-1})).$$

If $|N_a| < |\Gamma|$, $|N_b| < |\Gamma|$, and $|N_a \cap N_b| = 0$, then

$$\dim \operatorname{Ker} A = 0 \ \ or \ \ \dim \operatorname{Ker} A^* = 0$$

If $a, b \in L^\infty(\Gamma)$ and $aP_\Gamma + bQ_\Gamma$ is Fredholm, then a and b can be shown to be invertible in $L^\infty(\Gamma)$ (this is an extension of Theorem 6.20). Thus, Theorem 10.8 implies the following.

Corollary 10.9. *Let Γ be a composed Carleson curve, $p \in (1, \infty)$, and $w \in A_p(\Gamma)$. Suppose a and b are functions in $L^\infty(\Gamma)$. If the operator $A := aP_\Gamma + bQ_\Gamma$ is Fredholm on $L^p(\Gamma, w)$, then*

$$\dim \operatorname{Ker} A = \dim \operatorname{Coker} A^* = \max\{0, \operatorname{Ind} A\},$$
$$\dim \operatorname{Coker} A = \dim \operatorname{Ker} A^* = \max\{0, -\operatorname{Ind} A\}.$$

In particular, A is invertible on $L^p(\Gamma, w)$ if and only if A is Fredholm of index zero.

For composed Lyapunov curves, Theorem 10.8 was established by Gohberg and Krupnik [89, Chapter 7.5]. In the general case, it first appeared in [10]. The proof of [10] makes use of the arguments of [89, Chapter 7.5] and the following Theorem 10.10, which in turn extends Theorem 9.13.

Given a Jordan curve Γ, denote by $D(\Gamma)$ the bounded connected component of $\mathbf{C} \setminus \Gamma$ and let $\overline{D(\Gamma)} := D(\Gamma) \cup \Gamma$. Recall that we always assume that a Jordan curve is oriented counter-clockwise. We call a composed curve G a *loop curve* if it may be represented in the form $G = \Gamma_1 \cup \ldots \cup \Gamma_m$ where $\Gamma_1, \ldots, \Gamma_m$ are Jordan curves such that $\overline{D(\Gamma_j)}$ and $\overline{D(\Gamma_k)}$ have at most finitely many points in common whenever $j \neq k$. Obviously, flowers are examples of loop curves. If G is a loop curve then $S_G^2 = I$ and hence P_G is a projection.

Theorem 10.10. *Let Γ be a composed Carleson curve, $p \in (1, \infty)$, and $w \in A_p(\Gamma)$. There exists a Carleson loop curve G and a weight $W \in A_p(G)$ such that $\Gamma \subset G$ and $W|\Gamma = w$.*

The main ideas of the proof of this theorem are in [10].

Corollary 10.9 divides the problem of deciding whether an operator of the form $aP_\Gamma + bQ_\Gamma$ is invertible into two "halves": into finding out whether it is Fredholm and into computing its index. Unfortunately, results like Theorem 10.8 or Corollary 10.9 are neither true for operators with matrix coefficients nor for generic operators in the algebra alg $(S_\Gamma, PC(\Gamma))$.

10.4   Spectrum of the Cauchy singular integral

Let Γ be a composed Carleson curve, $p \in (1, \infty)$, and $w \in A_p(\Gamma)$. Since $S_\Gamma - \lambda I = (1 - \lambda)P_\Gamma + (-1 - \lambda)Q_\Gamma$, we can employ Corollary 10.9. to deduce that

$$\operatorname{sp} S_\Gamma = \operatorname{sp}_{\operatorname{ess}} S_\Gamma \cup \{\lambda \in \mathbf{C} \setminus \operatorname{sp}_{\operatorname{ess}} S_\Gamma : \operatorname{Ind}(S_\Gamma - \lambda I) \neq 0\}. \qquad (10.13)$$

The essential spectrum is known from Theorem 9.28. We are so left with looking at what Theorem 10.5 gives in the special case where $A = S_\Gamma - \lambda I$.

Write $\Gamma = \eta_1 \cup \ldots \cup \eta_m$ as in Section 10.2 and denote by $T = \{t_1, \ldots, t_n\}$ the set of nodes of Γ. For $t \in T$, define $\mathcal{L}_{t,k,n}$ by (9.71) and denote by $\mathcal{L}_{t,k,n}^\#$ any simple arc joining 0 to 1 and lying entirely in $\mathcal{L}_{t,k,n}$. In Section 10.2 we saw that it is always possible to find a (possibly degenerated) logarithmic double spiral $\mathcal{L}_{t,k,n}^\# \subset \mathcal{L}_{t,k,n}$ joining 0 to 1. Given $z_1, z_2 \in \mathbf{C}$, we put

$$\mathcal{L}_{t,k,n}(z_1, z_2) := \big\{(1 - x)z_1 + xz_2 : x \in \mathcal{L}_{t,k,n}\big\},$$
$$\mathcal{L}_{t,k,n}^\#(z_1, z_2) := \big\{(1 - x)z_1 + xz_2 : x \in \mathcal{L}_{t,k,n}^\#\big\},$$

and we think of $\mathcal{L}_{t,k,n}^\#(z_1, z_2)$ as being oriented from z_1 to z_2.

Let T^* be the set of all nodes $t \in T$ with valency $\varepsilon(t) \neq 0$. If $T^* = \emptyset$, then Theorem 9.28 shows that $\operatorname{sp}_{\operatorname{ess}} S_\Gamma = \{-1, 1\}$, and since $\operatorname{Ind}(S_\Gamma - \lambda I)$ is constant on $\mathbf{C} \setminus \{-1, 1\}$ and vanishes for large $|\lambda|$, we infer from (10.13) that $\operatorname{sp} S_\Gamma = \{-1, 1\}$.

So assume $T^* \neq \emptyset$. Theorem 9.28 shows that

$$\operatorname{sp}_{\operatorname{ess}} S_\Gamma = \bigcup_{t \in T^*} \bigcup_{k=0}^{|\varepsilon(t)|-1} \mathcal{L}_{t,k,|\varepsilon(t)|}\big(-\operatorname{sign}\varepsilon(t), \operatorname{sign}\varepsilon(t)\big).$$

Consider the curve

$$C_\Gamma^\# := C_{\Gamma,p,w}^\# := \bigcup_{t \in T^*} \bigcup_{k=0}^{|\varepsilon(t)|-1} \mathcal{L}_{t,k,|\varepsilon(t)|}^\#\big(-\operatorname{sign}\varepsilon(t), \operatorname{sign}\varepsilon(t)\big).$$

The curve $C_\Gamma^\#$ consists of $\sum_{t \in T^*} |\varepsilon(t)|$ oriented simple arcs. Since $\sum_{t \in T^*} \varepsilon(t) = 0$, the number of arcs joining -1 to 1 is equal to the number of arcs going from 1 to -1. Thus, the arcs of $C_\Gamma^\#$ may be traced out in a succession so that $C_\Gamma^\#$ becomes a closed, continuous, and oriented curve. As obviously $C_\Gamma^\# \subset \operatorname{sp}_{\operatorname{ess}} S_\Gamma$, the winding number $\operatorname{wind}(C_\Gamma^\#, \lambda)$ of $C_\Gamma^\#$ about each point $\lambda \in \mathbf{C} \setminus \operatorname{sp}_{\operatorname{ess}} S_\Gamma$ is well-defined.

Theorem 10.11. *Let Γ be a composed Carleson curve, $p \in (1, \infty)$, and $w \in A_p(\Gamma)$. If $T^* \neq \emptyset$, then*

$$\text{Ind}\,(S_\Gamma - \lambda I) = -\text{wind}\,(C_\Gamma^\#, \lambda)$$

for every $\lambda \in \mathbf{C} \setminus \text{sp}_{\text{ess}}\, S_\Gamma$ and hence,

$$\text{sp}\, S_\Gamma = \text{sp}_{\text{ess}}\, S_\Gamma \cup \{\lambda \in \mathbf{C} \setminus \text{sp}_{\text{ess}}\, S_\Gamma : \text{wind}\,(C_\Gamma^\#, \lambda) \neq 0\}.$$

If $T^* = \emptyset$, then $\text{sp}\, S_\Gamma = \text{sp}_{\text{ess}}\, S_\Gamma = \{-1, 1\}$.

This theorem was established in [10]. We confine ourselves to illustrate it by three examples.

Example 10.12. Suppose first $\Gamma = \eta_1 \cup \eta_2 \cup \eta_3$ where η_1, η_2, η_3 are simple arcs with exactly one common endpoint and with no other common points. Denote the common endpoint by t_0, and let t_j be the other endpoint of η_j. Assume η_j is oriented from t_0 to t_j. Then $\varepsilon(t_0) = 3$ and $\varepsilon(t_1) = \varepsilon(t_2) = \varepsilon(t_3) = -1$. Let α_t and β_t be the indicator functions of Γ, p, w at $t \in \Gamma$. In accordance with (9.71) for $n = 3$, we put

$$\alpha_t^{(k)}(x) = \frac{1}{3}\left(k - \frac{2}{p} + \alpha_t(3x)\right), \quad \beta_t^{(k)}(x) = \frac{1}{3}\left(k - \frac{2}{p} + \beta_t(3x)\right).$$

Then define

$$\begin{aligned}
\mathcal{L}_{t_0,k,3}(-1,1) &= \mathcal{L}\big(-1,1; p, \alpha_{t_0}^{(k)}, \beta_{t_0}^{(k)}\big) \quad (k = 0, 1, 2),\\
\mathcal{L}_{t_j,0,1}(1,-1) &= \mathcal{L}\big(1,-1; p, \alpha_{t_j}, \beta_{t_j}\big) \quad (j = 1, 2, 3),\\
\mathcal{L}_{t_0,k,3}^\#(-1,1) &= \mathcal{L}\big(-1,1; p, \alpha_{t_0}^{(k)}, \alpha_{t_0}^{(k)}\big) \quad (k = 0, 1, 2),\\
\mathcal{L}_{t_j,0,1}^\#(1,-1) &= \mathcal{L}\big(1,-1; p, \alpha_{t_j}, \alpha_{t_j}\big) \quad (j = 1, 2, 3).
\end{aligned}$$

By Theorem 9.28,

$$\text{sp}_{\text{ess}}\, S_\Gamma = \bigcup_{k=0}^{2} \mathcal{L}_{t_0,k,3}(-1,1) \cup \bigcup_{j=1}^{3} \mathcal{L}_{t_j,0,1}(1,-1),$$

and the curve $C_\Gamma^\#$ in Theorem 10.11 is

$$\begin{aligned}
C_\Gamma^\# =\ & \mathcal{L}_{t_0,0,3}^\#(-1,1) \cup \mathcal{L}_{t_1,0,1}^\#(1,-1) \cup \mathcal{L}_{t_0,1,3}^\#(-1,1)\\
& \cup \mathcal{L}_{t_2,0,1}^\#(1,-1) \cup \mathcal{L}_{t_0,2,3}^\#(-1,1) \cup \mathcal{L}_{t_3,0,1}^\#(1,-1),
\end{aligned}$$

where the arrangement of the arcs in the latter union indicates the succession in which they are traversed. Obviously, if the six arcs of $C_\Gamma^\#$ are circular arcs, then the possible values of $\text{Ind}\,(S_\Gamma - \lambda I)$ are $0, \pm 1, \pm 2, \pm 3$. $\quad\square$

Example 10.13. Now suppose $\Gamma = \eta$ is an arbitrary simple Carleson arc. Denote the starting point of η by t_1 and the terminating point by t_2. Theorem 9.28 shows that

$$\text{sp}_{\text{ess}} \, S_\eta = \mathcal{L}(-1, 1; p; \alpha_{t_1}, \beta_{t_1}) \cup \mathcal{L}(1, -1; p; \alpha_{t_2}, \beta_{t_2}) \qquad (10.14)$$

and $\text{Ind}\,(S_\eta - \lambda I)$ can be determined from Theorem 10.11 with

$$C_\eta^{\#} = \mathcal{L}(-1, 1; p; \alpha_{t_1}, \alpha_{t_1}) \cup \mathcal{L}(1, -1; p; \alpha_{t_2}, \alpha_{t_2}). \qquad (10.15)$$

If the two leaves in (10.14) are horns, then the possible values of $\text{Ind}\,(S_\eta - \lambda I)$ are $-1, 0, 1$. However, if one of the leaves is a logarithmic double spiral and the other one is a circular arc, then for each $\varkappa \in \mathbf{Z}$ there exists a point $\lambda \in \mathbf{C} \setminus \text{sp}_{\text{ess}} \, S_\eta$ such that $\text{Ind}\,(S_\eta - \lambda I) = \varkappa$. $\qquad\qquad\Box$

Example 10.14. Finally, suppose we are given a Carleson star Γ comprised of four arcs. Denote the center of the star by t and let t_1, t_2, t_3, t_4 stand for the other endpoints of the star. Assume t has the valency 4 and the four points t_1, t_2, t_3, t_4 have the valency -1. Let the local spectrum $\text{sp}\,\pi_t(S_\Gamma)$ be as in Figure 19b. In order to decide which of the three holes in Figure 19b belong to the spectrum of S_Γ, we need know more about the curve and the weight at the points t_j $(j = 1, \ldots, 4)$. In Figures 23a and 23b, we see four arcs $\mathcal{L}_{t,k,4}^{\#}(-1, 1)$ entirely contained in the leaves $\mathcal{L}_{t,k,4}(-1, 1)$ $(k = 1, \ldots, 4)$.

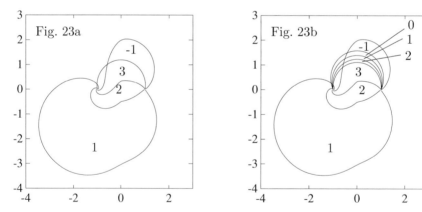

Figures 23a and 23b illustrate the determination of the spectrum of S_Γ in case the local spectrum $\text{sp}\,\pi_t(S_\Gamma)$ is as in Figure 19b.

If the curve is nice at the points t_j and if the weight is equivalent to a constant weight at these points, then each of points t_j $(j = 1, \ldots, 4)$ contributes one and the same circular arc from 1 to -1 to the curve $C_\Gamma^{\#}$. This situation is shown in Figure 23a: we there see four arcs from -1 to 1 and one circular arc from 1 to -1. The latter circular arc must be traversed four times when determining wind $(C_\Gamma^\#, \lambda)$. The corresponding winding numbers are indicated in Figure 23a. Consequently, in this case all the three holes of Figure 19b are contained in the spectrum of S_Γ.

Now assume again that the curve is nice at the points t_j, but assume the weight is equivalent to a power weight $|t - t_j|^{\lambda_j}$ at these points. If the numbers λ_j $(j = 1, \ldots, 4)$ are distinct but close to each other, then the four points t_j produce four circular arcs from 1 to -1 which are close to each other. This is shown in Figure 23b. Determining the corresponding winding numbers, we arrive at the conclusion that the spectrum of S_Γ does not contain the piece of the upper hole of Figure 19b which is included between the most upper two circular arcs (labeled by 0 in Figure 23b), but that the rest of the upper hole and the two lower holes of Figure 19b are contained in the spectrum of S_Γ. $\qquad\square$

10.5 Orlicz spaces

The results of this section are due to Alexei Yu. Karlovich and are taken from his papers [110], [111]. They extend part of the above results on the Lebesgue spaces $L^p(\Gamma)$ to reflexive Orlicz spaces $L^M(\Gamma)$. For the sake of simplicity, we assume that Γ is a Jordan curve.

The theory of Orlicz spaces is well presented in [128], [8], [140], [141]. We here restrict ourselves to citing a few basic concepts and results.

A *Young function* is a continuous and convex function $M : [0, \infty) \to [0, \infty)$ such that $M(0) = 0$, $M(x) > 0$ for $x > 0$,

$$\lim_{x \to 0} \frac{M(x)}{x} = 0, \quad \lim_{x \to \infty} \frac{M(x)}{x} = +\infty.$$

Given a Young function M, one defines a new function $N : [0, \infty) \to [0, \infty)$ by

$$N(x) := \max_{y \geq 0} \left(xy - M(y) \right).$$

It turns out that N is also a Young function. It is called the *complementary Young function* of M.

Now let Γ be a rectifiable Jordan curve and equip Γ with Lebesgue length measure $|d\tau|$. Let further M be a Young function and let N be the complementary Young function. The *Orlicz space* $L^M(\Gamma)$ is defined as the linear space of all measurable functions $f : \Gamma \to \mathbf{C}$ for which there is a $\lambda = \lambda(f) > 0$ such that

$$\int_\Gamma M(|f(\tau)|/\lambda) \, |d\tau| < \infty.$$

If $f \in L^M(\Gamma)$, then the two numbers

$$\|f\|_{(M)} := \inf \left\{ \lambda > 0 : \int_\Gamma M(|f(\tau)|/\lambda) \, |d\tau| \leq 1 \right\}$$

and

$$\|f\|_M := \sup \left\{ \int_\Gamma |f(\tau)g(\tau)| \, |d\tau| : \int_\Gamma N(|g(\tau)|) \, |d\tau| \leq 1 \right\}$$

are finite and are referred to as the *Luxemburg* and *Orlicz norms* of f, respectively. These two norms are equivalent, $\|f\|_{(M)} \leq \|f\|_M \leq 2\|f\|_{(M)}$ for all $f \in L^M(\Gamma)$, and $L^M(\Gamma)$ is a Banach space with anyone of these norms.

Let M be a Young function and denote by $M^{-1} : [0, \infty) \to [0, \infty)$ the inverse function of M. Define a new function $\varrho : (0, \infty) \to (0, \infty]$ by

$$\varrho(x) := \limsup_{y \to \infty} \frac{M^{-1}(y)}{M^{-1}(y/x)}, \quad x \in (0, \infty).$$

One can show that ϱ is a regular submultiplicative function in the sense of Section 1.4. The lower and upper indices of ϱ are called the *Boyd indices* of $L^M(\Gamma)$ (see [28]) and are denoted by α_M and β_M, respectively. By Theorem 1.13,

$$\alpha_M = \lim_{x \to 0} \frac{\log \varrho(x)}{\log x}, \quad \beta_M = \lim_{x \to \infty} \frac{\log \varrho(x)}{\log x}.$$

One always has

$$0 \leq \alpha_M \leq \beta_M \leq 1, \quad \alpha_M + \beta_N = 1, \quad \alpha_N + \beta_M = 1.$$

The Boyd indices are important characteristics of an Orlicz space. For example, it can be shown (see, e.g., [140, Theorem 3.2(b)]) that $L^M(\Gamma)$ is *reflexive* (i.e. canonically isomorphic to its second dual) if and only if

$$0 < \alpha_M \leq \beta_M < 1. \tag{10.16}$$

L^p spaces fit in with Orlicz spaces as follows. The function $M(x) = x^p/p$ is a Young function for every $p \in (1, \infty)$. Its complementary Young function is $N(x) = x^q/q$ where $1/p + 1/q = 1$. In this case $L^M(\Gamma)$ is the Lebesgue space $L^p(\Gamma)$ and we have

$$\|f\|_{(M)} = p^{-1/p} \left(\int_\Gamma |f(\tau)|^p |d\tau| \right)^{1/p}, \quad \|f\|_M = q^{1/q} \left(\int_\Gamma |f(\tau)|^p |d\tau| \right)^{1/p}.$$

Since $M^{-1}(y) = (py)^{1/p}$, we get $\varrho(x) = x^{1/p}$ and thus $\alpha_M = \beta_M = 1/p$.

If $1 \leq r < 1/\beta_M \leq 1/\alpha_M < s \leq \infty$, then $L^s(\Gamma) \subset L^M(\Gamma) \subset L^r(\Gamma)$, the inclusion maps being continuous. Thus, if $L^M(\Gamma)$ is a reflexive Orlicz space then

$$L^\infty(\Gamma) \subset L^M(\Gamma) \subset L^1(\Gamma)$$

by virtue of (10.16).

Theorem 10.15. *Let Γ be a rectifiable Jordan curve and let $L^M(\Gamma)$ be a reflexive Orlicz space. The Cauchy singular integral operator S_Γ generates a bounded operator $\tilde{S}_\Gamma$ on $L^M(\Gamma)$ if and only if Γ is a Carleson curve. In that case for every $f \in L^M(\Gamma)$ the limit*

$$(S_\Gamma f)(t) := \lim_{\varepsilon \to 0} \frac{1}{\pi i} \int\limits_{|\tau - t| > \varepsilon} \frac{f(\tau)}{\tau - t}\, d\tau$$

exists and coincides with $(\tilde{S}_\Gamma f)(t)$ for almost all $t \in \Gamma$.

A derivation of this theorem from David's Theorem 4.17 is in [110].

We now come to the Orlicz space version of the U_t^0 transform introduced in Section 3.2. Let $t \in \Gamma$. For $0 < R_1 \le R_2 < \infty$, put

$$\Gamma(t, R_1, R_2) := \left\{ \tau \in \Gamma : R_1 \le |\tau - t| < R_2 \right\}$$

and let χ_{t, R_1, R_2} denote the characteristic function of $\Gamma(t, R_1, R_2)$. Fix a number $\kappa \in (0, 1)$. Given a continuous function $\psi : \Gamma \setminus \{t\} \to (0, \infty)$, we define the function $U_t^M \psi : (0, \infty) \to (0, \infty]$ by

$$(U_t^M \psi)(\xi) := \limsup_{R \to 0} \frac{\|\psi \chi_{t, \kappa \xi R, \xi R}\|_M \, \|\psi^{-1} \chi_{t, \kappa R, R}\|_N}{|\Gamma(t, \kappa R, R)|} \tag{10.17}$$

for $\xi \in (0, \infty)$. Define $\eta_t^x : \Gamma \setminus \{t\} \to (0, \infty)$ by $\eta_t^x(\tau) = e^{-x \arg(\tau - t)}$ where $\arg(\tau - t)$ is any continuous branch of the argument on $\Gamma \setminus \{t\}$. One can show that the function $U_t^M \eta_t^x : (0, \infty) \to (0, \infty)$ is regular and submultiplicative for every $x \in \mathbf{R}$. Thus, by Theorem 1.13, the lower and upper indices

$$\alpha(U_t^M \eta_t^x) := \lim_{\xi \to 0} \frac{\log(U_t^M \eta_t^x)(\xi)}{\log \xi}, \quad \beta(U_t^M \eta_t^x) := \lim_{\xi \to \infty} \frac{\log(U_t^M \eta_t^x)(\xi)}{\log \xi}$$

exist and are finite for every $x \in \mathbf{R}$. We put

$$\alpha_t^M(x) := \alpha(U_t^M \eta_t^x), \quad \beta_t^M(x) := \beta(U_t^M \eta_t^x)$$

and call α_t^M and β_t^M the *modified outer indicator functions* of $L^M(\Gamma)$ at t.

In order to compare the definitions in the preceding paragraph with those of Chapter 3, let $M(x) = x^p/p$ and thus $L^M(\Gamma) = L^p(\Gamma)$ $(1 < p < \infty)$. In that case the right-hand side of (10.17) is

$$\varrho_1(\xi) := \limsup_{R \to 0} \frac{p^{1/p} q^{1/q} \|\psi \chi_{t, \kappa \xi R, \xi R}\|_p \, \|\psi^{-1} \chi_{t, \kappa R, R}\|_q}{|\Gamma(t, \kappa R, R)|},$$

while the right-hand side of (3.11) gives

$$\varrho_2(\xi) := \limsup_{R \to 0} \frac{\|\psi \chi_{t, \kappa \xi R, \xi R}\|_p \, \|\psi^{-1} \chi_{t, \kappa R, R}\|_q}{|\Gamma(t, \kappa \xi R, \xi R)|^{1/p} |\Gamma(t, \kappa R, R)|^{1/q}}.$$

Since

$$\frac{(1-\kappa)\xi}{C_\Gamma} = \frac{(1-\kappa)\xi R}{C_\Gamma R} \leq \frac{|\Gamma(t,\kappa\xi R,\xi R)|}{|\Gamma(t,\kappa R,R)|} \leq \frac{C_\Gamma \xi R}{(1-\kappa)R} = \frac{C_\Gamma \xi}{1-\kappa},$$

we have

$$p^{1/p}q^{1/q}\left(\frac{(1-\kappa)\xi}{C_\Gamma}\right)^{1/p}\varrho_2(\xi) \leq \varrho_1(\xi) \leq p^{1/p}q^{1/q}\left(\frac{C_\Gamma \xi}{1-\kappa}\right)^{1/p}\varrho_2(\xi)$$

for all $\xi \in (0,\infty)$. Hence,

$$\lim_{\xi\to 0}\frac{\log\varrho_1(\xi)}{\log\xi} = \lim_{\xi\to 0}\frac{\log\varrho_2(\xi)}{\log\xi} + \frac{1}{p}, \quad \lim_{\xi\to\infty}\frac{\log\varrho_1(\xi)}{\log\xi} = \lim_{\xi\to\infty}\frac{\log\varrho_2(\xi)}{\log\xi} + \frac{1}{p},$$

and letting $\psi := \eta_t^x$, we obtain

$$\alpha_t^M(x) = \alpha(U_t^0 \eta_t^x) + 1/p, \quad \beta_t^M(x) = \beta(U_t^0 \eta_t^x) + 1/p$$

for all $x \in \mathbf{R}$. From Corollary 3.12 (with $w = 1$ and $\gamma = -ix$), (3.24), and (3.67) we therefore get

$$\alpha_t^M(x) = \alpha_t^*(x) + 1/p, \quad \beta_t^M(x) = \beta_t^*(x) + 1/p. \tag{10.18}$$

Thus, if $L^M(\Gamma) = L^p(\Gamma)$, then the modified outer indicator functions α_t^M, β_t^M can be expressed via (10.18) in terms of the outer indicator functions α_t^*, β_t^*.

Here is the analogue of Theorem 3.31 in the case $w = 1$. The numbers δ_t^-, δ_t^+ are the spirality indices of Γ at $t \in \Gamma$ and α_M, β_M are the Boyd indices of $L^M(\Gamma)$. As we have no weight on Γ, we define

$$\alpha_t(y) := \min\{\delta_t^- y, \delta_t^+ y\}, \quad \beta_t(y) := \max\{\delta_t^- y, \delta_t^+ y\}$$

in accordance with Theorem 3.30.

Theorem 10.16. *Let Γ be a Carleson Jordan curve and suppose $L^M(\Gamma)$ is reflexive. If $t \in \Gamma$, then*

(a) $-\infty < \alpha_t^M(x) \leq \beta_t^M(x) < +\infty$ *for all $x \in \mathbf{R}$;*

(b) $\alpha_t^M(0) = \alpha_M$, $\beta_t^M(0) = \beta_M$;

(c) α_t^M *is concave and β_t^M is convex;*

(d) *for all $x, y \in \mathbf{R}$ one has*

$$\alpha_t^M(x) + \alpha_t(y) \leq \alpha_t^M(x+y) \leq \min\left\{\alpha_t^M(x) + \beta_t(y),\ \beta_t^M(x) + \alpha_t(y)\right\},$$
$$\beta_t^M(x) + \beta_t(y) \geq \beta_t^M(x+y) \geq \max\{\alpha_t^M(x) + \beta_t(y),\ \beta_t^M(x) + \alpha_t(y)\}.$$

Using Theorem 10.15 one can show that if Γ is a Carleson Jordan curve and $L^M(\Gamma)$ is reflexive, then $S_\Gamma^2 f = f$ for all $f \in L^M(\Gamma)$. Hence $P_\Gamma := (I + S_\Gamma)/2$

is a projection, and one can define the Hardy subspace of $L^M(\Gamma)$ as $L^M_+(\Gamma) := P_\Gamma L^M(\Gamma)$. If $a \in L^\infty(\Gamma)$, then the multiplication operator aI is easily seen to be bounded on $L^M(\Gamma)$. Given $a \in L^\infty(\Gamma)$, we may therefore define the Toeplitz operator $T(a) \in \mathcal{B}(L^M_+(\Gamma))$ by $T(a) : f \mapsto P_\Gamma(af)$.

In Section 7.3, we defined the leaves $\mathcal{L}(z_1, z_2; p, \alpha_t, \beta_t)$. Since now no p is present, we have to modify this definition. The easiest way to remove the $1/p$ occuring in the definitions is to put $p = \infty$. So let

$$Y\left(\infty, \alpha^M_t, \beta^M_t\right) := \left\{\gamma = x + iy \in \mathbf{C} : \alpha^M_t(x) \le y \le \beta^M_t(x)\right\}$$

and then put

$$\mathcal{L}\left(z_1, z_2; \infty; \alpha^M_t, \beta^M_t\right) := \left\{M_{z_1, z_2}(e^{2\pi\gamma}) : \gamma \in Y\left(\infty, \alpha^M_t, \beta^M_t\right)\right\} \cup \{z_1, z_2\}$$

where $M_{z_1, z_2}(\zeta) := (z_2\zeta - z_1)/(\zeta - 1)$. Equivalently, we might define

$$\mathcal{L}\left(z_1, z_2; \infty; \alpha^M_t, \beta^M_t\right) := \mathcal{L}\left(z_1, z_2; p; \alpha^M_t - 1/p, \beta^M_t - 1/p\right), \tag{10.19}$$

where p is any number. Clearly, the right-hand side of (10.19) is independent of p.

Theorem 10.17. *Let Γ be a Carleson Jordan curve and suppose $L^M(\Gamma)$ is reflexive. Assume further that*

$$\alpha^M_t(x) = \alpha_M + \min\{\delta^-_t x, \delta^+_t x\}, \quad \beta^M_t(x) = \beta_M + \max\{\delta^-_t x, \delta^+_t x\} \tag{10.20}$$

for all $t \in \Gamma$ and all $x \in \mathbf{R}$. If $a \in PC(\Gamma)$, then the essential spectrum of $T(a)$ on $L^M(\Gamma)$ is given by

$$\mathrm{sp}_{\mathrm{ess}} T(a) = \mathcal{R}(a) \cup \bigcup_{t \in \Lambda_a} \mathcal{L}\left(a(t-0), a(t+0); \infty; \alpha^M_t, \beta^M_t\right), \tag{10.21}$$

where

$$\Lambda_a = \{t \in \Gamma : a(t-0) \ne a(t+0)\},$$

$\mathcal{R}(a)$ is the essential range of a, and α^M_t, β^M_t are the modified outer indicator functions of $L^M(\Gamma)$ at t. If $T(a)$ is Fredholm, then its index is

$$\mathrm{Ind}\, T(a) = -\mathrm{wind}\left(a^\#, 0\right)$$

where $a^\# := a^\#_{\Gamma, M}$ is the closed, continuous, and naturally oriented curve

$$a^\# := \mathcal{R}(a) \cup \bigcup_{t \in \Lambda_a} \mathcal{L}\left(a(t-0), a(t+0); \infty; \psi_t, \psi_t\right),$$

ψ_t being any continuous function such that $\alpha^M_t \le \psi_t \le \beta^M_t$ on $\mathbf{R}$.

In case where Γ is a piecewise Lyapunov curve, the inclusion

$$\mathrm{sp}_{\mathrm{ess}} T(a) \subset \mathcal{R}(a) \cup \bigcup_{t \in \Lambda_a} \mathcal{L}\big(a(t-0), a(t+0); \infty; \alpha_t^M, \beta_t^M\big) \qquad (10.22)$$

was (in completely other terms) already established in [89, Section 9.11]. In the form cited here the theorem is proved in [110], [111]. The reverse inclusion in (10.22) is actually the hard part of the problem. The proof of A.Yu. Karlovich is based on finding conditions for the boundedness of the operator

$$\varphi_{t,\gamma} S_\Gamma \varphi_{t,\gamma}^{-1} I$$

on $L^M(\Gamma)$, where $\varphi_{t,\gamma}(\tau) := |(\tau - t)^\gamma|$ is as in Section 3.1, in terms of the Boyd indices α_M, β_M and the modified outer indicator functions α_t^M, β_t^M.

Theorem 10.16 implies that (10.20) is automatically satisfied in two interesting cases, namely, if

(i) the Boyd indices coincide, i.e. $\alpha_M = \beta_M$,

or

(ii) the spirality indices coincide everywhere on Γ, i.e. $\delta_t^- = \delta_t^+$ for all $t \in \Gamma$.

Thus, in the case $\alpha_M = \beta_M$, Theorem 10.17 contains Theorems 7.4 and 7.14 (with $w = 1$) and is a theorem for general Carleson Jordan curves. In case the curve Γ is spiralic (i.e. $\delta_t^- = \delta_t^+$ for all $t \in \Gamma$), Theorem 10.17 covers all reflexive Orlicz spaces.

We note that "halos" do not appear in Theorem 10.17. The leaves in (10.21) are pure logarithmic leaves due to the assumption (10.20). In the notation of Section 7.5, we have

$$\mathcal{L}\big(a(t-0), a(t+0); \infty; \alpha_t^M, \beta_t^M\big)$$
$$= \mathcal{L}^0\big(a(t-0), a(t+0); \delta_t^-, \delta_t^+; \alpha_M, \alpha_M, \beta_M, \beta_M\big).$$

Theorem 10.17 and (10.18) also indicate that calling $\alpha_t(x), \beta_t(x)$ the indicator functions of Γ, p, w (or of the Lebesgue space $L^p(\Gamma, w)$) is perhaps not very fortunate. It would be more convenient to refer to $\alpha_t(x) + 1/p$ and $\beta_t(x) + 1/p$ as the indicator functions of Γ, p, w. We then could write

$$\mathcal{L}(z_1, z_2; p, \alpha_t, \beta_t) = \mathcal{L}(z_1, z_2; \infty; \alpha_t + 1/p, \beta_t + 1/p),$$

and instead of the three "parameters" p, α_t, β_t we had only the two "parameters" $\alpha_t + 1/p, \beta_t + 1/p$.

Finally, under the hypotheses of Theorem 10.17, one can without difficulty establish Orlicz space analogues of Theorems 8.20, 10.7, Corollary 8.26, and Theorem 10.6 (for $w = 1$ and for Carleson Jordan curves Γ). Precise results are formulated in [110], [111].

10.6 Mellin techniques

This section is based on the papers [19], [20], [21] by V.S. Rabinovich and the authors and exhibits an entirely different approach to singular integral operators on composed Carleson curves.

Let $\mathbf{R}_+ := \{x \in \mathbf{R} : x > 0\}$ and abbreviate $S_{\mathbf{R}_+}$ to S_+. As far as we know, it was Alexander Dynin who first made the following simple but fundamental observation: on writing

$$(S_+ f)(x) := \frac{1}{\pi i} \int\limits_{\mathbf{R}_+} \frac{f(y)}{y - x}\, dy = \frac{1}{\pi i} \int\limits_{\mathbf{R}_+} \frac{f(y)}{1 - x/y}\frac{dy}{y} \quad (x \in \mathbf{R}_+),$$

we see that S_+ is a convolution operator on the multiplicative group $\mathbf{R}_+$ with its invariant measure $d\mu(y) := dy/y$; thus, to find the spectrum of S_+ on $L^2(\mathbf{R}_+, d\mu)$ we have only to evaluate Fk, where F is the Fourier transform associated with $L^2(\mathbf{R}_+, d\mu)$ and k is the convolution kernel of S_+, i.e. $k(x) = 1/(\pi i(1 - x))$.

Formally computing, one gets $(Fk)(\xi) = \coth(-\pi\xi)$ and then arrives at the conclusion that the spectrum of S_+ on $L^2(\mathbf{R}_+, d\mu)$ is $(-\infty, -1] \cup [1, \infty)$. This is an unbounded set. The point is that, in the notation of Section 4.1, $L^2(\mathbf{R}_+, d\mu)$ is $L^2(\mathbf{R}_+, w)$ with $w(y) = y^{-1/2}$ and that w does not belong to $A_2(\mathbf{R}_+)$. Thus, S_+ is unbounded on $L^2(\mathbf{R}_+, d\mu)$. However, we will see that by means of a simple trick (see (10.26) below) the above argument gives the spectrum of S_+ on $L^2(\mathbf{R}_+, w)$ where w is any weight of the form $w(y) = y^\lambda$ with $\lambda \in (-1/2, 1/2)$.

The half-line $\mathbf{R}_+$ is a locally compact abelian group with the group operation $x * y = xy$ and the invariant measure $d\mu(y) := dy/y$. A convolution operator K on $L^2(\mathbf{R}_+, d\mu)$ can formally be written as

$$(Kf)(x) = \int\limits_{\mathbf{R}_+} k(x * y^{-1}) f(y)\, d\mu(y) = \int\limits_{\mathbf{R}_+} k\left(\frac{x}{y}\right) f(y)\frac{dy}{y} \quad (x \in \mathbf{R}_+). \qquad (10.23)$$

All characters $\chi : \mathbf{R}_+ \to \mathbf{T}$ are given by $\chi_\xi(x) = x^{i\xi}$ with $\xi \in \mathbf{R}$, and hence the dual group $\mathbf{R}_+^*$ may be identified with the additive group $\mathbf{R}$. The invariant measure on $\mathbf{R}$ is $d\nu(\xi) := d\xi$. The Fourier transform $F : L^2(\mathbf{R}_+, d\mu) \to L^2(\mathbf{R}, d\nu)$ is formally given by

$$(Fk)(\xi) = \int\limits_{\mathbf{R}_+} k(x)\chi_\xi(x)\, d\mu(x) = \int\limits_{\mathbf{R}_+} f(x) x^{i\xi}\frac{dx}{x} \quad (\xi \in \mathbf{R}). \qquad (10.24)$$

This transform is known under the name *Mellin transform*, and therefore the methods employed in this section are usually called Mellin techniques. Following the general practice, we henceforth denote the transform (10.24) by M.

The Mellin transform is an isomorphism of $L^2(\mathbf{R}_+, d\mu)$ onto $L^2(\mathbf{R}, d\nu)$. An operator $K \in \mathcal{B}(L^2(\mathbf{R}_+, d\mu))$ is said to be a *Mellin convolution* if it can be represented in the form $K = M^{-1} a M$ with some function $a \in L^\infty(\mathbf{R})$. In that case a

is referred to as the *Mellin symbol* of K and one abbreviates $M^{-1}aM$ to $M^0(a)$. For example, if K is given by (10.23) with $k \in L^1(\mathbf{R}_+, d\mu)$, then $K = M^0(a)$ with $a = Mk$.

We let $L^2_n(\mathbf{R}_+, d\mu)$ stand for the direct sum of n copies of $L^2(\mathbf{R}_+, d\mu)$ and we think of $L^2_n(\mathbf{R}_+, d\mu)$ as a space of column vectors. By $L^\infty_{n \times n}(\mathbf{R})$ we denote the $n \times n$ matrix functions with entries in $L^\infty(\mathbf{R})$. For $a \in L^\infty_{n \times n}(\mathbf{R})$, the operator $M^0(a)$ is defined on $L^2_n(\mathbf{R}_+, d\mu)$ in the natural manner.

Theorem 10.18. *Let $a \in L^\infty_{n \times n}(\mathbf{R})$. Then $M^0(a)$ is invertible on $L^2_n(\mathbf{R}_+, d\mu)$ if and only if a is invertible in $L^\infty_{n \times n}(\mathbf{R})$.*

This is a well-known result whose proof is in every text on harmonic analysis.

Passage from $L^2(\mathbf{R}_+, d\mu)$ to $L^p(\mathbf{R}_+, d\mu)$ leads to a multiplier problem. So let $p \in (1, \infty)$. The functions $a \in L^\infty(\mathbf{R})$ for which there exists a constant $C \in (0, \infty)$ such that

$$\|M^0(a)f\|_{L^p(\mathbf{R}_+, d\mu)} \leq C\|f\|_{L^p(\mathbf{R}_+, d\mu)} \text{ for all } f \in L^p(\mathbf{R}_+, d\mu) \cap L^2(\mathbf{R}_+, d\mu)$$

are called *Mellin multipliers* of $L^p(\mathbf{R}_+, d\mu)$. The set of all Mellin multipliers of $L^p(\mathbf{R}_+, d\mu)$ is denoted by M_p. If $a \in M_p$, then $M^0(a)$ extends from $L^p(\mathbf{R}_+, d\mu) \cap L^2(\mathbf{R}_+, d\mu)$ to a bounded operator on $L^p(\mathbf{R}_+, d\mu)$, which is also denoted by $M^0(a)$. The set M_p is a Banach algebra with the norm

$$\|a\|_{M_p} := \|M^0(a)\|_{\mathcal{B}(L^p(\mathbf{R}_+, d\mu))}.$$

One can show that $a \in M_p$ for all $p \in (1, \infty)$ whenever a $(\in L^\infty(\mathbf{R}))$ has bounded total variation $V_1(a)$ and that

$$\|a\|_{M_p} \leq C_p\big(\|a\|_\infty + V_1(a)\big) \tag{10.25}$$

with some $C_p < \infty$ in this case. The estimate (10.25) is known as *Stechkin's inequality* (see, e.g., [57, Theorem 2.11]).

Let $\dot{\mathbf{R}} = \mathbf{R} \cup \{\infty\}$ be the usual compactification of $\mathbf{R}$ by one point at infinity. In what follows, we will denote by $PC(\dot{\mathbf{R}})$ the functions $a \in L^\infty(\mathbf{R})$ which have finite limits $a(\xi \pm 0)$ at every $\xi \in \dot{\mathbf{R}}$; by definition $a(\infty - 0) = a(+\infty)$, $a(\infty + 0) = a(-\infty)$, where $a(-\infty)$ and $a(+\infty)$ have the usual meaning. The closure in M_p of all functions in $PC(\dot{\mathbf{R}})$ with finite total variation and with at most finitely many jumps is denoted by $PC_p(\dot{\mathbf{R}})$. One can show that $PC_2(\dot{\mathbf{R}}) = PC(\dot{\mathbf{R}})$ and that $PC_p(\dot{\mathbf{R}})$ is continuously embedded in $PC(\dot{\mathbf{R}})$ for every $p \in (1, \infty)$. Notice that $PC_p(\dot{\mathbf{R}})$ is a proper subset of $PC(\dot{\mathbf{R}}) \cap M_p$ if $p \neq 2$.

The following result saves Theorem 10.18 for multipliers in $PC_p(\dot{\mathbf{R}})$. The notations $L^p_n(\mathbf{R}_+, d\mu)$, $PC^{n \times n}_p(\dot{\mathbf{R}})$, etc. have their usual meaning.

Theorem 10.19. *Let $a \in PC_p^{n \times n}(\dot{\mathbf{R}})$. Then $M^0(a)$ is invertible on the space $L_{n \times n}^p(\mathbf{R}_+, d\mu)$ if and only if a is invertible in $L_{n \times n}^\infty(\mathbf{R})$.*

For a proof see, e.g., [57, Theorem 2.18].

We now return to the Cauchy singular integral operator S_+ on $L^p(\mathbf{R}_+, w_\lambda)$ where $1 < p < \infty$, $w_\lambda(y) = y^\lambda$, and $-1/p < \lambda < 1/q$. Note that $w_\lambda \in A_p(\mathbf{R}_+)$ if and only if $-1/p < \lambda < 1/q$. The map

$$C : L^p(\mathbf{R}_+, w_\lambda) \to L^p(\mathbf{R}_+, d\mu), \quad (Ch)(x) := x^{1/p+\lambda} h(x)$$

is an isometric isomorphism and we have

$$(CS_+C^{-1}f)(x) = \lim_{\varepsilon \to 0} \frac{1}{\pi i} \int_{|\log(x/y)| > \varepsilon} \frac{(x/y)^{1/p+\lambda}}{1 - x/y} f(y) \frac{dy}{y} \quad (x \in \mathbf{R}_+). \quad (10.26)$$

Thus, the spectrum of S_+ on $L^p(\mathbf{R}_+, w_\lambda)$ coincides with the spectrum of the convolution operator with the kernel

$$k(x) = \frac{1}{\pi i} \frac{x^{1/p+\lambda}}{1 - x} \quad (10.27)$$

on $L^p(\mathbf{R}_+, d\mu)$. A formal calculation gives

$$(Mk)(\xi) = \frac{1}{\pi i} \int_{\mathbf{R}_+} \frac{x^{1/p+\lambda}}{1 - x} x^{i\xi} \frac{dx}{x} = \coth \pi \left(i \left(\frac{1}{p} + \lambda \right) - \xi \right). \quad (10.28)$$

Here is a precise result.

Theorem 10.20. *Given $\eta \in (0, 1)$ and a complex number ϱ such that $\operatorname{Re} \varrho \geq 1$, define*

$$a_{\eta, \varrho}(\xi) := \coth \left(\pi \left(\frac{i\eta - \xi}{\varrho} \right) \right) \quad (\xi \in \mathbf{R}). \quad (10.29)$$

Then $a_{\eta, \varrho} \in PC_p(\dot{\mathbf{R}}) \cap C(\mathbf{R})$ for all $p \in (1, \infty)$ and $M^0(a_{\eta, \varrho})$ acts on the space $L^p(\mathbf{R}_+, d\mu)$ by the rule

$$\left(M^0(a_{\eta, \varrho}) f \right)(x) = \lim_{\varepsilon \to 0} \frac{\varrho}{\pi i} \int_{|\log(x/y)| > \varepsilon} \frac{(x/y)^\eta}{1 - (x/y)^\varrho} f(y) \frac{dy}{y} \quad (x \in \mathbf{R}_+).$$

For $\varrho = 1$, a proof is in [57, pp. 24–25, 51–52] and [174, pp. 12-13]. For $\operatorname{Re} \varrho \geq 1$, see [19]. Note that $a_{\eta, \varrho}$ actually has finite total variation.

Letting $\eta = 1/p + \lambda$ and $\varrho = 1$ in the preceding theorem, we infer from (10.26) that S_+ on $L^p(\mathbf{R}_+, w_\lambda)$ is similar to $M^0(a_{1/p+\lambda, 1})$ on $L^p(\mathbf{R}_+, d\mu)$. Consequently, by Theorem 10.19, the spectrum of S_+ on $L^p(\mathbf{R}_+, w_\lambda)$ is $a_{1/p+\lambda,1}(\overline{\mathbf{R}})$, where $\overline{\mathbf{R}} = \mathbf{R} \cup \{\pm\infty\}$. We have

$$\coth\left(\pi\left(\frac{i\eta - \xi}{\varrho}\right)\right) = M_{-1,1}\left(\exp\left(2\pi\left(\frac{i\eta - \xi}{\varrho}\right)\right)\right) \tag{10.30}$$

with $M_{-1,1}(\zeta) := (\zeta + 1)/(\zeta - 1)$. Thus, $a_{1/p+\lambda,1}(\overline{\mathbf{R}})$ equals

$$\left\{M_{-1,1}\left(\exp\left(2\pi\left(i\left(\frac{1}{p}+\lambda\right)-\xi\right)\right)\right) : \xi \in \overline{\mathbf{R}}\right\}$$
$$= \left\{M_{-1,1}\left(e^{2\pi(x+iy)}\right) : y = 1/p + \lambda\right\},$$

and we know from Example 7.6 that this is the circular arc

$$\mathcal{L}(-1, 1; p; \lambda, \lambda) = \mathcal{A}\left(-1, 1; 1/p + \lambda\right). \tag{10.31}$$

At this point we have identified the spectrum of S_+ on the half-line $\mathbf{R}_+$ (with the weight $w_\lambda(y) = y^\lambda$).

What is the simplest curve beyond the half-line $\mathbf{R}_+$? We made essential use of the fact that $\mathbf{R}_+$ is a multiplicative group. So let us look for the curves in the plane which are subgroups of the multiplicative group $\mathbf{C} \setminus \{0\}$, i.e. let us look for the nontrivial Lie subgroups of $\mathbf{C} \setminus \{0\}$. The exponential $z \mapsto e^z$ takes additive subgroups of $\mathbf{C}$ to multiplicative subgroups of $\mathbf{C} \setminus \{0\}$. The linear subspaces of $\mathbf{C}$ are the lines

$$L_\delta := \{x + iy : y = -\delta x\} \quad (\delta \in \mathbf{R}) \text{ and } L_\infty := i\mathbf{R}.$$

This easily implies that the multiplicative Lie subgroups of $\mathbf{C} \setminus \{0\}$ are

$$g_\delta := \exp(L_\delta) = \{e^{(1-i\delta)x} : x \in \mathbf{R}\} = \{x^{1-i\delta} : x \in \mathbf{R}_+\} \quad (\delta \in \mathbf{R}) \tag{10.32}$$

and $g_\infty := \exp(i\mathbf{R}) = \mathbf{T}$. Since $g_0 = \mathbf{R}_+$, we can say that it is logarithmic spirals which are the simplest curves beyond $\mathbf{R}_+$. The unit circle $\mathbf{T} = g_\infty$ is regarded as the boundary case of a logarithmic spiral in this philosophy.

Let $\delta \in \mathbf{R}$ and consider the logarithmic spiral g_δ with the orientation from the origin to infinity. We know from Example 1.6 that g_δ is a Carleson curve and it can be verified straightforwardly that the weight w_λ defined by $w_\lambda(\tau) = |\tau|^\lambda$ ($\tau \in g_\delta$) belongs to $A_p(g_\delta)$ if and only if $-1/p < \lambda < 1/q$. Thus, let $\lambda \in (-1/p, 1/q)$. In order to study S_{g_δ} on $L^p(g_\delta, w_\lambda)$ we could repeat the above constructions with g_δ in place of $\mathbf{R}_+$, i.e. we could do elementary "harmonic analysis on logarithmic spirals". However, it is easier to make direct use of the isomorphy of g_δ and $\mathbf{R}_+$. The map

$$C_\delta : L^p(g_\delta, w_\lambda) \to L^p(\mathbf{R}_+, d\mu), \quad (C_\delta h)(x) = |1 - i\delta|^{1/p} x^{1/p+\lambda} h(x^{1-i\delta})$$

is an isometric isomorphism. Formally we get

$$(C_\delta S_{g_\delta} C_\delta^{-1} f)(x) = \frac{1 - i\delta}{\pi i} \int\limits_{\mathbf{R}_+} \frac{(x/y)^{1/p+\lambda}}{1 - (x/y)^{1-i\delta}} f(y) \frac{dy}{y} \quad (x \in \mathbf{R}_+),$$

and Theorem 10.20 shows that really $C_\delta S_{g_\delta} C_\delta^{-1} = M^0(a_{1/p+\lambda,1-i\delta})$. Thus, by Theorem 10.19 and by (10.30) the spectrum of S_{g_δ} on $L^p(g_\delta, w_\lambda)$ is

$$a_{1/p+\lambda,1-i\delta}(\overline{\mathbf{R}}) = \left\{ M_{-1,1}\left(\exp\left(2\pi \left(\frac{i(1/p+\lambda) - \xi}{1 - i\delta} \right) \right) \right) : \xi \in \overline{\mathbf{R}} \right\}$$
$$= \left\{ M_{-1,1}\left(e^{2\pi(x+iy)} \right) : y = 1/p + \lambda + \delta x \right\},$$

and from Example 7.8 we infer that this is the logarithmic double spiral

$$\mathcal{L}(-1, 1; p; \alpha, \beta) = \mathcal{S}(-1, 1; \delta; 1/p + \lambda) \tag{10.33}$$

where $\alpha(x) = \beta(x) = \lambda + \delta x$. In summary, we have proved the following result.

Theorem 10.21. *Let $\delta \in \mathbf{R}$, $p \in (1, \infty)$, $\lambda \in (-1/p, 1/q)$, denote by g_δ the logarithmic spiral (10.32) with the orientation from 0 to infinity, and let $w_\lambda(\tau) = |\tau|^\lambda$ for $\tau \in g_\delta$. Then the spectrum of S_{g_δ} on $L^p(g_\delta, w_\lambda)$ is a (possibly degenerated) logarithmic double spiral and coincides with the values assumed by the function*

$$\mathbf{C} \setminus i\mathbf{Z} \to \mathbf{C}, \quad z \mapsto \coth(\pi z) = M_{-1,1}(e^{2\pi z}) \tag{10.34}$$

on the line of slope δ through $i(1/p + \lambda)$ compactified by two points at infinity.

In view of this theorem, the function (10.34) may be called the *complex Mellin symbol* of the Cauchy singular integral operator on the logarithmic spirals g_δ ($\delta \in \mathbf{R}$).

Now pick $\delta \in \mathbf{R}$ and consider the *logarithmic star*

$$\gamma_\delta := \{0\} \cup \bigcup_{j=1}^N e^{i\theta_j} g_\delta$$

where $\theta_1, \ldots, \theta_N$ are real numbers satisfying $0 \le \theta_1 < \theta_2 < \ldots < \theta_N < 2\pi$. Provide each of the logarithmic spirals $e^{i\theta_j} g_\delta$ with an orientation, put $\varepsilon_j = +1$ if 0 is the starting point of the spiral, and set $\varepsilon_j = -1$ if 0 is the terminating point. Let further $p \in (1, \infty)$ and $\lambda \in (-1/p, 1/q)$. Define the weight w_λ on γ_δ by $w_\lambda(\tau) = |\tau|^\lambda$ ($\tau \in \gamma_\delta$).

The space $L^p(\gamma_\delta, w_\lambda)$ is isometrically isomorphic to the direct sum of the spaces $L^p(e^{i\theta_j} g_\delta, w_\lambda)$ and thus to $L^p_N(g_\delta, w_\lambda)$. The space $L^p_N(g_\delta, w_\lambda)$ in turn may be isometrically and isomorphically mapped onto $L^p_N(\mathbf{R}_+, d\mu)$ as above. Accordingly, the Cauchy singular integral operator S_{γ_δ} on $L^p(\gamma_\delta, w_\lambda)$ is similar to an operator matrix $(S_{jk})_{j,k=1}^{N}$ on $L_{N}^{p}(\mathbf{R}_{+}, d\mu)$. A direct computation shows that S_{jk} acts by the rule

$$(S_{jk}f)(x) = \frac{(1 - i\delta)\varepsilon_{k}}{\pi i} \int_{\mathbf{R}_{+}} \frac{(x/y)^{1/p+\lambda}}{1 - e^{i(\theta_{j}-\theta_{k})}(x/y)^{1-i\delta}} f(y)\, \frac{dy}{y},$$

the integral understood in a principal sense as in (10.26). Thus, S_{jk} is a multiplicative convolution on $L^{p}(\mathbf{R}_{+}, d\mu)$ whose kernel is (up to the ε_{k}) of the form

$$l_{\eta,\varrho,\beta}(x) = \frac{\varrho}{\pi i} \frac{x^{\eta}}{1 - e^{i\beta}x^{\varrho}}$$

with $\eta \in (0,1)$, $\operatorname{Re}\varrho \geq 1$, $\beta \in (-2\pi, 2\pi)$. For $z \in \mathbf{C} \setminus i\mathbf{Z}$, we define

$$b_{\beta}(z) := \begin{cases} e^{(-\beta+\pi)z}/\sinh(\pi z) & \text{if} \quad \beta \in (0, 2\pi), \\ \coth(\pi z) & \text{if} \quad \beta = 0, \\ e^{(-\beta-\pi)z}/\sinh(\pi z) & \text{if} \quad \beta \in (-2\pi, 0). \end{cases}$$

Note that

$$b_{0}(z) = \frac{1}{2}\left(\lim_{\beta\to 0+0} b_{\beta}(z) + \lim_{\beta\to 0-0} b_{\beta}(z) \right).$$

Finally, put

$$a_{\eta,\varrho,\beta}(z) := b_{\beta}\left(\frac{i\eta - z}{\varrho} \right).$$

It is clear that $a_{\eta,\varrho,0}$ is the function $a_{\eta,\varrho}$ given by (10.29) on the real axis. Here is a generalization of Theorem 10.20 to the case $\beta \neq 0$.

Theorem 10.22. *Let $\eta \in (0,1)$, $\beta \in (-2\pi, 2\pi)$, $\operatorname{Re}\varrho \geq 1$. Then the restriction of $a_{\eta,\varrho,\beta}$ to $\mathbf{R}$ belongs to $PC_{p}(\dot{\mathbf{R}}) \cap C(\mathbf{R})$ for all $p \in (1,\infty)$ and $M^{0}(a_{\eta,\varrho,\beta})$ acts on $L^{p}(\mathbf{R}_{+}, d\mu)$ by the formula*

$$\left(M^{0}(a_{\eta,\varrho,\beta})f\right)(x) = \lim_{\varepsilon\to 0} \frac{\varrho}{\pi i} \int_{|\log(x/y)|>\varepsilon} \frac{(x/y)^{\eta}}{1 - e^{i\beta}(x/y)^{\varrho}} f(y)\, \frac{dy}{y} \quad (x \in \mathbf{R}_{+}).$$

A proof based upon Theorem 10.20 is in [19]. Again notice that $a_{\eta,\varrho,\beta}$ has in fact finite total variation.

Combining Theorems 10.22 and 10.20 we arrive at the following result, which extends Theorem 10.21 to logarithmic stars.

Theorem 10.23. *Let γ_{δ} $(\delta \in \mathbf{R})$ and w_{λ} $(-1/p < \lambda < 1/q)$ be as above. Then the spectrum of $S_{\gamma_{\delta}}$ on $L^{p}(\gamma_{\delta}, w_{\lambda})$ is the union of the eigenvalues of the matrices*

$$\sigma(z) := \left(\varepsilon_{k}b_{\theta_{j}-\theta_{k}}(z)\right)_{j,k=1}^{N}$$

as z ranges over the line of slope δ through the point $i(1/p + \lambda)$ compactified by two points at infinity.

Denote by $\varepsilon := \varepsilon_1 + \ldots + \varepsilon_N$ the valency of γ_δ at the origin. Evaluating the determinant $\det(\sigma(z) - \lambda I)$ as in the proof of Theorem 9.28 one obtains that the spectrum of S_{γ_δ} on $L^p(\gamma_\delta, w_\lambda)$ is

$$\{-1, 1\} \text{ if } \varepsilon = 0,$$

$$\bigcup_{k=0}^{\varepsilon-1} \mathcal{S}\big(-1, 1; \delta; (1/p + \lambda + k)/\varepsilon\big) \text{ if } \varepsilon > 0,$$

$$\bigcup_{k=0}^{|\varepsilon|-1} \mathcal{S}\big(1, -1; \delta; (1/p + \lambda + k)/|\varepsilon|\big) \text{ if } \varepsilon < 0,$$

where $\mathcal{S}(z_1, z_2; \delta; \varphi)$ is as in Example 7.8. One can show that the essential spectrum of S_{γ_δ} coincides with the spectrum of S_{γ_δ}.

Let γ be a curve which is homeomorphic to some logarithmic star γ_δ, let $\psi : \gamma_\delta \to \gamma$ be a homeomorphism, and put $t := \psi(0)$. We say that Γ is a *deformed logarithmic star* at the point t if there exist open bounded neighborhoods $U \subset \mathbf{C}$ of t and $V \subset \mathbf{C}$ of 0 as well as a homeomorphism

$$\varphi : \gamma_\delta \cap V \to \gamma \cap U =: \{t\} \cup \bigcup_{j=1}^{N} (\gamma_j \cap U)$$

such that the following hold:

(i) for each j, the restriction $\varphi|e^{i\theta_j} g_\delta \cap V \to \gamma_j \cap U$ is differentiable and the derivative satisfies a Hölder condition, i.e.

$$\big|\varphi'(\tau_1) - \varphi'(\tau_2)\big| \leq C|\tau_1 - \tau_2|^\lambda \text{ for all } \tau_1, \tau_2 \in e^{i\theta_j} g_\delta \cap V$$

with some $C < \infty$ and some $\lambda > 0$;

(ii) for each j, the derivative $\varphi'(\tau)$ converges to 1 as τ approaches 0 along the arc $e^{i\theta_j} g_\delta \cap V$.

Compactifying a deformed logarithmic star γ by a single point at infinity and proceeding as in Section 8.2, one can introduce the local algebras $\mathcal{B}_t(\gamma)$ and $\mathcal{A}_t(\gamma)$ at the "center" t of γ (note that $\gamma \cap U$ is a Carleson star). Employing Roch and Silbermann's method [174, pp. 40-46] one can show that the local spectrum of S_γ in $\mathcal{B}_t(\gamma)$ and $\mathcal{A}_t(\gamma)$ coincides with the local spectrum of S_{γ_δ} in $\mathcal{B}_0(\gamma_\delta)$ and $\mathcal{A}_0(\gamma_\delta)$, respectively. The latter two local spectra in turn coincide with the spectrum of S_{γ_δ}.

Finally, a bounded curve Γ is called a *curve composed of locally deformed logarithmic stars* if for each point $t \in \Gamma$ there exists a deformed logarithmic star $\gamma_t = \varphi_t(\gamma_{\delta_t})$ and an open neighborhood $U \subset \mathbf{C}$ of t such that $\Gamma \cap U = \gamma_t \cap U$. The result of the preceding paragraph in conjunction with Theorem 8.2 and constructions as in Section 9.4 yield a symbol calculus for operators in $\mathrm{alg}\,(S_\Gamma, PC(\Gamma))$ on

$L^p(\Gamma, w)$ provided w is a power weight,

$$w(\tau) = \prod_{j=1}^{n} |\tau - t_j|^{\lambda_j} \quad \left(t_j \in \Gamma \text{ distinct}, \ \lambda_j \in (-1/p, 1/q)\right),$$

and Γ is composed of locally deformed logarithmic stars. Note that composed Lyapunov curves are nothing but curves composed by locally deformed straight stars (which means that $\delta_t = 0$ for all $t \in \Gamma$).

Curves which locally do not look like a deformed logarithmic star but like a "slowly varying" logarithmic star and weights which are locally "slowly varying" weights can also be tackled by the methods outlined above. However, in this case one has to replace Mellin convolutions by Mellin pseudodifferential operators. For details, we refer to [163], [164] ($\delta = 0$) and [19], [167] ($\delta \in \mathbf{R}$). We remark that the results of these papers obtained with the help of Mellin techniques do not yield heavy leaves (i.e. leaves with a nonempty set of interior points). Thus, some time ago it seemed that the Mellin approach cannot go beyond some barrier.

Only in our recent papers [20], [21] with V.S. Rabinovich we were able to move essentially ahead this barrier. Moreover, the approaches of [20], [21] led to completely new insights into the nature of leaves.

Let us call a subset Y of $\mathbf{C}$ a *preleaf* if there exist two continuous functions $\alpha : \mathbf{R} \to \mathbf{R}$ and $\beta : \mathbf{R} \to \mathbf{R}$ such that

$$Y = \tilde{Y}(\alpha, \beta) := \left\{ x + iy \in \mathbf{C} : \alpha(x) \le y \le \beta(x) \right\} \tag{10.35}$$

and

(i) $-\infty < \alpha(x) \le \beta(x) < +\infty$ for all $x \in \mathbf{R}$;
(ii) $0 < \alpha(0) \le \beta(0) < 1$;
(iii) α is concave and β is convex;
(iv) $\alpha(x)$ and $\beta(x)$ have asymptotes as $x \to \pm\infty$ and the (convex) regions $\{x+iy \in \mathbf{C} : y < \alpha(x)\}$ and $\{x+iy \in \mathbf{C} : y > \beta(x)\}$ are separated by these asymptotes.

A moment's thought reveals that a preleaf can always be represented as a union of straight lines

$$\Pi(\omega, \delta) := \left\{ x + iy \in \mathbf{C} : y = \omega + \delta x \right\}. \tag{10.36}$$

The following result of [20] characterizes those unions of straight lines of the form (10.36) which are really preleaves.

Theorem 10.24. *A subset Y of $\mathbf{C}$ is a preleaf if and only if there exists a nonempty compact convex subset $\mathcal{N}$ of $(0,1) \times \mathbf{R}$ such that*

$$Y = \bigcup_{(\omega, \delta) \in \mathcal{N}} \Pi(\omega, \delta). \tag{10.37}$$

Note that if $Y = \tilde{Y}(\alpha, \beta)$ is represented by (10.37) then the functions α and β in (10.35) are given by

$$\alpha(x) = \min_{(\omega,\delta)\in\mathcal{N}} (\omega + \delta x), \quad \beta(x) = \max_{(\omega,\delta)\in\mathcal{N}} (\omega + \delta x). \tag{10.38}$$

The set $Y(p, \alpha_t, \beta_t)$ introduced in Section 7.3 is a preleaf:

$$Y(p, \alpha_t, \beta_t) = \tilde{Y}(1/p + \alpha_t, 1/p + \beta_t).$$

We therefore arrive at the question of determining the set $\mathcal{N} = \mathcal{N}_t(\Gamma, p, w)$ in the representation (10.37) of $Y(p, \alpha_t, \beta_t)$. This set was identified in [20] for a large class of curves and weights. Before citing the result, we need one more definition.

Let Γ be an oriented simple arc with the starting point t and let w be a weight on Γ. We write $(\Gamma, w) \in A_p^0$ $(1 < p < \infty)$ if the following conditions are satisfied: the arc Γ can be given by

$$\Gamma = \left\{ \tau = t + re^{i\theta(r)} : 0 \le r \le s \right\}$$

and the weight w is of the form

$$w(t + re^{i\theta(r)}) = e^{v(r)} \quad (0 < r < s)$$

where θ and v are real-valued infinitely differentiable functions on $(0, s)$ such that

$$\sup_{r\in(0,s)} \left| (rD_r)^j \theta(r) \right| < \infty, \quad \sup_{r\in(0,s)} \left| (rD_r)^j v(r) \right| < \infty \text{ for all } j \ge 1,$$

$$\lim_{r\to 0} (rD_r)^2 \theta(r) = 0, \quad \lim_{r\to 0} (rD_r)^2 v(r) = 0,$$

$$-1/p < \liminf_{r\to 0} \left(rv'(r) \right) \le \limsup_{r\to 0} \left(rv'(r) \right) < 1/q.$$

Taking into account that Proposition 1.19 remains true under the hypothesis (1.55), one can easily verify that $w \in A_p(\Gamma)$ whenever $(\Gamma, w) \in A_p^0$.

If $(\Gamma, w) \in A_p^0$, a similarity like (10.26) transforms S_Γ on $L^p(\Gamma, w)$ into a pseudodifferential operator on $L^p(\mathbf{R}_+, d\mu)$ which is covered by the theory of [166]. What results is that a point $z \in \mathbf{C}$ belongs to the local spectrum $sp\pi_t(S_\Gamma)$ if and only if

$$\lim_{\varepsilon\to 0} \inf_{(r,\lambda)\in(0,\varepsilon)\times\mathbf{R}} \left| \coth\left(\pi \frac{\lambda + i(1/p + rv'(r))}{1 + ir\theta'(r)} \right) - z \right| = 0;$$

notice that $\coth(\pi\zeta) = M_{-1,1}(e^{2\pi\zeta})$. On the basis of this observation, the following result is proved in [20].

Theorem 10.25. *Let* $(\Gamma, w) \in A_p^0$. *Denote by* $\mathcal{P} = \mathcal{P}_t(\Gamma, p, w)$ *the set of the partial limits of the map*

$$(0, s) \to (0, 1) \times \mathbf{R}, \quad r \mapsto \left(1/p + rv'(r), r\theta'(r)\right) \tag{10.39}$$

as $r \to 0$ *and let* α_t, β_t *be the indicator functions of* Γ, p, w *at* t. *Then*

$$Y(p, \alpha_t, \beta_t) = \bigcup_{(\omega, \delta) \in \text{conv } \mathcal{P}} \Pi(\omega, \delta),$$

where conv $\mathcal{P}$ *is the convex hull of* $\mathcal{P}$.

This theorem gives $Y(p, \alpha_t, \beta_t)$ and thus the leaf $\mathcal{L}(-1, 1; p, \alpha_t, \beta_t)$ entirely in terms of the partial limits of the map (10.39) as $r \to 0$. In particular, the sole knowledge of $\mathcal{P}$ provides us with the indicator functions $\alpha_t(x), \beta_t(x)$, the spirality indices $\delta_t^\pm$, the indices of powerlikeness μ_t, ν_t, and their "splittings" $\mu_t^\pm, \nu_t^\pm$ as simple geometric characteristics of the set conv $\mathcal{P}$ (see Fig. 24):

$$\frac{1}{p} + \alpha_t(x) = \min_{(\omega, \delta) \in \text{conv } \mathcal{P}} (\omega + \delta x), \quad \frac{1}{p} + \beta_t(x) = \max_{(\omega, \delta) \in \text{conv } \mathcal{P}} (\omega + \delta x),$$

$$\delta_t^- = \min_{(\omega, \delta) \in \text{conv } \mathcal{P}} \delta, \qquad \delta_t^+ = \max_{(\omega, \delta) \in \text{conv } \mathcal{P}} \delta,$$

$$\frac{1}{p} + \mu_t = \min_{(\omega, \delta) \in \text{conv } \mathcal{P}} \omega, \qquad \frac{1}{p} + \nu_t = \max_{(\omega, \delta) \in \text{conv } \mathcal{P}} \omega,$$

$$\frac{1}{p} + \mu_t^\pm = \min_{(\omega, \delta_t^\pm) \in \text{conv } \mathcal{P}} \omega, \qquad \frac{1}{p} + \nu_t^\pm = \max_{(\omega, \delta_t^\pm) \in \text{conv } \mathcal{P}} \omega.$$

It should not be hidden that, despite its elegance, Theorem 10.25 is only applicable to curves and weights in A_p^0 and thus not (yet) to general curves and weights. On the other hand, the construction of Section 3.8 implies that if Γ is any Carleson arc with the starting point t and w is any weight in $A_p(\Gamma)$, then there exists a pair $(\Gamma_0, w_0) \in A_p^0$ such that $Y(p, \alpha_t, \beta_t) = Y(p, \alpha_t^0, \beta_t^0)$, where α_t^0, β_t^0 are the indicator functions of Γ_0, p, w_0 at t. Equivalently: every possible local spectrum sp$\pi_t(S_\Gamma)$ of S_Γ on $L^p(\Gamma, w)$ is attained at curves and weights in A_p^0.

The paper [20] also contains a proof of the following result.

Theorem 10.26. *Given a nonempty compact convex subset* $\mathcal{N}$ *of* $(0, 1) \times \mathbf{R}$, *there exists a pair* $(\Gamma, w) \in A_p^0$ *such that* $\mathcal{N} = $ conv $\mathcal{P}$, *where* $\mathcal{P}$ *is as in Theorem 10.25.*

The proof of this theorem given in [20] employs arguments different from those of Section 3.8 and is thus an alternative proof of Theorem 3.37 and Corollary 3.38.

Our paper [21] with Rabinovich is based on a combination of the so-called method of limit operators (see [165], [169] and the references given there) and of Mellin techniques and can be regarded as a third approach to the spectral theory of singular integral operators. We there consider S_Γ on $L^2(\Gamma, w)$ under the assumption that $(\Gamma, w) \in A_2^0$.

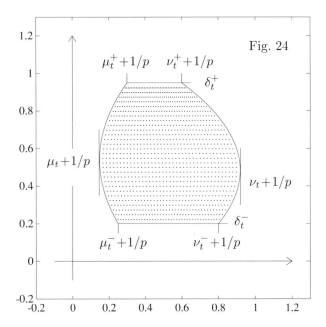

Fig. 24

Looking at the set conv $\mathcal{P}$ (dotted), we see the parameters $\delta_t^{\pm}$, μ_t, ν_t, $\mu_t^{\pm}$, $\nu_t^{\pm}$ at a glance. In the case of a power-like weight ($\mu_t = \nu_t$) the set conv $\mathcal{P}$ is a vertical line segment, while in the case of a spiralic curve ($\delta_t^{-} = \delta_t^{+}$) the set conv $\mathcal{P}$ degenerates to a horizontal line segment. The emergence of arbitrary convex sets convincingly demonstrates the "interference" of the curve and the weight.

A function $a \in C^{\infty}(0, s) \cap L^{\infty}(0, s)$ is said to be *slowly oscillating at the origin* if

$$\lim_{r \to 0} |(rD_r)^j a(r)| = 0 \text{ for all } j \geq 1.$$

Note that if $f \in C^{\infty}(\mathbf{R})$ and f as well as all its derivatives are bounded, then

$$a(r) = f(\log(-\log r)), \quad r \in (0, 1),$$

is slowly oscillating at the origin.

Let SO be the closure in $L^{\infty}(0, s)$ of the set of all $a \in C(0, s]$ which are slowly oscillating at the origin. Then SO is a C^*-subalgebra of $L^{\infty}(0, s)$, and we denote by $M_0(SO)$ the set (fiber) of all ξ in the maximal ideal space $M(SO)$ of SO such that $\xi(\varphi) = \varphi(0)$ whenever $\varphi \in C[0, s]$.

In [21], the problem of determining the local spectrum $\mathrm{sp}\pi_t(S_\Gamma)$ for (Γ, w) in the class A_2^0 is completely localized in the following sense: we associate with

$S_\Gamma \in \mathcal{B}(L^2(\Gamma, w))$ a family of limit operators

$$\{S_{\Gamma_\xi}\}_{\xi \in M_0(SO)}, \quad S_{\Gamma_\xi} \in \mathcal{B}(L^2(\Gamma_\xi, w_\xi))$$

such that Γ_ξ is a pure logarithmic spiral, w_ξ is a pure power weight, and

$$\mathrm{sp}\pi_t(S_\Gamma) = \bigcup_{\xi \in M_0(SO)} \mathrm{sp}\pi_t(S_{\Gamma_\xi}). \tag{10.40}$$

Since $\mathrm{sp}\pi_t(S_{\Gamma_\xi})$ is a logarithmic double spiral (Theorem 10.21), we arrive at the conclusion that the interpretation of a leaf as a union of logarithmic double spirals (= the interpretation of a preleaf as a union of straight lines) is equivalent to the equality (10.40).

The method of [21] is also applicable to singular integral operators with slowly oscillating coefficients. Here is a sample result.

Theorem 10.27. *Suppose $(\Gamma, w) \in A_2^0$ and let a, b be slowly oscillating at the origin. Put*

$$a_\Gamma(t + re^{i\theta(r)}) := a(r), \quad b_\Gamma(t + re^{i\theta(r)}) := b(r)$$

and consider $A := a_\Gamma I + b_\Gamma S_\Gamma \in \mathcal{B}(L^2(\Gamma, w))$. Denote by $\mathcal{P}_A$ the partial limits of the map

$$(0, s) \to \mathbf{C}^2 \times \mathbf{R} \times (-1/2, 1/2), \quad r \mapsto \Big(a(r), b(r), r\theta'(r), rv'(r)\Big)$$

as $r \to 0$. Then the local spectrum of A at the starting point t of Γ is given by

$$\mathrm{sp}\pi_t(A) = \bigcup_{(\alpha, \beta, \delta, \gamma) \in \mathcal{P}_A} \big(\alpha + \beta\, \mathcal{S}(-1, 1; \delta, 1/2 + \gamma)\big),$$

where $\mathcal{S}(-1, 1; \delta, 1/2 + \gamma)$ is as in Example 7.8. Equivalently,

$$\mathrm{sp}\pi_t(A) = \bigcup_{\xi \in M_0(SO)} \big(a(\xi) + b(\xi)\, \mathcal{S}(-1, 1; \xi\theta'(\xi), 1/2 + \xi v'(\xi))\big),$$

where $a(\xi), b(\xi), \xi\theta'(\xi), \xi v'(\xi)$ stand for the values of the functional $\xi \in M_0(SO)$ at the slowly oscillating functions $a(r), b(r), r\theta'(r), rv'(r)$, respectively.

Actually, the paper [21] contains a symbol calculus for operators in the algebra $\mathrm{alg}\,(S_\Gamma, SO(\Gamma))$ provided $(\Gamma, w) \in A_2^0$. Thus, [21] is a step towards one of the great challenges of the spectral theory of singular integral operators: to a theory which unifies the three "forces" determining the spectra, namely the oscillation of the Carleson curve, the oscillation of the Muckenhoupt weight, and the oscillation of the coefficients. Theorem 10.27 demonstrates that the shape of the spectra depends very sensitively on the interference of these three kinds of oscillation.

To conclude this section we wish to remark that the name "Mellin techniques", though in general use, is somewhat misleading. The map

$$E : L^p(\mathbf{R}_+, d\mu) \to L^p(\mathbf{R}) := L^p(\mathbf{R}, d\nu), \quad (Eh)(x) = h(e^x)$$

is an isometric isomorphism. Taking into account (10.26) we obtain that

$$(ECS_+C^{-1}E^{-1}f)(x) = \lim_{\varepsilon \to 0} \frac{1}{\pi i} \int_{|x-y|>\varepsilon} \frac{e^{(x-y)(1/p+\lambda)}}{1-e^{x-y}} f(y)\, dy \quad (x \in \mathbf{R}).$$

Hence, the Cauchy singular integral operator S_+ on $L^p(\mathbf{R}_+, w_\lambda)$ is not only similar to the multiplicative convolution with the kernel (10.27) but is also similar to the additive convolution with the kernel

$$\tilde{k}(x) = \frac{1}{\pi i} \frac{e^{x(1/p+\lambda)}}{1-e^x}.$$

Therefore the spectrum of S_+ on $L^p(\mathbf{R}_+, w_\lambda)$ may be determined by taking the usual Fourier transform of $\tilde{k}$:

$$(F\tilde{k})(\xi) := \int_{\mathbf{R}} \tilde{k}(x)e^{i\xi x} dx = \frac{1}{\pi i} \int_{\mathbf{R}} \frac{e^{x(1/p+\lambda)}}{1-e^x} e^{i\xi x}\, dx \quad (\xi \in \mathbf{R}).$$

The substitution $e^x = t$ gives

$$(F\tilde{k})(\xi) = \frac{1}{\pi i} \int_{\mathbf{R}_+} \frac{t^{1/p+\lambda}}{1-t} t^{i\xi} \frac{dt}{t},$$

and hence, we arrive at the same integral as in (10.28). In other terms: $F\tilde{k} = Mk$, or more general, $M = FE$. Thus, everything we are doing with the Mellin transform on $L^p(\mathbf{R}_+, d\mu)$ can be translated into an equivalent action with the usual Fourier transform. The concrete integrals one has to evaluate nevertheless remain the same. In other words, there is no result obtained by means of "Mellin techniques" which cannot be gotten with the help of "Fourier techniques", and vice versa.

10.7 Wiener-Hopf integral operators

In this section we cite the main result of the paper [26] by I. Spitkovsky and one of the authors, which describes the essential spectrum of Wiener-Hopf integral operators with piecewise continuous symbols on $L^p(\mathbf{R}_+, w)$ for $p \in (1, \infty)$ and $w \in A_p(\mathbf{R}_+)$, and which generalizes the results of Duduchava [55] and Schneider [185] to the most general setting.

As in Section 10.6, we first have to deal with a multiplier problem. Let $w \in A_p(\mathbf{R}_+)$ and extend w symmetrically to a weight in $A_p(\mathbf{R})$ (see Section 2.4).

Chapter 10. Further results

Thus, the extended weight is an even function. We denote it again by w. Let $F : L^2(\mathbf{R}) \to L^2(\mathbf{R})$ be the *Fourier transform*,

$$(Ff)(x) := \int_{\mathbf{R}} e^{ix\xi} f(x)\, dx \quad (x \in \mathbf{R}) \text{ for } f \in L^2(\mathbf{R}),$$

and denote by F^{-1} the inverse Fourier transform. A function $a \in L^\infty(\mathbf{R})$ is called a *Fourier multiplier* on $L^p(\mathbf{R}, w)$ if the map $f \mapsto F^{-1}aF$ extends from $L^2(\mathbf{R}) \cap L^p(\mathbf{R}, w)$ to a bounded operator on all of $L^p(\mathbf{R}, w)$. The latter operator is then usually denoted by $W^0(a)$. We let $M_{p,w}$ stand for the set of all Fourier multipliers on $L^p(\mathbf{R}, w)$. One can easily show (see, e.g., [174, Proposition 12.5] or the equality $F = ME^{-1}$ of the last paragraph of Section 10.6) that if w is identically 1, then $M_{p,w}$ coincides with the set M_p of all Mellin multipliers introduced in Section 10.6. The set $M_{p,w}$ is a Banach algebra with the norm

$$\|a\|_{M_{p,w}} := \|W^0(a)\|_{\mathcal{B}(L^p(\mathbf{R}, w))}.$$

Bounded functions with finite total variation belong to $M_{p,w}$ and *Stechkin's inequality* holds:

$$\|a\|_{M_{p,w}} \leq C_{p,w}\big(\|a\|_\infty + V_1(a)\big) \text{ with some } C_{p,w} < \infty.$$

Let $PC(\dot{\mathbf{R}})$ be as in Section 10.6. We denote by $PC_{p,w}(\dot{\mathbf{R}})$ the closure in $M_{p,w}$ of the collection of all functions in $PC(\dot{\mathbf{R}})$ having finite total variation and at most finitely many jumps. One can show (see [174, Proposition 12.2]) that $PC_{p,w}(\dot{\mathbf{R}})$ is continuously embedded in $PC(\dot{\mathbf{R}})$.

The analogues of Theorems 10.18 and 10.19 are valid for the operators $W^0(a)$: if $a \in L^\infty_{n \times n}(\mathbf{R})$, then $W^0(a)$ is invertible on $L^2_n(\mathbf{R})$ if and only if a invertible in $L^\infty_{n \times n}(\mathbf{R})$; if $a \in PC^{n \times n}_{p,w}(\dot{\mathbf{R}})$, then $W^0(a)$ is invertible on $L^p_n(\mathbf{R}, w)$ if and only if a is invertible in $L^\infty_{n \times n}(\mathbf{R})$.

The *Wiener-Hopf integral operator* $W(a)$ generated by a function $a \in M_{p,w}$, its so-called *symbol*, is the compression of $W^0(a)$ to the positive half-line $\mathbf{R}_+ = (0, \infty)$. Thus, $W(a)$ is the bounded operator on $L^p(\mathbf{R}_+, w)$ ($= L^p(\mathbf{R}_+, w|\mathbf{R}_+)$) acting by the rule $f \mapsto (W^0(a)\tilde{f})|\mathbf{R}_+$ where $\tilde{f}$ is the extension of f by zero to all of $\mathbf{R}$. Let χ_+ be the characteristic function of $\mathbf{R}_+$. The space $L^p(\mathbf{R}_+, w)$ may be identified with $\chi_+ L^p(\mathbf{R}, w)$ and consequently, we may also think of $W(a)$ as the operator $\chi_+ W^0(a)|\operatorname{Im}\chi_+ I$.

If a is of the form $a = c + Fk$ with $c \in \mathbf{C}$ and $k \in L^1(\mathbf{R})$, then $a \in M_{p,w}$ for all $p \in (1, \infty)$ and all $w \in A_p(\mathbf{R})$. In this case $W(a)$ can be written as

$$(W(a)f)(x) = c\, f(x) + \int_0^\infty k(x - y) f(y)\, dy \quad (x > 0).$$

The Cauchy singular integral operator S_+ on $\mathbf{R}_+$ is the archetypal example of a Wiener-Hopf integral operator with a piecewise continuous symbol: we have $S_+ = W(\sigma)$ where $\sigma(\xi) := -\operatorname{sign}\xi$. More general, if

$$a(\xi) = -\sum_{j=1}^{m} c_j \operatorname{sign}(\xi - \alpha_j) \quad (\xi \in \mathbf{R})$$

with $c_j \in \mathbf{C}$ and $\alpha_j \in \mathbf{R}$, then

$$\big(W(a)f\big)(x) = \sum_{j=1}^{m} \frac{c_j}{\pi i} \int_0^\infty \frac{e^{i\alpha_j(y-x)}}{y-x}\, dy \quad (x > 0),$$

the integrals understood in the Cauchy principal value sense.

A *power weight* on $\mathbf{R}_+$ is a weight ϱ of the form

$$\varrho(x) = |x - i|^{\lambda_\infty} \prod_{j=1}^{n} \left| \frac{x - \xi_j}{x - i} \right|^{\lambda_j} \tag{10.41}$$

where $\xi_1, \ldots, \xi_n \in [0, \infty)$ are distinct points and $\lambda_1, \ldots, \lambda_n, \lambda_\infty$ are real numbers. A straightforward computation reveals that $\varrho \in A_p(\mathbf{R}_+)$ if and only if $\lambda_1, \ldots, \lambda_n, \lambda_\infty$ all belong to $(-1/p, 1/q)$. Given a general weight $w \in A_p(\mathbf{R}_+)$, we call the sets

$$N_\xi(p, w) := \left\{ \gamma \in \mathbf{C} : \left| \left(\frac{x - \xi}{x - i} \right)^\gamma \right| w(x) \in A_p(\mathbf{R}_+) \right\} \quad (\xi \in [0, \infty)),$$

$$N_\infty(p, w) := \left\{ \gamma \in \mathbf{C} : \left| (x - i)^\gamma \right| w(x) \in A_p(\mathbf{R}_+) \right\}$$

the indicator sets of p, w at $\xi \in [0, \infty) \cup \{\infty\}$. Since $\arg((x - \xi)/(x - i))$ and $\arg(x - i)$ are bounded on $\mathbf{R}$, it is clear that

$$N_\xi(p, w) = I_\xi(p, w) + i\mathbf{R}$$

with

$$I_\xi(p, w) := \left\{ \gamma \in \mathbf{R} : \left| \frac{x - \xi}{x - i} \right|^\gamma w(x) \in A_p(\mathbf{R}_+) \right\} \quad (\xi \in [0, \infty)), \tag{10.42}$$

$$I_\infty(p, w) := \left\{ \gamma \in \mathbf{R} : |x - i|^\gamma w(x) \in A_p(\mathbf{R}_+) \right\}. \tag{10.43}$$

In [26] it was proved that for every $\xi \in [0, \infty) \cup \{\infty\}$ the set $I_\xi(p, w)$ is an open interval of length at most 1 which contains the origin. Thus,

$$I_\xi(p, w) = \left(-1/p - \mu_\xi, 1/q - \nu_\xi \right)$$

with $-1/p < \mu_\xi \le \nu_\xi < 1/q$. We refer to the numbers μ_ξ and ν_ξ as the *indices of powerlikeness* of the weight w at ξ. Note that if ϱ is the power weight (10.41), then

$$\mu_{\xi_j} = \nu_{\xi_j} = \lambda_j \ (j = 1, \ldots, n) \text{ and } \mu_\infty = \nu_\infty = \lambda_\infty,$$

and $\mu_\xi = \nu_\xi = 0$ for all remaining ξ.

Given $z_1, z_2 \in \mathbf{C}$ and a number $\varphi \in (0,1)$, we define the circular arc $\mathcal{A}(z_1, z_2; \varphi)$ as in Example 7.6. We think of $\mathcal{A}(z_1, z_2; \varphi)$ as being oriented from z_1 to z_2. If $0 < \delta \leq \gamma < 1$, we let $\mathcal{H}(z_1, z_2; \delta, \gamma)$ stand for the *horn* given by

$$\mathcal{H}(z_1, z_2; \delta, \gamma) := \bigcup_{\varphi \in [\delta, \gamma]} \mathcal{A}(z_1, z_2; \varphi),$$

which is in accordance with Example 7.7. For a function $a \in PC(\dot{\mathbf{R}})$, we denote by Λ_a the points on $\mathbf{R}$ at which it has a jump, i.e.

$$\Lambda_a := \{\xi \in \mathbf{R} : a(\xi - 0) \neq a(\xi + 0)\}.$$

As usual, $\mathcal{R}(a)$ is the essential range of a.

Theorem 10.28. *Let $p \in (1, \infty)$, $w \in A_p(\mathbf{R}_+)$, and $a \in PC_{p,w}(\dot{\mathbf{R}})$. The essential spectrum of the Wiener-Hopf operator $W(a)$ on $L^p(\mathbf{R}_+, w)$ is*

$$\mathrm{sp}_{\mathrm{ess}}\, W(a) \;=\; \mathcal{R}(a) \cup \bigcup_{\xi \in \Lambda_a} \mathcal{H}\big(a(\xi - 0), a(\xi + 0); 1/q - \nu_\infty, 1/q - \mu_\infty\big)$$

$$\cup \, \mathcal{H}\big(a(+\infty), a(-\infty); 1/p + \mu_0, 1/p + \nu_0\big), \tag{10.44}$$

where μ_0, ν_0 and μ_∞, ν_∞ are the indices of powerlikeness of w at 0 and ∞, respectively. If $W(a)$ is Fredholm, then its index on $L^p(\mathbf{R}_+, w)$ is minus the winding number of the closed, continuous, and oriented curve

$$\mathcal{R}(a) \cup \bigcup_{\xi \in \Lambda_a} \mathcal{A}\big(a(\xi - 0), a(\xi + 0); \varphi_\infty\big) \cup \mathcal{A}\big(a(+\infty), a(-\infty); \varphi_0\big) \tag{10.45}$$

where $\varphi_\infty \in [1/q - \nu_\infty, 1/p - \mu_\infty]$ and $\varphi_0 \in [1/p + \mu_0, 1/p + \nu_0]$ are arbitrary numbers.

This theorem is due to Duduchava [55] for spaces without weight and to Schneider [185] for spaces with power weights. In the form quoted here, the theorem was established in [26].

The proof of Theorem 10.28 given in [26] follows Roch and Silbermann's approach [174] and is based on localization techniques, by means of which the question can be reduced to studying local spectra of operators in the algebra $\mathrm{alg}\,(S_\mathbf{R}, PC(\mathbf{R}))$ on $L^p(\mathbf{R}, w)$, where w is the symmetric extension of w. The map

$$B : L^p(\mathbf{R}, w) \to L^p(\mathbf{T}, \eta), \quad (B\varphi)(t) := \frac{1}{t-1} \varphi\left(i\frac{t+1}{t-1}\right)$$

is an isomorphism in case

$$\eta(t) := w\left(i\frac{t+1}{t-1}\right)|t-1|^{1-2/p} \quad (t \in \mathbf{T}).$$

Since $BS_{\mathbf{R}}B^{-1} = -S_{\mathbf{T}}$, it follows that $w \in A_p(\mathbf{R})$ if and only if $\eta \in A_p(\mathbf{T})$. In the end one arrives at Toeplitz operators with $PC(\mathbf{T})$ symbols on $L_+^p(\mathbf{T}, \eta)$, the local spectra of which are at our disposal due to Theorem 7.4 (and were available in the case at hand from Spitkovsky's paper [200] since 1992). Since the spirality indices of $\mathbf{T}$ are zero, we understand in particular that in Theorem 10.28 the metamorphosis of line segments ends up with horns.

There are some interesting differences between the Toeplitz case (Theorems 7.4 and 7.14) and the Wiener-Hopf case (Theorem 10.28). First, in (10.44) the horns we have to fill in between $a(\xi - 0)$ and $a(\xi + 0)$ for $\xi \in \mathbf{R}$ do only depend on p and the behavior of the weight at infinity, while the horn between $a(+\infty)$ and $a(-\infty)$ is completely determined by p and the behavior of the weight at the origin. The indices of powerlikeness μ_ξ, ν_ξ for $\xi \in (0, \infty)$ do not play any role. Secondly, the indices of powerlikeness μ_∞, ν_∞ and μ_0, ν_0 enter (10.44) in a different way. It is easily seen that

$$\mathcal{H}(z_1, z_2; 1/q - \nu, 1/q - \mu) = \mathcal{H}(z_2, z_1; 1/p + \mu, 1/p + \nu). \tag{10.46}$$

Thus, we could rewrite (10.44) as

$$\mathrm{sp}_{\mathrm{ess}} W(a) \ = \ \mathcal{R}(a) \cup \bigcup_{\xi \in \Lambda_a} \mathcal{H}(a(\xi + 0), a(\xi - 0); 1/p + \mu_\infty, 1/p + \nu_\infty)$$

$$\cup \ \mathcal{H}(a(+\infty), a(-\infty); 1/p + \mu_0, 1/p + \nu_0). \tag{10.47}$$

Notice, however, that accordingly rewriting (10.45) changes the orientations of the arcs between $a(\xi - 0)$ and $a(\xi + 0)$ for $\xi \in \mathbf{R}$.

Application of (10.44) to the Cauchy singular integral $S_+ = W(\sigma)$ gives

$$\mathrm{sp}_{\mathrm{ess}} S_+ = \mathcal{H}(1, -1; 1/q - \nu_\infty, 1/q - \mu_\infty) \cup \mathcal{H}(-1, 1; 1/p + \mu_0, 1/p + \nu_0),$$

while (10.47) yields

$$\mathrm{sp}_{\mathrm{ess}} S_+ = \mathcal{H}(-1, 1; 1/p + \mu_\infty, 1/p + \nu_\infty) \cup \mathcal{H}(-1, 1; 1/p + \mu_0, 1/p + \nu_0). \tag{10.48}$$

By (10.46), both are the same. Thus, the essential spectrum of S_+ on $L^p(\mathbf{R}_+, w)$ is always the union of two horns between -1 and 1. If w is the power weight

$$w(x) = x^\lambda = \left| \frac{x}{x - i} \right|^\lambda |x - i|^\lambda,$$

then $\mu_\infty = \nu_\infty = \mu_0 = \nu_0 = \lambda$, and (10.48) degenerates to

$$\mathrm{sp}_{\mathrm{ess}} S_+ = \mathcal{A}(-1, 1; 1/p + \lambda),$$

which is in accordance with (10.31).

10.8   Zero-order pseudodifferential operators

Both the operators in the algebra $\mathrm{alg}\,(S_{\mathbf{R}}, PC(\dot{\mathbf{R}}))$ and Wiener-Hopf integral operators with piecewise continuous symbols are special pseudodifferential operators. The *zero-order pseudodifferential operator* Ψ_σ on the real line $\mathbf{R}$ generated by a function $\sigma(x, \xi)$ of two real variables x and ξ is formally defined by

$$(\Psi_\sigma f)(x) = \frac{1}{2\pi} \int_{\mathbf{R}} e^{-i\xi x} \sigma(x, \xi) \left(\int_{\mathbf{R}} e^{i\xi y} f(y)\, dy \right) d\xi \quad (x \in \mathbf{R}).$$

This is frequently also written in the form $\Psi_\sigma = F^{-1}_{\xi \to x} \sigma(x, \xi) F_{x \to \xi}$.

Let $p \in (1, \infty)$ and $w \in A_p(\mathbf{R})$. Suppose σ is a finite sum

$$\sigma(x, \xi) = \sum_{j=1}^{n} a_j(x)\, b_j(\xi) \quad (x \in \mathbf{R}, \xi \in \mathbf{R}) \tag{10.49}$$

where $a_j \in PC(\dot{\mathbf{R}})$ and $b_j \in PC_{p,w}(\dot{\mathbf{R}})$. In that case we can write

$$\Psi_\sigma = \sum_{j=1}^{n} a_j W^0(b_j), \tag{10.50}$$

and from the definition of $PC_{p,w}(\dot{\mathbf{R}})$ given in Section 10.7 we infer that (10.50) is a bounded operator in $L^p(\mathbf{R}, w)$. Let us denote by $\Psi(p, w)$ the smallest closed subalgebra of $\mathcal{B}(L^p(\Gamma, w))$ containing all operators of the form (10.50) with $a_j \in PC(\dot{\mathbf{R}})$ and $b_j \in PC_{p,w}(\dot{\mathbf{R}})$.

Here are two important examples. Let first $\sigma(x, \xi) = c(x) + d(x)\,(-\operatorname{sign}\xi)$. We then get
$$\Psi_\sigma = cW^0(1) + dW^0(\sigma) = cI + dS_{\mathbf{R}},$$
and hence, $\mathrm{alg}\,(S_{\mathbf{R}}, PC(\dot{\mathbf{R}}))$ is contained in $\Psi(p, w)$. Secondly, let χ_+ and χ_- denote the characteristic functions of $(0, \infty)$ and $(-\infty, 0)$, respectively. If $\sigma(x, \xi)$ is $a(\xi)$, then $\Psi_\sigma = W^0(a)$. Since $\chi_{\pm} I$ are obviously in $\Psi(p, w)$, it follows that $\Psi(p, w)$ contains the operators,

$$\chi_+ W^0(a)\chi_+ I + \chi_- I \quad \big(a \in PC_{p,w}(\dot{\mathbf{R}})\big),$$

which may be identified with Wiener-Hopf integral operators.

A classical result on pseudodifferential operators says that if the functions a_j and b_j are continuous on $\dot{\mathbf{R}}$ (and the b_j's are subject to a multiplier condition), then the operator Ψ_σ given by (10.50) is Fredholm on $L^p(\mathbf{R}_+, w)$ if and only if

$$\sigma(x, \infty) \neq 0, \quad \sigma(\infty, \xi) \neq 0, \quad \sigma(\infty, \infty) \neq 0$$

for all $x \in \mathbf{R}$ and all $\xi \in \mathbf{R}$. Equivalently, on putting

$$M := (\dot{\mathbf{R}} \times \dot{\mathbf{R}}) \setminus (\mathbf{R} \times \mathbf{R}) = (\mathbf{R} \times \{\infty\}) \cup (\{\infty\} \times \mathbf{R}) \cup \{(\infty, \infty)\},$$

we see that Ψ_σ is Fredholm if and only if $\sigma(x, \xi) \neq 0$ for all $(x, \xi) \in M$.

In the case where a_j and b_j are allowed to be piecewise continuous or, more general, for operators in $\Psi(p, w)$, a symbol calculus for deciding whether the operator is Fredholm was developed by Cordes [41], Duduchava [55], [56], [57], [58], Schneider [185], and by Roch and Silbermann [174] provided w is a power weight. General weights w in $A_p(\mathbf{R})$ were considered in the paper [27] by Spitkovsky and one of the authors. The result is as follows.

Replacing in (10.42), (10.43) the class $A_p(\mathbf{R}_+)$ by $A_p(\mathbf{R})$, we obtain the definition of the *indices of powerlikeness* μ_ξ, ν_ξ of the weight $w \in A_p(\mathbf{R})$ for every point $\xi \in \dot{\mathbf{R}}$. We now associate a certain horn $\mathcal{H}_m$ with each point $m = (x, \xi) \in M$: for $m = (x, \infty)$ with $x \in \mathbf{R}$, we let

$$\mathcal{H}_m = \mathcal{H}_{(x,\infty)} := \mathcal{H}(0, 1; 1/q - \nu_x, 1/q - \mu_x),$$

if $m = (\infty, \xi)$ with $\xi \in \mathbf{R}$, we put

$$\mathcal{H}_m = \mathcal{H}_{(\infty,\xi)} := \mathcal{H}(0, 1; 1/p + \mu_\infty, 1/q + \nu_\infty),$$

and if $m = (\infty, \infty)$, we set

$$\mathcal{H}_m = \mathcal{H}_{(\infty,\infty)} := \{0, 1\};$$

thus, $\mathcal{H}_{(\infty,\infty)}$ is in fact not a horn but the doubleton $\{0, 1\}$. The *horn bundle* of $\Psi(p, w)$ is defined as

$$\mathcal{M} := \mathcal{M}_{p,w} := \bigcup_{m \in M} (\{m\} \times \mathcal{H}_m).$$

Finally, for $a \in PC(\dot{\mathbf{R}})$, $b \in PC_{p,w}(\dot{\mathbf{R}})$, $\mu \in \mathcal{M}$, we define 2×2 matrices $\mathrm{Sym}_\mu aI$ and $\mathrm{Sym}_\mu W^0(b)$. There are three types of points $\mu \in \mathcal{M}$:

$$(x, \infty, t) \text{ with } x \in \mathbf{R}, \quad (0, \xi, t) \text{ with } \xi \in \mathbf{R}, \quad (\infty, \infty, j) \text{ with } j \in \{0, 1\}.$$

Let

$$E(\alpha, \beta, t) := \begin{pmatrix} \alpha t + \beta(1 - t) & (\beta - \alpha)t(1 - t) \\ \beta - \alpha & \beta t + \alpha(1 - t) \end{pmatrix}$$

and put

$$
\begin{aligned}
\mathrm{Sym}_{(x,\infty,t)} aI &= \mathrm{diag}\,(a(x + 0),\, a(x - 0)), \\
\mathrm{Sym}_{(0,\xi,t)} aI &= \mathrm{diag}\,(a(+\infty),\, a(-\infty)), \\
\mathrm{Sym}_{(\infty,\infty,0)} aI &= \mathrm{diag}\,(a(+\infty),\, a(-\infty)), \\
\mathrm{Sym}_{(\infty,\infty,1)} aI &= \mathrm{diag}\,(a(-\infty),\, a(+\infty)), \\
\mathrm{Sym}_{(x,\infty,t)} W^0(b) &= E\big(b(-\infty),\, b(+\infty), t\big), \\
\mathrm{Sym}_{(\infty,\xi,t)} W^0(b) &= E\big(b(\xi - 0),\, b(\xi + 0), t\big), \\
\mathrm{Sym}_{(\infty,\infty,0)} W^0(b) &= \mathrm{diag}\,(b(-\infty),\, b(+\infty)), \\
\mathrm{Sym}_{(\infty,\infty,1)} W^0(b) &= \mathrm{diag}\,(b(+\infty),\, b(-\infty)).
\end{aligned}
$$

Theorem 10.29. *Let $p \in (1, \infty)$ and $w \in A_p(\mathbf{R})$.*

(a) *For each $\mu \in \mathcal{M}$ the map*

$$\mathrm{Sym}_\mu : \big\{aI : a \in PC(\dot{\mathbf{R}})\big\} \cup \big\{W^0(b) : b \in PC_{p,w}(\dot{\mathbf{R}})\big\} \to \mathbf{C}^{2\times 2}$$

extends to a Banach algebra homomorphism $\mathrm{Sym}_\mu : \Psi(p, w) \to \mathbf{C}^{2\times 2}$.

(b) *An operator $A \in \Psi(p, w)$ is Fredholm on $L^p(\mathbf{R}, w)$ if and only if $\mathrm{Sym}_\mu A$ is invertible for every $\mu \in \mathcal{M}$.*

(c) *If $A \in \Psi(p, w)$ is Fredholm, then A has a regularizer in $\Psi(p, w)$.*

A proof of this theorem is in [27].

10.9 Conformal welding and Haseman's problem

This section is based on the papers [2], [3], [4] by A.V. Aizenshtat, G.S. Litvinchuk, and one of the authors. Its purpose is to show that singular integral operators on spiralic Carleson curves arise very naturally when studying singular integral operators with a shift on Lyapunov arcs.

The problem of *conformal welding* consists in the following. We are given an oriented simple arc η and a homeomorphism $\alpha : \eta \to \eta$ which preserves the orientation of the arc η. What we are looking for is another simple arc Γ and a conformal (i.e. analytic and bijective) map $\omega : \dot{\mathbf{C}} \setminus \eta \to \dot{\mathbf{C}} \setminus \Gamma$ such that

$$\omega^+\big(\alpha(t)\big) = \omega^-(t) \quad \text{for all } t \in \eta;$$

here $\dot{\mathbf{C}} := \mathbf{C} \cup \{\infty\}$, and $\omega^+(\xi)$ and $\omega^-(\xi)$ are the limits of $\omega(z)$ as $z \in \mathbf{C} \setminus \eta$ approaches $\xi \in \eta$ from the left and from the right, respectively. Note that the limits $\omega^\pm(\xi)$ exist for all $\xi \in \eta$; by definition,

$$\omega^+(t_j) := \omega^-(t_j) := \lim_{z \to t_j, z \in \mathbf{C} \setminus \eta} \omega(z)$$

at the endpoints t_j of η.

Example 10.30. Let $\eta_0 = [0, \infty)$ be the nonnegative real line with the orientation from 0 to ∞ and let $\alpha_0 : [0, \infty) \to [0, \infty)$ be given by $\alpha_0(t) = \mu t$ with $\mu > 0$ and $\mu \neq 1$. Put

$$\delta := \frac{1}{2\pi} \log \mu, \quad \Gamma_0 := \{r^{1-i\delta} : r > 0\} \cup \{0\}.$$

Notice that Γ_0 is a logarithmic spiral and that the spirality indices of Γ_0 at the origin are just δ. Let $\log z$ stand for the branch of the logarithm which is analytic in $\mathbf{C} \setminus [0, \infty)$ and assumes the value $-i\pi$ at $z = -1$. Define

$$\omega_0(z) := z^{1/(1+i\delta)} := \exp\left(\frac{1}{1+i\delta} \log z\right). \tag{10.51}$$

We claim that Γ_0 and w_0 solve the problem of conformal welding for η_0 and α_0, i.e. that w_0 maps $\mathbf{C} \setminus [0, \infty)$ conformally onto $\mathbf{C} \setminus \Gamma_0$ and that $w_0^+ (\mu t) = w_0^- (t)$ for all $t \in [0, \infty)$.

Put $g(z) := \log z$. Clearly, g maps $\mathbf{C} \setminus [0, \infty)$ conformally onto the horizontal stripe $\{\zeta \in \mathbf{C} : -2\pi < \operatorname{Im} \zeta < 0\}$ and we have

$$g^+ (\mu t) = \log(\mu t) - 2\pi i, \quad g^- (t) = \log t \quad \text{for all } t \in (0, \infty). \tag{10.52}$$

Thus, the map h defined by $h(z) := g(z)/(1 + i\delta)$ maps $\mathbf{C} \setminus [0, \infty)$ conformally onto the stripe $\Pi_\delta := \{\zeta \in \mathbf{C} : -2\pi - \delta \operatorname{Re} \zeta < \operatorname{Im} \zeta < -\delta \operatorname{Re} \zeta\}$, and since $\log \mu = 2\pi\delta$, we infer from (10.52) that

$$h^+ (\mu t) = \frac{\log \mu + \log t - 2\pi i}{1 + i\delta} = \frac{\log t - 2\pi i(1 + i\delta)}{1 + i\delta} = h^- (t) - 2\pi i \tag{10.53}$$

for all $t \in (0, \infty)$. Finally, let $f(\zeta) := e^\zeta$. The map f maps Π_δ conformally onto $\mathbf{C} \setminus \Gamma_0$ and the identity $f(\zeta) = f(\zeta - 2\pi i)$ holds. The map $f \circ h$ is just the map w_0 given by (10.51). Consequently, w_0 maps $\mathbf{C} \setminus [0, \infty)$ conformally onto $\mathbf{C} \setminus \Gamma_0$, and from (10.53) we see that

$$w_0^+ (\mu t) = f(h^+ (\mu t)) = f(h^- (t) - 2\pi i) = f(h^- (t)) = w_0^- (t)$$

for all $t \in (0, \infty)$. This proves our claim.

With regard to Theorem 10.32, we remark that w_0 enjoys the following properties. Because $w_0'(z) = (1 + i\delta)^{-1} z^{-1} w_0(z)$ and w_0 is continuously extendible to $(0, \infty)$ from each of the two sides, the derivative w_0' also extends continuously to $(0, \infty)$ from each side. Obviously,

$$|w_0(z)| \simeq |z|^{1/(1+\delta^2)}, \quad |w_0'(z)| \simeq |z|^{-\delta^2/(1+\delta^2)}$$

as $z \in \mathbf{C} \setminus [0, \infty)$ approaches the origin. □

Example 10.31. Now let η be the line segment $[0, 1]$, orient η from 0 to 1, and let $\alpha : [0, 1] \to [0, 1]$ be the homeomorphism given by

$$\alpha(t) := \frac{\mu t}{\mu t + 1 - t} \quad (\mu > 0, \ \mu \neq 1).$$

Consider the Möbius transform $\varphi : \dot{\mathbf{C}} \to \dot{\mathbf{C}}$, $z \mapsto z/(1 - z)$ and let α_0, Γ_0, w_0 be as in Example 10.30. Obviously, $\varphi([0, 1]) = [0, \infty) \cup \{\infty\}$ and $\alpha = \varphi^{-1} \circ \alpha_0 \circ \varphi$. Thus,

$$\omega := \varphi^{-1} \circ w_0 \circ \varphi : \dot{\mathbf{C}} \setminus [0, 1] \to \dot{\mathbf{C}} \setminus \varphi^{-1}(\Gamma_0)$$

is a conformal map and for $t \in (0, 1)$,

$$\begin{aligned} \omega^+ (\alpha(t)) &= (\varphi^{-1} \circ w_0^+ \circ \varphi)((\varphi^{-1} \circ \alpha_0 \circ \varphi)(t)) \\ &= \varphi^{-1} (w_0^+ (\alpha_0(\varphi(t)))) = \varphi^{-1} (w_0^- (\varphi(t))) \\ &= (\varphi^{-1} \circ w_0^- \circ \varphi)(t) = \omega^- (t). \end{aligned}$$

This shows that $\Gamma := \varphi^{-1}(\Gamma_0)$ and ω solve the problem of conformal welding for η and α. We have

$$
\begin{aligned}
\Gamma \setminus \{0,1\} &= \left\{ \varphi^{-1}\left(e^{2\pi x(1-i\delta)}\right) : x \in \mathbf{R} \right\} = \left\{ \frac{e^{2\pi x(1-i\delta)}}{e^{2\pi x(1-i\delta)}+1} : x \in \mathbf{R} \right\} \\
&= \left\{ \frac{e^{2\pi(x(1-i\delta)+i/2)}}{e^{2\pi(x(1-i\delta)+i/2)}-1} : x \in \mathbf{R} \right\} \\
&= \left\{ M_{0,1}(e^{2\pi\gamma}) : \gamma = x + i(1/2 - \delta x) \text{ with } x \in \mathbf{R} \right\} \\
&= \mathcal{S}\left(0,1; -\delta; 1/2\right) \setminus \{0,1\},
\end{aligned}
$$

and hence, Γ is a logarithmic double spiral between 0 and 1. The spirality indices of Γ at its endpoints 0 and 1 are δ.

It can be straightforwardly verified that ω' is continuously extendible to $(0,1)$ from each of the two sides, that

$$
|\omega(z)| \simeq |z|^{1/(1+\delta^2)}, \quad |\omega'(z)| \simeq |z|^{-\delta^2/(1+\delta^2)}
$$

as $z \in \mathbf{C} \setminus [0,1]$ approaches zero, and that

$$
|\omega(z)-1| \simeq |z-1|^{1/(1+\delta^2)}, \quad |\omega'(z)| \simeq |z-1|^{-\delta^2/(1+\delta^2)}
$$

as $z \in \mathbf{C} \setminus [0,1]$ goes to 1. $\qquad\qquad\qquad\qquad\qquad\qquad\qquad\qquad\square$

The preceding examples exhibit some characteristic features of conformal welding. The following theorem concerns more general cases.

Theorem 10.32. *Let η be a simple Lyapunov arc with two endpoints, t_0 and t_1. Suppose η is oriented from t_0 to t_1. Let $\alpha : \eta \to \eta$ be an orientation-preserving diffeomorphism whose derivative satisfies a Hölder condition on η. Then there exist a simple Carleson arc Γ with two endpoints $\tau_0, \tau_1 \in \mathbf{C}$ and a conformal map $\omega : \dot{\mathbf{C}} \setminus \eta \to \dot{\mathbf{C}} \setminus \Gamma$ such that $\omega^+(\alpha(t)) = \omega^-(t)$ for all $t \in \eta \setminus \{t_0, t_1\}$ and such that the following hold.*

(a) *If τ', τ'' are any two points of $\Gamma \setminus \{\tau_0, \tau_1\}$, then the subarc of Γ between τ' and τ'' is a Lyapunov curve.*

(b) *The boundary functions $\omega^\pm$ satisfy a Hölder condition on η.*

(c) *If $z \in \mathbf{C} \setminus \eta$ approaches t_j $(j = 0, 1)$ then $|\omega(z) - t_j| \simeq |z - t_j|^{1/(1+\delta_j^2)}$ with*

$$
\delta_j := (-1)^j \frac{1}{2\pi} \log |\alpha'(t_j)|. \tag{10.54}
$$

(d) *The derivative ω' is continuously extendible to $\eta \setminus \{t_0, t_1\}$ from the left and from the right, and we have*

$$
|\omega'(z)| \simeq |z - t_j|^{-\delta_j^2/(1+\delta_j^2)}
$$

as $z \in \mathbf{C} \setminus \eta$ goes to t_j $(j = 0, 1)$, where δ_j is given by (10.54).

(e) *For every continuous argument* $\tau \mapsto \arg(\tau - \tau_j)$ $(j = 0,1)$ *on* $\Gamma \setminus \{\tau_0, \tau_1\}$ *the asymptotic relation*

$$\arg(\tau - \tau_j) = -\delta_j \log|\tau - \tau_j| + O(1) \quad (\tau \in \Gamma, \tau \to \tau_j)$$

is valid. In particular, Γ *is spiralic at its endpoints and the spirality indices are given by* (10.54).

We now turn to the Haseman boundary value problem. Let η and α be as in Theorem 10.32. If $f \in L^1(\eta)$, then the Cauchy integral

$$(Cf)(z) := \frac{1}{2\pi i} \int_\eta \frac{f(t)}{t - z} \, dt \quad (z \in \dot{\mathbf{C}} \setminus \eta)$$

defines an analytic function in $\dot{\mathbf{C}} \setminus \eta$ and this function has nontangential limits $(Cf)^+(t)$ and $(Cf)^-(t)$ from the left and from the right, respectively, for almost all $t \in \eta$. The Sokhotski-Plemelj formulas say that

$$(Cf)^\pm(t) = \pm\frac{1}{2}f(t) + \frac{1}{2}(S_\eta f)(t) \tag{10.55}$$

for almost all $t \in \eta$. Fix $p \in (1, \infty)$ and let ϱ be a power weight of the form

$$\varrho(t) = |t - t_0|^{\beta_0}|t - t_1|^{\beta_1} \quad (t \in \eta \setminus \{t_0, t_1\})$$

with $\beta_0, \beta_1 \in (-1/p, 1/q)$. The *Haseman problem* is as follows: given functions $G \in C(\eta \cup \{t_0, t_1\})$ and $g \in L^p(\eta, \varrho)$, find all functions $f \in L^p(\eta, \varrho)$ such that

$$(Cf)^+\big(\alpha(t)\big) = G(t)(Cf)^-(t) + g(t) \text{ for almost all } t \in \eta.$$

In the case where α is the identity map, $\alpha(t) = t$ for all $t \in \eta$, Haseman's problem is known as the *Riemann-Hilbert problem* or simply as the *Riemann problem*.

The composition operator W given by $W\varphi = \varphi \circ \alpha$ is easily seen to be bounded on $L^p(\eta, \varrho)$. Taking into account (10.55) we therefore see that the Haseman problem is equivalent to solving the equation $Af = g$ in $L^p(\eta, \varrho)$ where $A := WP_\eta + GQ_\eta$ with $P_\eta := (I + S_\eta)/2$, $Q_\eta := (I - S_\eta)/2$.

By means of conformal welding, the Haseman problem may be reduced to a Riemann-Hilbert problem. Let Γ and ω be as in Theorem 10.32. One can show that $G \circ (\omega^-)^{-1}$ is continuous on $\Gamma \cup \{\tau_0, \tau_1\}$. With δ_j $(j = 0,1)$ given by (10.54), put

$$m_j := \big[(1/p + \beta_j)(1 + \delta_j^2)\big],$$

$[\cdot]$ standing for the integral part, and set

$$\gamma_j := (1/p + \beta_j)(1 + \delta_j^2) - m_j - 1/p.$$

If $(1/p + \beta_j)(1 + \delta_j^2) \notin \mathbf{Z}$ then $\gamma_j \in (-1/p, 1/q)$. Finally, define the power weight w on Γ by

$$w(\tau) := |\tau - \tau_0|^{\gamma_0}|\tau - \tau_1|^{\gamma_1} \quad \big(\tau \in \Gamma \setminus \{\tau_0, \tau_1\}\big)$$

and let $m := m_0 + m_1$.

Theorem 10.33. *Suppose*

$$(1/p + \beta_j)(1 + \delta_j^2) \notin \mathbf{Z} \ for \ j \in \{0, 1\}. \tag{10.56}$$

Then the operator $A = WP_\eta + GQ_\eta$ *is Fredholm on* $L^p(\eta, \varrho)$ *if and only if the operator* $B := P_\Gamma + (G \circ (\omega^-)^{-1})Q_\Gamma$ *is Fredholm on* $L^p(\Gamma, w)$*. In that case*

$$\dim \operatorname{Ker} A = \max \{0, \operatorname{Ind} B + m\},$$

$$\dim \operatorname{Coker} A = \max \{0, -\operatorname{Ind} B - m\},$$

and thus, $\operatorname{Ind} A = \operatorname{Ind} B + m$.

This theorem shows that the Haseman problem on a Lyapunov arc is equivalent to a Riemann-Hilbert problem on a spiralic Carleson arc. Using Theorems 9.27 and 10.5, one gets a Fredholm criterion and an index formula for the operator B and thus also for A. We confine ourselves to formulating the final result.

If t traverses η from t_1 to t_0 (i.e. opposite to the orientation of η), then $G(t)$ describes a curve Σ from $G(t_1)$ to $G(t_0)$. Join $G(t_0)$ to 1 by the logarithmic double spiral

$$\Sigma_0 := \mathcal{S}\big(G(t_0), 1; \delta_0, 1; \delta_0, 1/p + \gamma_0\big)$$

and then join 1 to $G(t_1)$ by the logarithmic double spiral

$$\Sigma_1 := \mathcal{S}\big(1, G(t_1); \delta_1, 1/p + \gamma_1\big).$$

The union $\Sigma \cup \Sigma_0 \cup \Sigma_1$ is a closed, continuous, and oriented curve which will be denoted by $G_{\eta,\alpha}^\#$.

Theorem 10.34. *Suppose* (10.56) *holds. Then the operator* $A = WP_\eta + GQ_\eta$ *is Fredholm on* $L^p(\eta, \varrho)$ *if and only if* $0 \notin G_{\eta,\alpha}^\#$*. In that case*

$$\dim \operatorname{Ker} A = \max \{0, -\operatorname{wind} G_{\eta,\alpha}^\# + m\},$$
$$\dim \operatorname{Coker} A = \max \{0, \operatorname{wind} G_{\eta,\alpha}^\# - m\},$$

and, in particular, $\operatorname{Ind} A = -\operatorname{wind} G_{\eta,\alpha}^\# + m$.

We remark that in the papers [113], [114], [115], [116] the operator $A = WP_\eta + GQ_\eta$ was studied with the help of different methods and without the extra assumption (10.56). These papers contain a Fredholm criterion and an index formula for A in "analytic" language (like in Proposition 7.3) and not via the "geometric" information of $G_{\eta,\alpha}^\#$. However, in these papers no formulas for the kernel and cokernel dimensions of A are given.

10.10   Notes and comments

1.1. Theorems 10.1 to 10.4 are commented in the text. The most influential early work on Toeplitz and Wiener-Hopf operators with matrix-valued symbols was certainly the Gohberg-Krein paper [81]. The monographs Böttcher, Silbermann [23], Clancey, Gohberg [33], Gohberg, Goldberg, Kaashoek [80], Litvinchuk, Spitkovsky [137], Mikhlin, Prössdorf [147] all deal with phenomena caused by the matrix case. For operator-valued symbols we refer to the recent articles [24] and [63] and the literature cited there.

Passage from piecewise continuous matrix-valued symbols to semi-almost periodic (SAP) matrix-valued symbols is accompanied with overcoming serious obstacles. In this connection see our papers [117], [118], [119], [22] with Spitkovsky as well as the paper [113] and the references there.

10.2–10.3. References are in the text.

10.4. It had been well known for a long time that if Γ is a half-line, then the spectrum of S_Γ on $L^p(\Gamma, w)$ with a power weight w is the union of two circular arcs and the set encircled by these two arcs; see Widom [209], Shamir [189], Duduchava [57]. Theorem 10.11 was established only very recently by Bishop, Spitkovsky, and the authors in [10]. We remark that this theorem is even new in the case where Γ is a composed Lyapunov curve and w is a power weight.

10.5. References are in the text. We want to add that examples of Young functions which generate Orlicz spaces with distinct Boyd indices are in the book by Maligranda [141] and the paper by Aslanov and one of the authors [7]. On the basis of Lindberg's paper [135], Maligranda considered a Young function which for sufficiently large $x > 0$ is given by

$$M(x) = \exp\left\{\left(p + \kappa \sin(\log(-\log x))\right)\right\}$$

with $\kappa > 0$ and $p > 1 + \sqrt{2}$; he showed that the Boyd indices are

$$\alpha_M = 1/(p + \kappa\sqrt{2}) \quad \text{and} \quad \beta_M = 1/(p - \kappa\sqrt{2}).$$

This is what we had in mind in the notes and comments to Sections 1.5 and 1.6 when saying that functions like $\sin(\log(-\log x))$ are good friends of workers in Orlicz spaces.

10.6. A. Dynin and G.I. Eskin were probably the first to understand the relevancy of the Mellin calculus to singular and convolution operators. Dynin has never published his observation cited in the text, Eskin has the book [64]. The book by Duduchava [57] and the report by Roch and Silbermann [174] pay due attention to the delicacies of the Mellin approach and the L^p theory of one-dimensional convolution operators and are both excellent introductions to as well as encyclopaedic sources of the topic.

Mellin techniques for studying pseudodifferential operators with slowly oscillating data were developed in Rabinovich's papers [163], [164], [165], [166], [167],

[168]. As already said, while Wiener-Hopf techniques are strongly tied down to one dimension, Mellin techniques are the most promising apparatus for handling higher dimensions.

10.7–10.8. For Wiener-Hopf integral and zero-order pseudodifferential operators on the spaces $L^p(\mathbf{R}, w)$ with power weights w we refer to the books by Böttcher, Silbermann [23], Duduchava [57], and the report by Roch, Silbermann [174]. The pioneering papers in this field include Widom [209], Cordes [41], Duduchava [55], [56], [57], [58].

10.9. The Haseman problem was first studied by Haseman [100]. The idea of applying conformal welding to the Haseman problem on Lyapunov Jordan curves goes back to Mandzhavidze and Khvedelidze [142] and was also employed by Simonenko [193], Antontsev and Monakhov [6], and Zverovich [212]. A detailed discussion of results on Haseman's problem can be found in the books by Litvinchuk [136] and by Kravchenko and Litvinchuk [129].

We here consider Haseman's problem on a simple Lyapunov arc η under the assumption that the derivative $\alpha'(t_j)$ of the diffeomorphism $\alpha : \eta \to \eta$ at the endpoints t_j of the arc is not equal to 1. In this case the solution of the corresponding problem of conformal welding is based on invoking results from the theory of quasi-conformal mappings (see [4]). An analogous approach was also used by Lyubarski [138], [139] in order to study different problems.

The operator $A := WP_\Gamma + GQ_\Gamma$ associated with Haseman's problem is a singular integral operator with a shift. In the case where this operator is considered on $L^2_n(\Gamma)$ over a composed Lyapunov curve Γ and the coefficient G belongs to $PC_n(\Gamma)$, the Fredholm theory of A was worked out by Kravchenko and one of the authors [114], [115], [116]. For the spaces $L^p_n(\Gamma, w)$, Γ being a composed smooth curve and w being a power weight, and coefficients G with entries in SAP, a Fredholm theory of A was established in [113] (under some additional hypotheses in case $p \neq 2$). The results of Section 10.9 are all from the papers [2], [3], [4] by Aizenshtat, Litvinchuk and one of the authors. We should emphasize that these results do not only contain Fredholm criteria and index formulas for the operator A but also formulas for the kernel and cokernel dimensions.

Bibliography

[1] L.V. Ahlfors: Zur Theorie der Überlagerungsflächen. *Acta Math.* **65** (1935), 157–194.

[2] A.V. Aizenshtat, Yu.I. Karlovich, and G.S. Litvinchuk: On a theorem on conformal gluing and its application. *Dokl. Rasshir. Zased. Semin. Inst. Prikl. Matem. (Tbilisi)* **3** (1988), 9–12 [Russian].

[3] A.V. Aizenshtat, Yu.I. Karlovich, and G.S. Litvinchuk: On the defect numbers of the Kveselava-Vekua operator with discontinuous derivative of the shift. *Soviet Math. Dokl.* **43** (1991), 633–638.

[4] A.V. Aizenshtat, Yu.I. Karlovich, and G.S. Litvinchuk: The method of conformal gluing for the Haseman boundary value problem on an open contour. *Complex Variables* **28** (1996), 313–346.

[5] G.R. Allan: Ideals of vector-valued functions. *Proc. London Math. Soc.*, 3rd ser., **18** (1968), 193–216.

[6] S.N. Antontsev and V.N. Monakhov: On the solvability of a class of conjugation problems with shift. *Dokl. Akad. Nauk SSSR* **205** (1972), 263–266 [Russian].

[7] V.D. Aslanov and Yu.I. Karlovich: One-sided invertibility of functional operators in reflexive Orlicz spaces. *Dokl. Akad. Nauk Azerb. SSR* **45**, No. 10–11 (1989), 3–7 [Russian].

[8] C. Bennett and R. Sharpley: *Interpolation of Operators.* Academic Press, Boston 1988.

[9] A.S. Besicovitch: A general form of the covering principle and relative differentiation of additive functions. *Proc. Cambridge Philos. Soc.* **41** (1945), 103–110 and **42** (1946), 1–10.

[10] C.J. Bishop, A. Böttcher, Yu.I. Karlovich, and I. Spitkovsky: Local spectra and index of singular integral operators with piecewise continuous coefficients on composed curves. *Math. Nachrichten* (to appear).

[11] C.J. Bishop and P.W. Jones: Harmonic measure and arclength. *Ann. Math.* **132** (1990), 511-547.

[12] C.J. Bishop and P.W. Jones: Harmonic measure, L^2 estimates and the Schwarzian derivative. *J. d'Analyse Math.* **62** (1994), 77–113.

[13] A. Böttcher: Toeplitz operators with piecewise continuous symbols - a neverending story ? *Jahresber. d. Deutschen Mathematiker-Vereinigung* **97** (1995), 115–129.

[14] A. Böttcher and S.M. Grudsky: On the composition of Muckenhoupt weights and inner functions. *J. London Math. Soc.* (to appear)

[15] A. Böttcher and Yu.I. Karlovich: Toeplitz and singular integral operators on Carleson curves with logarithmic whirl points. *Integral Equations and Operator Theory* **22** (1995), 127–161.

[16] A. Böttcher and Yu.I. Karlovich: Toeplitz and singular integral operators on general Carleson Jordan curves. *Operator Theory: Advances and Applications* **90** (1996), 119–152.

[17] A. Böttcher and Yu.I. Karlovich: Submultiplicative functions and spectral theory of Toeplitz operators. *Integral Transforms and Special Functions* **4** (1996), 181-202.

[18] A. Böttcher and Yu.I. Karlovich: Toeplitz operators with PC symbols on general Carleson Jordan curves with arbitrary Muckenhoupt weights. *Trans. Amer. Math. Soc.* (to appear).

[19] A. Böttcher, Yu.I. Karlovich, and V.S. Rabinovich: Emergence, persistence, and disappearance of logarithmic spirals in the spectra of singular integral operators. *Integral Equations and Operator Theory* **25** (1996), 406–444.

[20] A. Böttcher, Yu.I. Karlovich, and V.S. Rabinovich: Mellin pseudodifferential operators with slowly varying symbols and singular integrals on Carleson curves with Muckenhoupt weights. *Math. Annalen* Preprint, TU Chemnitz 1997.

[21] A. Böttcher, Yu.I. Karlovich, and V.S. Rabinovich: The method of limit operators for one-dimensional singular integrals with slowly oscillating data. Preprint, TU Chemnitz 1997.

[22] A. Böttcher, Yu.I. Karlovich, and I. Spitkovsky: Toeplitz operators with semi-almost periodic symbols on spaces with Muckenhoupt weight. *Integral Equations and Operator Theory* **18** (1994), 261–276.

[23] A. Böttcher and B. Silbermann: *Analysis of Toeplitz Operators*. Akademie-Verlag, Berlin 1989 and Springer-Verlag, Berlin, Heidelberg, New York 1990.

[24] A. Böttcher and B. Silbermann: Infinite Toeplitz and Hankel matrices with operator-valued entries. *SIAM J. Math. Analysis* **27** (1996), 805–822.

[25] A. Böttcher and I. Spitkovsky: Toeplitz operators with PQC symbols on weighted Hardy spaces. *J. Funct. Analysis* **97** (1991), 194–214.

[26] A. Böttcher and I. Spitkovsky: Wiener-Hopf integral operators with PC symbols on spaces with Muckenhoupt weight. *Revista Matemática Iberoamericana* **9** (1993), 257–279.

[27] A. Böttcher and I. Spitkovsky: Pseudodifferential operators with heavy spectrum. *Integral Equations and Operator Theory* **19** (1994), 251–269.

[28] D.W. Boyd: Indices for the Orlicz spaces. *Pacific J. Math.* **38** (1971), 315–323.

[29] A. Brown and P. Halmos: Algebraic properties of Toeplitz operators. *J. Reine Angew. Math.* **231** (1963), 89–102.

[30] D.L. Burkholder and R.F. Gundy: Distribution function inequalities for the area integral. *Studia Math.* **44** (1972), 527–544.

[31] A.P. Calderón: Inequalities for the maximal function relative to a metric. *Studia Math.* **57** (1976), 297–306.

[32] A.P. Calderón: Cauchy integrals on Lipschitz curves and related operators. *Proc. Nat. Acad. Sci. USA* **74** (1977), 1324–1327.

[33] K.F. Clancey and I. Gohberg: *Factorization of Matrix Functions and Singular Integral Operators.* Birkhäuser Verlag, Basel 1981.

[34] L.A. Coburn: Weyl's theorem for non-normal operators. *Michigan Math. J.* **13** (1966), 285–286.

[35] R.R. Coifman: Distribution function inequalities for singular integrals. *Proc. Nat. Acad. Sci. USA* **69** (1972), 2838–2839.

[36] R.R. Coifman and C. Fefferman: Weighted norm inequalities for maximal functions and singular integrals. *Studia Math.* **51** (1974), 241–249.

[37] R.R. Coifman, P.W. Jones, and S. Semmes: Two elementary proofs of the L^2 boundedness of Cauchy integrals on Lipschitz curves. *J. Amer. Math. Soc.* **2** (1989), 553–564.

[38] R.R. Coifman, A. McIntosh, and Y. Meyer: L'intégrale de Cauchy définit un opérateur borné sur L^2 pour les courbes Lipschitziennes. *Ann. of Math.* **116** (1982), 361–388.

[39] R.R. Coifman and G. Weiss: *Analyse harmonique non-commutative sur certains espaces homogénes.* Lecture Notes Math., Vol. 242, Springer-Verlag, Berlin 1971.

[40] R.R. Coifman and G. Weiss: Extensions of Hardy spaces and their use in analysis. *Bull. Amer. Math. Soc.* **83** (1977), 569–645.

[41] H.O. Cordes: Pseudodifferential operators on a half-line. *J. Mech. Math.* **18** (1969), 893–908.

[42] M. Costabel: Singular integral operators on curves with corners. *Integral Equations and Operator Theory* **3** (1980), 323–349.

[43] M. Costabel: An inverse of the Gohberg-Krupnik symbol map. *Proc. Royal Soc. Edinburgh* **87A** (1980), 153–165.

[44] E.A. Danilov: The Riemann boundary value problem on contours with unbounded distortion. *Cand. Dissertation*, Odessa 1984 [Russian].

[45] I.I. Danilyuk: *Irregular boundary Value Problems in the Plane.* Nauka, Moscow 1975 [Russian].

[46] G. David: L'intégrale de Cauchy sur les courbes rectifiables. *Prepublication Univ. Paris-Sud*, Dept. Math. 82T05, 1982.

[47] G. David: Opérateurs integraux singuliers sur certaines courbes du plan complexe. *Ann. Sci. École Norm. Sup.* **17** (1984), 157–189.

[48] G. David and J.L. Journé: A boundedness criterion for general Calderón-Zygmund operators. *Ann. Math.* **120** (1984), 371–397.

[49] G. David, J.L. Journé, and S. Semmes: Opérateurs de Calderón-Zygmund, fonctions para acrétives et interpolation. *Revista Matemática Iberoamericana* **4** (1985), 1–56.

[50] G. David and S. Semmes: *Analysis of and on Uniformly Rectifiable Sets.* Math. Surveys and Monographs, Vol. 38, Amer. Math. Soc., Providence, R.I., 1993.

[51] A. Devinatz: Toeplitz operators on H^2 spaces. *Trans. Amer. Math. Soc.* **112** (1964), 304–317.

[52] R.G. Douglas: *Banach Algebra Techniques in Operator Theory.* Academic Press, New York 1972.

[53] R.G. Douglas: Local Toeplitz operators. *Proc. London Math. Soc.*, 3rd ser., **36** (1978), 234-276.

[54] R.G. Douglas and D. Sarason: Fredholm Toeplitz operators. *Proc. Amer. Math. Soc.* **26** (1970), 117–120.

[55] R.V. Duduchava: Wiener-Hopf integral operators with discontinuous symbols. *Soviet Math. Dokl.* **14** (1973), 1001–1005.

[56] R.V. Duduchava: On convolution integral operators with discontinuous coefficients. *Math. Nachrichten* **79** (1977), 75–98 [Russian].

[57] R.V. Duduchava: *Integral Equations with Fixed Singularities.* B.G. Teubner Verlagsgesellschaft, Leipzig 1979.

[58] R.V. Duduchava: On algebras generated by convolutions and discontinuous functions. *Integral Equations and Operator Theory* **10** (1987), 505–530.

[59] P. Duren: *Theory of H^p Spaces.* Academic Press, New York 1970.

[60] E.M. Dynkin: Methods of the theory of singular integrals (Hilbert transform and Calderon-Zygmund theory). In: *Itogi Nauki Tekh., Sovr. Probl. Matem., Fund. Napravl.*, Vol. 15, Moscow 1987, pp. 197–292 [Russian].

[61] E.M. Dynkin: Methods of the theory of singular integrals II (Littlewood-Paley theory and its applications). In: *Itogi Nauki Tekh., Sovr. Probl. Matem., Fund. Napravl.*, Vol. 42, Moscow 1989, pp. 105–198 [Russian].

[62] E.M. Dynkin and B.P. Osilenker: Weighted norm estimates for singular integrals and their applications. *J. Soviet Math.* **30** (1985), 2094–2154.

[63] T. Ehrhardt, S. Roch, and B. Silbermann: Symbol calculus for singular integral operators with operator-valued PQC-coefficients. *Operator Theory: Advances and Applications* **90** (1996), 182–203.

[64] G.I. Eskin: *Boundary Value Problems for Elliptic Pseudodifferential Operators.* Nauka, Moscow 1971 [Russian].

[65] J.L. Fernández, J. Heinonen, and O. Martio: Quasilines and conformal mappings. *J. d'Analyse Math.* **52** (1989), 117–132.

[66] J.L. Fernández and M. Zinsmeister: Ensembles de niveau des représentations conformes. *Comptes Rendus de l'Acad. Sci.* **305** (1987), 449-452.

[67] T. Finck and S. Roch: Banach algebras with matrix symbol of bounded order. *Integral Equations and Operator Theory* **18** (1994), 427–434.

[68] T. Finck, S. Roch, and B.Silbermann: Two projections theorems and symbol calculus for operators with massive local spectra. *Math. Nachrichten* **162** (1993), 167–185.

[69] G.B. Folland and E.M. Stein: *Hardy Spaces on Homogeneous Groups.* Princeton Univ. Press, Princeton 1982.

[70] D. Gaier: *Lectures on Complex Approximation*. Birkhäuser Verlag, Basel, Boston, Stuttgart 1987.

[71] F.D. Gakhov: On Riemann's boundary value problem. *Matem. Sbornik* **2 (44)** (1937), 673–683 [Russian].

[72] F.D. Gakhov: *Boundary Value Problems*. Pergamon Press, London 1966. [Extended Russian edition: Nauka, Moscow 1977]

[73] J. Garcia-Cuerva and J.L. Rubio de Francia: *Weighted Norm Inequalities and Related Topics*. Elsevier Sci. Publ. B.V., North-Holland, Amsterdam, New York, Oxford 1985.

[74] J.B. Garnett: *Bounded Analytic Functions*. Academic Press, New York 1981.

[75] J.B. Garnett, F. Gehring, and P.W. Jones: Conformally invariant length sums. *Indiana Univ. Math. J.* **32** (1983), 809–829.

[76] F.W. Gehring: The L^p integrability of partial derivatives of a quasiconformal mapping. *Acta Math.* **130** (1973), 265–277.

[77] I. Gohberg: On an application of the theory of normed rings to singular integral equations. *Uspekhi Mat. Nauk.* **7** (1952), 149–156 [Russian].

[78] I. Gohberg: On the number of solutions of homogeneous singular equations with continuous coefficients. *Dokl. Akad. Nauk SSSR* **122** (1958), 327–330 [Russian].

[79] I. Gohberg: On Toeplitz matrices constituted by the Fourier coefficients of piecewise continuous functions. *Funkts. Anal. Prilozh.* **1** (1967), 91–92 [Russian].

[80] I. Gohberg, S. Goldberg, and M.A. Kaashoek: *Classes of Linear Operators*. Vol. 1: Birkhäuser Verlag, Basel, Boston, Berlin 1990; Vol. 2: Birkhäuser Verlag, Basel, Boston, Berlin 1994.

[81] I. Gohberg and M.G. Krein: Systems of integral equations on a half-line with kernel depending upon the difference of the arguments. *Amer. Math. Soc. Transl.* (2) **14** (1960), 217–287.

[82] I. Gohberg and N. Krupnik: The spectrum of one-dimensional singular integral operators with piecewise continuous coefficients. *Matem. Issled.* **3** (1968), 16–30 [Russian].

[83] I. Gohberg and N. Krupnik: On the spectrum of singular integral operators on the space L^p. *Studia Math.* **31** (1968), 347–362 [Russian].

[84] I. Gohberg and N. Krupnik: On the spectrum of singular integral operators on weighted L^p spaces. *Dokl. Akad. Nauk SSSR* **185** (1969), 745–749 [Russian].

[85] I. Gohberg and N. Krupnik: Systems of singular integral equations in weighted L^p spaces. *Soviet Math. Dokl.* **10** (1969), 688–691.

[86] I. Gohberg and N. Krupnik: On singular integral operators on a non-simple contour. *Soobshzh. Akad. Nauk Gruz. SSR* **64** (1971), 21–24 [Russian].

[87] I. Gohberg and N. Krupnik: Singular integral operators with piecewise continuous coefficients and their symbols. *Math. USSR Izv.* **5** (1971), 955–979.

[88] I. Gohberg and N. Krupnik: On the symbol of singular integral operators on a composed contour. In: *Trudy Simpoz. Mech. Sploshnych Sred*, Vol. 1, pp. 46–59, Tbilisi 1973 [Russian].

[89] I. Gohberg and N. Krupnik: *One-Dimensional Linear Singular Integral Equations.*
      Vols. I and II. Birkhäuser Verlag, Basel, Boston, Berlin 1992.

[90] I. Gohberg and N. Krupnik: Extension theorems for invertibility symbols in Banach
      algebras. *Integral Equations and Operator Theory* **15** (1992), 991–1010.

[91] I. Gohberg, N. Krupnik, and I. Spitkovsky: Banach algebras of singular integral op-
      erators with piecewise continuous coefficients. General contour and weight. *Integral
      Equations and Operator Theory* **17** (1993), 322–337.

[92] I. Gohberg and A.A. Sementsul: Toeplitz matrices composed by the Fourier coef-
      ficients of functions with discontinuities of almost periodic type. *Matem. Issled.* **5**
      (1970), 63–83 [Russian].

[93] S.M. Grudsky: Singular integral equations and the Riemann boundary value prob-
      lem with infinite index in the space $L^p(\Gamma, \varrho)$. *Izv. Akad. Nauk SSSR*, Ser. Mat. **49**
      (1985), 55–80 [Russian].

[94] S.M. Grudsky and A.B. Khevelev: On invertibility in $L^2(\mathbf{R})$ of singular integral
      operators with periodic coefficients and a shift. *Soviet Math. Dokl.* **27** (1983) 486-
      489.

[95] M. Guzman: *Differentiation of Integrals in* $\mathbf{R}^n$. Lecture Notes Math., Vol. 481,
      Springer-Verlag, Berlin, Heidelberg, New York 1975.

[96] R. Hagen, S. Roch, and B. Silbermann: *Spectral Theory of Approximation Methods
      for Convolution Equations.* Birkhäuser Verlag, Basel, Boston, Berlin 1995.

[97] P. Halmos: Two projections. *Trans. Amer. Math. Soc.* **144** (1969), 381–389.

[98] P. Hartman: On completely continuous Hankel operators. *Proc. Amer. Math. Soc.*
      9 (1958), 862–866.

[99] P. Hartman and A. Wintner: The spectra of Toeplitz's matrices. *Amer. J. Math.*
      76 (1954), 867–882.

[100] C. Haseman: *Anwendung der Theorie der Integralgleichungen auf einige Randwert-
      aufgaben.* Göttingen 1907.

[101] V.P. Havin: Boundary properties of Cauchy type integrals and harmonically con-
      jugated functions in domains with rectifiable boundary. *Matem. Sbornik* **68** (1965),
      499–517 [Russian].

[102] W.K. Hayman and J.-M. Wu: Level sets of univalent functions. *Comment. Math.
      Helv.* **56** (1981), 366–403.

[103] A.Ya. Helemskii: *Banach and Locally Convex Algebras.* Clarendon Press, Oxford
      1993.

[104] H. Helson and G. Szegö: A problem in prediction theory. *Ann. Mat. Pura Appl.*
      51 (1960), 107–138.

[105] E. Hille and R.S. Phillips: *Functional Analysis on Semi-Groups.* Amer. Math. Soc.
      Coll. Publ., Vol. 37, revised edition, Providence, R.I., 1957.

[106] R.A. Horn, C.A. Johnson: *Matrix analysis.* Cambridge University Press, Cambridge
      1986.

[107] R. Hunt, B. Muckenhoupt, and R. Wheeden: Weighted norm inequalities for the
      conjugate function and Hilbert transform. *Trans. Amer. Math. Soc.* **176** (1973),
      227–251.

[108] P.W. Jones: Square functions, Cauchy integrals, analytic capacity, and harmonic measure. In: *Harmonic Analysis and Partial Differential Equations* (J. Garcia-Cuerva, ed.), Lecture Notes Math., Vol. 1384, Springer-Verlag, Berlin 1989.

[109] J.-L. Journé: *Calderón-Zygmund Operators, Pseudo-Differential Operators and the Cauchy Integral of Calderón.* Lecture Notes Math., Vol. 994, Springer-Verlag, Berlin, Heidelberg, New York, Tokyo 1983.

[110] A.Yu. Karlovich: Algebras of singular integral operators with piecewise continuous coefficients on reflexive Orlicz spaces. *Math. Nachrichten* **179** (1996), 187–222.

[111] A.Yu. Karlovich: Singular integral operators with piecewise continuous coefficients in reflexive rearrangement-invariant spaces. Preprint 1996.

[112] A.Yu. Karlovich: On the index of singular integral operators in reflexive Orlicz spaces. *Matem. Zametki* (submitted) [Russian].

[113] Yu.I. Karlovich: On the Haseman problem. *Demonstratio Mathematica* **26** (1993), 581–595.

[114] Yu.I. Karlovich and V.G. Kravchenko: On a singular integral operator with non-Carleman shift on an open contour. *Soviet Math. Dokl.* **18** (1977), 1263–1267.

[115] Yu.I. Karlovich and V.G. Kravchenko: Singular integral equations with non-Carleman translations on an open contour. *Differential Equations* **17** (1982), 1408–1417.

[116] Yu.I. Karlovich and V.G. Kravchenko: An algebra of singular integral operators with piecewise continuous coefficients and piecewise smooth shift on a composite contour. *Math USSR Izv.* **23** (1984), 307–352.

[117] Yu.I. Karlovich and I. Spitkovsky: Factorization of almost periodic matrix-valued functions and Fredholm theory of Toeplitz operators with semi-almost periodic matrix symbols. *Math. USSR Izv.* **34** (1990), 281–316.

[118] Yu.I. Karlovich and I. Spitkovsky: (Semi-)Fredholmness of convolution operators on the spaces of Bessel potentials. *Operator Theory: Advances and Applications* **71** (1994), 122–152.

[119] Yu.I. Karlovich and I. Spitkovsky: Semi-Fredholm properties of certain singular integral operators. *Operator Theory: Advances and Applications* **90** (1996), 264–287.

[120] S.S. Kazarian: Integral inequalities for the conjugate function in weighted reflexive Orlicz spaces. *Soviet J. Contemp. Math. Anal., Armen. Acad. Sci.* **25** (1990), 45–57.

[121] C. Kenig: Weighted H^p spaces on Lipschitz domains. *Amer. J. Math.* **102** (1980), 129–163.

[122] B.V. Khvedelidze: Linear discontinuous boundary value problems of function theory, singular integral equations, and some of their applications. *Trudy Tbil. Matem. Inst. Akad. Nauk Gruz. SSR* **23** (1956), 3–158 [Russian].

[123] B.V. Khvedelidze: A remark on my paper "Linear discontinuous boundary value problems ...". *Soobshzh. Akad. Nauk Gruz. SSR* **21** (1958), 129–138 [Russian].

[124] B.V. Khvedelidze: The method of Cauchy type integrals for discontinuous boundary value problems of the theory of holomorphic functions of one complex variable. *J. Sov. Math.* **7** (1977), 309–414.

[125] V.M. Kokilashvili: On singular integrals and maximal operators with Cauchy kernel. *Dokl. Akad. Nauk SSSR* **223** (1975), 555–558 [Russian].

[126] A.N. Kolmogorov and S.V. Fomin: *Reele Funktionen und Funktionalanalysis.* Deutscher Verlag d. Wiss., Berlin 1982.

[127] P. Koosis: *Introduction to H^p spaces.* Cambridge Univ. Press, Cambridge 1980.

[128] M.A. Krasnoselskii and Ya.B. Rutitskii: *Convex Functions and Orlicz Spaces.* Noordhoff, Groningen 1961.

[129] V.G. Kravchenko and G.S. Litvinchuk: *Introduction to the Theory of Singular Integral Operators with Shift.* Kluwer Academic Publishers, Dordrecht, Boston, London 1994.

[130] M.G. Krein: Integral equations on a half-line with kernel depending upon the difference of the arguments. *Amer. Math. Soc. Transl.* (2) **22** (1962), 163-288.

[131] S.G. Krein, Yu.I. Petunin, and E.M. Semenov: *Interpolation of Linear Operators.* Transl. Math. Monographs **54**, Amer. Math. Soc., Providence, R.I., 1982.

[132] N. Krupnik: On singular integral operators with matrix coefficient. *Matem. Issled.* **45** (1977), 93–100 [Russian].

[133] N. Krupnik: Conditions for the existence of an n-symbol and of a sufficient supply of n-dimensional representations of a Banach algebra. *Matem. Issled.* **54** (1980), 84–97 [Russian].

[134] N. Krupnik: *Banach Algebras with Symbol and Singular Integral Operators.* Birkhäuser Verlag, Basel 1987.

[135] K. Lindberg: On subspaces of Orlicz sequence spaces. *Studia Math.* **45** (1973), 119–143.

[136] G.S. Litvinchuk: *Boundary Value Problems and Singular Integral Equations with Shift.* Nauka, Moscow 1977 [Russian].

[137] G.S. Litvinchuk and I. Spitkovsky: *Factorization of Measurable Matrix Functions.* Akademie-Verlag, Berlin 1987 and Birkhäuser Verlag, Basel, Boston 1987.

[138] Yu.I. Lyubarski: Conformal gluing for a Carleman shift having discontinuous derivative. *Dokl. Akad. Nauk Ukr. SSR*, Ser. A, **5** (1987), 18–20 [Ukrainian].

[139] Yu.I. Lyubarski: Properties of systems of linear combinations of powers. *Algebra i Analiz* **1** (1989), 1–69 [Russian].

[140] L. Maligranda: Indices and interpolation. *Dissert. Math.* **49** (1985), 1–49.

[141] L. Maligranda: *Orlicz Spaces and Interpolation.* Sem. Math. 5, Dept. Mat., Univ. Estadual de Campinas, Campinas SP, Brazil, 1989.

[142] G.F. Mandzhavidze and B.V. Khvedelidze: On the Riemann-Privalov problem with continuous coefficients. *Dokl. Akad. Nauk SSSR* **123** (1958), 791–794 [Russian].

[143] P. Mattila, M.S. Melnikov, and J.Verdera: The Cauchy integral, analytic capacity, and uniform rectifiability. *Annals of Math.* **144** (1996), 127–136.

[144] M.S. Melnikov: Analytic capacity: discrete approach and curvature of measure. *Sbornik: Mathematics* **186** (1995), 827–846.

[145] M.S. Melnikov and J. Verdera: A geometric proof of the L^2 boundedness of the Cauchy integral on Lipschitz graphs. *Intern. Math. Research Notices* **7** (1995), 325–331.

[146] S.G. Mikhlin: Singular integral equations. *Uspekhi Mat. Nauk* **3** (25) (1948), 29–112 [Russian].

[147] S.G. Mikhlin and S. Prössdorf: *Singular Integral Operators.* Akademie-Verlag, Berlin 1986.

[148] B. Muckenhoupt: Weighted norm inequalities for the Hardy maximal function. *Trans. Amer. Math. Soc.* **165** (1972), 207–226.

[149] T. Murai: Boundedness of singular integral operators of Calderon type. *Proc. Jap. Acad. Sci.* (A) **59** (1983), 364–367.

[150] T. Murai: *A Real Variable Method for Cauchy Transform, and Analytic Capacity.* Lecture Notes Math., Vol. 1307, Springer-Verlag, Berlin 1988.

[151] N.I. Muskhelishvili: *Singuläre Integralgleichungen.* Akademie-Verlag, Berlin 1965. [Extended Russian edition: Nauka, Moscow 1968]

[152] N.K. Nikolski: *Treatise on the Shift Operator.* Springer-Verlag, Berlin, Heidelberg 1986.

[153] V.I. Nyaga: Matrix singular integral operators on composed curves. *Matem. Zametki* **30** (1981), 553–560 [Russian].

[154] K. Øyma: Harmonic measure and conformal length. *Proc. Amer. Math. Soc.* **115** (1992), 687–689.

[155] V.A. Paatashvili and G.A. Khuskivadze: On the boundedness of the Cauchy singular integral on Lebesgue spaces in the case of non-smooth contours. *Trudy Tbil. Matem. Inst. Akad. Nauk GSSR* **69** (1982), 93–107 [Russian].

[156] G.K. Pedersen: Measure theory in C^*-algebras.II. *Math. Scand.* **22** (1968), 63–74.

[157] B.A. Plamenevski and V.N. Senichkin: On C^*-algebras of singular integral operators with discontinuous coefficients on composed curves. Part I: *Izv. Vyssh. Uchebn. Zaved., Mat.* **84/1** (1984), 25–33; Part II: *ibidem* **84/4** (1984), 37–46 [Russian].

[158] B.A. Plamenevski and V.N. Senichkin: Representations of C^*-algebras generated by pseudodifferential operators in spaces with weight. In: *Probl. Mat. Fiziki*, pp. 165–189, Izd. Leningrad. Univ., Leningrad 1987 [Russian].

[159] J. Plemelj: Ein Ergänzungssatz zur Cauchy'schen Integraldarstellung analytischer Funktionen, Randwerte betreffend. *Monatshefte Math. Phys.* **19** (1908), 205–210.

[160] C. Pommerenke: *Univalent Functions.* Vandenhoeck and Ruprecht, Göttingen 1975.

[161] S.C. Power: *Hankel Operators on Hilbert Space.* Pitman Research Notes, Vol. 64, Pitman, Boston, London, Melbourne 1982.

[162] I.I. Privalov: *Randeigenschaften analytischer Funktionen.* Deutscher Verlag der Wiss., Berlin 1956.

[163] V.S. Rabinovich: Singular integral operators on a composed contour with oscillating tangent and pseudodifferential Mellin operators. *Soviet Math. Dokl.* **44** (1992), 791–796.

[164] V.S. Rabinovich: Singular integral operators on composed contours and pseudod-ifferential operators. *Math. Notes* **58** (1995), 722–734.

[165] V.S. Rabinovich: Pseudofifferential operators with operator symbols – local invert-ibility and limit operators. In: *Proceedings of the Seminar on Linear Topological Spaces and Complex Analysis*, pp. 58–73, Metu-Tübitak, Ankara 1994.

[166] V.S. Rabinovich: Pseudodifferential operators with analytic symbols and some of their applications. In: *Proceedings of the Seminar on Linear Topological Spaces and Complex Analysis*, pp. 79–98, Metu-Tübitak, Ankara 1995.

[167] V.S. Rabinovich: Algebras of singular integral operators on composed contours with knots that are logarithmic whirl points. *Izv. Ross. Akad. Nauk, Ser. Mat.* **60** (1996), 169–200 [Russian].

[168] V.S. Rabinovich: Mellin pseudodifferential operator techniques in the theory of singular integral operators on some Carleson curves. In: *Proceedings of the IWOTA 1995* (to appear).

[169] V.S. Rabinovich, S. Roch, and B. Silbermann: Fredholm theory and finite section method for band-dominated operators. *Integral Equations and Operator Theory* (to appear).

[170] I. Raeburn and A.M. Sinclair: The C^*-algebra generated by two projections. *Math. Scand.* **65** (1989), 278–290.

[171] M. Reed and B. Simon: *Methods of Modern Mathematical Physics.* Vol. 1, Academic Press, New York 1972.

[172] S. Roch et al.: Banach algebras generated by N idempotents and applications. *Operator Theory: Advances and Applications* **90** (1996), 19–54.

[173] S. Roch and B. Silbermann: Algebras generated by idempotents and the symbol calculus for singular integral operators. *Integral Equations and Operator Theory* **11** (1988), 385–419.

[174] S. Roch and B. Silbermann: *Algebras of Convolution Operators and Their Image in the Calkin Algebra.* Report R-Math-05/90, Karl-Weierstrass-Inst. f. Math., Berlin 1990.

[175] S. Roch and B. Silbermann: Representations of noncommutative Banach algebras by continuous functions. *St. Petersburg Math. J.* **3** (1992), 865–879.

[176] S. Roch and B. Silbermann: The structure of algebras of singular integral operators. *J. Integral Equations Appl.* **4** (1992), 421–442.

[177] R.T. Rockafellar: *Convex Analysis.* Princeton Univ. Press, Princeton 1970.

[178] L.H. Rowen: *Polynomial Identities in Ring Theory.* Academic Press, New York 1980.

[179] W. Rudin: *Principles of Mathematical Analysis.* McGraw-Hill, Inc., New York 1964.

[180] W. Rudin: *Real and Complex Analysis.* McGraw-Hill, Inc., New York 1970.

[181] W. Rudin: *Functional Analysis.* McGraw-Hill, Inc., New York 1991.

[182] Yu.S. Samoilenko: *Spectral Theory of Families of Selfadjoint Operators.* Kluwer Academic Publishers, Dordrecht, Boston, London 1991.

[183] E.T. Sawyer: A characterization of a two-weight norm inequality for maximal operators. *Studia Math.* **75** (1982), 1–11.

[184] E.T. Sawyer: Two-weight norm inequalities for certain maximal and integral operators. In: *Harmonic Analysis* (F. Ricci and G. Weiss, eds.), Lecture Notes Math., Vol. 908, pp. 102–127, Springer-Verlag, Berlin 1982.

[185] R. Schneider: Integral equations with piecewise continuous coefficients in L^p spaces with weight. *J. Integral Equations* **9** (1985), 135–152.

[186] R.K. Seifullayev: The Riemann boundary value problem on a non-smooth open curve. *Matem. Sbornik* **112** (1980), 147–161 [Russian].

[187] V.N. Semenyuta and A.V. Khevelev: A local principle for special classes of Banach algebras. *Izv. Severo-Kavkazkogo Nauchn. Zentra Vyssh. Shkoly, Ser. Estestv. Nauk* **1/1977** (1977), 15–17 [Russian].

[188] S. Semmes: Square function estimates and the $T(b)$ theorem. *Proc. Amer. Math. Soc.* **110** (1990), 721–726.

[189] E. Shamir: The solution of a Riemann-Hilbert system with piecewise continuous coefficients. *Dokl. Akad. Nauk SSSR* **167** (1966), 1000–1003 [Russian].

[190] I.Ya. Shneiberg: On the solvability of linear equations in interpolation families of Banach spaces. *Dokl. Akad. Nauk SSSR* **212** (1973), 57–59 [Russian].

[191] I.B. Simonenko: The Riemann boundary value problem with continuous coefficients. *Dokl. Akad. Nauk SSSR* **124** (1959), 279–281 [Russian].

[192] I.B. Simonenko: The Riemann boundary value problem with measurable coefficients. *Dokl. Akad. Nauk SSSR* **135** (1960), 538–541 [Russian].

[193] I.B. Simonenko: The Riemann and Riemann-Haseman boundary value problems with continuous coefficients. In: *Issled. po Sovrem. Probl. Teor. Funkts. Kompl. Perem.*, Fizmatgiz, Moscow 1961, pp. 380–389 [Russian].

[194] I.B. Simonenko: The Riemann boundary value problem for n pairs of functions with measurable coefficients and its application to the investigation of singular integrals in the spaces L^p with weight. *Izv. Akad. Nauk SSSR, Ser. Mat.* **28** (1964), 277–306 [Russian].

[195] I.B. Simonenko: A new general method of studying linear operator equations of the type of singular integral equations. Part I: *Izv. Akad. Nauk SSSR, Ser. Mat.* **29** (1965), 567–586; Part II: *ibidem*, 757–782. [Russian]

[196] I.B. Simonenko: Some general questions of the theory of the Riemann boundary value problem. *Math. USSR Izv.* **2** (1968), 1091–1099.

[197] I.B. Simonenko: Example of a function which satifies the Muckenhoupt condition but is not a weight function for the Cauchy singular integral in the case of a contour with cusps. Deposited in VINITI, No. 2659-83, Rostov-on-Don 1983. [Russian]

[198] I.B. Simonenko: Stability of weight properties of functions with respect to the singular integral. *Matem. Zametki* **33** (1983), 409–416 [Russian].

[199] I.B. Simonenko and Chin Ngok Min: *The local method in the theory of one-dimensional singular integral equations with piecewise continuous coefficients. Noethericity.* Izd. Rostov-on-Don State Univ. 1986 [Russian].

[200] I. Spitkovsky: Singular integral operators with PC symbols on the spaces with general weights. *J. Funct. Analysis* **105** (1992), 129–143.

[201] I. Spitkovsky: Once more on algebras generated by two projections. *Lin. Algebra Appl.* **208/209** (1994), 377–395.

[202] E.M. Stein: *Singular Integrals and Differentiability Properties of Functions.* Princeton Univ. Press, Princeton 1970.

[203] J.-O. Strömberg and A. Torchinsky: *Weighted Hardy Spaces.* Lecture Notes in Math., Vol. 1381, Springer-Verlag, Berlin, Heidelberg, New York 1989.

[204] N.B. Vasil'ev: C^*-algebras with finite-dimensional representations. *Uspekhi Mat. Nauk* **1** (1966), 135–154 [Russian].

[205] N.L. Vasilevski and I. Spitkovsky: On the algebra generated by two projections. *Dokl. Akad. Nauk Ukrain. SSR*, Ser. A, **8** (1981), 10–13 [Ukrainian].

[206] I.N. Vekua: *Generalized Analytic Functions.* Nauka, Moscow 1988 [Russian].

[207] H. Widom: Inversion of Toeplitz matrices.II. *Illinois J. Math.* **4** (1960), 88–99.

[208] H. Widom: Inversion of Toeplitz matrices.III. *Notices Amer. Math. Soc.* **7** (1960), p. 63.

[209] H. Widom: Singular integral equations on L^p. *Trans. Amer. Math. Soc.* **97** (1960), 131–160.

[210] K. Zhu: *Operator Theory in Function Spaces.* Marcel Dekker, Inc., New York and Basel 1990.

[211] M. Zinsmeister: Les domains de Carleson. *Michigan Math. J.* **36** (1989), 213–220.

[212] E.I. Zverovich: Two-element boundary value problems and the method of locally conformal gluing. *Sibirsk. Matem. Zh.* **14** (1973), 64–85 [Russian].

[213] A. Zygmund: *Trigonometric Series.* Cambridge Univ. Press, London and New York 1968.

Index

Titles previously published in the series

OPERATOR THEORY: ADVANCES AND APPLICATIONS
BIRKHÄUSER VERLAG

Edited by
I. Gohberg,
.School of Mathematical Sciences, Tel-Aviv University, Ramat Aviv, Israel

This series is devoted to the publication of current research in operator theory, with particular emphasis on applications to classical analysis and the theory of integral equations, as well as to numerical analysis, mathematical physics and mathematical methods in electrical engineering.

MMA • Monographs in Mathematics

Managing Editors
H. Amann / J.-P. Bourguignon / K. Grove / P.-L. Lions

Editorial Board
H. Araki / J. Ball / F. Brezzi / K.C. Chang / N. Hitchin / H. Hofer / H. Knörrer / K. Masuda / D. Zagier

The foundations of this outstanding book series were laid in 1944. Until the end of the 1970s, a total of 77 volumes appeared, including works of such distinguished mathematicians as Carathéodory, Nevanlinna and Shafarevich, to name a few. The series came to its name and present appearance in the 1980s. According to its well-established tradition, only monographs of excellent quality will be published in this collection. Comprehensive, in-depth treatments of areas of current interest are presented to a readership ranging from graduate students to professional mathematicians. Concrete examples and applications both within and beyond the immediate domain of mathematics illustrate the import and consequences of the theory under discussion.

Progress in Mathematics

Edited by:

H. Bass
Columbia University
New York
10027
U.S.A.

J. Oesterlé
Dépt. de Mathématiques
Université de Paris VI
4, Place Jussieu
75230 Paris Cedex 05, France

A. Weinstein
Dept. of Mathematics
University of CaliforniaNY
Berkeley, CA 94720
U.S.A.

Progress in Mathematics is a series of books intended for professional mathematicians and scientists, encompassing all areas of pure mathematics. This distinguished series, which began in 1979, includes authored monographs, and edited collections of papers on important research developments as well as expositions of particular subject areas.

We encourage preparation of manuscripts in such form of TeX for delivery in camera-ready copy which leads to rapid publication, or in electronic form for interfacing with laser printers or typesetters.

Proposals should be sent directly to the editors or to: Birkhäuser Boston, 675 Massachusetts Avenue, Cambridge, MA 02139, U.S.A.